VOLUME SEVENTY THREE

CURRENT TOPICS IN MEMBRANES

Exchangers

CURRENT TOPICS IN MEMBRANES, VOLUME 73

Series Editors

ROBERT BALABAN
National Heart, Lung and Blood Institute
National Institutes of Health
Bethesda, Maryland, USA

SIDNEY A. SIMON
Department of Neurobiology
Duke University Medical Centre
Durham, North Carolina, USA

VOLUME SEVENTY THREE

CURRENT TOPICS IN MEMBRANES

Exchangers

Edited by

MARK O. BEVENSEE

Department of Cell, Developmental and Integrative Biology
University of Alabama at Birmingham
Birmingham, AL, USA

ELSEVIER

AMSTERDAM • BOSTON • HEIDELBERG • LONDON
NEW YORK • OXFORD • PARIS • SAN DIEGO
SAN FRANCISCO • SINGAPORE • SYDNEY • TOKYO
Academic Press is an imprint of Elsevier

Academic Press is an imprint of Elsevier
225 Wyman Street, Waltham, MA 02451, USA
525 B Street, Suite 1800, San Diego, CA 92101-4495, USA
The Boulevard, Langford Lane, Kidlington, Oxford, OX5 1GB, UK
32 Jamestown Road, London NW1 7BY, UK
Radarweg 29, PO Box 211, 1000 AE Amsterdam, The Netherlands

First edition 2014

ISBN: 978-0-12-800223-0
ISSN: 1063-5823

14 15 16 11 10 9 8 7 6 5 4 3 2 1

DEDICATION

This volume on exchangers follows its companion volume (70) on cotransporters, and the two as a pair are dedicated to the late Dr. Dale J. Benos, Endowed Professor and former Chair of the Department of Physiology and Biophysics at the University of Alabama at Birmingham (UAB). Dale's sudden death near the end of 2010 was an enormous loss for both UAB and the general scientific community. However, Dale's dedication and commitment to advancing science and encouraging others lives on through the many individuals he inspired.

CONTENTS

CONTRIBUTORS

Christine Anne
Université Paris Descartes, Sorbonne Paris Cité, Centre National de la Recherche Scientifique, Unité Mixte de Recherche 8192, Centre Universitaire des Saints-Pères, Paris, France

Janice Y. Chou
Section on Cellular Differentiation, Program on Developmental Endocrinology and Genetics, Eunice Kennedy Shriver National Institute of Child Health and Human Development, National Institutes of Health, Bethesda, Maryland, USA

Emmanuelle Cordat
Department of Physiology, University of Alberta, Edmonton, Alberta, Canada

Bruno Gasnier
Université Paris Descartes, Sorbonne Paris Cité, Centre National de la Recherche Scientifique, Unité Mixte de Recherche 8192, Centre Universitaire des Saints-Pères, Paris, France

Bruno Hagenbuch
Department of Pharmacology, Toxicology and Therapeutics, The University of Kansas Medical Center, Kansas City, Kansas, USA

Ruth Hendus-Altenburger
Section for Biomolecular Sciences, and Section for Cell and Developmental Biology, Department of Biology, University of Copenhagen, Copenhagen, Denmark

Zhanjun Hou
Department of Oncology, Wayne State University School of Medicine, and Molecular Therapeutics Program, Barbara Ann Karmanos Cancer Institute, Detroit, Michigan, USA

Ali H. Jalloul
Department of Physiology and Pharmacology, Hotchkiss Brain Institute, University of Calgary, Calgary, Alberta, Canada

Martin Kolisek
Institute of Veterinary Physiology, Free University Berlin, Berlin, Germany

Birthe B. Kragelund
Section for Biomolecular Sciences, Department of Biology, University of Copenhagen, Copenhagen, Denmark

Guohong Liu
Department of Physiology and Pharmacology, Hotchkiss Brain Institute, University of Calgary, Calgary, Alberta, Canada

Brian C. Mansfield
Section on Cellular Differentiation, Program on Developmental Endocrinology and Genetics, Eunice Kennedy Shriver National Institute of Child Health and Human Development, National Institutes of Health, Bethesda, and Foundation Fighting Blindness, Columbia, Maryland, USA

Larry H. Matherly
Departments of Oncology and Pharmacology, Wayne State University School of Medicine, Molecular Therapeutics Program, Barbara Ann Karmanos Cancer Institute, Detroit, Michigan, USA

Magnus Monné
Department of Biosciences, Biotechnology and Biopharmaceutics, Laboratory of Biochemistry and Molecular Biology, University of Bari, Bari, and Department of Sciences, University of Basilicata, Potenza, Italy

Ferdinando Palmieri
Department of Biosciences, Biotechnology and Biopharmaceutics, Laboratory of Biochemistry and Molecular Biology, University of Bari, Bari, Italy

Stine Falsig Pedersen
Section for Cell and Developmental Biology, Department of Biology, University of Copenhagen, Copenhagen, Denmark

Ryan M. Pelis
Department of Pharmacology, Dalhousie University, Halifax, Nova Scotia, Canada

Reinhart A.F. Reithmeier
Department of Biochemistry, 1 King's College Circle, University of Toronto, Toronto, Ontario, Canada

Paul P.M. Schnetkamp
Department of Physiology and Pharmacology, Hotchkiss Brain Institute, University of Calgary, Calgary, Alberta, Canada

Monika Schweigel-Röntgen
Institute for Muscle Biology & Growth, Leibniz Institute for Farm Animal Biology, Dummerstorf, Germany

Bruno Stieger
Department of Clinical Pharmacology and Toxicology, University Hospital, Zürich, Switzerland

Robert T. Szerencsei
Department of Physiology and Pharmacology, Hotchkiss Brain Institute, University of Calgary, Calgary, Alberta, Canada

Stephen H. Wright
Department of Physiology, University of Arizona, Tucson, Arizona, USA

PREFACE

This volume on exchangers in conjunction with the previous volume 70 in this series entitled "Co-transport Systems" completes the planned coverage of secondary active transporters, which play pivotal roles in the reabsorption/secretion of solutes by epithelia and the regulation of intracellular ions and molecules by cells in general. A more detailed description of the importance and categorization of secondary active transporters are presented in the Preface of the cotransporter volume. The format of the current volume parallels that of the previous one. For example, I have focused on covering the exchangers of the SoLute Carrier (SLC) gene series with an emphasis on transported species; thus, some chapters include multiple SLCs. As in the previous volume, there is an emphasis on presenting the latest biophysical and structural characteristics of the transporters.

A major goal of this two-volume project was to examine all the relevant SLCs comprehensively. However, due to unavoidable circumstances, there are a couple of unfortunate omissions in this exchanger volume. Although amino acid cotransporters were covered in the previous volume and vesicular neurotransmitter transporters are presented in this volume, the SLC1 glutamate exchangers are not examined. Furthermore, although there is a chapter on the potassium-dependent sodium–calcium exchangers in the current volume, the SLC8 potassium-independent sodium–calcium exchangers are only minimally discussed. I apologize for these unplanned omissions but direct the reader to excellent reviews on SLC1 by Kanai *et al.* and SLC8 by Khananshvili in a special SLC issue of Molecular Aspects of Medicine, volume 34 (2013) edited by Dr. Matthias H. Hediger. Overall, however, these two volumes examine the vast majority of SLC secondary active transporters, and readers will be amazed at the extensive advances made in characterizing the structure, function, and regulation of these important proteins. Even so, the biological function and characteristics of many SLC family members—particularly those presented in the second half of this exchanger volume—are still poorly understood. I hope the excitement of recent advances conveyed in these chapters will drive further groundbreaking research.

I am indebted to many people for making this exchanger volume a success. The contributing authors were a pleasure to work with, and I thank them for their enthusiasm and dedication to this project. I owe special thanks to Dr. Monika Schweigel-Röntgen who submitted two chapters comprising a particularly large cluster of transporters (SLC30, 31, 39, and 41). I am also grateful to many individuals who timely reviewed submissions and provided valuable feedback to me and the authors. Finally, I thank the series editors (Drs. Sidney Simon and Robert Balaban, and the late Dr. Dale Benos) for initiating and supporting this project, as well as the Elsevier team (editorial project manager Helene Kabes and book production project manager Preeta Kumaraguruparan) for routine advice and assistance.

MARK O. BEVENSEE

PREVIOUS VOLUMES IN SERIES

Current Topics in Membranes and Transport

Volume 23 Genes and Membranes: Transport Proteins and Receptors* (1985)
Edited by Edward A. Adelberg and Carolyn W. Slayman

Volume 24 Membrane Protein Biosynthesis and Turnover (1985)
Edited by Philip A. Knauf and John S. Cook

Volume 25 Regulation of Calcium Transport across Muscle Membranes (1985)
Edited by Adil E. Shamoo

Volume 26 Na^+–H^+ Exchange, Intracellular pH, and Cell Function* (1986)
Edited by Peter S. Aronson and Walter F. Boron

Volume 27 The Role of Membranes in Cell Growth and Differentiation (1986)
Edited by Lazaro J. Mandel and Dale J. Benos

Volume 28 Potassium Transport: Physiology and Pathophysiology* (1987)
Edited by Gerhard Giebisch

Volume 29 Membrane Structure and Function (1987)
Edited by Richard D. Klausner, Christoph Kempf, and Jos van Renswoude

Volume 30 Cell Volume Control: Fundamental and Comparative Aspects in Animal Cells (1987)
Edited by R. Gilles, Arnost Kleinzeller, and L. Bolis

Volume 31 Molecular Neurobiology: Endocrine Approaches (1987)
Edited by Jerome F. Strauss, III, and Donald W. Pfaff

Volume 32 Membrane Fusion in Fertilization, Cellular Transport, and Viral Infection (1988)
Edited by Nejat Düzgünes and Felix Bronner

Volume 33 Molecular Biology of Ionic Channels* (1988)
Edited by William S. Agnew, Toni Claudio, and Frederick J. Sigworth

---

*Part of the series from the Yale Department of Cellular and Molecular Physiology

Volume 34 Cellular and Molecular Biology of Sodium Transport* (1989)
Edited by Stanley G. Schultz

Volume 35 Mechanisms of Leukocyte Activation (1990)
Edited by Sergio Grinstein and Ori D. Rotstein

Volume 36 Protein–Membrane Interactions* (1990)
Edited by Toni Claudio

Volume 37 Channels and Noise in Epithelial Tissues (1990)
Edited by Sandy I. Helman and Willy Van Driessche

Current Topics in Membranes

Volume 38 Ordering the Membrane Cytoskeleton Trilayer* (1991)
Edited by Mark S. Mooseker and Jon S. Morrow

Volume 39 Developmental Biology of Membrane Transport Systems (1991)
Edited by Dale J. Benos

Volume 40 Cell Lipids (1994)
Edited by Dick Hoekstra

Volume 41 Cell Biology and Membrane Transport Processes* (1994)
Edited by Michael Caplan

Volume 42 Chloride Channels (1994)
Edited by William B. Guggino

Volume 43 Membrane Protein–Cytoskeleton Interactions (1996)
Edited by W. James Nelson

Volume 44 Lipid Polymorphism and Membrane Properties (1997)
Edited by Richard Epand

Volume 45 The Eye's Aqueous Humor: From Secretion to Glaucoma (1998)
Edited by Mortimer M. Civan

Volume 46 Potassium Ion Channels: Molecular Structure, Function, and Diseases (1999)
Edited by Yoshihisa Kurachi, Lily Yeh Jan, and Michel Lazdunski

Volume 47 Amiloride-Sensitive Sodium Channels: Physiology and Functional Diversity (1999)
Edited by Dale J. Benos

Volume 63 Membrane Protein Crystallization
Edited by Larry DeLucas

Volume 64 Leukocyte Adhesion
Edited by Klaus Ley

Volume 65 Claudins
Edited by Alan S. L. Yu

Volume 66 Structure and Function of Calcium Release Channels
Edited by Irina I. Serysheva

Volume 67 Advances in Adrenergic Receptor Biology
Edited by Qin Wang

Volume 68 Membrane Fusion
Edited by Leonid V. Chernomordik and Michael M. Kozlov

Volume 69 Metal Transporters
Edited by Svetlana Lutsenko and José M. Argüello

Volume 70 Co-Transport Systems
Edited by Mark O. Bevensee

Volume 71 Store-Operated Calcium Channels
Edited by Murali Prakriya

Volume 72 Functional Organization of Vertebrate Plasma Membrane
Edited by Vann Bennett

CHAPTER ONE

Structure, Function, and Trafficking of SLC4 and SLC26 Anion Transporters

Emmanuelle Cordat*, Reinhart A.F. Reithmeier†,1

*Department of Physiology, University of Alberta, Edmonton, Alberta, Canada

†Department of Biochemistry, 1 King's College Circle, University of Toronto, Toronto, Ontario, Canada

[1]Corresponding author: e-mail address: r.reithmeier@utoronto.ca

Contents

Current Topics in Membranes, Volume 73
ISSN 1063-5823
http://dx.doi.org/10.1016/B978-0-12-800223-0.00001-3

Abstract

The structure and function of the red cell anion exchanger 1 (AE1, Band 3, SLC4A1), the truncated kidney anion exchanger 1 (kAE1), and the other members of the SLC4 family of bicarbonate transporters are reviewed. Mutations in the *AE1* gene cause human diseases like Southeast Asian ovalocytosis and hereditary spherocytosis in the red cell and distal renal tubular acidosis in the kidney. These mutations affect the folding, trafficking, and functional expression of these membrane glycoproteins. In the SLC26 family of anion transporters, mutations also cause trafficking defects and human disease. Membrane glycoproteins are cotranslationally N-glycosylated in the endoplasmic reticulum (ER) and when properly folded, traffic via the secretory pathway to their final destination such as the plasma membrane. Misfolded glycoproteins are retained in ER and are targeted for degradation by the proteasome following retrotranslocation and ubiquitinylation. ER chaperones, like membrane-bound calnexin, interact transiently with glycoproteins and are part of the quality control system that monitors the folding of glycoproteins during their biosynthesis. Recent results have indicated that it is possible to "correct" trafficking defects caused by some mutations in the SLC4 and 26 families through the use of small molecules that interfere with the interaction of glycoproteins with the components of the quality control system. This review summarizes the current knowledge on structure and function of anion transporters from the SLC4 and SLC26 families, and the effect of mutations on their trafficking and functional expression.

ABBREVIATIONS

AE1 anion exchanger 1
CAII carbonic anhydrase II
CDA congenital dyserythropoietic anemia type II
cdAE1 cytosolic domain of AE1
dRTA distal renal tubular acidosis
ER endoplasmic reticulum
GAPDH glyceraldehyde-3-phosphate dehydrogenase
HEMPAS hereditary erythrocytic multinuclearity with positive acidified serum lysis
HS hereditary spherocytosis
HT high transport
kAE1 kidney anion exchanger 1
mdAE1 membrane domain of AE1
SAO Southeast Asian Ovalocytosis
SLC solute carrier
TM transmembrane

1. INTRODUCTION

The secretory pathway, beginning from the initial targeting of secreted and membrane proteins to the endoplasmic reticulum (ER), transit through the Golgi, and finally on to organelles like the lysosome or the plasma membrane, has been well characterized in mammalian cells and in simple organisms like yeast (Schekman, 2010). Membrane proteins typically contain N-linked oligosaccharides (Landoltmarticorena & Reithmeier, 1994) that are involved in protein folding and quality control during the biosynthetic process (Benyair, Ron, & Lederkremer, 2011; Helenius, 1994). Some single-span membrane proteins, like red cell Glycophorin A or membrane-anchored enzymes, contain multiple short O-linked oligosaccharide chains in the external juxta-membrane region as well. Oligosaccharides can play a role in the mature protein by increasing the solubility of secreted proteins and protecting them from proteolysis (Elbein, 1991). The role of N-linked oligosaccharides on glycoproteins—including those in membranes—varies from essential to dispensable. In membrane glycoproteins, the oligosaccharides face the cell exterior or the lumen of organelles, "sugar coating" the membrane surface. This review focuses on studies of the effect of disease-causing mutations in erythroid and kidney anion exchanger 1 (AE1 and kAE1, respectively), other members of the SLC4 family of bicarbonate transports, and the SLC26 family of anion transporters, on the folding, trafficking, and functional expression of these membrane glycoproteins.

2. ANION EXCHANGER 1 STRUCTURE AND FUNCTION

AE1 or Band 3 is the major glycoprotein of the red blood cell membrane, present at over 1 million copies per cell (Fairbanks, Steck, & Wallach, 1971; Reithmeier, 1993; Reithmeier et al., 1989). It is the founding member of the solute carrier SLC4 family of bicarbonate transporters that includes bicarbonate/chloride exchangers and sodium-coupled bicarbonate transporters (Alper, 2009; Cordat & Casey, 2009; Pushkin & Kurtz, 2006; Romero, 2005; Romero, Fulton, & Boron, 2004). Thus, these exchangers and cotransporters share a common ancestor and are related in sequence and structure. They likely operate by a similar transport mechanism with a central substrate-binding site alternately facing the cytosolic or extra-cytosolic side of the membrane (Jardetzky, 1966).

2.1. AE1 function

AE1 catalyzes the rapid electroneutral exchange of chloride and bicarbonate anions across the red cell membrane (Fig. 1.1A). In the tissues, carbon dioxide that enters the red cell by diffusion or perhaps through the aquaporin (AQP1) water channel is hydrated by intracellular carbonic anhydrase II (CA II) that is associated with the C-terminus of AE1 (Vince & Reithmeier, 1998) to produce bicarbonate, which is then in turn transported out of the cell in exchange for chloride (Reithmeier, 2001; Sterling, Reithmeier, & Casey, 2001). In the lungs, the system reverses. Bicarbonate that enters the red cell via AE1 is converted by CAII to carbon dioxide, which then diffuses across the plasma membrane. This important part of the respiratory system increases the blood's capacity to transport carbon dioxide as soluble plasma bicarbonate. A truncated form of AE1 (kAE1) is expressed in the basolateral membrane (BLM) of α-intercalated cells of the kidney where it mediates the reabsorption of bicarbonate produced by intracellular CAII into blood in exchange for chloride (Fig. 1.1B). The proton produced by CAII in turn is pumped across the apical membrane by a V-ATPase resulting in acidification of the urine. See Section 7 for a more complete description of kAE1.

Anion transport in red blood cells can be inhibited by a number of organic anions (Cabantchik, Knauf, & Rothstein, 1978), most notably by 4,4′-diisothiocyano-2,2′-stilbene disulfonate (DIDS) or its reduced dihydro derivative (H_2DIDS). H_2DIDS, which binds to an outward-facing inhibitor site (Jennings & Nicknish, 1985; Landoltmarticorena, Casey, & Reithmeier, 1995; Lieberman & Reithmeier, 1983; Rao, Martin, Reithmeier, & Cantley, 1979), reacts irreversibly with Lys 539 at neutral pH and crosslinks to Lys 851 under alkaline conditions, thereby bridging the ~15 Å distance between these two residues within each AE1 monomer (Jennings & Nicknish, 1985; Okubo, Kang, Hamasaki, & Jennings, 1994). Mutation of Lys 539, however, does not affect the anion transport activity of AE1 (Garcia & Lodish, 1989). Inhibitor binding induces a conformational change in AE1 that locks the protein into an outward-facing state, consistent with an alternating sites model for transport (Falke & Chan, 1985; Macara, Kuo, & Cantley, 1983).

2.2. AE1 structure

Human AE1 consists of 911 residues arranged in the following two major domains: an N-terminal cytosolic domain (cdAE1) from residues 1–360

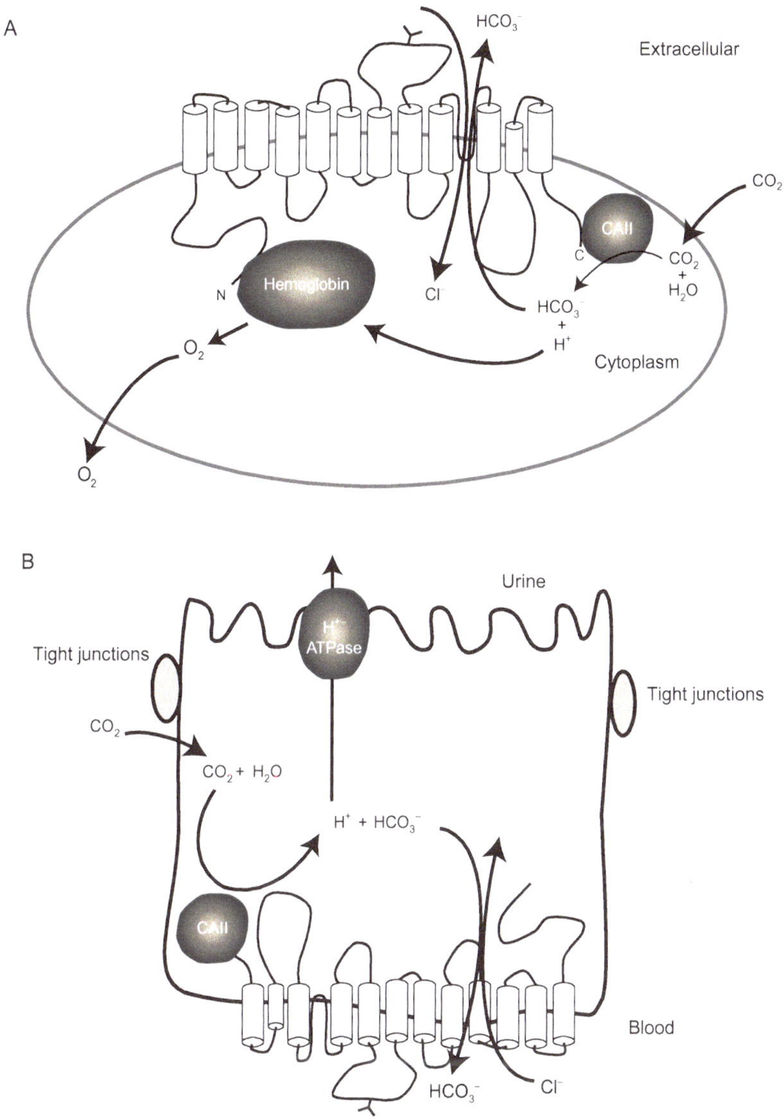

Figure 1.1 Role of Cl^-/HCO_3^- anion exchanger 1 (AE1) in (A) red blood cells and (B) in acid-secreting α-intercalated cells of the kidney. (A) In the tissues, carbon dioxide that diffuses through the red blood cell membrane (represented as red trace) or enters via aquaporin channels is rapidly converted with water to bicarbonate and a proton by the action of carbonic anhydrase II that is associated with the anion exchanger. The bicarbonate is transported out of the cell in exchange for chloride and the proton is buffered by hemoglobin, which lowers its affinity for oxygen delivering it to the tissues. In the lungs, the system reverses (not shown); bicarbonate enters the cell in exchange for chloride via the anion exchanger, bicarbonate and a proton are converted to carbon dioxide

(Continued)

and a C-terminal membrane domain (mdAE1) from residues 361–911 (Lux, John, Kopito, & Lodish, 1989; Steck, Ramos, & Strapazon, 1976; Tanner, Martin, & High, 1988). The two domains can be separated by mild proteolytic treatment with trypsin cleavage at Lys 360 (Reithmeier, 1979; Steck, Koziarz, Singh, Reddy, & Kohler, 1978; Steck et al., 1976). The cytosolic domain interacts with hemoglobin (Walder et al., 1984), glycolytic enzymes (Campanella, Chu, & Low, 2005; Chu & Low, 2006), and cytoskeletal proteins including ankyrin (Chang & Low, 2003; Stefanovic et al., 2007), protein 4.1 (Lombardo, Willardson, & Low, 1992), and protein 4.2 (Low, 1986). The crystal structure of cdAE1 has been determined under low pH conditions (Zhang, Kiyatkin, Bolin, & Low, 2000) and more recently (Shnitsar et al., 2013) under neutral pH conditions (Fig. 1.2B). It consists of a symmetric dimer with an N-terminal interaction domain and a C-terminal arm that mediates dimer formation. This arrangement allows the protein to undergo dramatic pH-dependent conformational changes, with a more asymmetric structure at neutral pH compared to the compact structure at acidic pH (Appell & Low, 1981; Low, Westfall, Allen, & Appell, 1984). The N-terminal 54 residues of AE1 form an extended intrinsically disordered region not visible in the crystal structure.

The membrane domain is responsible for the transport function of AE1 and is functional in the absence of the cytosolic domain when produced by proteolysis (Grinstein, Ship, & Rothstein, 1978) or by expression of mdAE1 in transfected cells (Shnitsar et al., 2013). Extensive studies of AE1 topology using sequence analysis (Kopito & Lodish, 1985a, 1985b), proteolysis (Abe, Chaen, Jin, Hamasaki, & Hamasaki, 2004; Steck et al., 1976), chemical labeling (Kawano et al., 1988), scanning N-glycosylation (Cheung, Li, & Reithmeier, 2005; Cheung & Reithmeier, 2007; Popov, Li, & Reithmeier, 1999; Popov, Tam, Li, & Reithmeier, 1997), and scanning cysteine mutagenesis (Fujinaga, Tang, & Casey, 1999; Zhu & Casey, 2004) have revealed that the

Figure 1.1—Cont'd and water by carbonic anhydrase II, the carbon dioxide then diffuses out of the red blood cell, and is expired by the lungs. (B) In the α-intercalated cells of the kidney, carbon dioxide that enters the cell is converted by carbonic anhydrase II to bicarbonate and a proton. The bicarbonate is transported into the blood by the kidney anion exchanger 1 (kAE1) in exchange for chloride, while the proton is pumped into the urine by a H^+ATPase. The result is acidification of the urine and delivery of bicarbonate into the blood. Defects in kAE1 cause an abnormal urine acidification and result in the disease distal renal tubular acidosis (dRTA). (See the color plate.)

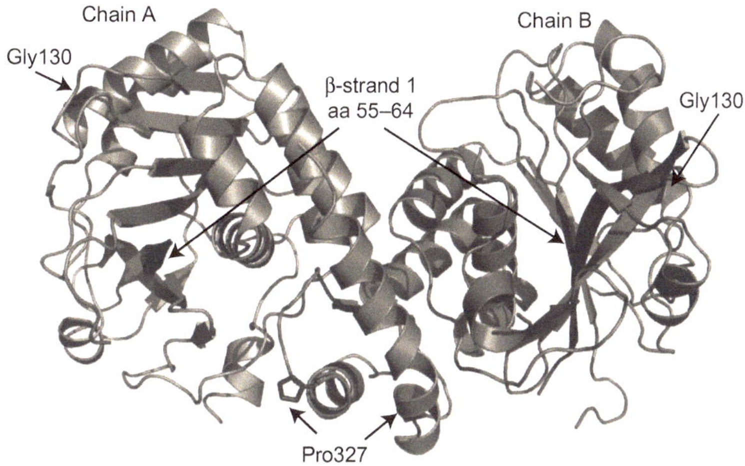

Figure 1.2 Crystal structure of the cytosolic domain of human anion exchanger 1 (cdΔ54AE1, residues 55–356) determined at neutral pH (Shnitsar et al., 2013). The cytosolic domain of AE1 exists as a symmetrical dimer with N-terminal interaction domains and C-terminal dimerization arms. The first β-strand (residues 55–64, colored in red) is missing in kAE1, thereby yielding a more open structure than full length AE1. Also not present in this structure is the intrinsically disordered acidic N-terminal extension (residues 1–54) involved in protein interactions. The positions of two HS mutations (G130R, P327R) that are impaired in Protein 4.2 binding are indicated. A third HS mutation (E40K) is located in the N-terminal extension. All three HS mutations result in the introduction of a basic residue into an otherwise acidic cdAE1. (See the color plate.)

membrane domain consists of some 14 transmembrane (TM) segments, some as mobile reentrant loops that may not span the membrane completely as helical segments and are likely involved in the transport mechanism. Low-resolution electron microscopic studies of negatively stained two-dimensional crystals have shown that mdAE1 has a dimeric structure (Wang, Kuhlbrandt, Sarabia, & Reithmeier, 1993; Wang, Sarabia, Reithmeier, & Kuhlbrandt, 1994) in agreement with previous hydrodynamic (Reithmeier, 1979) and other measurements (Jennings, 1984). More recent studies (Yamaguchi et al., 2010) using unstained samples of mdAE1 with bound H_2DIDS revealed the presence of long-tilted V-shaped TM helices reminiscent of those found in ClC chloride-proton cotransporters. A model of AE1 based on the similarity with *Escherichia coli* ClC has been proposed (Bonar, Schneider, Becker, Deitmer, & Casey,

2013; Hirai, Hamasaki, Yamaguchi, & Ikeda, 2011) and agrees remarkably well with spatial constraints based on past biochemical characterizations. The membrane domain of AE1 has also been modeled on the *E. coli* UraA uracil permease (Barneaud-Rocca, Etchebest, & Guizouarn, 2013), which is a proton-uracil symporter. A search using the human AE1 sequence on the Phyre2 Protein Fold Recognition Server (Kelley & Sternberg, 2009) found only UraA (Reithmeier, unpublished). The resulting folding model is shown in Fig. 1.3A. The protein contains 14 TM segments and correctly predicts most of the information on the topology of AE1, including the extracellular location of blood group antigens and other topological markers (Fig. 1.3B). The N and C termini are in the cytosol and the N-glycosylation site (N642) faces the cell exterior. The model is supported with recent experimental findings of TM segments (TM3–5 and 8) involved in ion permeation (Barneaud-Rocca et al., 2013). TM segments 3 and 10 consist of a central β-strand joined to a short helical segment. This unusual feature creates a positive helix dipole at the N-terminal ends of the short helices, which may create an anion-binding site as in the case of bacterial periplasmic sulfate-binding proteins (He & Quiocho, 1993). Small movements of these two helices could dispose the anion to alternating outside or inside-facing conformations. A more complete description of the AE1 transport mechanism awaits a structure of this membrane protein obtained from crystals that diffract to higher resolution than those obtained to date (Lemieux, Reithmeier, & Wang, 2002).

AE1 exists as a mixture of dimers and tetramers in the membrane and in detergent solutions (Casey & Reithmeier, 1993; Jennings, 1984; Steck, 1972). In contrast, the membrane domain does not form tetramers (Reithmeier, 1979). It is a symmetrical dimer with cross-sectional dimensions of 110 × 55 Å and a height of 80 Å with considerable mass extending beyond the bilayer on the cytosolic side of the membrane (Reithmeier, 1979; Wang et al., 1993, 1994). The tetrameric form of AE1 is responsible for the interaction with ankyrin and linkage to the spectrin/actin cytoskeleton (Fig. 1.4) (Thevenin & Low, 1990; Thevenin, Willardson, & Low, 1989; Van Dort, Moriyama, & Low, 1998). Transport studies have shown that blocking one subunit with an inhibitor does not prevent transport by the other subunit (Macara & Cantley, 1981a, 1981b). Thus, although AE1 exists as a dimer, the two subunits can operate independently from one another, even when one subunit is inactive. This feature is important in the consideration of AE1 dimers produced in heterozygotes with a defective *AE1* allele.

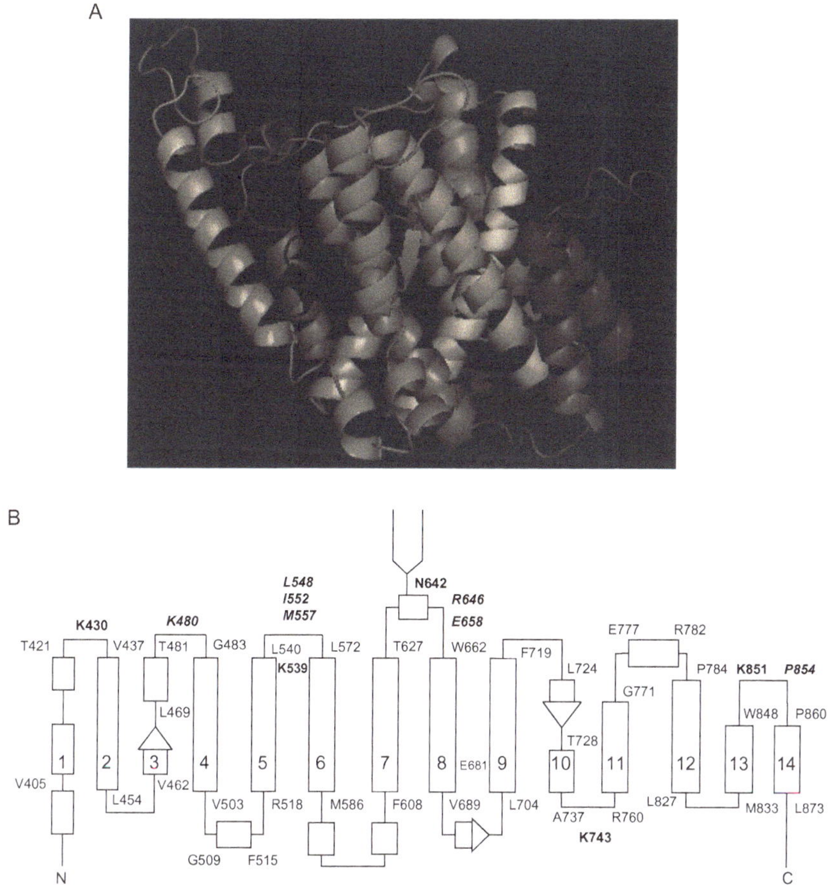

Figure 1.3 The homology model for the membrane domain of AE1 based on the crystal structure of the *E. coli* UraA uracil permease (Barneaud-Rocca et al., 2013). (A) A three-dimensional model for the membrane domain of the human AE1 monomer based on a Phyre2 homology search that identified UraA as the sole membrane protein in the PDB with a similar fold. The model contains 14 TM segments colored from N-terminus (blue) to C-terminus (red). The protein has an inverse topology with TMs 1–7 and TMs 8–14 having opposite orientations in the bilayer. TM segments 3 and 10 (orange β-strand in TM10 visible) are not completely helical, each containing a central short β-strand in anti-parallel orientation that likely form part of a central anion-binding site. The green loop at the top-right hand side of the model contains the single site of N-glycosylation (N642) and the top of the model therefore corresponds to the outside of the red cell. The bottom of the model has the N- and C-termini of AE1 facing the cytosol as expected. AE1 exists as a stable dimer, which is not shown in this model as UraA is a monomer. (B) A topology model of the membrane domain of AE1 based on the UraA fold. The residues that form the approximate limits of the TM segments are indicated by the rectangles that denote predicted TM helical segments. The residues indicated in italics are external

(Continued)

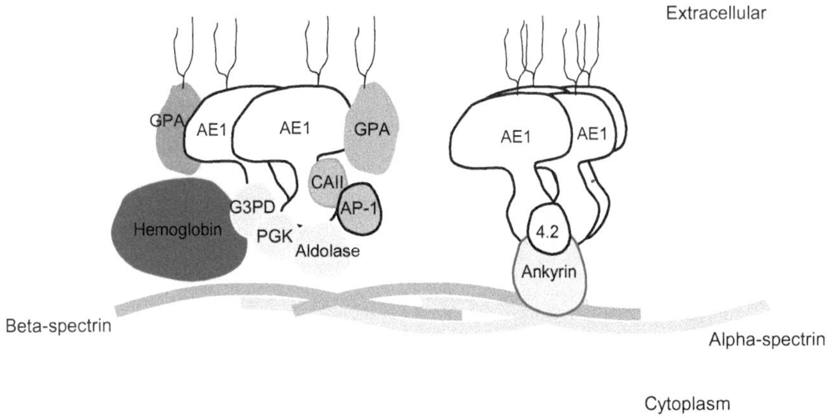

Figure 1.4 Schematic representation of the dimeric and tetrameric populations of the Band 3 protein and its interacting partners. In the dimeric state, AE1 interacts via its N-terminus with glycolytic enzymes such as glyceraldehyde 3-phosphate dehydrogenase (GAPDH), aldolase, and phosphoglycerate kinase (PGK), as well as with deoxyhemoglobin. The C-terminus of AE1 binds to carbonic anhydrase II (CAII), GAPDH, and the μ subunit of adaptor protein 1 (AP-1). Additionally, AE1 interacts with Glycophorin A (GPA), which together with AE1 forms the Wr^b blood group. In the tetrameric form, AE1 interacts with protein 4.2 and ankyrin, which anchor the protein complex to the underlying spectrin/actin cytoskeleton. The oligosaccharides are schematically represented as a Y shape on the extracellular face of the protein. The lipid bilayer is shown in light grey.

3. AE1 BIOSYNTHESIS AND N-GLYCOSYLATION

3.1. AE1 biosynthesis

AE1, like other mammalian glycoproteins, is cotranslationally inserted into the ER membrane with the first TM segment acting as a stable signal sequence (Kopito & Lodish, 1985a; Lux et al., 1989; Sabban, Marchesi, Adesnik, & Sabatini, 1981). More distal TM segments can act as signal

Figure 1.3—Cont'd blood group antigen polymorphisms due to single amino acid changes at the indicated sites. The single site of N-glycosylation is at an external site located at N642. K539 and K851 are the sites crosslinked from the outside of the red cell with H_2DIDS. K430 is an external site modified by eosin maleimide. K743 is an internal trypsin cleavage site. TM segments 3 and 10 are each composed of a short β-strand (arrows) linked to a short α-helix. The N termini of the helices in TM3 and 10 create a $\delta+$ve dipole that may form part of a central anion-binding site. E681 in TM8 is also essential for transport and is likely located close to this anion-binding site. (See the color plate.)

sequences and stop transfer sequences (Tam, Loo, Clarke, & Reithmeier, 1994), and it is possible to insert complementary portions of AE1 into the membrane and to reconstitute a functional transporter from several parts (Groves & Tanner, 1995, 1999a, 1999b; Groves, Wang, & Tanner, 1998; Wang, Groves, Mawby, & Tanner, 1997). AE1 forms a dimer in the ER that then moves through the Golgi to the plasma membrane. In red cell precursors, the oligosaccharide chain is modified from the original high-mannose form to a complex structure containing a long polylactosaminyl chain on a fraction of Band 3 (Landolt-Marticorena, Charuk, & Reithmeier, 1998). In transfected human embryonic kidney (HEK 293) cells, AE1 largely retains a high-mannose oligosaccharide structure, although it traffics to the plasma membrane and is functional (Casey, Ding, & Kopito, 1995; Quilty & Reithmeier, 2000). Insertion of a novel N-glycosylation site (–NXS/T–) into the larger preceding EC loop 3 allows the oligosaccharide chain to be processed to a complex form, even as the native N-linked oligosaccharide on EC loop 4 retains a high-mannose structure (Tam, Landolt-Marticorena, & Reithmeier, 1996). This shows that AE1 has trafficked through the Golgi, but its oligosaccharide escapes processing in transfected HEK cells. N-glycosylation sites are commonly found on loops greater than 35 residues in size with the acceptor site spaced 12 and 14 residues away from the proximal and distal TM segments, respectively (Landoltmarticorena & Reithmeier, 1994). Interestingly, removal of the cytosolic domain allows the oligosaccharide on mdAE1 to be efficiently modified to a complex structure (Shnitsar et al., 2013). How changes on the cytosolic side of the membrane can affect oligosaccharide processing on the opposite side of the membrane remains unknown, but it may have to do with oligomer formation.

3.2. AE1 N-glycosylation

AE1 contains a single site of N-glycosylation at Asn^{642}; however, the oligosaccharide attached to this site is heterogeneous. Extending beyond the core ($GlcNAc_2Man_3$) structure are two (bi-antennary) long polylactosaminyl chains consisting of repeating GlcNAcGal units (Fukuda, Dell, Oates, & Fukuda, 1984; Tsuji, Irimura, & Osawa, 1980). AE1 consists of two glyco-populations: one with long polylactosaminyl chains and the other with short oligosaccharide chains (Landolt-Marticorena et al., 1998; Tsuji, Irimura, & Osawa, 1981). Some of the long oligosaccharide chains have a branched tetra-antennary structure. A small fraction of AE1 present in mature red blood cells contains shorter oligosaccharide chains that can bind the lectin Concanavalin A (Findlay, 1974). The heterogeneity of the oligosaccharide attached to AE1 may be due to the massive amount of this glycoprotein made late during red cell development that may overwhelm the processing enzymes.

The complex oligosaccharide on AE1 is not essential for the functional expression of this membrane glycoprotein. When expressed in transfected HEK-293 cells, AE1 retains a high-mannose oligosaccharide and its anion transport activity (Casey et al., 1995). Mutation of the oligosaccharide acceptor site (N642D) does not impair the trafficking of AE1 to the cell surface or its transport function (Groves & Tanner, 1994a). Likewise, enzymatic removal of the oligosaccharide chain from mature AE1 using N-glycanase F does not affect its transport activity (Casey, Pirraglia, & Reithmeier, 1992).

4. ERYTHROID AE1 PROTEIN INTERACTIONS

4.1. Glycophorin A

Glycophorin A is a type I single-span membrane protein with an extracellular N-terminal domain that contains 1 N-linked (Asn 15) and 15 O-linked (Thr and Ser) oligosaccharide chains terminating in sialic acid residues that display negative charges on the red cell surface. Glycophorin A is made with an amino-terminal signal sequence that is cotranslationally cleaved during biosynthesis in the ER, and then traffics to the cell surface (Jokinen, Gahmberg, & Andersson, 1979; Jokinen, Ulmanen, Andersson, Kaariainen, & Gahmberg, 1981). Glycophorin A is expressed endogenously in human K562 erythroleukemic cells (Gahmberg, Jokinen, & Andersson, 1979). Treatment of these cells with tunicamycin, which blocks N-glycosylation, does not affect the cell surface expression of Glycophorin A (Gahmberg, Jokinen, Karhi, & Andersson, 1980), showing that N-glycosylation is not essential for trafficking to the plasma membrane.

Glycophorin A is about equally abundant to AE1 in the human red cell membrane. These two proteins interact with each other in the mature red cell membrane forming the Wright (Wr) blood group antigen. Patients with the Wr (a+b−) blood group have an E658K mutation in AE1 (Bruce et al., 1995). Glu 658 of AE1 is proposed to interact with Arg 61 in the extracellular domain of Glycophorin A to form the Wr antigen. In red cells from symptomless En(a−) individuals that are devoid of Glycophorin A, AE1 becomes heavily sialylated, compensating for the loss of these terminal sugars on Glycophorin A (Tanner & Anstee, 1976; Tanner, Jenkins, Anstee,

& Clamp, 1976). Experiments in transfected cells have shown that Glycophorin A can form a complex with AE1 in the ER, suggesting that these two proteins traffic together to the cell surface (Auffray et al., 2001; Pang & Reithmeier, 2009; Young, Beckmann, Toye, & Tanner, 2000). When AE1 is expressed in *Xenopus* oocytes, movement to the cell surface is facilitated by Glycophorin A, although Glycophorin A is not essential for AE1 trafficking (Groves & Tanner, 1992, 1994b). Knockdown of Glycophorin A expression in human erythroleukemic K562 cells had no effect on the cell surface expression of AE1 (Pang & Reithmeier, 2009). Indeed, the presence of normal amounts of AE1 in the En(a−) red cells shows that Glycophorin A is not essential for the cell surface expression of normal AE1 during erythropoiesis (Tanner & Anstee, 1976). Some changes in the stability and transport properties of AE1 in these cells have, however, been noted in the absence of Glycophorin A (Bruce, Groves, Okubo, Thilaganathan, & Tanner, 1994; Bruce et al., 2004).

Glycophorin A can rescue the trafficking of some AE1 mutants to the cell surface (see below), including AE1 Southeast Asian ovalocytosis (SAO) (Patterson et al., 2009), certain hereditary spherocytosis (HS) mutants (Toye et al., 2008), and in kAE1, dominant mutations causing distal renal tubular acidosis (dRTA) (Williamson & Toye, 2008; Young et al., 2000). AE1 HS mutants that are not rescued by Glycophorin A may have lost the ability to interact with this protein partner due to misfolding. The rescue of dominant and often functional dRTA mutants by Glycophorin A allows their expression in the red cells of patients with this kidney disease (Williamson & Toye, 2008). This explains why some dRTA patients with a mutation in the gene that encodes both AE1 and kAE1 have normal red cells, yet a kidney defect.

4.2. Proteins that interact with the cytosolic domain of AE1

The N-terminal cytosolic domain of AE1 (cdAE1) mediates a bewildering array of direct interactions with cytosolic proteins including hemoglobin (Walder et al., 1984), glycolytic enzymes like glyceraldehyde 3-phosphate dehydrogenase (GAPDH) (Kant & Steck, 1973) and aldolase (Jenkins, Madden, & Steck, 1984; Strapazon & Steck, 1976), and the cytoskeletal proteins like ankyrin (Thevenin & Low, 1990), Protein 4.1, and Protein 4.2 (Low, 1986). The binding sites for glycolytic enzymes (Chu & Low, 2006; Murthy, Liu, Kaul, Kohler, & Steck, 1981), ankyrin (Chang & Low, 2003; Stefanovic et al., 2007), and Protein 4.1 (Lombardo et al.,

1992) have been localized to specific regions of cdAE1. The binding of glycolytic enzymes is regulated by phosphorylation of key tyrosine residues (Tyr 8 and Tyr 21) in the amino-terminal region of cdAE1 (Harrison, Rathinavelu, Arese, Geahlen, & Low, 1991; Low, Rathinavelu, & Harrison, 1993). Indeed, a complex of glycolytic enzymes is associated with the inner surface of the red cell membrane via an interaction with AE1, providing a local pool of ATP that is used by the sodium pump (Campanella et al., 2005, 2008; Puchulu-Campanella et al., 2013). Deoxyhemoglobin binds to cdAE1 with high affinity (Chu, Breite, Ciraolo, Franco, & Low, 2008), linking AE1 to the oxygenation state of the red cell. Denatured hemoglobin (hemichrome) causes clustering of associated AE1 (Kannan, Labotka, & Low, 1988), providing a possible inside-to-outside signal for red cell senescence (Kannan, Yuan, & Low, 1991; Low, 1991).

The assembly of the cytoskeleton onto the membrane during erythropoiesis has been studied in a number of systems. Early studies using cultured human erythroid precursors indicated a staged assembly process, whereby Protein 4.2 was the last protein to be assembled onto the cytoskeleton (Wada et al., 1999). Spectrin was detected early on, followed by ankyrin and Protein 4.1, then AE1, and finally Protein 4.2 (Wada et al., 1999; Wickrema, Koury, Dai, & Krantz, 1994). Early work using chicken erythroblasts showed synthesis and assembly of the cytoskeletal proteins (spectrin, ankyrin, and Protein 4.1) well before AE1 (Cox, Stack, & Lazarides, 1987; Woods, Boyer, Vogt, & Lazarides, 1986). Other work has challenged whether *in transformed cells*, the staged assembly of the cytoskeleton followed by AE1 occurs during normal erythropoiesis where synchronous synthesis of AE1 and the cytoskeletal components occurs (Nehls, Zeitler-Zapf, & Drenckhahn, 1993). AE1 is made late in the production of erythroid cells isolated from rabbit bone marrow (Foxwell & Tanner, 1981). A cytoskeleton can assemble in the absence of AE1 in a mouse knockout model, but the resulting red cells are extremely fragile (Peters et al., 1996).

Studies of primary chicken erythroblasts have shown that AE1 first traffics as a dimer to the plasma membrane without processing of its N-linked oligosaccharide. It is then endocytosed and traffics back to the Golgi where it associates with the cytoskeletal protein ankyrin, likely as a tetramer. Its oligosaccharide is then modified to a complex form (Ghosh, Cox, & Cox, 1999). The AE1–ankyrin complex then moves to the plasma membrane where ankyrin can interact with spectrin/actin, locking AE1 to the underlying cytoskeleton. It is not known whether the same assembly mechanism is in place for human AE1.

4.3. Carbonic anhydrase interaction with the C-terminal tail of AE1

The identical C-terminal tails of AE1 and kAE1 can bind CAII via an acidic motif (–LDADD–), thereby forming a bicarbonate transport metabolon within red cells and kidney cells, respectively (Reithmeier, 2001; Sowah & Casey, 2011; Sterling et al., 2001; Vince, Carlsson, & Reithmeier, 2000; Vince & Reithmeier, 1998, 2000). This interaction promotes bicarbonate transport across the membrane by rapidly interconverting CO_2+H_2O and $HCO_3^-+H^+$ close to the site of bicarbonate transport across the plasma membrane, although this effect has been disputed (Boron, 2009). Bicarbonate may be "channeled" directly between CAII and AE1 creating a local pool of bicarbonate that is not in equilibrium with bulk bicarbonate. Indeed, there is evidence that the bicarbonate transport metabolon creates a pH micro-domain in the vicinity of AE1 (Johnson & Casey, 2011). CO_2 that enters the cell is rapidly converted to bicarbonate by CAII lining the inner surface of the membrane that is then immediately transported out of the cell. This process essentially converts an extracellular molecule of carbon dioxide into an extracellular molecule of bicarbonate. Evidence has been presented that membrane-transport metabolons that link transporters and enzymes may be common throughout Nature (Moraes & Reithmeier, 2012).

5. ERYTHROID AE1-ASSOCIATED DISEASES

Mutations in the *AE1* gene are linked to a number of human diseases and conditions (Table 1.1) (Almomani, Chu, & Cordat, 2011; Bruce et al., 2000; Cordat & Casey, 2009; Tanner, 2002; Wrong, Bruce, Unwin, Toye, & Tanner, 2002). Many of these mutations affect the folding of AE1 and its subsequent trafficking from its site of synthesis in the ER, through the Golgi and to its final destination in the plasma membrane (Almomani et al., 2011). Some also directly affect its transport function. It is important to recognize that AE1 as a membrane glycoprotein is made in red cell precursors prior to the loss of the ER and other components of the secretory pathway. Once inserted in the plasma membrane of the red cell, AE1 does not turnover but is present for the life of the red cell. Defective red cells as in HS have a shorter half-life than the 120 days of a typical erythrocyte. Kidney AE1 has a very different life story as detailed in Section 7.1. Like AE1, it is made in the ER and traffics to the cell surface where it likely associates with the underlying cytoskeleton, which stabilizes it at the level of

Table 1.1 Mutations in human AE1 linked to red cell diseases and conditions

Mutation	Name	Structural context	Comments	References
Blood group antigens				
R432W	ELO	Defines extracellular region (EC1)		(Jarolim, Rubin, Zakova, Storry, & Reid, 1998)
P548L	Diego, Rba	Defines extracellular region (EC3)		(Jarolim, Murray, Rubin, Smart, & Moulds, 1997)
K551N	Diego, Tra	Defines extracellular region (EC3)		(Jarolim, Murray, Rubin, Smart, et al., 1997)
T552I	Warrior	Defines extracellular region (EC3)		(Jarolim, Murray, Rubin, Coghlan, & Zelinski, 1997)
Y555H	VanVugt, Vga	Defines extracellular region (EC3)		(Jarolim et al., 1998)
V557M	Diego, Wda	Defines extracellular region (EC3)		(Bruce, Zelinski, Ridgwell, & Tanner, 1996; Jarolim, Murray, Rubin, Smart, et al., 1997)
P561S	BOW	Defines extracellular region (EC3)		(Jarolim et al., 1998)
G565A	Wulfsberg (Wu)	Defines extracellular region (EC3)		(Jarolim et al., 1998)
N569K	Bishop (Bpa)	Defines extracellular region (EC3)		(Jarolim et al., 1998)
R656C	Hughes (Hga)	Within N-glycosylated EC4		(Jarolim et al., 1998)
R656H	Moen (Moa)	Within N-glycosylated EC4		(Jarolim et al., 1998)

Table 1.1 Mutations in human AE1 linked to red cell diseases and conditions—cont'd

Mutation	Name	Structural context	Comments	References
E658K	Wright, $Wra^{+}b^{-}$	Within N-glycosylated EC4	Interaction with GPA	(Bruce et al., 1995)
Band 3 Memphis				
K56E	Memphis	Altered mobility on SDS gels	Linked to disease-causing mutations	(Jarolim, Murray, Rubin, Smart, et al., 1997; Mueller & Morrison, 1977)
P854L	Memphis II Diago (Dia)	Altered DIDS reactivity		(Bruce, Anstee, Spring, & Tanner, 1994)
Band 3 High Transport (HT)				
P868L	Band 3 HT	Altered DIDS reactivity	Higher V_{max} for transport	(Bruce, Kay, Lawrence, & Tanner, 1993)
Southeast Asian Ovalocytosis (SAO)				
Del400-408	Band 3 SAO	Misfolded	Misfolded and inactive protein, localized to ER in transfected cells, present in red cell membrane as heterodimer	(Cheung et al., 2005b; Jarolim et al., 1991; Schofield, Martin, Spillett, & Tanner, 1994; Schofield, Tanner, et al., 1992)
Hereditary spherocytosis (HS)—cytosolic domain				
E40K		In N-terminal disordered region	Affects Protein 4.2 binding	(Low, Zhang, & Bolin, 2001)
G130R		Exposed in middle of helix	Affects Protein 4.2 binding	(Low et al., 2001)
R180H			Affects ankyrin binding site	(Van Zwieten et al., 2013)
A285D	Boston		Destabilizes Band 3 protein	(Jarolim et al., 1996)

Continued

Table 1.1 Mutations in human AE1 linked to red cell diseases and conditions—cont'd

Mutation	Name	Structural context	Comments	References
P327R	Tuscaloosa	First turn of helix, minor effect on protein stability	Affects Protein 4.2 binding	(Jarolim, Palek, et al., 1992; Low et al., 2001; Zhou et al., 2007)
Hereditary spherocytosis (HS)—membrane domain				
G455E	Prague V			(Jarolim et al., 1996)
C479W	Edmonton		Misfolded, recessive for dRTA	(Chu et al., 2009)
V488M	Coimbra		Recessive for dRTA	(Alloisio et al., 1997)
R490C	Bicetre			(Dhermy et al., 1997)
R518C	Dresden			(Eber et al., 1996)
S667F	Courcouronnes		Recessive for dRTA, only partially rescued by GPA	(Toye et al., 2008)
L707P			Retained in the ER	(Jarolim et al., 1996; Quilty & Reithmeier, 2000)
G714R	Okinawa			(Kanzaki et al., 1997)
R760Q/W			Retained in the ER	(Quilty & Reithmeier, 2000)
G771D	Chur			(Maillet, Alloisio, Morle, & Delaunay, 1996)
I783N	Napoli II			(Miraglia del Giudice et al., 1997)
R808C/H	Jablonec, Nara			(Jarolim et al., 1996; Kanzaki et al., 1997; Quilty & Reithmeier, 2000)

Table 1.1 Mutations in human AE1 linked to red cell diseases and conditions—cont'd

Mutation	Name	Structural context	Comments	References
H834P	Birmingham			(Jarolim et al., 1996; Quilty & Reithmeier, 2000)
T837M/A	Philadelphia, Tokyo			(Iwase et al., 1998; Jarolim et al., 1996; Quilty & Reithmeier, 2000)
A858D			Recessive, also causes dRTA in the homozygous state and incomplete dRTA in the heterozygous state	(Shmukler et al., 2010; Sinha, Agarwal, Bawazir, & Bruce, 2013)
R870W	Prague III		Retained in the ER	(Jarolim et al., 1996; Quilty & Reithmeier, 2000)

This table does not include polymorphisms and frameshift/truncations.

the BLM. Unlike AE1 in red cells, kAE1 is subject to endocytosis and turnover (Cordat, unpublished; Williamson, Brown, Mawby, & Toye, 2008). Furthermore, red cells express Glycophorin A, which as explain above, can facilitate the cell surface expression of AE1 (Groves & Tanner, 1992, 1994b; Young et al., 2000) and certain AE1 mutants (Pang & Reithmeier, 2009; Toye et al., 2008). Finally, human red cell precursors selectively lose the ER chaperone calnexin and other components of the protein quality control system of nucleated cells prior to the maximum expression of membrane glycoproteins like AE1 (Patterson et al., 2009). This loss of protein quality control may allow mutant proteins like AE1 SAO to escape the ER and traffic to the plasma membrane, thereby accounting for their presence in mature red cells.

5.1. AE1 Memphis

The Memphis AE1 variant was described by Mueller and Morrison (1977) and is the first polymorphism found in AE1. It is due to a K56E point mutation located at the beginning of the first β strand in cdAE1 (Jarolim, Rubin, et al., 1992). Treatment of intact red blood cells with Pronase or chymotrypsin produces a 60 kDa N-terminal fragment and a complementary ~35 kDa C-terminal N-glycosylated fragment (Steck, Fairbanks, & Wallach, 1971). In heterozygotes with the Memphis mutation, proteolytic treatment of AE1 produces a doublet of 63-kDa and 60-kDa polypeptide bands of equal intensity on SDS gels, with the upper band carrying the Memphis point mutation. This gel shift assay provides a convenient biochemical assay for the presence of AE1 Memphis, which is often linked to other mutations.

A high transport (HT) variant (P868L) also carries the Memphis mutation. AE1 HT exhibits increased anion transport activity in red cells and results in an abnormal red cell shape (acanthocytosis) (Bruce et al., 1993). It was originally discovered by Marguerite Kay as an AE1 protein with a slower than normal migration rate on SDS gels (Kay, Bosman, & Lawrence, 1988). The mutation is located within the last TM segment of AE1, suggesting that this region plays an important role in the transport mechanism. The proline to leucine mutation is expected to remove a kink or bend in a helical segment.

5.2. AE1-associated blood group antigens

AE1 displays various blood group antigens that are often due to single amino acid changes. The Diego blood (Dia) group antigen was the first to be ascribed to an amino acid change in the extracellular portion of AE1 (Spring, Bruce, Anstee, & Tanner, 1992). As such, these epitopes define regions of AE1 that face the cell exterior, thereby providing valuable topological information (Fig. 1.3B). The Diego blood group is associated with a P854L mutation in the last extracellular loop of AE1 (Bruce, Anstee, Spring, & Tanner, 1994) and is associated with the Memphis K56E mutation. A T552I mutation gives rise to the Warrior blood group antigen and is designated part of the Diego blood group system (Jarolim, Murray, Rubin, Coghlan, & Zelinski, 1997).

5.3. Congenital dyserythropoietic anemia, type II (CDA II)

CDAII or HEMPAS is a human disease that results in impaired erythroid development with red cells displaying multinuclearity and a "double membrane" with the inner membrane derived from the ER (Alloisio et al., 1996). AE1 in these red cells contains a high-mannose oligosaccharide rather than the normal complex structure, suggesting a defect in trafficking and oligosaccharide processing (Kameh, Landolt-Marticorena, Charuk, Schachter, & Reithmeier, 1998). HEMPAS is due to a defect in the SEC23B component of the COPII complex (Dancourt & Barlowe, 2010) that is responsible for the budding of vesicles from the ER that carries cargo to the Golgi (Schwarz et al., 2009). A defect in SEC23B expression in red cell precursors results in retention of remnant ER in the mature red cell and a defect in the processing of the N-linked oligosaccharide to the complex form. The red cell pathway depends on SEC23B, while other cells can use either SEC23A or B for this key step in the secretory pathway. A zebrafish mutant (retsina), devoid of AE1, shows a defect in cytokinesis and exhibits multinuclearity similar to CDA II, suggesting a role of AE1 in erythroid-specific cell division (Paw et al., 2003).

5.4. Southeast Asian Ovalocytosis

SAO is due to a 9-amino acid deletion at the boundary of the cytosolic domain and the beginning of transmembrane segment 1 (TM1) (Jarolim et al., 1991; Schofield, Reardon, & Tanner, 1992). This deletion results in removal of a bend at Pro 403 and part of the hydrophobic region of TM1 (Chambers, Bloomberg, Ring, & Tanner, 1999) and impairs the insertion of the mutant protein into the membrane (Cheung & Reithmeier, 2005). AE1 SAO is prone to aggregation and exhibits tighter binding to ankyrin and the cytoskeleton (Liu et al., 1995, 1990). Not surprisingly, the mutant is misfolded and is not capable of mediating anion transport or binding inhibitors (Sarabia, Casey, & Reithmeier, 1993). It has a normal secondary structure but does not undergo a cooperative thermal melt (Moriyama, Ideguchi, Lombardo, Van Dort, & Low, 1992). This is reminiscent of a "molten globule" state of a protein that has not collapsed to its final folded state. What is surprising, however, is that AE1 SAO—a misfolded membrane protein—is present in about equal amounts to the wild-type AE1 in the red cells of heterozygotes. SAO in the homozygous state is embryonic lethal as no homozygotes have yet been identified in families with SAO (Jarolim et al., 1991; Wrong et al., 2002). In order to survive, homozygotes would need blood transfusions, beginning early in life.

How misfolded AE1 SAO escapes the quality control system during biosynthesis in red cell precursors is an important question. Expression of AE1 SAO in transfected HEK or kAE1 SAO in MDCK cells results in retention of this mutant in the ER (Cheung, Cordat, & Reithmeier, 2005b). In red cell precursors however, AE1 SAO can traffic to the cell surface. Heterodimer formation with wild-type AE1 may facilitate the ER exit of AE1 SAO and its movement to the plasma membrane (Cheung et al., 2005b). Indeed, evidence for heterodimers of wild-type AE1 and SAO AE1 in mature red cells has been obtained (Jennings & Gosselink, 1995). The wild-type AE1 may rescue the trafficking of the SAO mutant by heterodimer formation, while the SAO dimer is subject to ER retention and rapid degradation. AE1 SAO traffics to the cell surface in human K562 cells that express endogenous Glycophorin A (Patterson et al., 2009). This suggests that Glycophorin A facilitates the trafficking of AE1 SAO and may account for the presence of the mutant protein in mature red cells. In kidney cells devoid of Glycophorin A, kAE1 SAO would be retained in the ER and act in a dominant-negative manner in heterodimers with wild-type kAE1 (Cheung et al., 2005b). The SAO mutation in TM1 may not preclude the interaction of Glycophorin A with AE1 that is centered at the extracellular proximal end of TM8 (Bruce et al., 1995). It would certainly be worthwhile studying the biosynthesis and trafficking of AE1 in red cell progenitors from patients with SAO, and the effect of knockdown of Glycophorin A.

5.5. Hereditary spherocytosis

Single amino acid changes in AE1 can cause HS, a hemolytic anemia resulting from smaller misshapen red cells. Most of these dominant mutations cause AE1 to misfold and be retained in the ER when expressed in transfected cells (Quilty & Reithmeier, 2000). Because the ER is lost during red cell development, the HS mutant protein never makes it to the cell surface and is therefore not present in mature red cells. This results in membrane instability and blebbing off of membrane vesicles resulting in smaller spheroidal red cells. Even more important is the finding that dominant HS mutants can form heterodimers with the wild-type protein and retain the wild-type protein in the ER (Quilty & Reithmeier, 2000). In contrast, recessive HS mutants can traffic to the cell surface on their own, albeit sometimes poorly, and more efficiently when in a heterodimer with wild-type AE1. In this regard, the wild-type protein acts in a dominant-positive fashion to facilitate the trafficking of the mutant in a heterodimer. Recessive HS mutants typically have a severely impaired transport function. Dominant HS mutants in contrast are typically functional. Yet these dominant mutants are retained in the ER. This is similar to the situation in cystic fibrosis (CF) where the common delta F508 mutant of the cystic fibrosis transmembrane regulator (CFTR) chloride channel is functional but retained in the ER. Clearly, the quality control system is sensing changes in the proteins that do not necessarily affect their ability to function. A cellular survey of the plethora of functions mediated by membrane proteins for quality control purposes is impossible. Rather, cells likely monitor the exposure of hydrophobic regions common in misfolded proteins.

HS mutations are found in both the membrane domain (Quilty & Reithmeier, 2000) and the cytosolic domain of AE1 (Low et al., 2001). The cytosolic mutations affect the interaction of AE1 with the protein components of the underlying cytoskeleton that leads to membrane instability. Band 3 Tuscaloosa (P327R) is associated with a 30% deficiency of protein 4.2 in the heterozygous state (Jarolim, Palek, et al., 1992). This mutation is associated with the asymptomatic Memphis mutation (K56E). Three HS mutations (E40K, G130R, and P327R) in cdAE1 do not affect the stability of the domain or the expression of AE1, but cause a diminished binding to protein 4.2 (Bustos & Reithmeier, 2006, 2010). Mutations in the cytosolic domain do not have a dramatic effect on the structure of the domain (Zhou et al., 2007). Furthermore, the three mutations do not cluster, but they are located on the same surface of the protein (Fig. 1.2). The three mutations all involve the introduction of positive charges into a largely acidic protein surface. This suggests that the interaction of protein 4.2 with the cytosolic domain is electrostatic and may involve a large surface area (Bustos & Reithmeier, 2010).

Disruption of the murine erythroid *AE1* gene caused spherocytosis due to the absence of AE1 in the plasma membrane. There was also a complete loss of Protein 4.2, while the levels of spectrin, actin, ankyrin, and protein 4.1 were normal (Southgate, Chishti, Mitchell, Yi, & Palek, 1996). This indicates that the stability of protein 4.2 is dependent upon its interaction with AE1. Interestingly, there was also a complete deficiency of Glycophorin A in the red cells from these mice (Hassoun et al., 1998). Clearly, AE1 is part of a complex of proteins that includes Glycophorin A and Protein 4.2 that are interdependent upon one another for their proper assembly and stability at the plasma membrane (Fig. 1.4).

There are also mutations in protein 4.2 that lead to HS in the homozygous or compound heterozygous state (Cohen, Dotimas, & Korsgren, 1993). Protein 4.2^{NIPPON} contains an A142T mutation that results in a dramatic decrease in the amount of this protein in red cells (Bouhassira et al., 1992). Red cells from a compound heterozygote of the Nippon mutation with the A317C Shiga mutation also had a nearly complete deficiency of protein 4.2 (Kanzaki et al., 1995). A homozygous Tozeur mutation R310Q is associated with a lack of protein 4.2 in red cells due to the instability of the protein rather than a defect in binding to AE1 (Hayette et al., 1995). Similarly, the Komatsu mutation (D175Y) results in a complete absence of protein 4.2 in the red cells of the homozygous patient, but normal amounts in the heterozygous parents. Expression studies in *Xenopus* oocytes have shown that the Tozeur and Komatsu, but not Nippon, protein 4.2 mutants exhibited an impaired interaction with co-expressed AE1 (Toye et al., 2005). Clearly, a disruption of the interaction of Protein 4.2 with AE1 leads to membrane instability and HS.

6. RETENTION AND RESCUE OF AE1 MUTANTS

The following questions arise: How are mutant forms of AE1 retained in the ER? How do misfolded mutants like AE1 SAO escape the quality control system of the ER in red cell precursors? Is it possible to rescue mutants by interfering with the retention and quality control system of the cell to regain functional expression of AE1 mutants at the plasma membrane?

6.1. Role of Glycophorin A in AE1 trafficking (Williamson & Toye, 2008)

Patients with dRTA rarely display a red cell defect. One example is the G701D dRTA mutant, which exists in both AE1 and kAE1 because these two membrane proteins are encoded by the same gene. Expression studies using *Xenopus* oocytes have shown that coexpression of Glycophorin A (GPA) facilitates the cell surface expression of AE1 (Tanphaichitr et al., 1998). In transfected HEK cells, the G701D mutant can escape the ER and move to the Golgi where it is retained. AE1 forms a complex with GPA in the ER (Pang & Reithmeier, 2009), and this association may allow the G701D mutant to escape retention in the Golgi. The ability of GPA to rescue the G701D mutant accounts for the lack of a red cell defect because GPA associated with AE1 in the erythrocyte to form the Wright blood group antigen (Bruce et al., 1995; Pang & Reithmeier, 2009). Kidney cells do not express GPA; hence, the G701D mutant remains localized intracellularly, likely in the Golgi. En(a−) individuals with a complete absence of GPA in the red blood are healthy with normal amounts of AE1 at the cell surface; therefore, GPA is not essential for AE1 trafficking. En(a−) patients with a dRTA mutation like G701D in AE1 are predicted to have red cell as well as kidney defects.

6.2. AE1 biosynthesis during erythropoiesis

AE1 is produced at the late stage of erythropoiesis, but before the reticulocyte stage. As mentioned earlier, previous studies using chicken erythroblasts indicated a sequential expression of red cell membrane proteins (Cox et al., 1987; Lazarides & Moon, 1984; Lazarides & Woods, 1989; Woods et al., 1986). Studies using primary human CD34+ cells reinforced this work (Wickrema et al., 1994). More recent studies using cultured human erythroblasts (Satchwell et al., 2011) have indicated that AE1 and protein 4.1 are synthesized at the same time and interact early during red cell development. We examined the role of chaperones during red cell development and discovered that calnexin and several other components like the disulfide isomerase ERp57 were selectively removed from the ER during the late stages of red cell development and before the maximal levels of AE1 and GPA were reached (Patterson et al., 2009). This startling finding shows that calnexin is not required for the biosynthesis and folding of these major glycoproteins during red cell development. Because AE1 contains no disulfide bonds, and GPA contains no cysteine residues, these proteins are not required to engage the calnexin–disulfide isomerase complex in order to fold properly. The loss of quality control machinery during the late stages of red cell development may facilitate the massive synthesis and trafficking of major glycoproteins like AE1 and GPA to the plasma membrane. The lack of quality control may also permit misfolded proteins like AE1 SAO to exit the ER in partnership with GPA during red cell development, and to be present in the plasma membrane of mature erythrocytes. Thus, one could hypothesize that HS mutants absent from the erythroid cell surface may be unable to interact with GPA. They would be retained in the ER and lost during red cell development. It would be useful to compare the trafficking of AE1 in CD34+ cells from patients with normal and mutant forms of AE1, and to determine the effect of reintroducing calnexin.

7. KIDNEY AE1 BIOGENESIS, INTERACTORS, AND ASSOCIATED PATHOLOGIES

7.1. Kidney AE1 biosynthesis

The use of an alternative promoter and splicing produces the kidney version of AE1 (kAE1) that is missing the first 65 residues in AE1, but is otherwise identical in amino acid sequence (Alper, Kopito, Libresco, & Lodish, 1988; Brosius, Alper, Garcia, & Lodish, 1989). This protein is found in the BLM of acid-secreting α-intercalated cells where it mediates bicarbonate reabsorption into the blood in exchange for chloride. Like AE1, kAE1 is made in the ER and traffics to the basolateral surface where it likely associates with the underlying cytoskeleton, which stabilizes it at the level of the BLM. While the interacting partners for cdAE1 are well known, the function of the cytosolic domain of kAE1 is poorly understood. It may play a role in the retention of kAE1 in the BLM of polarized α-intercalated cells through interactions with the underlying cytoskeleton, or it could be involved in the endocytosis and turnover of kAE1. Key tyrosine residues within the C-terminal tail of kAE1 (Y904) and the N-terminal domain are involved in targeting to and endocytosis of the protein at the BLM, likely in a phosphorylation-dependent manner (Toye, Banting, & Tanner, 2004).

7.2. Kidney AE1 interactors

The interacting partners for kAE1 are less well characterized than for AE1. The cytosolic domain of kAE1 is missing the first 65 amino acids present in AE1. The missing domain includes the acidic disordered amino-terminal region and the first β-strand in the domain (Fig. 1.2). The lack of the β-strand results in a more open structure for the kidney cytosolic domain (Pang, Bustos, & Reithmeier, 2008). In contrast to its erythroid counterpart, kAE1 does not bind ankyrin (Ding, Casey, & Kopito, 1994). It was reported to bind kanadaptin, a protein expressed in intracellular vesicles in kidney cells that may serve as an adaptor for proper targeting of kAE1 to the BLM (Chen, VijayaKumar, Li, & Al-Awqati, 1998). However, later studies reported the nuclear localization of kanadaptin making it an unlikely specific adaptor to kAE1 (Hubner, Bahr, Gossmann, Efthymiadis, & Drenchahn, 2003). Furthermore, the physical interaction was undetectable in transfected HEK 293 cells (Kittanakom, Keskanokwong, Akkarapatumwong, Yenchitsomanus, & Reithmeier, 2004).

More recently, GAPDH was found to interact with the short carboxyl-terminal domain of kAE1 in kidney homogenates and immortalized Madin-Darby canine kidney (MDCK) cells (Su Y et al., 2011). Knockdown of endogenous GAPDH resulted in intracellular retention of kAE1 proteins in MDCK cells. However, it is unclear whether this intracellular retention is directly due to the lack of interaction, or indirectly due to the abnormal glycolytic state of the cells.

According to the results of a yeast two-hybrid assay, the adaptor protein complex 1 (AP-1A) subunit μA (μ-1A) interacts with the C-terminal tail of kAE1, and this interaction was observed in transfected HEK293 cells (Sawasdee et al., 2010). This interaction has been later confirmed by co-immunoprecipitation studies in the renal epithelial MDCK cell line and mouse kidney extracts. Knockdown of AP-1A dramatically destabilized kAE1 in MDCK cells and prevented kAE1 trafficking to the plasma membrane that caused degradation by the lysosome (Almomani et al., 2012). This study highlighted the unsuspected role of ubiquitous AP-1A in basolateral targeting of the exchanger to the BLM.

Williamson and colleagues demonstrated that kAE1 basolateral targeting is regulated by the phosphorylation status of the following two critical cytosolic tyrosine residues: the amino-terminal Tyr 359 and the carboxyl-terminal Tyr 904 (Williamson et al., 2008). Hyperosmolarity, hypertonicity, and nonphysiological concentrations of extracellular bicarbonate caused phosphorylation of these tyrosine amino acids and internalization of kAE1. It was thus proposed that the phosphorylation status of these two tyrosine residues controls the level of plasma membrane expression of kAE1. A physiological trigger for this trafficking cycle remains to be identified.

Wu and colleagues reported that kAE1 is expressed at a low level in podocytes in the kidney (Wu et al., 2010; Wu, Satchwell, & Toye, 2011). The exchanger interacts with nephrin, which is one of the proteins involved in maintenance of the glomerular slit diaphragm (Wu et al., 2010). The interaction of kAE1 with integrin-linked kinase (Keskanokwong et al., 2007) together with the interaction with nephrin is proposed to maintain the integrity of the slit diaphragm by participating in the linkage of the podocyte's plasma membrane to the cytoskeleton (Wu et al., 2011).

Membrane yeast two-hybrid screens using tagged AE1 and kAE1 as baits, and a kidney cDNA library as prey identified an array of potential interacting partners for these membrane proteins (Reithmeier, unpublished). Some of the putative kAE1 interactors identified in this screen include the following: glycolytic enzymes (aldolase, GAPDH, lactate dehydrogenase, phosphofructokinase, and pyruvate kinase), cytoskeletal proteins (actin and myosin), and proteins involved in protein folding and trafficking (adaptin AP1μ2, kinesin, and Hsp70). It is important to show that the putative partner proteins interact with kAE1 in a cellular context, and that the interaction is physiologically important. This is typically done by "knocking down" the interacting partner and examining the effect on the functional expression of kAE1, as was done for AP-1A (Almomani et al., 2012).

7.3. Distal renal tubular acidosis (Yenchitsomanus et al., 2003)

Mutations in the *AE1* gene also cause a kidney disease, dRTA (Cordat, 2006; Yenchitsomanus, Kittanakom, Rungroj, Cordat, & Reithmeier, 2005; Yenchitsomanus et al., 2003) (Fig. 1.5, Table 1.2). Kidney AE1 is expressed in acid-secreting α-intercalated cells and is found at the BLM of these polarized epithelial cells, where it mediates bicarbonate reabsorption into the blood in exchange for chloride (Fig. 1.1B). Protons are transported into the urine by an apical V-type H^+ ATPase. Mutations in the ATPase can also cause dRTA due to a loss of an interaction with the glycolytic enzyme phosphofructokinase (Su et al., 2008). Patients with dRTA are defective in acid secretion and display an elevated urine pH even after an acid challenge. They suffer from kidney stones and rickets and are small in stature. A paper published in the December 1992 issue of American Journal of Diseases of Children (Lewis, 1992) proposed that Tiny Tim, a fictional character in

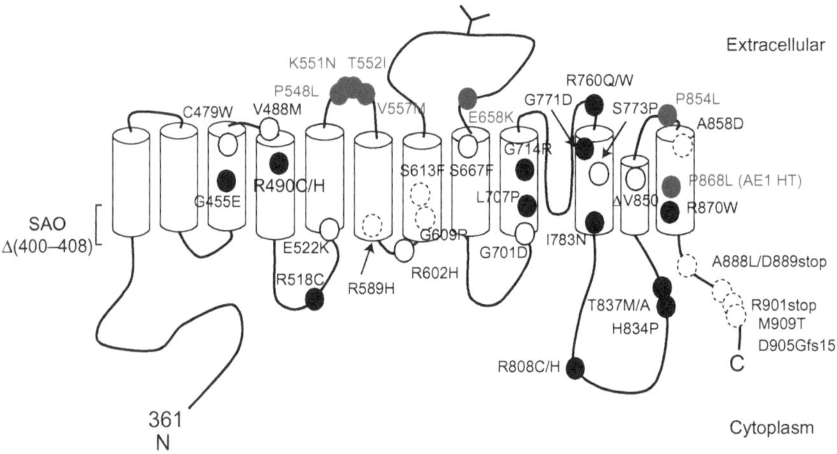

Figure 1.5 A folding model of the membrane domain of human AE1 (mdAE1, residues 361–911) showing the location of mutations associated with the blood group (dark grey circles), SAO, HS (black circles), dRTA (white circles with solid lines show recessive mutations; dashed lines show dominant mutations), and other mutations. Blood group antigens such as Diego (Di^a, P854L) face the cell exterior as expected. The Wright blood group antigen (Wr^{a+b-}) is due to a mutation (E658K) at the extracellular end of TM8 in AE1 that affects its interaction with Glycophorin A (dark grey). The SAO mutation is a 9-amino acid deletion (residues 400–408) at the cytosolic end of TM1. Another deletion is the dRTA mutant Band 3 Walton (R901STOP) that is missing the C-terminal 11 amino acids (residues 901–911). HS mutations are located throughout the membrane domain and are also present in the cytosolic N-terminal domain (not shown), while dRTA mutations are exclusively found in the membrane domain and the short C-terminal domain.

Table 1.2 Mutations in kAE1 linked to distal renal tubular acidosis (dRTA)

Mutation	Type/name	Structural context	Comments	References
V488M	Recessive, Coimbra		Causes HS as well in the homozygous state	(Alloisio et al., 1997)
C479W	Recessive, Edmonton	C479 labeled by eosin maleimide, misfolded	Retained in ER, dominantly causes HS	(Chu et al., 2009)
G494S	Recessive			(Zhang et al., 2012)
E522K	Recessive		20% at the plasma membrane compared to control	(Chang et al., 2009)
R589H/S/C	Dominant	Functional	Retained in ER	(Bruce et al., 1997; Cordat et al., 2006; Quilty et al., 2002b; Toye et al., 2004)
R602H	Recessive		Basolateral	(Cordat, unpublished results)
G609R	Dominant		Apical and basolateral	(Rungroj et al., 2004)
S613F	Dominant		Intracellular	(Bruce et al., 1997; Cordat et al., 2006; Toye et al., 2004)
S667F	Recessive for incomplete dRTA		Retained in the ER in MDCK cells	(Toye et al., 2008)
G701D	Recessive		Localized to Golgi, normal transport in red cells	(Bruce et al., 2000; Cordat et al., 2006)
S773P	Recessive		Impaired trafficking to cell surface	(Kittanakom, Cordat, Akkarapatumwong, Yenchitsomanus, & Reithmeier, 2004)

Continued

Table 1.2 Mutations in kAE1 linked to distal renal tubular acidosis (dRTA)—cont'd

Mutation	Type/name	Structural context	Comments	References
DelV850	Recessive		Near normal transport in red cells	(Bruce et al., 2000)
A858D	Mild Dominant		Causes incomplete dRTA	(Bruce et al., 2000)
A888L/ D889X	Dominant			(Cheidde, Vieira, Lima, Saad, & Heilberg, 2003)
R901STOP	Dominant, Walton		Near normal transport in red cells, missorted to apical membrane, retained in ER in MDCK cells	(Quilty, Cordat, & Reithmeier, 2002; Toye et al., 2004; Toye, Bruce, Unwin, Wrong, & Tanner, 2002b)
D905dup	Dominant			(Zhang et al., 2012)
D905Gfs15	Dominant, Qingdoa		Insertion of single base results in frameshift and novel 15-amino acid C-terminal sequence to residue 919	(Shao et al., 2010)
M909T	Dominant	Functional	Mislocalized to both apical and basolateral membranes	(Fry et al., 2012)

the Dickens novel A Christmas Carol, suffered from dRTA. Homozygous mice lacking AE1 exhibited symptoms of dRTA, while heterozygote mice were normal (Stehberger et al., 2007).

One of the first dRTA mutations to be identified and studied is a C-terminal truncation missing the last 11 amino acids in the protein (Band 3 Walton, 901STOP) (Jarolim et al., 1994). This mutation causes impaired exit of the protein from the ER and missorting to the apical membrane (Devonald, Smith, Poon, Ihrke, & Karet, 2003; Quilty, Li, & Reithmeier, 2002a; Toye, Bruce, Unwin, Wrong, & Tanner, 2002a). Further studies of a series of truncation mutants showed that deletion of as little as five amino acids from the C-terminus of AE1 impaired ER exit, while deletion of 20 residues profoundly decreased protein stability and expression (Cordat, Li, & Reithmeier, 2003). In agreement with these findings, some of the latest dRTA mutations identified are either point or frameshift mutations within the short carboxyl-terminus of AE1 (Fry et al., 2012; Shao et al., 2010; Zhang et al., 2012). Additionally, they are all dominantly inherited and do not cause erythroid symptoms. It is tempting to postulate that the short carboxyl-terminal tail of kAE1 plays an essential role in basolateral targeting, likely via a physical interaction with kidney-specific targeting proteins like members of the adaptin family.

The kidney disease, dRTA, presents in dominant and recessive forms (Cordat, 2006; Cordat et al., 2006). In the dominant disease, kAE1 typically retains function but is retained in the ER. In recessive dRTA, kAE1 has poor transport function and impaired trafficking to the BLM in polarized epithelial (MDCK) cells. Dominant mutants of kAE1 like R589Q can form heterodimers with wild-type AE1 and retain the wild-type protein in the ER (Quilty, Li, & Reithmeier, 2002b). Recessive mutants can also form heterodimers with wild-type kAE1, but in this case, the wild-type protein promotes trafficking of the heterodimer to the cell surface (Quilty et al., 2002b). Incomplete dRTA has also been described. In these patients, dRTA symptoms only appear after an acid challenge. Some dRTA mutations that cause incomplete dRTA have been characterized (Cordat & Reithmeier, 2006), but the molecular mechanisms resulting in the symptoms are less well understood than in complete dRTA.

Compound heterozygote recessive mutants such as G701D and A858D can form heterodimers that are predominantly retained in the ER (Ungsupravate et al., 2010), while heterodimers with the wild-type protein can traffic to the plasma membrane in transfected HEK cells. Compound heterozygotes can also carry one dominant and one recessive mutation. In the case of the A858D/delV850 compound heterozygote, the recessive and inactive delV850, which by itself is retained in the ER, reached the BLM when co-expressed with the dominant, mildly misfolded A858D (Cordat & Reithmeier, 2006). There are also compound heterozygotes of dRTA mutations with SAO. Recall that the SAO mutant is completely nonfunctional and retained in the ER in most transfected cells in the absence of Glycophorin A. SAO can form a heterodimer with the wild type or with dRTA mutants and retain the partner subunit in the ER (Cheung, Cordat, & Reithmeier, 2005a; Kittanakom, Cordat, & Reithmeier, 2008). In these examples, AE1 SAO exhibits a dominant-negative effect on its partner subunit, preventing its exit from the ER and trafficking to the cell surface. Thus, SAO AE1 in combination with a dRTA mutation can cause not only SAO in red cells, but severe dRTA in the kidney.

7.4. AE1 knockout mice develop dRTA

Two independently generated lines of AE1-knockout mice have been described, one lacking both erythroid AE1 and kAE1 (Peters et al., 1996); the other lacking only erythroid AE1 (Southgate et al., 1996). They display similar phenotypes, with 85% of the animals dying within 2 weeks of birth. Disruption of the murine erythroid *AE1* gene resulted in spherocytosis, as well as complete loss of protein 4.2, but normal levels of spectrin, actin, ankyrin, and protein 4.1 (Southgate et al., 1996). This indicates that the stability of protein 4.2 is dependent upon its interaction with AE1. Interestingly, there was also a complete deficiency of Glycophorin A in the red cells from these mice (Hassoun et al., 1998).

The surviving $AE1^{-/-}$ animals also developed dRTA associated with nephrocalcinosis, and displayed hypercalciuria, hyperphosphaturia, and hypocitraturia in addition to metabolic acidosis (Stehberger et al., 2007). In addition, these animals displayed tubular calcification of the renal medulla that is reminiscent of nephrocalcinosis that is seen in dRTA patients. Because both animal lines lack the erythroid AE1 and become very sick, further studies of acid–base imbalance, nephrocalcinosis, and dRTA are difficult to perform.

7.5. Rescuing dRTA mutants

A major question remains: "How are mutant forms of kAE1 retained in the ER?" Some clues come from studies of kAE1 expressed in MDCK cells. In these polarized cells, kAE1 traffics efficiently to the BLM. AE1 can interact with the membrane-bound ER chaperone, calnexin, but not with luminal calreticulin (Popov & Reithmeier, 1999). Castanospermine is an inhibitor of glucosidases I and II and blocks the interaction of calnexin with client glycoproteins that retain the full $Glc_3Man_9GlcNAc$ precursor structure. Interestingly, castanospermine treatment of cells resulted in movement of some dominant but not recessive dRTA kAE1 mutants to the BLM (Patterson & Reithmeier, 2010). Mutation of the N-glycosylation site (N642D) that precludes an interaction with the lectin calnexin had the same effect. The severely misfolded SAO mutant, however, was not rescued. Thus, some dominant dRTA mutants are retained in the ER of polarized MDCK by their interaction with calnexin. It has been shown (Chu, King, Berrini, Alexander, & Cordat, 2013) that it is possible to rescue the trafficking and partially the function of the recessive G701D mutant to the BLM of MDCK cells using a chemical chaperone, while the dominant R589H or C479W mutants were not rescued. It would be worthwhile to perform high-throughput screens of chemical libraries to identify small molecules that can perform a similar rescue function (Romisch, 2004), without the toxic properties of drugs like castanospermine. Misfolded dRTA mutants are also prematurely degraded (Patterson & Reithmeier, 2010). It is thus imaginable that slowing down their premature degradation by blocking the proteasome pathway may help rescue functional chloride/bicarbonate exchange in type-A intercalated cells of these patients.

8. SLC4 FAMILY OF ANION TRANSPORTERS

Band 3 is the founding and best-characterized member of the SLC4 family of anion transport proteins (Reithmeier, 1993). In humans, the following 11 SLC4 members exist: SLC4A1-5 and SLC4A7-11, as well as SLC4A6, a pseudogene (Choi, 2012; Romero, Chen, Parker, & Boron, 2013). SLC4A2 and SLC4A3 are closely related to SL4CA1 and are typically designated as AE2 and AE3, respectively; they all mediate anion (Cl^-/HCO_3^-) exchange (Alper, 2006, 2009; Lee, Gunn, & Kopito, 1991). The second group of SLC4 transporters is the electrogenic sodium-bicarbonate cotransporters, of which there are two members: NBCe1 and NBCe2 that share about 50% sequence identity (Bevensee, Schmitt, Choi, Romero, & Boron, 2000; Romero, 2001; Romero & Boron, 1999; Romero et al., 2004). The third group is the electroneutral sodium-coupled bicarbonate transporters that includes NBCn1, NBCn2, and the sodium-dependent chloride/bicarbonate exchanger (NDCBE) that are highly related with over 70% sequence identity (Romero et al., 2004). Two other members of the human SLC4 family have been identified based on sequence homology, and these have been designated AE4 and BTR1. BTR1 shows less than 20% sequence identity to other members of the SLC4 family. The roles of these other two members remain poorly described.

8.1. AE2 and AE3 structure, function, and disease

Like AE1, AE2 and AE3 have a similar two-domain structure with an N-terminal cytosolic domain and a C-terminal membrane domain made up of ~14 TM segments (Alper, 2006, 2009; Romero, 2005). Overall, they share about 50% sequence identity with AE1, most notably in the membrane domain. There are some differences however. First, the N-terminal domains of AE2 and AE3 are bigger than AE1 with N-terminal extensions that vary in size. *AE2* and *AE3* genes are subject to the use of tissue-specific alternative promoters and differential splicing. The human *AE2* gene contains three alternative promoters that lead to at least three different transcripts (AE2a, b, c). The human AE2a protein is predicted to contain 1240 residues compared to 911 residues in human AE1. The role of the N-terminal extension is undefined, but this region could certainly provide sites of interaction with other proteins. A second difference between AE1 and AE2/3 is the location of the sites of N-glycosylation. While AE1 contains a single site of N-glycosylation in extracellular (EC) loop 4, AE2 and AE3 contain N-glycosylation sites in an expanded EC loop 3 (the so-called Z-loop).

AE2 is widely expressed, consistent with its role in the regulation of intracellular pH and volume (Lindsey et al., 1990). As a bicarbonate exporter, AE2 could help cells recover from alkaline pH. Coupled with a Na^+/H^+ exchanger (e.g., NHE1), it could mediate net NaCl uptake into cells that would result in an increase in cell volume due to osmotic water flow. AE2 is found in the BLM of epithelial cells where, like kAE1, it mediates chloride/bicarbonate exchange. In acid-secreting parietal cells, where AE2 is highly expressed, it is essential for acid secretion; reabsorbing bicarbonate into the blood while the H^+ P-type ATPase pumps protons produced by cytosolic CAII into the lumen of the gastric gland. Not surprisingly, AE2 contains a CAII binding site in its C-terminal tail that could localize the enzyme to the BLM. The activity of AE2 is stimulated by both alkaline pH and hypertonicity; the activity is regulated by a conserved motif in the N-terminal cytosolic domain, although the membrane domain also plays a role (Alper, Chernova, & Stewart, 2002; Humphreys, Jiang, Chernova, & Alper, 1995; Stewart, Chernova, Kunes, & Alper, 2001; Stewart, Kerr, Chernova, Alper, & Vaughan-Jones, 2004; Stewart, Kurschat, Vaughan-Jones, & Alper, 2009; Zhang, Chernova, Stuart-Tilley, Jiang, & Alper, 1996).

Human AE3 has two major splice variants (AE3a, b) derived from two alternative promoters that result in different lengths of the N-terminal region. AE3a was first cloned by Ron Kopito from a human brain library and was designated a neuronal form (Kopito et al., 1989). This longer AE3a form (1232 residues) is abundantly expressed in brain (also designated the brain form, bAE3) (Kudrycki, Newman, & Shull, 1990), retina

(Kobayashi, Morgans, Casey, & Kopito, 1994), and smooth muscle (Brosius et al., 1997), while the much shorter AE3b form is expressed in the heart (also designated the cardiac form, cAE3) (Linn, Kudrycki, & Shull, 1992; Yannoukakos et al., 1994). Like AE2, the activity of AE3 is stimulated by alkaline pH. AE3 transcripts have been detected in other tissues, so this transporter may play a role in pH regulation and bicarbonate transport beyond the brain and heart (Sterling & Casey, 1999).

Polymorphisms in the *AE2* gene that reduce transporter expression have been linked to primary biliary cirrhosis due to decreased expression of this transporter (Aiba et al., 2011; Juran, Atkinson, Larson, Schlicht, & Lazaridis, 2009; Medina, 2011). AE2-deficient mice also develop primary biliary cirrhosis (Salas et al., 2008). A targeted disruption of the *AE2* gene in mice leads to achlorhydria, a condition where gastric content is abnormally alkaline (Gawenis et al., 2004). This finding confirms the essential role that AE2 plays in acid production in the stomach. An A867D mutation in EC loop 3 of human AE3 has been linked to idiopathic generalized epilepsy (Sander et al., 2002), but the molecular basis of this effect has not been determined. Mice with a targeted knockout of the *AE3* gene show a reduced seizure threshold (Hentschke et al., 2006). A deficiency in AE3 also causes blindness in mice due to loss of the protein in the retina (Alvarez et al., 2007).

8.2. Sodium-bicarbonate cotransporters

In humans, the SLC4 sodium-bicarbonate cotransporters (SLCA4-A11) are expressed in a tissue-specific manner (Damkier, Nielsen, & Praetorius, 2007). The SLC4 family contains two well-characterized electrogenic sodium-bicarbonate cotransporters, SLC4A4 (NBCe1) and SLC4A5 (NBCe2) (Choi, 2012; Romero et al., 2013). There are three splice variants of NBCe1: the renal form (NBCe1A); a wide-spread form (NBCe1B) expressed in the pancreas and other organs; and a brain form (NBCe1c). The N-terminal 41 residues in the "A" form are replaced by an 85-residue sequence in the "B" form. The "C" form has a different C-terminal region with a 61-residue sequence replacing the 46-residue sequence found in the other two forms. NBCe1 is found in the BLM of polarized cells where it mediates bicarbonate uptake into the blood. This uptake is important in the proximal tubule of the kidney and plays a major role in the reabsorption of bicarbonate from the lumen to the blood. On the apical membrane, the sodium/proton exchanger 3 (NHE3) imports sodium and exports protons, which are converted by an extracellular carbonic anhydrase IV with bicarbonate to carbon dioxide and water. CO_2 moves into the cell from the lumen to be converted to bicarbonate, which is in turn exported across the BLM by NBCe1 in a sodium-dependent process. NBCe1 operates electrogenically with a 1:3 Na^+:bicarbonate stoichiometry in renal preparations, but the cloned version behaves differently when expressed in oocytes (Lee, Boron, & Parker, 2013).

Mutations in NBCe1 may be expected to cause proximal tubular acidosis due to a defect in bicarbonate reabsorption and acid secretion (Gawenis et al., 2007; Igarashi et al., 1999). Indeed, acidosis is seen with a dozen mutations reported to date (Dinour et al., 2004). Not only do the patients exhibit metabolic acidosis, with a lowering of the blood pH close to 7, but they also exhibit visual impairments including blindness (Igarashi et al., 2001). Mutations in human NBCe1 are located in the membrane domain and include the following: S427L (Demirci, Chang, Mah, Romero, & Gorin, 2006; Dinour et al., 2004; Horita et al., 2005; Igarashi et al., 1999; Parker, Qin, Williamson, Toye, & Boron, 2012; Suzuki et al., 2008), T485S, G486, R510H, L522P, A799V, R881C, plus some missense mutations like Q29X (Igarashi et al., 2001) and W516X (Lo et al., 2011) that would not produce intact protein products. Polymorphisms such as K558R decrease transport activity (Yamazaki et al., 2011), as do the R298S, S427L, and A799V mutations (Dinour et al., 2004; Parker et al., 2012). The T485S, G486R, R510H, W516X, L5522P, A799V, and R881C mutations cause trafficking defects, likely caused by protein misfolding (Demirci et al., 2006; Horita et al., 2005; Parker et al., 2012; Suzuki et al., 2008; Toye et al., 2006). For example, the R881C mutant was localized to the ER in MDCK cells, and it also has reduced cell surface expression in oocytes with a concomitant decrease in transport activity. Similar to the AE1 mutants described above, NBCe1 mutants exhibit ER retention of the proteins and a decrease in cell surface expression, although the protein itself may still retain transport activity.

R298S is an interesting NBCe1 mutation that is located in the cytosolic domain. This mutation does not impair trafficking but does decrease transport activity. Modeling studies using the cytosolic domain of AE1 as a template showed an interaction of Arg298 with Glu91. Mutation of this acidic residue causes a dramatic decrease in transport activity prompting the intriguing suggestion that the cytosolic domain contains a bicarbonate tunnel that funnels substrate to the membrane domain (Chang, DiPiero, Sonnichsen, & Romero, 2008). Mutations in the equivalent conserved residues in AE1 had no effect on transport activity, but did cause instability in the isolated cytosolic domain due to disruption of an extensive hydrogen-bonding network involving Arg298 (Shnitsar et al., 2013). Thus, a substrate-access tunnel in the cytosolic domain is not an essential feature of the SLC4 family of bicarbonate transporters.

Initial reports of variants of NBCe2 (SLC4A5) were likely due to incomplete mRNA splicing (Groger et al., 2012; Romero et al., 2013). NBCe2, variant c consists of 1121 residues and operates with a sodium:bicarbonate stoichiometry of 1:2. This form is expressed predominantly in liver, spleen, and testis, but also in the kidney, stomach, lung, heart, and brain. SLC4A5 has been linked to hypertension and the *slc4a5* mouse knockout exhibits hypertension, as well as dRTA, and vision and neurological defects (Groger et al., 2012; Kao et al., 2011).

The SLC4 family contains a number of electroneutral sodium-coupled bicarbonate transporters (Romero et al., 2013), namely SLC4A7 (NBCn1/NBC3), SLC4A8 (NDCBE), and SLC4A10 (NBCn2). NBCn1 is expressed in the heart, skeletal muscle, and also in the kidney where it is found in the BLM of the thick ascending loop of Henle. SLC4A8 catalyzes the functional exchange of one extracellular sodium and two bicarbonates for one intracellular chloride (Grichtchenko et al., 2001; Romero et al., 2000). It is expressed in the brain and contributes to pH regulation in neurons and CSF secretion by the choroid plexus (Chen, Haddad, & Boron, 2008; Chen, Kelly, et al., 2008). Knockout mice exhibit neurological defects where SlcA8 is proposed to be a key regulator of presynaptic pH and modulator of glutamine release (Sinning et al., 2011). It is also expressed in type-B intercalated cells in the cortical collecting duct of the kidney where together with SLC26A4 (pendrin), it regulates sodium and chloride reabsorption in a thiazide-sensitive way (Jacques et al., 2013; Leviel et al., 2010). NBCn2 is an electroneutral sodium-bicarbonate cotransporter (Parker et al., 2008) that is expressed in the brain where it contributes to CSF secretion (Wang, Yano, Nagashima, & Seino, 2000). The SLC4A9 and A11 forms are poorly characterized, although A11 is unique in the SLC4 family as it transports borate (BOH_4^-) rather than bicarbonate (Ogando, Jalimarada, Zhang, Vithana, & Bonanno, 2011). Mutations in SLC26A11 (BTR1) cause congenital, hereditary endothelial dystrophy (CHED2), Fuchs dystrophy, and Harboyan syndrome (Desir et al., 2007; Hemadevi et al., 2008; Kumar, Bhattacharjee, Prakash, & Sadanand, 2007; Riazuddin et al., 2011; Shah et al., 2008; Vithana et al., 2006). These mutations have been discovered to induce trafficking defects resulting in a decrease in active protein at the cell surface and a defect in water flux (Groger et al., 2010; Vilas et al., 2011; Vilas et al., 2013; Vilas, Morgan, Loganathan, Quon, & Casey, 2011; Vithana et al., 2008).

9. THE SLC26 (SULP) FAMILY OF ANION TRANSPORTERS

Another family of membrane proteins that transport anions is the SLC26 or SulP family (Alper & Sharma, 2013). The SLC designation tends to be used for human/animal members, while SulP is used to describe members from other organisms, especially bacteria and plants that transport sulfate (Felce & Saier, 2004). The proteins consist of an N-terminal membrane domain and a C-terminal STAS domain (Aravind & Koonin, 2000). STAS stands for "**s**ulfate **t**ransporter and **a**nti-**s**igma factor antagonist". The SpoIIAA anti-sigma factor from *Bacillus subtilis* acts to antagonize the activity of anti-sigma factors that in turn act to repress the activity of sigma factors in sporulation. Remarkably, STAS domains, even in human SLC26 family members, are homologous to SpoIIAA and have a similar fold and three-dimensional structure (Babu et al., 2010; Pasqualetto et al., 2010; Sharma, Ye, et al., 2011). The bacterial SLC26 members contain about 550 amino acid residues, while the eukaryotic members are larger—some up to about 900 residues, with extensive loops in the STAS domain.

9.1. SLC26 structure and function

The human SLC26 family of anion transporters (Table 1.3) consists of 10 functional members, SLC26A1 to A11, with A10 being a pseudogene (Alper & Sharma, 2013). These transporters are expressed in polarized cells in a tissue-specific manner with high levels found in organs like the intestine, pancreas, and kidney. The substrate specificity (Cl^-,HCO_3^-,I^-,SO_4^-, oxalate, formate, glyoxylate) and mode of transport (commonly anion exchange but some as channel) vary among family members (Moseley et al., 1999; Ohana, Shcheynikov, Yang, So, & Muallem, 2011; Ohana, Yang, Shcheynikov, & Muallem, 2009). The SLC26A1 and A2 members transport sulfate and oxalate in an exchange mode with chloride (Ohana, Shcheynikov, Park, & Muallem, 2012; Schnedler, Burckhardt, & Burckhardt, 2011; Xie, Welch, Mercado, Romero, & Mount, 2002). The other members carry out primarily chloride/bicarbonate exchange as in the case of SLC26A3. SLC26A3 works together with NHE3 to mediate NaCl transport across the apical membrane in the intestine. This functional association is facilitated by a physical association of SLC26A3 and NHE3 with PDZ adapter proteins via their C-terminal PDZ motifs (Alper &

Table 1.3 Mutations in the human SLC26 family of anion transporters linked to human disease

SLC26 family member	Substrate and transport mode	Tissue expression	Disease and mutations	References
SLC26A1	Anion exchange SO_4^{2-}, oxalate, Cl^-	Intestine (basolateral membrane), kidney (proximal tubule), liver hepatocytes	None reported, but mouse knockout has oxalate urolithiasis, hepatotoxicity	(Alper & Sharma, 2013)
SLC26A2 (DTDST)	Anion exchange SO_4^{2-}, oxalate, Cl^-	Chondrocytes, kidney proximal tubule, intestine and pancreatic duct (apical membrane)	Chondrodysplasias Most frequent is R279W Diastrophic Dysplasia: R279W, V340del, N77G, D111Y, L132P, Q135R, G166R, G255E, G259V, R279W, A386V, N425D, Q454P, L483P, G484D, R492W, C653S, H665P, G678V, T689S, A715V, I751A Atelosteogenesis type II: R279W, delta V340, Q454P, C653S, G678V, A715V Achondrogensis type R279W A715V, Q454P, C653S, G678V, I751A combined with N425D Multiple epiphyseal dysplasia type 4: R279W, C653S	(Rossi & Superti-Furga, 2001) (Karniski, 2001)
SLC26A3 (DRA, CLD)	Anion exchange Cl^-, HCO_3^-, oxalate	Intestine enterocyte, epididymis (apical membrane)	Congenital chloride losing diarrhea: G120S, H124L, P129L, P131R/L, M136I, Y204D, S206P, H220P, C343Y, G379A, S398F, D468V, L496R, Y520C, K521N, delY527, I544N, D652N	(Makela, Kere, Holmberg, & Hoglund, 2002; Wedenoja et al., 2011)

Continued

Table 1.3 Mutations in the human SLC26 family of anion transporters linked to human disease—cont'd

SLC26 family member	Substrate and transport mode	Tissue expression	Disease and mutations	References
SLC26A4 (PDS, Pendrin)	Anion exchange I^-, Cl^-, HCO_3^-	Cochlear, vestibular epithelial cells, thyrocytes, Type-B intercalated cells and airway epithelial cell (apical membrane)	Nonsyndromic deafness (with enlargement of the vestibular aqueduct), Pendred Syndrome, DFNB4: Most prevalent mutations are L236P and T416P	
			Pendred Syndrome: G139A, G209V, L236P, D271H, G384E, F383S, F400S, R409H, T416P, C565Y, F634S, H723R	(Van Hauwe et al., 1998)
			Pendred Syndrome: E384G, Y530H	(Coyle et al., 1998)
			Nonsyndromic deafness: G209V, K369E, A372V, T721M, H723R	(Tsukamoto et al., 2003)
			PDS and DFNB4: E29Q, Y105C, A106D, V138F, G139A, T193I, G209V, L236P, D271H, F335L, K369E, A372V, E384G, R409H, T410M, T416P, L445W, V480D, G497S, T508N, Y530H, Y556C, C565Y, L597S, V653A, F667C, G672E, T721M, H723R	(Campbell et al., 2001)
SLC26A5	Sensor, transport inactive	Cochlear hair cells	None reported, but linked to deafness	(Alper & Sharma, 2013)

SLC26A6 (CFEX, PAT1)	Anion exchange Cl^-, HCO_3^-, OH^-, oxalate, formate	Intestine enterocyte, pancreas, kidney proximal tubule, heart myocytes, sperm	None reported, but mouse knockout has reduced bicarbonate secretion	(Alper & Sharma, 2013)
SLC26A7 (SUT2)	Anion exchange Cl^-, HCO_3^-, OH^-, sulfate Anion channel Cl^-	Stomach gastric parietal cells, kidney type-A intercalated cells, endothelial cells (apical membrane and endosome)	None reported, but mouse knockout has gastric hypochlorohydria, dRTA	(Alper & Sharma, 2013)
SLC26A8 (TAT1)	Anion exchange Cl^-, HCO_3^-, OH^-	Sperm	Mouse knockout is male infertile R87Q, E812K and R954C	(Alper & Sharma, 2013) (Dirami et al., 2013)
SLC26A9	Anion exchange Cl^-, HCO_3^- Anion channel Cl^-, HCO_3^-	Lung airway epithelial cells, stomach gastric parietal cells, kidney, brain	None reported, but mouse knockout has gastric hypochlorohydria	(Alper & Sharma, 2013)
SLC26A10	Pseudogene	N/A	N/A	(Alper & Sharma, 2013)
SLC26A11 (SUT1, KBAT)	Anion exchange Cl^-, HCO_3^-, sulfate, oxalate	Kidney intercalated cells, pancreas, brain, endothelial cells, brain	None reported	(Alper & Sharma, 2013)

Sharma, 2013). SLC26A3 protein has been shown to interact with the R domain of the CFTR chloride channel via its STAS domain, and this interaction regulates the activity of both CFTR and SLC26A3 proteins (Chernova et al., 2003; Ko et al., 2002). SLC26A4 is a chloride/bicarbonate exchanger, but it can also transport iodide (Reimold et al., 2011; Shcheynikov et al., 2008). SLCA5 is a nonfunctional transporter and acts as a sensor in cochlear hair cells (Zheng et al., 2000). In addition to exchange activity, SLC26A9 and A11 have chloride channel activity.

The human SLC26 transporters are predicted to be glycoproteins and consist of an N-terminal membrane domain that spans the membrane ~12 times and a C-terminal cytosolic STAS domain (Fig. 1.6). The proteins

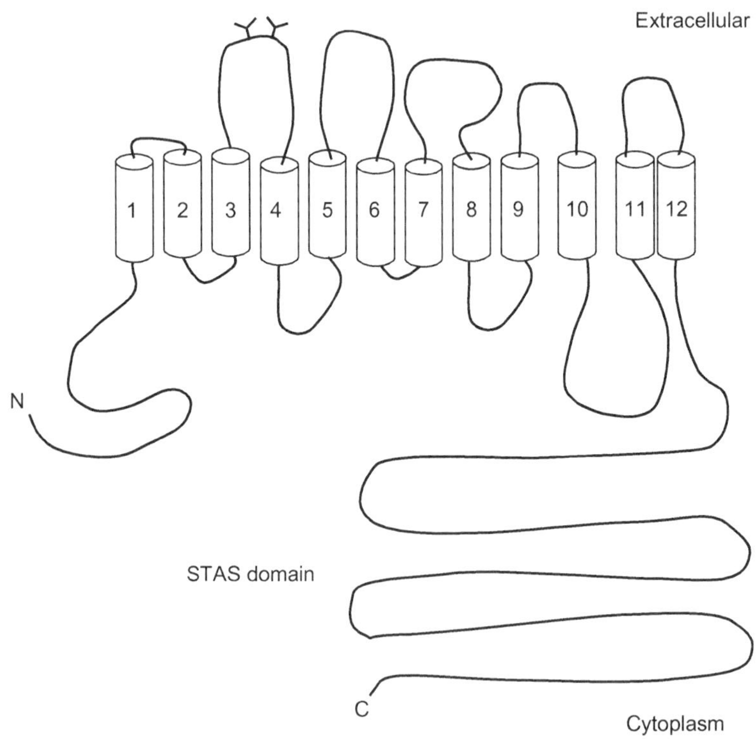

Figure 1.6 A folding model of a human SLC26 protein showing the location of the two putative N-glycosylation sites on the second extracellular loop, represented as Y symbols. In this model, there are 12 putative transmembrane segments with both N- and C-termini located in the cytosol. The C-terminal sulfate transporter and anti-sigma factor agonist (STAS) domain—common to all human SLC26 proteins—is located in the cytosol.

are oligomers, likely operating as dimers. Evidence has accumulated that SLC26A5 (Prestin) is a tetramer (Hallworth & Nichols, 2011). N-glycosylation of SLC26A3 has been reported to be required for efficient trafficking to the cell surface, but not for its transport activity (Hayashi & Yamashita, 2012). N-glycosylation may be important during the folding of these membrane proteins in the ER as part of protein quality control. Like AE1, the interaction of mutant forms of SLC26 transporters with the chaperone calnexin may provide a mechanism for retention of misfolded proteins in the ER.

The SLC26 transporters are members of a larger family of anion transport proteins including the bacterial SulP family of sulfate transporters. Indeed, SulP transporters are widely distributed in Nature and are found not only in animals, but also in archaea, bacteria, fungi, and plants. The BicA transporter from a cyanobacterium has been reported to be a sodium-bicarbonate cotransporter (Price & Howitt, 2011; Price, Woodger, Badger, Howitt, & Tucker, 2004). Topology studies on this protein have established a 12 TM model (Shelden, Howitt, & Price, 2009). The Rv3273 SulP transporter from *Mycobacterium tuberculosis* contains a C-terminal β-carbonic anhydrase rather than a typical STAS domain (Nishimori et al., 2010), suggesting that this protein is a bicarbonate transporter (Felce & Saier, 2004). The plant Arabidopsis contains at least 14 SulP genes, with most of these proteins functioning as sulfate transporters. The STAS domain of Sultr1;2 is required for proper targeting to the plasma membrane, and it binds and activates *O*-acetylserine-lyase/cysteine synthetase (Shibagaki & Grossman, 2004, 2006, 2010)—an example of a membrane-transport metabolon (Moraes & Reithmeier, 2012).

The structures of the C-terminal STAS domain from several organisms have been determined. It appears that the function of STAS domains is likely to provide interacting platforms to other proteins (Sharma, Rigby, & Alper, 2011). *E. coli* has only one SLC26 family member, YchM. The STAS domain of YchM was isolated as a complex with acyl carrier protein, suggesting a link between this transporter and fatty acid metabolism (Babu et al., 2010). YchM has also been shown to transport C-4 dicarboxylic acids like succinate and has been designated DauA (Karinou, Compton, Morel, & Javelle, 2013). The STAS domain of human SLC26A5—deleted of its disordered loops—has also been determined (Pasqualetto et al., 2010), as has the NMR structure of the STAS domain from *Mycobacterium tuberculosis* (Sharma, Ye, et al., 2011). They all resemble the SpoIIA and consist of a core of beta strands packed between α-helices. Also as mentioned earlier, the

STAS domain of SLC26A3 has been shown to interact with the R domain of CFTR (Ko et al., 2004). The association of these two proteins stimulates their mutual transport activities.

9.2. SLC26 family and disease

Mutations in some of the genes encoding SLC26 family members have been linked to a variety of human diseases (Table 1.3). Mutations in SLC26A2 cause chrondrodysplasias, mutations in SLC26A3 cause chronic chloride diarrhea, and mutations in SLC26A4 cause enlargement of the vestibular aqueduct in the inner ear and deafness. Mouse knockout models have revealed the physiological consequences of losing Slc26 family members including Slc26a1, 5, 6, 7, 8, 9. No (or only a few in the case of SLC26A8) human mutations associated with these genes have been clearly linked to disease to date.

In contrast, there is a plethora of mutations identified in SLC26A2, 3, and 4 linked to human disease. Very little is known about the consequences of these mutations on the structure, folding, and functional expression of the proteins. These mutations may be in the membrane domain or the STAS domain. As with AE1, the mutations likely cause protein misfolding and defects in the trafficking of the mutants from the ER to the cell surface. Mutations introduced into the conserved region of SLC26A2 can cause transport defects such as Glu417 that is purported to lie on the permeation pathway (Ohana et al., 2012). Diastrophic dysplasia sulfate transporter (DTDST) mutants varied in their transport properties when expressed in *Xenopus* oocytes. G255E, L483P, and N425D mutants had minimal sulfate transport activity, delV340 and R279W had transport rates of 17% and 32% respectively, while A715V, C653S, Q454P, and G678V had normal transport (Karniski, 2001). When expressed in HEK cells, the cell surface and transport levels of SLC26A2 mutants (R279W, delV340, Q454P, C635S, G678V, A715V) were lower than normal protein, indicating impaired trafficking. In particular, G678V mutant was entirely intracellular (Karniski, 2004).

In the case of SLC26A3, an I544N mutation causes the STAS domain to misfold and results in ER retention (Dorwart et al., 2008). The Tyr527 STAS deletion mutant, on the other hand, exhibits transport activity when expressed in cultured cells (Chernova et al., 2003; Dorwart et al., 2008). The Val318 deletion found in conjunction with a C307W polymorphism in the

Finnish population results in a protein devoid of transport activity, yet with normal levels of expression in the colon (Hoglund et al., 1996; Moseley et al., 1999). The equivalent mouse deletion mutant is retained in the ER, as a likely consequence of its misfolding (Ko et al., 2002; Wedenoja et al., 2011). The human L489R mutant is expressed at the plasma membrane where SLC26A3 normally interacts with CFTR (Ko et al., 2002).

Mutations in SLC26A4 (Pendrin) gene cause Pendred syndrome, characterized by deafness and the development of goiter. Its overexpression results in chloride-sensitive hypertension (Alper & Sharma, 2013). These mutations can also cause trafficking defects, but some mutations can diminish intrinsic function as well (Choi et al., 2009; Dai et al., 2009). Interestingly, the trafficking of certain SLC26A4 mutants can be "rescued" at lower temperature and by chemical chaperones (Shepshelovich et al., 2005; Yoon et al., 2008).

In 2013, mutations in testis-specific human SLC26A8 were identified that cause male infertility (Dirami et al., 2013) in agreement with the observation that Slc26A8 knockout mice are not fertile (Toure et al., 2007). SLC26A8 transports chloride, sulfate, and oxalate and its STAS domain interacts with a RacGap, a GTPase accelerating protein (Toure et al., 2001).

10. CONCLUSIONS AND FUTURE PROSPECTS

Mutations in the gene encoding AE1 and the truncated kAE1 typically cause protein misfolding and trafficking defects. How the mutations cause misfolding has not been characterized at the protein level. According to the results from an NMR study with synthetic peptides, the SAO deletion in AE1 would change the relative orientation of the cytosolic and membrane domains (Chambers et al., 1999). A crystal structure of AE1 would reveal much more information by putting the mutations in a structural context. Preliminary "threading" experiments using Phyre^2 have suggested that AE1 and other members of the SLC4 family may have a similar fold to the *E. coli* UraA uracil permease that works by an alternating access mode of transport, providing a useful structural model (Fig. 1.3).

Studies of red cell progenitors (CD34+ cells) from patients with AE1 mutations would provide a better experimental system than the transient transfection experiments routinely performed using transformed cells such as HEK-293. At the other extreme, studies of the mutant forms of the AE1 homologues in yeast (Jennings, Howren, Cui, Winters, & Hannigan, 2007; Zhao & Reithmeier, 2001) or *C. elegans* could provide powerful model systems amenable to genetic manipulation. Mouse models of human red cell diseases would be very useful. A mouse AE1 knockout produced fragile red cells due to loss of the connection of the cytoskeleton to the plasma membrane mediated by AE1 (Peters et al., 1996). These mice were anemic, failed to thrive, and also displayed symptoms of dRTA (Stehberger et al., 2007). It is possible to culture primary cells derived from the spleens of anemic wild-type and mutant mice and to study AE1 biosynthesis. Studies using primary cells from mice with various red cell diseases would complement the whole-animal studies.

Research on the effect of disease-causing mutations on the functional expression of other members of the SLC4 family and the SLC26 family of anion transporters lags behind the studies done with AE1. An advantage of the SLC26 family is that these transporters are expressed in all kingdoms of life, including bacteria. Great success has been achieved over the last decade in determining the crystal structures of bacterial membrane-transport proteins that have homologues in higher organisms, including humans. Once a structure has been obtained, it will lead to an understanding of the molecular basis of action of the transport protein and will also put the mutations into a structural context.

REFERENCES

Abe, Y., Chaen, T., Jin, X. R., Hamasaki, T., & Hamasaki, N. (2004). Mass spectrometric analyses of transmembrane proteins in human erythrocyte membrane. *Journal of Biochemistry, 136*, 97–106.

Aiba, Y., Nakamura, M., Joshita, S., Inamine, T., Komori, A., Yoshizawa, K., et al. (2011). Genetic polymorphisms in CTLA4 and SLC4A2 are differentially associated with the pathogenesis of primary biliary cirrhosis in Japanese patients. *Journal of Gastroenterology, 46*, 1203–1212.

Alloisio, N., Texier, P., Denoroy, L., Berger, C., Miraglia del Giudice, E., Perrotta, S., et al. (1996). The cisternae decorating the red blood cell membrane in congenital dyserythropoietic anemia (type II) originate from the endoplasmic reticulum. *Blood, 87*, 4433–4439.

Alloisio, N., Texier, P., Vallier, A., Ribeiro, M. L., Morle, L., Bozon, M., et al. (1997). Modulation of clinical expression and band 3 deficiency in hereditary spherocytosis. *Blood, 90*, 414–420.

Almomani, E. Y., Chu, C. Y., & Cordat, E. (2011). Mis-trafficking of bicarbonate transporters: Implications to human diseases. *Biochemistry and Cell Biology, 89*, 157–177.

Almomani, E. Y., King, J. C., Netsawang, J., Yenchitsomanus, P. T., Malasit, P., Limjindaporn, T., et al. (2012). Adaptor protein 1 complexes regulate intracellular

trafficking of the kidney anion exchanger 1 in epithelial cells. *American Journal of Physiology. Cell Physiology, 303*, C554–C566.
Alper, S. L. (2006). Molecular physiology of SLC4 anion exchangers. *Experimental Physiology, 91*, 153–161.
Alper, S. L. (2009). Molecular physiology and genetics of Na+-independent SLC4 anion exchangers. *The Journal of Experimental Biology, 212*, 1672–1683.
Alper, S. L., Chernova, M. N., & Stewart, A. K. (2002). How pH regulates a pH regulator: A regulatory hot spot in the N-terminal cytoplasmic domain of the AE2 anion exchanger. *Cell Biochemistry and Biophysics, 36*, 123–136.
Alper, S. L., Kopito, R. R., Libresco, S. M., & Lodish, H. F. (1988). Cloning and characterization of a murine band 3-related cDNA from kidney and from a lymphoid cell line. *The Journal of Biological Chemistry, 263*, 17092–17099.
Alper, S. L., & Sharma, A. K. (2013). The SLC26 gene family of anion transporters and channels. *Molecular Aspects of Medicine, 34*, 494–515.
Alvarez, B. V., Gilmour, G. S., Mema, S. C., Martin, B. T., Shull, G. E., Casey, J. R., et al. (2007). Blindness caused by deficiency in AE3 chloride/bicarbonate exchanger. *PLoS ONE, 2*, e839.
Appell, K. C., & Low, P. S. (1981). Partial structural characterization of the cytoplasmic domain of the erythrocyte membrane protein, band 3. *The Journal of Biological Chemistry, 256*, 11104–11111.
Aravind, L., & Koonin, E. V. (2000). The STAS domain—A link between anion transporters and antisigma-factor antagonists. *Current Biology, 10*, R53–R55.
Auffray, I., Marfatia, S., de Jong, K., Lee, G., Huang, C. H., Paszty, C., et al. (2001). Glycophorin A dimerization and band 3 interaction during erythroid membrane biogenesis: In vivo studies in human glycophorin A transgenic mice. *Blood, 97*, 2872–2878.
Babu, M., Greenblatt, J. F., Emili, A., Strynadka, N. C., Reithmeier, R. A., & Moraes, T. F. (2010). Structure of a SLC26 anion transporter STAS domain in complex with acyl carrier protein: Implications for E. coli YchM in fatty acid metabolism. *Structure, 18*, 1450–1462.
Barneaud-Rocca, D., Etchebest, C., & Guizouarn, H. (2013). Structural model of the anion exchanger 1 (SLC4A1) and identification of transmembrane segments forming the transport site. *The Journal of Biological Chemistry, 288*, 26372–26384.
Benyair, R., Ron, E., & Lederkremer, G. Z. (2011). Protein quality control, retention, and degradation at the endoplasmic reticulum. *International Review of Cell and Molecular Biology, 292*, 197–280.
Bevensee, M. O., Schmitt, B. M., Choi, I., Romero, M. F., & Boron, W. F. (2000). An electrogenic Na(+)-HCO(-)(3) cotransporter (NBC) with a novel COOH-terminus, cloned from rat brain. *American Journal of Physiology. Cell Physiology, 278*, C1200–C1211.
Bonar, P., Schneider, H. P., Becker, H. M., Deitmer, J. W., & Casey, J. R. (2013). Three-dimensional model for the human Cl/HCO exchanger, AE1, by homology to the E. coli ClC protein. *Journal of Molecular Biology, 425*, 2591–2608.
Boron, W. F. (2009). Evaluating the role of carbonic anhydrases in the transport of HCO3-related species. *Biochimica et Biophysica Acta, 1804*, 410–421.
Bouhassira, E. E., Schwartz, R. S., Yawata, Y., Ata, K., Kanzaki, A., Qiu, J. J., et al. (1992). An alanine-to-threonine substitution in protein 4.2 cDNA is associated with a Japanese form of hereditary hemolytic anemia (protein 4.2NIPPON). *Blood, 79*, 1846–1854.
Brosius, F. C., 3rd., Alper, S. L., Garcia, A. M., & Lodish, H. F. (1989). The major kidney band 3 gene transcript predicts an amino-terminal truncated band 3 polypeptide. *The Journal of Biological Chemistry, 264*, 7784–7787.

Brosius, F. C., 3rd., Pisoni, R. L., Cao, X., Deshmukh, G., Yannoukakos, D., Stuart-Tilley, A. K., et al. (1997). AE anion exchanger mRNA and protein expression in vascular smooth muscle cells, aorta, and renal microvessels. *The American Journal of Physiology, 273*, F1039–F1047.

Bruce, L. J., Anstee, D. J., Spring, F. A., & Tanner, M. J. (1994). Band 3 Memphis variant II. Altered stilbene disulfonate binding and the Diego (Dia) blood group antigen are associated with the human erythrocyte band 3 mutation Pro854– >Leu. *The Journal of Biological Chemistry, 269*, 16155–16158.

Bruce, L. J., Cope, D. L., Jones, G. K., Schofield, A. E., Burley, M., Povey, S., et al. (1997). Familial distal renal tubular acidosis is associated with mutations in the red cell anion exchanger (Band 3, AE1) gene. *The Journal of Clinical Investigation, 100*, 1693–1707.

Bruce, L. J., Groves, J. D., Okubo, Y., Thilaganathan, B., & Tanner, M. J. (1994). Altered band 3 structure and function in glycophorin A- and B-deficient (MkMk) red blood cells. *Blood, 84*, 916–922.

Bruce, L. J., Kay, M. M., Lawrence, C., & Tanner, M. J. (1993). Band 3 HT, a human red-cell variant associated with acanthocytosis and increased anion transport, carries the mutation Pro-868– >Leu in the membrane domain of band 3. *The Biochemical Journal, 293*(Pt 2), 317–320.

Bruce, L. J., Pan, R. J., Cope, D. L., Uchikawa, M., Gunn, R. B., Cherry, R. J., et al. (2004). Altered structure and anion transport properties of band 3 (AE1, SLC4A1) in human red cells lacking glycophorin A. *The Journal of Biological Chemistry, 279*, 2414–2420.

Bruce, L. J., Ring, S. M., Anstee, D. J., Reid, M. E., Wilkinson, S., & Tanner, M. J. (1995). Changes in the blood group Wright antigens are associated with a mutation at amino acid 658 in human erythrocyte band 3: A site of interaction between band 3 and glycophorin A under certain conditions. *Blood, 85*, 541–547.

Bruce, L. J., Wrong, O., Toye, A. M., Young, M. T., Ogle, G., Ismail, Z., et al. (2000). Band 3 mutations, renal tubular acidosis and South-East Asian ovalocytosis in Malaysia and Papua New Guinea: Loss of up to 95% band 3 transport in red cells. *The Biochemical Journal, 350*(Pt 1), 41–51.

Bruce, L. J., Zelinski, T., Ridgwell, K., & Tanner, M. J. (1996). The low-incidence blood group antigen, Wda, is associated with the substitution Val557– >Met in human erythrocyte band 3 (AE1). *Vox Sanguinis, 71*, 118–120.

Bustos, S. P., & Reithmeier, R. A. F. (2006). Structure and stability of hereditary spherocytosis mutants of the cytosolic domain of the erythrocyte anion exchanger 1 protein. *Biochemistry, 45*, 1026–1034.

Bustos, S. P., & Reithmeier, R. A. (2010). Protein 4.2 interaction with hereditary spherocytosis mutants of the cytoplasmic domain of human anion exchanger 1. *The Biochemical Journal, 433*, 313–322.

Cabantchik, Z. I., Knauf, P. A., & Rothstein, A. (1978). The anion transport system of the red blood cell. The role of membrane protein evaluated by the use of 'probes'. *Biochimica et Biophysica Acta, 515*, 239–302.

Campanella, M. E., Chu, H., & Low, P. S. (2005). Assembly and regulation of a glycolytic enzyme complex on the human erythrocyte membrane. *Proceedings of the National Academy of Sciences of the United States of America, 102*, 2402–2407.

Campanella, M. E., Chu, H., Wandersee, N. J., Peters, L. L., Mohandas, N., Gilligan, D. M., et al. (2008). Characterization of glycolytic enzyme interactions with murine erythrocyte membranes in wild-type and membrane protein knockout mice. *Blood, 112*, 3900–3906.

Campbell, C., Cucci, R. A., Prasad, S., Green, G. E., Edeal, J. B., Galer, C. E., et al. (2001). Pendred syndrome, DFNB4, and PDS/SLC26A4 identification of eight novel mutations and possible genotype-phenotype correlations. *Human Mutation, 17*, 403–411.

Casey, J. R., Ding, Y., & Kopito, R. R. (1995). The role of cysteine residues in the erythrocyte plasma membrane anion exchange protein, AE1. *The Journal of Biological Chemistry, 270,* 8521–8527.

Casey, J. R., Pirraglia, C. A., & Reithmeier, R. A. F. (1992). Enzymatic deglycosylation of human band-3, the anion transport protein of the erythrocyte-membrane—Effect on protein-structure and transport-properties. *Journal of Biological Chemistry, 267,* 11940–11948.

Casey, J. R., & Reithmeier, R. A. F. (1993). Detergent interaction with band-3, a model polytopic membrane-protein. *Biochemistry, 32,* 1172–1179.

Chambers, E. J., Bloomberg, G. B., Ring, S. M., & Tanner, M. J. (1999). Structural studies on the effects of the deletion in the red cell anion exchanger (band 3, AE1) associated with South East Asian ovalocytosis. *Journal of Molecular Biology, 285,* 1289–1307.

Chang, M. H., DiPiero, J., Sonnichsen, F. D., & Romero, M. F. (2008). Entry to "HCO3-tunnel" revealed by SLC4A4 human mutation and structural model. *The Journal of Biological Chemistry, 283,* 18402–18410.

Chang, S. H., & Low, P. S. (2003). Identification of a critical ankyrin-binding loop on the cytoplasmic domain of erythrocyte membrane band 3 by crystal structure analysis and site-directed mutagenesis. *The Journal of Biological Chemistry, 278,* 6879–6884.

Chang, Y. H., Shaw, C. F., Jian, S. H., Hsieh, K. H., Chiou, Y. H., & Lu, P. J. (2009). Compound mutations in human anion exchanger 1 are associated with complete distal renal tubular acidosis and hereditary spherocytosis. *Kidney International, 76,* 774–783.

Cheidde, L., Vieira, T. C., Lima, P. R., Saad, S. T., & Heilberg, I. P. (2003). A novel mutation in the anion exchanger 1 gene is associated with familial distal renal tubular acidosis and nephrocalcinosis. *Pediatrics, 112,* 1361–1367.

Chen, J., Vijayakumar, S., Li, X., & Al-Awqati, Q. (1998). Kanadaptin is a protein that interacts with the kidney but not the erythroid form of band 3. *The Journal of Biological Chemistry, 273,* 1038–1043.

Chen, L. M., Haddad, G. G., & Boron, W. F. (2008). Effects of chronic continuous hypoxia on the expression of SLC4A8 (NDCBE) in neonatal versus adult mouse brain. *Brain Research, 1238,* 85–92.

Chen, L. M., Kelly, M. L., Parker, M. D., Bouyer, P., Gill, H. S., Felie, J. M., et al. (2008). Expression and localization of Na-driven Cl-HCO(3)(-) exchanger (SLC4A8) in rodent CNS. *Neuroscience, 153,* 162–174.

Chernova, M. N., Jiang, L., Shmukler, B. E., Schweinfest, C. W., Blanco, P., Freedman, S. D., et al. (2003). Acute regulation of the SLC26A3 congenital chloride diarrhoea anion exchanger (DRA) expressed in Xenopus oocytes. *The Journal of Physiology, 549,* 3–19.

Cheung, J. C., Cordat, E., & Reithmeier, R. A. (2005a). Trafficking defects of the Southeast Asian ovalocytosis deletion mutant of anion exchanger 1 membrane proteins. *The Biochemical Journal, 392,* 425–434.

Cheung, J. C., Cordat, E., & Reithmeier, R. A. F. (2005b). Trafficking defects of the Southeast Asian ovalocytosis deletion mutant of anion exchanger 1 membrane proteins. *Biochemical Journal, 392,* 425–434.

Cheung, J. C., Li, J., & Reithmeier, R. A. F. (2005). Topology of transmembrane segments 1-4 in the human chloride/bicarbonate anion exchanger 1 (AE1) by scanning N-glycosylation mutagenesis. *Biochemical Journal, 390,* 137–144.

Cheung, J. C., & Reithmeier, R. A. F. (2005). Membrane integration and topology of the first transmembrane segment in normal and Southeast Asian ovalocytosis human erythrocyte anion exchanger 1. *Molecular Membrane Biology, 22,* 203–214.

Cheung, J. C., & Reithmeier, R. A. F. (2007). Scanning N-glycosylation mutagenesis of membrane proteins. *Methods, 41,* 451–459.

Choi, I. (2012). SLC4A transporters. *Current Topics in Membranes*, *70*, 77–103.

Choi, B. Y., Stewart, A. K., Madeo, A. C., Pryor, S. P., Lenhard, S., Kittles, R., et al. (2009). Hypo-functional SLC26A4 variants associated with nonsyndromic hearing loss and enlargement of the vestibular aqueduct: Genotype-phenotype correlation or coincidental polymorphisms? *Human Mutation*, *30*, 599–608.

Chu, H., Breite, A., Ciraolo, P., Franco, R. S., & Low, P. S. (2008). Characterization of the deoxyhemoglobin binding site on human erythrocyte band 3: Implications for O2 regulation of erythrocyte properties. *Blood*, *111*, 932–938.

Chu, C. Y., King, J. C., Berrini, M., Alexander, R. T., & Cordat, E. (2013). Functional rescue of a kidney anion exchanger 1 trafficking mutant in renal epithelial cells. *PLoS ONE*, *8*, e57062.

Chu, H., & Low, P. S. (2006). Mapping of glycolytic enzyme-binding sites on human erythrocyte band 3. *The Biochemical Journal*, *400*, 143–151.

Chu, C., Woods, N., Sawasdee, N., Guizouarn, H., Pellissier, B., Borgese, F., et al. (2009). Band 3 Edmonton I, a novel mutant of the anion exchanger 1 causing spherocytosis and distal renal tubular acidosis. *The Biochemical Journal*, *426*, 379–388.

Cohen, C. M., Dotimas, E., & Korsgren, C. (1993). Human erythrocyte membrane protein band 4.2 (pallidin). *Seminars in Hematology*, *30*, 119–137.

Cordat, E. (2006). Unraveling trafficking of the kidney anion exchanger 1 in polarized MDCK epithelial cells. *Biochemistry and Cell Biology*, *84*, 949–959.

Cordat, E., & Casey, J. R. (2009). Bicarbonate transport in cell physiology and disease. *The Biochemical Journal*, *417*, 423–439.

Cordat, E., Kittanakom, S., Yenchitsomanus, P. T., Li, J., Du, K., Lukacs, G. L., et al. (2006). Dominant and recessive distal renal tubular acidosis mutations of kidney anion exchanger 1 induce distinct trafficking defects in MDCK cells. *Traffic*, *7*, 117–128.

Cordat, E., Li, J., & Reithmeier, R. A. (2003). Carboxyl-terminal truncations of human anion exchanger impair its trafficking to the plasma membrane. *Traffic*, *4*, 642–651.

Cordat, E., & Reithmeier, R. A. (2006). Expression and interaction of two compound heterozygous distal renal tubular acidosis mutants of kidney anion exchanger 1 in epithelial cells. *American Journal of Physiology. Renal Physiology*, *291*, F1354–F1361.

Cox, J. V., Stack, J. H., & Lazarides, E. (1987). Erythroid anion transporter assembly is mediated by a developmentally regulated recruitment onto a preassembled membrane cytoskeleton. *The Journal of Cell Biology*, *105*, 1405–1416.

Coyle, B., Reardon, W., Herbrick, J. A., Tsui, L. C., Gausden, E., Lee, J., et al. (1998). Molecular analysis of the PDS gene in Pendred syndrome. *Human Molecular Genetics*, *7*, 1105–1112.

Dai, P., Stewart, A. K., Chebib, F., Hsu, A., Rozenfeld, J., Huang, D., et al. (2009). Distinct and novel SLC26A4/Pendrin mutations in Chinese and U.S. patients with nonsyndromic hearing loss. *Physiological Genomics*, *38*, 281–290.

Damkier, H. H., Nielsen, S., & Praetorius, J. (2007). Molecular expression of SLC4-derived-Na+-dependent anion transporters in selected human tissues. *American Journal of Physiology. Regulatory, Integrative and Comparative Physiology*, *293*, R2136–R2146.

Dancourt, J., & Barlowe, C. (2010). Protein sorting receptors in the early secretory pathway. *Annual Review of Biochemistry*, *79*, 777–802.

Demirci, F. Y., Chang, M. H., Mah, T. S., Romero, M. F., & Gorin, M. B. (2006). Proximal renal tubular acidosis and ocular pathology: A novel missense mutation in the gene (SLC4A4) for sodium bicarbonate cotransporter protein (NBCe1). *Molecular Vision*, *12*, 324–330.

Desir, J., Moya, G., Reish, O., Van Regemorter, N., Deconinck, H., David, K. L., et al. (2007). Borate transporter SLC4A11 mutations cause both Harboyan syndrome and non-syndromic corneal endothelial dystrophy. *Journal of Medical Genetics*, *44*, 322–326.

Devonald, M. A., Smith, A. N., Poon, J. P., Ihrke, G., & Karet, F. E. (2003). Non-polarized targeting of AE1 causes autosomal dominant distal renal tubular acidosis. *Nature Genetics, 33*, 125–127.

Dhermy, D., Galand, C., Bournier, O., Boulanger, L., Cynober, T., Schismanoff, P. O., et al. (1997). Heterogenous band 3 deficiency in hereditary spherocytosis related to different band 3 gene defects. *British Journal of Haematology, 98*, 32–40.

Ding, Y., Casey, J. R., & Kopito, R. R. (1994). The major kidney AE1 isoform does not bind ankyrin (Ank1) in vitro. An essential role for the 79 NH2-terminal amino acid residues of band 3. *The Journal of Biological Chemistry, 269*, 32201–32208.

Dinour, D., Chang, M. H., Satoh, J., Smith, B. L., Angle, N., Knecht, A., et al. (2004). A novel missense mutation in the sodium bicarbonate cotransporter (NBCe1/SLC4A4) causes proximal tubular acidosis and glaucoma through ion transport defects. *The Journal of Biological Chemistry, 279*, 52238–52246.

Dirami, T., Rode, B., Jollivet, M., Da Silva, N., Escalier, D., Gaitch, N., et al. (2013). Missense mutations in SLC26A8, encoding a sperm-specific activator of CFTR, are associated with human asthenozoospermia. *American Journal of Human Genetics, 92*, 760–766.

Dorwart, M. R., Shcheynikov, N., Baker, J. M., Forman-Kay, J. D., Muallem, S., & Thomas, P. J. (2008). Congenital chloride-losing diarrhea causing mutations in the STAS domain result in misfolding and mistrafficking of SLC26A3. *The Journal of Biological Chemistry, 283*, 8711–8722.

Eber, S. W., Gonzalez, J. M., Lux, M. L., Scarpa, A. L., Tse, W. T., Dornwell, M., et al. (1996). Ankyrin-1 mutations are a major cause of dominant and recessive hereditary spherocytosis. *Nature Genetics, 13*, 214–218.

Elbein, A. D. (1991). The role of N-linked oligosaccharides in glycoprotein function. *Trends in Biotechnology, 9*, 346–352.

Fairbanks, G., Steck, T. L., & Wallach, D. F. (1971). Electrophoretic analysis of the major polypeptides of the human erythrocyte membrane. *Biochemistry, 10*, 2606–2617.

Falke, J. J., & Chan, S. I. (1985). Evidence that anion transport by band 3 proceeds via a ping-pong mechanism involving a single transport site. A 35 Cl NMR study. *The Journal of Biological Chemistry, 260*, 9537–9544.

Felce, J., & Saier, M. H., Jr. (2004). Carbonic anhydrases fused to anion transporters of the SulP family: Evidence for a novel type of bicarbonate transporter. *Journal of Molecular Microbiology and Biotechnology, 8*, 169–176.

Findlay, J. B. (1974). The receptor proteins for concanavalin A and Lens culinaris phytohemagglutinin in the membrane of the human erythrocyte. *The Journal of Biological Chemistry, 249*, 4398–4403.

Foxwell, B. M., & Tanner, M. J. (1981). Synthesis of the erythrocyte anion-transport protein. Immunochemical study of its incorporation into the plasma membrane of erythroid cells. *The Biochemical Journal, 195*, 129–137.

Fry, A. C., Su, Y., Yiu, V., Cuthbert, A. W., Trachtman, H., & Karet Frankl, F. E. (2012). Mutation conferring apical-targeting motif on AE1 exchanger causes autosomal dominant distal RTA. *Journal of the American Society of Nephrology, 23*, 1238–1249.

Fujinaga, J., Tang, X. B., & Casey, J. R. (1999). Topology of the membrane domain of human erythrocyte anion exchange protein, AE1. *The Journal of Biological Chemistry, 274*, 6626–6633.

Fukuda, M., Dell, A., Oates, J. E., & Fukuda, M. N. (1984). Structure of branched lactosaminoglycan, the carbohydrate moiety of band 3 isolated from adult human erythrocytes. *The Journal of Biological Chemistry, 259*, 8260–8273.

Gahmberg, C. G., Jokinen, M., & Andersson, L. C. (1979). Expression of the major red cell sialoglycoprotein, glycophorin A, in the human leukemic cell line K562. *The Journal of Biological Chemistry, 254*, 7442–7448.

Gahmberg, C. G., Jokinen, M., Karhi, K. K., & Andersson, L. C. (1980). Effect of tunicamycin on the biosynthesis of the major human red cell sialoglycoprotein, glycophorin A, in the leukemia cell line K562. *The Journal of Biological Chemistry, 255*, 2169–2175.

Garcia, A. M., & Lodish, H. F. (1989). Lysine 539 of human band 3 is not essential for ion transport or inhibition by stilbene disulfonates. *The Journal of Biological Chemistry, 264*, 19607–19613.

Gawenis, L. R., Bradford, E. M., Prasad, V., Lorenz, J. N., Simpson, J. E., Clarke, L. L., et al. (2007). Colonic anion secretory defects and metabolic acidosis in mice lacking the NBC1 Na+/HCO3- cotransporter. *The Journal of Biological Chemistry, 282*, 9042–9052.

Gawenis, L. R., Ledoussal, C., Judd, L. M., Prasad, V., Alper, S. L., Stuart-Tilley, A., et al. (2004). Mice with a targeted disruption of the AE2 Cl-/HCO3- exchanger are achlorhydric. *The Journal of Biological Chemistry, 279*, 30531–30539.

Ghosh, S., Cox, K. H., & Cox, J. V. (1999). Chicken erythroid AE1 anion exchangers associate with the cytoskeleton during recycling to the Golgi. *Molecular Biology of the Cell, 10*, 455–469.

Grichtchenko, I. I., Choi, I., Zhong, X., Bray-Ward, P., Russell, J. M., & Boron, W. F. (2001). Cloning, characterization, and chromosomal mapping of a human electroneutral Na(+)-driven Cl-HCO3 exchanger. *The Journal of Biological Chemistry, 276*, 8358–8363.

Grinstein, S., Ship, S., & Rothstein, A. (1978). Anion transport in relation to proteolytic dissection of band 3 protein. *Biochimica et Biophysica Acta, 507*, 294–304.

Groger, N., Frohlich, H., Maier, H., Olbrich, A., Kostin, S., Braun, T., et al. (2010). SLC4A11 prevents osmotic imbalance leading to corneal endothelial dystrophy, deafness, and polyuria. *The Journal of Biological Chemistry, 285*, 14467–14474.

Groger, N., Vitzthum, H., Frohlich, H., Kruger, M., Ehmke, H., Braun, T., et al. (2012). Targeted mutation of SLC4A5 induces arterial hypertension and renal metabolic acidosis. *Human Molecular Genetics, 21*, 1025–1036.

Groves, J. D., & Tanner, M. J. (1992). Glycophorin A facilitates the expression of human band 3-mediated anion transport in Xenopus oocytes. *The Journal of Biological Chemistry, 267*, 22163–22170.

Groves, J. D., & Tanner, M. J. (1994a). Role of N-glycosylation in the expression of human band 3-mediated anion transport. *Molecular Membrane Biology, 11*, 31–38.

Groves, J. D., & Tanner, M. J. (1994b). The effects of glycophorin A on the expression of the human red cell anion transporter (band 3) in Xenopus oocytes. *The Journal of Membrane Biology, 140*, 81–88.

Groves, J. D., & Tanner, M. J. (1995). Co-expressed complementary fragments of the human red cell anion exchanger (band 3, AE1) generate stilbene disulfonate-sensitive anion transport. *The Journal of Biological Chemistry, 270*, 9097–9105.

Groves, J. D., & Tanner, M. J. (1999a). Structural model for the organization of the transmembrane spans of the human red-cell anion exchanger (band 3; AE1). *The Biochemical Journal, 344*(Pt 3), 699–711.

Groves, J. D., & Tanner, M. J. (1999b). Topology studies with biosynthetic fragments identify interacting transmembrane regions of the human red-cell anion exchanger (band 3; AE1). *The Biochemical Journal, 344*(Pt 3), 687–697.

Groves, J. D., Wang, L., & Tanner, M. J. (1998). Complementation studies with co-expressed fragments of human red cell band 3 (AE1): The assembly of the anion-transport domain in xenopus oocytes and a cell-free translation system. *The Biochemical Journal, 332*(Pt 1), 161–171.

Hallworth, R., & Nichols, M. G. (2011). Prestin in HEK cells is an obligate tetramer. *Journal of Neurophysiology, 107*, 5–11.

Harrison, M. L., Rathinavelu, P., Arese, P., Geahlen, R. L., & Low, P. S. (1991). Role of band 3 tyrosine phosphorylation in the regulation of erythrocyte glycolysis. *The Journal of Biological Chemistry, 266*, 4106–4111.

Hassoun, H., Hanada, T., Lutchman, M., Sahr, K. E., Palek, J., Hanspal, M., et al. (1998). Complete deficiency of glycophorin A in red blood cells from mice with targeted inactivation of the band 3 (AE1) gene. *Blood, 91*, 2146–2151.

Hayashi, H., & Yamashita, Y. (2012). Role of N-glycosylation in cell surface expression and protection against proteolysis of the intestinal anion exchanger SLC26A3. *American Journal of Physiology. Cell Physiology, 302*, C781–C795.

Hayette, S., Morle, L., Bozon, M., Ghanem, A., Risinger, M., Korsgren, C., et al. (1995). A point mutation in the protein 4.2 gene (allele 4.2 Tozeur) associated with hereditary haemolytic anaemia. *British Journal of Haematology, 89*, 762–770.

He, J. J., & Quiocho, F. A. (1993). Dominant role of local dipoles in stabilizing uncompensated charges on a sulfate sequestered in a periplasmic active transport protein. *Protein Science, 2*, 1643–1647.

Helenius, A. (1994). How N-linked oligosaccharides affect glycoprotein folding in the endoplasmic reticulum. *Molecular Biology of the Cell, 5*, 253–265.

Hemadevi, B., Veitia, R. A., Srinivasan, M., Arunkumar, J., Prajna, N. V., Lesaffre, C., et al. (2008). Identification of mutations in the SLC4A11 gene in patients with recessive congenital hereditary endothelial dystrophy. *Archives of Ophthalmology, 126*, 700–708.

Hentschke, M., Wiemann, M., Hentschke, S., Kurth, I., Hermans-Borgmeyer, I., Seidenbecher, T., et al. (2006). Mice with a targeted disruption of the Cl-/HCO3- exchanger AE3 display a reduced seizure threshold. *Molecular and Cellular Biology, 26*, 182–191.

Hirai, T., Hamasaki, N., Yamaguchi, T., & Ikeda, Y. (2011). Topology models of anion exchanger 1 that incorporate the anti-parallel V-shaped motifs found in the EM structure. *Biochemistry and Cell Biology, 89*, 148–156.

Hoglund, P., Haila, S., Socha, J., Tomaszewski, L., Saarialho-Kere, U., Karjalainen-Lindsberg, M. L., et al. (1996). Mutations of the Down-regulated in adenoma (DRA) gene cause congenital chloride diarrhoea. *Nature Genetics, 14*, 316–319.

Horita, S., Yamada, H., Inatomi, J., Moriyama, N., Sekine, T., Igarashi, T., et al. (2005). Functional analysis of NBC1 mutants associated with proximal renal tubular acidosis and ocular abnormalities. *Journal of the American Society of Nephrology, 16*, 2270–2278.

Hubner, S., Bahr, C., Gossmann, H., Efthymiadis, A., & Drenchahn, D. (2003). Mitochondrial and nuclear localization of kanadaptin. *European Journal of Cell Biology, 82*, 240–252.

Humphreys, B. D., Jiang, L., Chernova, M. N., & Alper, S. L. (1995). Hypertonic activation of AE2 anion exchanger in Xenopus oocytes via NHE-mediated intracellular alkalinization. *The American Journal of Physiology, 268*, C201–C209.

Igarashi, T., Inatomi, J., Sekine, T., Cha, S. H., Kanai, Y., Kunimi, M., et al. (1999). Mutations in SLC4A4 cause permanent isolated proximal renal tubular acidosis with ocular abnormalities. *Nature Genetics, 23*, 264–266.

Igarashi, T., Inatomi, J., Sekine, T., Seki, G., Shimadzu, M., Tozawa, F., et al. (2001). Novel nonsense mutation in the Na+/HCO3- cotransporter gene (SLC4A4) in a patient with permanent isolated proximal renal tubular acidosis and bilateral glaucoma. *Journal of the American Society of Nephrology, 12*, 713–718.

Iwase, S., Ideguchi, H., Takao, M., Horiguchi-Yamada, J., Iwasaki, M., Takahara, S., et al. (1998). Band 3 Tokyo: Thr837->Ala837 substitution in erythrocyte band 3 protein associated with spherocytic hemolysis. *Acta Haematologica, 100*, 200–203.

Jacques, T., Picard, N., Miller, R. L., Riemondy, K. A., Houillier, P., Sohet, F., et al. (2013). Overexpression of pendrin in intercalated cells produces chloride-sensitive hypertension. *Journal of the American Society of Nephrology, 24*, 1104–1113.

Jardetzky, O. (1966). Simple allosteric model for membrane pumps. *Nature, 211*, 969–970.

Jarolim, P., Murray, J. L., Rubin, H. L., Coghlan, G., & Zelinski, T. (1997). A Thr552 -> Ile substitution in erythroid band 3 gives rise to the Warrior blood group antigen. *Transfusion*, *37*, 398–405.

Jarolim, P., Murray, J. L., Rubin, H. L., Smart, E., & Moulds, J. M. (1997). Blood group antigens Rb(a), Tr(a), and Wd(a) are located in the third ectoplasmic loop of erythroid band 3. *Transfusion*, *37*, 607–615.

Jarolim, P., Murray, J. L., Rubin, H. L., Taylor, W. M., Prchal, J. T., Ballas, S. K., et al. (1996). Characterization of 13 novel band 3 gene defects in hereditary spherocytosis with band 3 deficiency. *Blood*, *88*, 4366–4374.

Jarolim, P., Palek, J., Amato, D., Hassan, K., Sapak, P., Nurse, G. T., et al. (1991). Deletion in erythrocyte band 3 gene in malaria-resistant Southeast Asian ovalocytosis. *Proceedings of the National Academy of Sciences of the United States of America*, *88*, 11022–11026.

Jarolim, P., Palek, J., Rubin, H. L., Prchal, J. T., Korsgren, C., & Cohen, C. M. (1992). Band 3 Tuscaloosa: Pro327–Arg327 substitution in the cytoplasmic domain of erythrocyte band 3 protein associated with spherocytic hemolytic anemia and partial deficiency of protein 4.2. *Blood*, *80*, 523–529.

Jarolim, P., Rubin, H. L., Cho, M. R., Brabec, V., Derick, L. H., Yi, S. J., et al. (1994). Duplication of 10 nucleotides in the erythroid band 3 (AE1) gene in a kindred with hereditary spherocytosis and band 3 protein deficiency (band 3PRAGUE). *The Journal of Clinical Investigation*, *93*, 121–130.

Jarolim, P., Rubin, H. L., Zakova, D., Storry, J., & Reid, M. E. (1998). Characterization of seven low incidence blood group antigens carried by erythrocyte band 3 protein. *Blood*, *92*, 4836–4843.

Jarolim, P., Rubin, H. L., Zhai, S., Sahr, K. E., Liu, S. C., Mueller, T. J., et al. (1992). Band 3 Memphis: A widespread polymorphism with abnormal electrophoretic mobility of erythrocyte band 3 protein caused by substitution AAG—GAG (Lys—Glu) in codon 56. *Blood*, *80*, 1592–1598.

Jenkins, J. D., Madden, D. P., & Steck, T. L. (1984). Association of phosphofructokinase and aldolase with the membrane of the intact erythrocyte. *The Journal of Biological Chemistry*, *259*, 9374–9378.

Jennings, M. L. (1984). Oligomeric structure and the anion transport function of human erythrocyte band 3 protein. *The Journal of Membrane Biology*, *80*, 105–117.

Jennings, M. L., & Gosselink, P. G. (1995). Anion exchange protein in Southeast Asian ovalocytes: Heterodimer formation between normal and variant subunits. *Biochemistry*, *34*, 3588–3595.

Jennings, M. L., Howren, T. R., Cui, J., Winters, M., & Hannigan, R. (2007). Transport and regulatory characteristics of the yeast bicarbonate transporter homolog Bor1p. *American Journal of Physiology. Cell Physiology*, *293*, C468–C476.

Jennings, M. L., & Nicknish, J. S. (1985). Localization of a site of intermolecular cross-linking in human red blood cell band 3 protein. *The Journal of Biological Chemistry*, *260*, 5472–5479.

Johnson, D. E., & Casey, J. R. (2011). Cytosolic H+microdomain developed around AE1 during AE1-mediated Cl-/HCO3- exchange. *The Journal of Physiology*, *589*, 1551–1569.

Jokinen, M., Gahmberg, C. G., & Andersson, L. C. (1979). Biosynthesis of the major human red cell sialoglycoprotein, glycophorin A, in a continuous cell line. *Nature*, *279*, 604–607.

Jokinen, M., Ulmanen, I., Andersson, L. C., Kaariainen, L., & Gahmberg, C. G. (1981). Cell-free synthesis and glycosylation of the major human-red-cell sialoglycoprotein, glycophorin A. *European Journal of Biochemistry*, *114*, 393–397.

Juran, B. D., Atkinson, E. J., Larson, J. J., Schlicht, E. M., & Lazaridis, K. N. (2009). Common genetic variation and haplotypes of the anion exchanger SLC4A2 in primary biliary cirrhosis. *The American Journal of Gastroenterology*, *104*, 1406–1411.

Kameh, H., Landolt-Marticorena, C., Charuk, J. H. M., Schachter, H., & Reithmeier, R. A. F. (1998). Structural and functional consequences of an N-glycosylation mutation (HEMPAS) affecting human erythrocyte membrane glycoproteins. *Biochemistry and Cell Biology-Biochimie et Biologie Cellulaire*, *76*, 823–835.
Kannan, R., Labotka, R., & Low, P. S. (1988). Isolation and characterization of the hemichrome-stabilized membrane protein aggregates from sickle erythrocytes. Major site of autologous antibody binding. *The Journal of Biological Chemistry*, *263*, 13766–13773.
Kannan, R., Yuan, J., & Low, P. S. (1991). Isolation and partial characterization of antibody- and globin-enriched complexes from membranes of dense human erythrocytes. *The Biochemical Journal*, *278*(Pt 1), 57–62.
Kant, J. A., & Steck, T. L. (1973). Specificity in the association of glyceraldehyde 3-phosphate dehydrogenase with isolated human erythrocyte membranes. *The Journal of Biological Chemistry*, *248*, 8457–8464.
Kanzaki, A., Hayette, S., Morle, L., Inoue, F., Matsuyama, R., Inoue, T., et al. (1997). Total absence of protein 4.2 and partial deficiency of band 3 in hereditary spherocytosis. *British Journal of Haematology*, *99*, 522–530.
Kanzaki, A., Yasunaga, M., Okamoto, N., Inoue, T., Yawata, A., Wada, H., et al. (1995). Band 4.2 Shiga: 317 CGC−>TGC in compound heterozygotes with 142 GCT−> ACT results in band 4.2 deficiency and microspherocytosis. *British Journal of Haematology*, *91*, 333–340.
Kao, L., Kurtz, L. M., Shao, X., Papadopoulos, M. C., Liu, L., Bok, D., et al. (2011). Severe neurologic impairment in mice with targeted disruption of the electrogenic sodium bicarbonate cotransporter NBCe2 (Slc4a5 gene). *The Journal of Biological Chemistry*, *286*, 32563–32574.
Karinou, E., Compton, E. L., Morel, M., & Javelle, A. (2013). The Escherichia coli SLC26 homologue YchM (DauA) is a C(4)-dicarboxylic acid transporter. *Molecular Microbiology*, *87*, 623–640.
Karniski, L. P. (2001). Mutations in the diastrophic dysplasia sulfate transporter (DTDST) gene: Correlation between sulfate transport activity and chondrodysplasia phenotype. *Human Molecular Genetics*, *10*, 1485–1490.
Karniski, L. P. (2004). Functional expression and cellular distribution of diastrophic dysplasia sulfate transporter (DTDST) gene mutations in HEK cells. *Human Molecular Genetics*, *13*, 2165–2171.
Kawano, Y., Okubo, K., Tokunaga, F., Miyata, T., Iwanaga, S., & Hamasaki, N. (1988). Localization of the pyridoxal phosphate binding site at the COOH-terminal region of erythrocyte band 3 protein. *The Journal of Biological Chemistry*, *263*, 8232–8238.
Kay, M. M., Bosman, G. J., & Lawrence, C. (1988). Functional topography of band 3: Specific structural alteration linked to functional aberrations in human erythrocytes. *Proceedings of the National Academy of Sciences of the United States of America*, *85*, 492–496.
Kelley, L. A., & Sternberg, M. J. (2009). Protein structure prediction on the Web: A case study using the Phyre server. *Nature Protocols*, *4*, 363–371.
Keskanokwong, T., Shandro, H., Johnson, D. E., Kittanakom, S., Vilas, G. L., Thorner, P., et al. (2007). Interaction of integrin-linked kinase with the kidney chloride/bicarbonate exchanger, kAE1. *The Journal of Biological Chemistry*, *282*, 23205–23218.
Kittanakom, S., Cordat, E., Akkarapatumwong, V., Yenchitsomanus, P., & Reithmeier, R. A. F. (2004). Trafficking defects of a novel autosomal recessive distal renal tubular acidosis mutant (S773P) of the human kidney anion exchanger (kAE1). *Journal of Biological Chemistry*, *279*, 40960–40971.
Kittanakom, S., Cordat, E., & Reithmeier, R. A. F. (2008). Dominant-negative effect of Southeast Asian ovalocytosis anion exchanger 1 in compound heterozygous distal renal tubular acidosis. *Biochemical Journal*, *410*, 271–281.

Kittanakom, S., Keskanokwong, T., Akkarapatumwong, V., Yenchitsomanus, P. T., & Reithmeier, R. A. (2004). Human kanadaptin and kidney anion exchanger 1 (kAE1) do not interact in transfected HEK 293 cells. *Molecular Membrane Biology*, *21*, 395–402.
Ko, S. B., Shcheynikov, N., Choi, J. Y., Luo, X., Ishibashi, K., Thomas, P. J., et al. (2002). A molecular mechanism for aberrant CFTR-dependent HCO(3)(-) transport in cystic fibrosis. *The EMBO Journal*, *21*, 5662–5672.
Ko, S. B., Zeng, W., Dorwart, M. R., Luo, X., Kim, K. H., Millen, L., et al. (2004). Gating of CFTR by the STAS domain of SLC26 transporters. *Nature Cell Biology*, *6*, 343–350.
Kobayashi, S., Morgans, C. W., Casey, J. R., & Kopito, R. R. (1994). AE3 anion exchanger isoforms in the vertebrate retina: Developmental regulation and differential expression in neurons and glia. *The Journal of Neuroscience*, *14*, 6266–6279.
Kopito, R. R., Lee, B. S., Simmons, D. M., Lindsey, A. E., Morgans, C. W., & Schneider, K. (1989). Regulation of intracellular pH by a neuronal homolog of the erythrocyte anion exchanger. *Cell*, *59*, 927–937.
Kopito, R. R., & Lodish, H. F. (1985a). Primary structure and transmembrane orientation of the murine anion exchange protein. *Nature*, *316*, 234–238.
Kopito, R. R., & Lodish, H. F. (1985b). Structure of the murine anion exchange protein. *Journal of Cellular Biochemistry*, *29*, 1–17.
Kudrycki, K. E., Newman, P. R., & Shull, G. E. (1990). cDNA cloning and tissue distribution of mRNAs for two proteins that are related to the band 3 Cl-/HCO3- exchanger. *The Journal of Biological Chemistry*, *265*, 462–471.
Kumar, A., Bhattacharjee, S., Prakash, D. R., & Sadanand, C. S. (2007). Genetic analysis of two Indian families affected with congenital hereditary endothelial dystrophy: Two novel mutations in SLC4A11. *Molecular Vision*, *13*, 39–46.
Landoltmarticorena, C., Casey, J. R., & Reithmeier, R. A. F. (1995). Transmembrane helix-helix interactions and accessibility of H2DIDS on labeled band-3, the erythrocyte anion-exchange protein. *Molecular Membrane Biology*, *12*, 173–182.
Landolt-Marticorena, C., Charuk, J. H. M., & Reithmeier, R. A. F. (1998). Two glycoprotein populations of band 3 dimers are present in human erythrocytes. *Molecular Membrane Biology*, *15*, 153–158.
Landoltmarticorena, C., & Reithmeier, R. A. F. (1994). Asparagine-linked oligosaccharides are localized to single extracytosolic segments in multispan membrane-glycoproteins. *Biochemical Journal*, *302*, 253–260.
Lazarides, E., & Moon, R. T. (1984). Assembly and topogenesis of the spectrin-based membrane skeleton in erythroid development. *Cell*, *37*, 354–356.
Lazarides, E., & Woods, C. (1989). Biogenesis of the red blood cell membrane-skeleton and the control of erythroid morphogenesis. *Annual Review of Cell Biology*, *5*, 427–452.
Lee, S. K., Boron, W. F., & Parker, M. D. (2013). Substrate specificity of the electrogenic sodium/bicarbonate cotransporter NBCe1-A (SLC4A4, variant A) from humans and rabbits. *American Journal of Physiology. Renal Physiology*, *304*, F883–F899.
Lee, B. S., Gunn, R. B., & Kopito, R. R. (1991). Functional differences among nonerythroid anion exchangers expressed in a transfected human cell line. *The Journal of Biological Chemistry*, *266*, 11448–11454.
Lemieux, M. J., Reithmeier, R. A. F., & Wang, D. N. (2002). Importance of detergent and phospholipid in the crystallization of the human erythrocyte anion-exchanger membrane domain. *Journal of Structural Biology*, *137*, 322–332.
Leviel, F., Hubner, C. A., Houillier, P., Morla, L., El Moghrabi, S., Brideau, G., et al. (2010). The Na+-dependent chloride-bicarbonate exchanger SLC4A8 mediates an electroneutral Na+reabsorption process in the renal cortical collecting ducts of mice. *The Journal of Clinical Investigation*, *120*, 1627–1635.
Lewis, D. W. (1992). What was wrong with Tiny Tim? *American Journal of Diseases of Children*, *146*, 1403–1407.

Lieberman, D. M., & Reithmeier, R. A. F. (1983). Characterization of the stilbenedisulfonate binding-site of the band-3 polypeptide of human-erythrocyte membranes. *Biochemistry, 22*, 4028–4033.
Lindsey, A. E., Schneider, K., Simmons, D. M., Baron, R., Lee, B. S., & Kopito, R. R. (1990). Functional expression and subcellular localization of an anion exchanger cloned from choroid plexus. *Proceedings of the National Academy of Sciences of the United States of America, 87*, 5278–5282.
Linn, S. C., Kudrycki, K. E., & Shull, G. E. (1992). The predicted translation product of a cardiac AE3 mRNA contains an N terminus distinct from that of the brain AE3 Cl-/HCO3- exchanger. Cloning of a cardiac AE3 cDNA, organization of the AE3 gene, and identification of an alternative transcription initiation site. *The Journal of Biological Chemistry, 267*, 7927–7935.
Liu, S. C., Palek, J., Yi, S. J., Nichols, P. E., Derick, L. H., Chiou, S. S., et al. (1995). Molecular basis of altered red blood cell membrane properties in Southeast Asian ovalocytosis: Role of the mutant band 3 protein in band 3 oligomerization and retention by the membrane skeleton. *Blood, 86*, 349–358.
Liu, S. C., Zhai, S., Palek, J., Golan, D. E., Amato, D., Hassan, K., et al. (1990). Molecular defect of the band 3 protein in southeast Asian ovalocytosis. *New England Journal of Medicine, 323*, 1530–1538.
Lo, Y. F., Yang, S. S., Seki, G., Yamada, H., Horita, S., Yamazaki, O., et al. (2011). Severe metabolic acidosis causes early lethality in NBC1 W516X knock-in mice as a model of human isolated proximal renal tubular acidosis. *Kidney International, 79*, 730–741.
Lombardo, C. R., Willardson, B. M., & Low, P. S. (1992). Localization of the protein 4.1-binding site on the cytoplasmic domain of erythrocyte membrane band 3. *The Journal of Biological Chemistry, 267*, 9540–9546.
Low, P. S. (1986). Structure and function of the cytoplasmic domain of band 3: Center of erythrocyte membrane-peripheral protein interactions. *Biochimica et Biophysica Acta, 864*, 145–167.
Low, P. S. (1991). Role of hemoglobin denaturation and band 3 clustering in initiating red cell removal. *Advances in Experimental Medicine and Biology, 307*, 173–183.
Low, P. S., Rathinavelu, P., & Harrison, M. L. (1993). Regulation of glycolysis via reversible enzyme binding to the membrane protein, band 3. *The Journal of Biological Chemistry, 268*, 14627–14631.
Low, P. S., Westfall, M. A., Allen, D. P., & Appell, K. C. (1984). Characterization of the reversible conformational equilibrium of the cytoplasmic domain of erythrocyte membrane band 3. *The Journal of Biological Chemistry, 259*, 13070–13076.
Low, P. S., Zhang, D., & Bolin, J. T. (2001). Localization of mutations leading to altered cell shape and anion transport in the crystal structure of the cytoplasmic domain of band 3. *Blood Cells, Molecules & Diseases, 27*, 81–84.
Lux, S. E., John, K. M., Kopito, R. R., & Lodish, H. F. (1989). Cloning and characterization of band 3, the human erythrocyte anion-exchange protein (AE1). *Proceedings of the National Academy of Sciences of the United States of America, 86*, 9089–9093.
Macara, I. G., & Cantley, L. C. (1981a). Interactions between transport inhibitors at the anion binding sites of the band 3 dimer. *Biochemistry, 20*, 5095–5105.
Macara, I. G., & Cantley, L. C. (1981b). Mechanism of anion exchange across the red cell membrane by band 3: Interactions between stilbenedisulfonate and NAP-taurine binding sites. *Biochemistry, 20*, 5695–5701.
Macara, I. G., Kuo, S., & Cantley, L. C. (1983). Evidence that inhibitors of anion exchange induce a transmembrane conformational change in band 3. *The Journal of Biological Chemistry, 258*, 1785–1792.
Maillet, P., Alloisio, N., Morle, L., & Delaunay, J. (1996). Spectrin mutations in hereditary elliptocytosis and hereditary spherocytosis. *Human Mutation, 8*, 97–107.

Makela, S., Kere, J., Holmberg, C., & Hoglund, P. (2002). SLC26A3 mutations in congenital chloride diarrhea. *Human Mutation, 20*, 425–438.

Medina, J. F. (2011). Role of the anion exchanger 2 in the pathogenesis and treatment of primary biliary cirrhosis. *Digestive Diseases, 29*, 103–112.

Miraglia del Giudice, E., Vallier, A., Maillet, P., Perrotta, S., Cutillo, S., Iolascon, A., et al. (1997). Novel band 3 variants (bands 3 Foggia, Napoli I and Napoli II) associated with hereditary spherocytosis and band 3 deficiency: Status of the D38A polymorphism within the EPB3 locus. *British Journal of Haematology, 96*, 70–76.

Moraes, T. F., & Reithmeier, R. A. (2012). Membrane transport metabolons. *Biochimica et Biophysica Acta, 1818*, 2687–2706.

Moriyama, R., Ideguchi, H., Lombardo, C. R., Van Dort, H. M., & Low, P. S. (1992). Structural and functional characterization of band 3 from Southeast Asian ovalocytes. *The Journal of Biological Chemistry, 267*, 25792–25797.

Moseley, R. H., Hoglund, P., Wu, G. D., Silberg, D. G., Haila, S., de la Chapelle, A., et al. (1999). Downregulated in adenoma gene encodes a chloride transporter defective in congenital chloride diarrhea. *The American Journal of Physiology, 276*, G185–G192.

Mueller, T. J., & Morrison, M. (1977). Detection of a variant of protein 3, the major transmembrane protein of the human erythrocyte. *The Journal of Biological Chemistry, 252*, 6573–6576.

Murthy, S. N., Liu, T., Kaul, R. K., Kohler, H., & Steck, T. L. (1981). The aldolase-binding site of the human erythrocyte membrane is at the NH2 terminus of band 3. *The Journal of Biological Chemistry, 256*, 11203–11208.

Nehls, V., Zeitler-Zapf, P., & Drenckhahn, D. (1993). Different sequences of expression of band 3, spectrin, and ankyrin during normal erythropoiesis and erythroleukemia. *American Journal of Pathology, 142*, 1565–1573.

Nishimori, I., Minakuchi, T., Maresca, A., Carta, F., Scozzafava, A., & Supuran, C. T. (2010). The beta-carbonic anhydrases from Mycobacterium tuberculosis as drug targets. *Current Pharmaceutical Design, 16*, 3300–3309.

Ogando, D. G., Jalimarada, S. S., Zhang, W., Vithana, E. N., & Bonanno, J. A. (2011). SLC4A11 is an EIPA-sensitive Na+permeable pHi regulator. *American Journal of Physiology. Cell Physiology, 305*, C716–C727.

Ohana, E., Shcheynikov, N., Park, M., & Muallem, S. (2012). Solute carrier family 26 member a2 (Slc26a2) protein functions as an electroneutral SOFormula/OH-/Cl- exchanger regulated by extracellular Cl. *The Journal of Biological Chemistry, 287*, 5122–5132.

Ohana, E., Shcheynikov, N., Yang, D., So, I., & Muallem, S. (2011). Determinants of coupled transport and uncoupled current by the electrogenic SLC26 transporters. *The Journal of General Physiology, 137*, 239–251.

Ohana, E., Yang, D., Shcheynikov, N., & Muallem, S. (2009). Diverse transport modes by the solute carrier 26 family of anion transporters. *The Journal of Physiology, 587*, 2179–2185.

Okubo, K., Kang, D., Hamasaki, N., & Jennings, M. L. (1994). Red blood cell band 3. Lysine 539 and lysine 851 react with the same H2DIDS (4,4'-diisothiocyanodihydrostilbene-2,2'-disulfonic acid) molecule. *The Journal of Biological Chemistry, 269*, 1918–1926.

Pang, A. J., Bustos, S. P., & Reithmeier, R. A. F. (2008). Structural characterization of the cytosolic domain of kidney chloride/bicarbonate anion exchanger 1 (kAE1). *Biochemistry, 47*, 4510–4517.

Pang, A. J., & Reithmeier, R. A. (2009). Interaction of anion exchanger 1 and glycophorin A in human erythroleukaemic K562 cells. *The Biochemical Journal, 421*, 345–356.

Parker, M. D., Musa-Aziz, R., Rojas, J. D., Choi, I., Daly, C. M., & Boron, W. F. (2008). Characterization of human SLC4A10 as an electroneutral Na/HCO3 cotransporter (NBCn2) with Cl- self-exchange activity. *The Journal of Biological Chemistry, 283*, 12777–12788.

Parker, M. D., Qin, X., Williamson, R. C., Toye, A. M., & Boron, W. F. (2012). HCO(3)(-)-independent conductance with a mutant Na(+)/HCO(3)(-) cotransporter (SLC4A4) in a case of proximal renal tubular acidosis with hypokalaemic paralysis. *The Journal of Physiology*, *590*, 2009–2034.

Pasqualetto, E., Aiello, R., Gesiot, L., Bonetto, G., Bellanda, M., & Battistutta, R. (2010). Structure of the cytosolic portion of the motor protein prestin and functional role of the STAS domain in SLC26/SulP anion transporters. *Journal of Molecular Biology*, *400*, 448–462.

Patterson, S. T., Li, J., Kang, J. A., Wickrema, A., Williams, D. B., & Reithmeier, R. A. (2009). Loss of specific chaperones involved in membrane glycoprotein biosynthesis during the maturation of human erythroid progenitor cells. *The Journal of Biological Chemistry*, *284*, 14547–14557.

Patterson, S. T., & Reithmeier, R. A. (2010). Cell surface rescue of kidney anion exchanger 1 mutants by disruption of chaperone interactions. *The Journal of Biological Chemistry*, *285*, 33423–33434.

Paw, B. H., Davidson, A. J., Zhou, Y., Li, R., Pratt, S. J., Lee, C., et al. (2003). Cell-specific mitotic defect and dyserythropoiesis associated with erythroid band 3 deficiency. *Nature Genetics*, *34*, 59–64.

Peters, L. L., Shivdasani, R. A., Liu, S. C., Hanspal, M., John, K. M., Gonzalez, J. M., et al. (1996). Anion exchanger 1 (band 3) is required to prevent erythrocyte membrane surface loss but not to form the membrane skeleton. *Cell*, *86*, 917–927.

Popov, M., Li, J., & Reithmeier, R. A. F. (1999). Transmembrane folding of the human erythrocyte anion exchanger (AE1, Band 3) determined by scanning and insertional N-glycosylation mutagenesis. *Biochemical Journal*, *339*, 269–279.

Popov, M., & Reithmeier, R. A. F. (1999). Calnexin interaction with N-glycosylation mutants of a polytopic membrane glycoprotein, the human erythrocyte anion exchanger 1 (band 3). *Journal of Biological Chemistry*, *274*, 17635–17642.

Popov, M., Tam, L. Y., Li, J., & Reithmeier, R. A. F. (1997). Mapping the ends of transmembrane segments in a polytopic membrane protein—Scanning N-glycosylation mutagenesis of extracytosolic loops in the anion exchanger, Band 3. *Journal of Biological Chemistry*, *272*, 18325–18332.

Price, G. D., & Howitt, S. M. (2011). The cyanobacterial bicarbonate transporter BicA: Its physiological role and the implications of structural similarities with human SLC26 transporters. *Biochemistry and Cell Biology*, *89*, 178–188.

Price, G. D., Woodger, F. J., Badger, M. R., Howitt, S. M., & Tucker, L. (2004). Identification of a SulP-type bicarbonate transporter in marine cyanobacteria. *Proceedings of the National Academy of Sciences of the United States of America*, *101*, 18228–18233.

Puchulu-Campanella, E., Chu, H., Anstee, D. J., Galan, J. A., Tao, W. A., & Low, P. S. (2013). Identification of the components of a glycolytic enzyme metabolon on the human red blood cell membrane. *The Journal of Biological Chemistry*, *288*, 848–858.

Pushkin, A., & Kurtz, I. (2006). SLC4 base (HCO3 -, CO3 2-) transporters: Classification, function, structure, genetic diseases, and knockout models. *American Journal of Physiology. Renal Physiology*, *290*, F580–F599.

Quilty, J. A., Cordat, E., & Reithmeier, R. A. F. (2002). Impaired trafficking of human kidney anion exchanger (kAE1) caused by hetero-oligomer formation with a truncated mutant associated with distal renal tubular acidosis. *Biochemical Journal*, *368*, 895–903.

Quilty, J. A., Li, J., & Reithmeier, R. A. (2002a). Impaired trafficking of distal renal tubular acidosis mutants of the human kidney anion exchanger kAE1. *American Journal of Physiology. Renal Physiology*, *282*, F810–F820.

Quilty, J. A., Li, J., & Reithmeier, R. A. (2002b). Impaired trafficking of distal renal tubular acidosis mutants of the human kidney anion exchanger kAE1. *American Journal of Physiology. Renal Physiology*, *282*, F810–F820.

Quilty, J. A., & Reithmeier, R. A. F. (2000). Trafficking and folding defects in hereditary spherocytosis mutants of the human red cell anion exchanger. *Traffic*, *1*, 987–998.

Rao, A., Martin, P., Reithmeier, R. A. F., & Cantley, L. C. (1979). Location of the stilbenedisulfonate binding-site of the human-erythrocyte anion-exchange system by resonance energy-transfer. *Biochemistry, 18*, 4505–4516.
Reimold, F. R., Heneghan, J. F., Stewart, A. K., Zelikovic, I., Vandorpe, D. H., Shmukler, B. E., et al. (2011). Pendrin function and regulation in Xenopus oocytes. *Cellular Physiology and Biochemistry, 28*, 435–450.
Reithmeier, R. A. F. (1979). Fragmentation of the band-3 polypeptide from human-erythrocyte membranes—Size and detergent binding of the membrane-associated domain. *Journal of Biological Chemistry, 254*, 3054–3060.
Reithmeier, R. A. F. (1993). The erythrocyte anion transporter (Band-3). *Current Opinion in Structural Biology, 3*, 515–523.
Reithmeier, R. A. F. (2001). A membrane metabolon linking carbonic anhydrase with chloride/bicarbonate anion exchangers. *Blood Cells, Molecules & Diseases, 27*, 85–89.
Reithmeier, R. A. F., Lieberman, D. M., Casey, J. R., Pimplikar, S. W., Werner, P. K., See, H., et al. (1989). Structure and function of the band-3 Cl-/Hco3- transporter. *Bicarbonate, Chloride, and Proton Transport Systems, 574*, 75–83.
Riazuddin, S. A., Vithana, E. N., Seet, L. F., Liu, Y., Al-Saif, A., Koh, L. W., et al. (2011). Missense mutations in the sodium borate cotransporter SLC4A11 cause late-onset Fuchs corneal dystrophy. *Human Mutation, 31*, 1261–1268.
Romero, M. F. (2001). The electrogenic Na+/HCO3- cotransporter, NBC. *JOP: Journal of the Pancreas, 2*, 182–191.
Romero, M. F. (2005). Molecular pathophysiology of SLC4 bicarbonate transporters. *Current Opinion in Nephrology and Hypertension, 14*, 495–501.
Romero, M. F., & Boron, W. F. (1999). Electrogenic Na+/HCO3- cotransporters: Cloning and physiology. *Annual Review of Physiology, 61*, 699–723.
Romero, M. F., Chen, A. P., Parker, M. D., & Boron, W. F. (2013). The SLC4 family of bicarbonate (HCO(3)(-)) transporters. *Molecular Aspects of Medicine, 34*, 159–182.
Romero, M. F., Fulton, C. M., & Boron, W. F. (2004). The SLC4 family of HCO 3—Transporters. *Pflügers Archiv, 447*, 495–509.
Romero, M. F., Henry, D., Nelson, S., Harte, P. J., Dillon, A. K., & Sciortino, C. M. (2000). Cloning and characterization of a Na+-driven anion exchanger (NDAE1). A new bicarbonate transporter. *The Journal of Biological Chemistry, 275*, 24552–24559.
Romisch, K. (2004). A cure for traffic jams: Small molecule chaperones in the endoplasmic reticulum. *Traffic, 5*, 815–820.
Rossi, A., & Superti-Furga, A. (2001). Mutations in the diastrophic dysplasia sulfate transporter (DTDST) gene (SLC26A2): 22 novel mutations, mutation review, associated skeletal phenotypes, and diagnostic relevance. *Human Mutation, 17*, 159–171.
Rungroj, N., Devonald, M. A., Cuthbert, A. W., Reimann, F., Akkarapatumwong, V., Yenchitsomanus, P. T., et al. (2004). A novel missense mutation in AE1 causing autosomal dominant distal renal tubular acidosis retains normal transport function but is mistargeted in polarized epithelial cells. *The Journal of Biological Chemistry, 279*, 13833–13838.
Sabban, E., Marchesi, V., Adesnik, M., & Sabatini, D. D. (1981). Erythrocyte membrane protein band 3: Its biosynthesis and incorporation into membranes. *The Journal of Cell Biology, 91*, 637–646.
Salas, J. T., Banales, J. M., Sarvide, S., Recalde, S., Ferrer, A., Uriarte, I., et al. (2008). Ae2a, b-deficient mice develop antimitochondrial antibodies and other features resembling primary biliary cirrhosis. *Gastroenterology, 134*, 1482–1493.
Sander, T., Toliat, M. R., Heils, A., Leschik, G., Becker, C., Ruschendorf, F., et al. (2002). Association of the 867Asp variant of the human anion exchanger 3 gene with common subtypes of idiopathic generalized epilepsy. *Epilepsy Research, 51*, 249–255.

Sarabia, V. E., Casey, J. R., & Reithmeier, R. A. F. (1993). Molecular characterization of the band-3 protein from Southeast-Asian ovalocytes. *Journal of Biological Chemistry, 268*, 10676–10680.

Satchwell, T. J., Bell, A. J., Pellegrin, S., Kupzig, S., Ridgwell, K., Daniels, G., et al. (2011). Critical band 3 multiprotein complex interactions establish early during human erythropoiesis. *Blood, 118*, 182–191.

Sawasdee, N., Junking, M., Ngaojanlar, P., Sukomon, N., Ungsupravate, D., Limjindaporn, T., et al. (2010). Human kidney anion exchanger 1 interacts with adaptor-related protein complex 1 mu1A (AP-1 mu1A). *Biochemical and Biophysical Research Communications, 401*, 85–91.

Schekman, R. (2010). Charting the secretory pathway in a simple eukaryote. *Molecular Biology of the Cell, 21*, 3781–3784.

Schnedler, N., Burckhardt, G., & Burckhardt, B. C. (2011). Glyoxylate is a substrate of the sulfate-oxalate exchanger, sat-1, and increases its expression in HepG2 cells. *Journal of Hepatology, 54*, 513–520.

Schofield, A. E., Martin, P. G., Spillett, D., & Tanner, M. J. (1994). The structure of the human red blood cell anion exchanger (EPB3, AE1, band 3) gene. *Blood, 84*, 2000–2012.

Schofield, A. E., Reardon, D. M., & Tanner, M. J. (1992). Defective anion transport activity of the abnormal band 3 in hereditary ovalocytic red blood cells. *Nature, 355*, 836–838.

Schofield, A. E., Tanner, M. J., Pinder, J. C., Clough, B., Bayley, P. M., Nash, G. B., et al. (1992). Basis of unique red cell membrane properties in hereditary ovalocytosis. *Journal of Molecular Biology, 223*, 949–958.

Schwarz, K., Iolascon, A., Verissimo, F., Trede, N. S., Horsley, W., Chen, W., et al. (2009). Mutations affecting the secretory COPII coat component SEC23B cause congenital dyserythropoietic anemia type II. *Nature Genetics, 41*, 936–940.

Shah, S. S., Al-Rajhi, A., Brandt, J. D., Mannis, M. J., Roos, B., Sheffield, V. C., et al. (2008). Mutation in the SLC4A11 gene associated with autosomal recessive congenital hereditary endothelial dystrophy in a large Saudi family. *Ophthalmic Genetics, 29*, 41–45.

Shao, L., Xu, Y., Dong, Q., Lang, Y., Yue, S., & Miao, Z. (2010). A novel SLC4A1 variant in an autosomal dominant distal renal tubular acidosis family with a severe phenotype. *Endocrine, 37*, 473–478.

Sharma, A. K., Rigby, A. C., & Alper, S. L. (2011). STAS domain structure and function. *Cellular Physiology and Biochemistry, 28*, 407–422.

Sharma, A. K., Ye, L., Baer, C. E., Shanmugasundaram, K., Alber, T., Alper, S. L., et al. (2011). Solution structure of the guanine nucleotide-binding STAS domain of SLC26-related SulP protein Rv1739c from Mycobacterium tuberculosis. *The Journal of Biological Chemistry, 286*, 8534–8544.

Shcheynikov, N., Yang, D., Wang, Y., Zeng, W., Karniski, L. P., So, I., et al. (2008). The Slc26a4 transporter functions as an electroneutral Cl-/I-/HCO3- exchanger: Role of Slc26a4 and Slc26a6 in I- and HCO3- secretion and in regulation of CFTR in the parotid duct. *The Journal of Physiology, 586*, 3813–3824.

Shelden, M. C., Howitt, S. M., & Price, G. D. (2009). Membrane topology of the cyanobacterial bicarbonate transporter, BicA, a member of the SulP (SLC26A) family. *Molecular Membrane Biology, 27*, 12–23.

Shepshelovich, J., Goldstein-Magal, L., Globerson, A., Yen, P. M., Rotman-Pikielny, P., & Hirschberg, K. (2005). Protein synthesis inhibitors and the chemical chaperone TMAO reverse endoplasmic reticulum perturbation induced by overexpression of the iodide transporter pendrin. *Journal of Cell Science, 118*, 1577–1586.

Shibagaki, N., & Grossman, A. R. (2004). Probing the function of STAS domains of the Arabidopsis sulfate transporters. *The Journal of Biological Chemistry, 279*, 30791–30799.

Shibagaki, N., & Grossman, A. R. (2006). The role of the STAS domain in the function and biogenesis of a sulfate transporter as probed by random mutagenesis. *The Journal of Biological Chemistry, 281*, 22964–22973.

Shibagaki, N., & Grossman, A. R. (2010). Binding of cysteine synthase to the STAS domain of sulfate transporter and its regulatory consequences. *The Journal of Biological Chemistry, 285*, 25094–25102.

Shmukler, B. E., Kedar, P. S., Warang, P., Desai, M., Madkaikar, M., Ghosh, K., et al. (2010). Hemolytic anemia and distal renal tubular acidosis in two Indian patients homozygous for SLC4A1/AE1 mutation A858D. *American Journal of Hematology, 85*, 824–828.

Shnitsar, V., Li, J., Li, X., Calmettes, C., Basu, A., Casey, J. R., et al. (2013). A substrate access tunnel in the cytosolic domain is not an essential feature of the SLC4 family of bicarbonate transporters. *The Journal of Biological Chemistry, 288*, 33848–33860.

Sinha, R., Agarwal, I., Bawazir, W. M., & Bruce, L. J. (2013). Distal renal tubular acidosis with hereditary spherocytosis. *Indian Pediatrics, 50*, 693–695.

Sinning, A., Liebmann, L., Kougioumtzes, A., Westermann, M., Bruehl, C., & Hubner, C. A. (2011). Synaptic glutamate release is modulated by the Na+-driven Cl-/HCO(3)(-) exchanger Slc4a8. *The Journal of Neuroscience, 31*, 7300–7311.

Southgate, C. D., Chishti, A. H., Mitchell, B., Yi, S. J., & Palek, J. (1996). Targeted disruption of the murine erythroid band 3 gene results in spherocytosis and severe haemolytic anaemia despite a normal membrane skeleton. *Nature Genetics, 14*, 227–230.

Sowah, D., & Casey, J. R. (2011). An intramolecular transport metabolon: Fusion of carbonic anhydrase II to the COOH terminus of the Cl(-)/HCO(3)(-)exchanger, AE1. *American Journal of Physiology. Cell Physiology, 301*, C336–C346.

Spring, F. A., Bruce, L. J., Anstee, D. J., & Tanner, M. J. (1992). A red cell band 3 variant with altered stilbene disulphonate binding is associated with the Diego (Dia) blood group antigen. *The Biochemical Journal, 288*(Pt 3), 713–716.

Steck, T. L. (1972). Cross-linking the major proteins of the isolated erythrocyte membrane. *Journal of Molecular Biology, 66*, 295–305.

Steck, T. L., Fairbanks, G., & Wallach, D. F. (1971). Disposition of the major proteins in the isolated erythrocyte membrane. Proteolytic dissection. *Biochemistry, 10*, 2617–2624.

Steck, T. L., Koziarz, J. J., Singh, M. K., Reddy, G., & Kohler, H. (1978). Preparation and analysis of seven major, topographically defined fragments of band 3, the predominant transmembrane polypeptide of human erythrocyte membranes. *Biochemistry, 17*, 1216–1222.

Steck, T. L., Ramos, B., & Strapazon, E. (1976). Proteolytic dissection of band 3, the predominant transmembrane polypeptide of the human erythrocyte membrane. *Biochemistry, 15*, 1153–1161.

Stefanovic, M., Markham, N. O., Parry, E. M., Garrett-Beal, L. J., Cline, A. P., Gallagher, P. G., et al. (2007). An 11-amino acid beta-hairpin loop in the cytoplasmic domain of band 3 is responsible for ankyrin binding in mouse erythrocytes. *Proceedings of the National Academy of Sciences of the United States of America, 104*, 13972–13977.

Stehberger, P. A., Shmukler, B. E., Stuart-Tilley, A. K., Peters, L. L., Alper, S. L., & Wagner, C. A. (2007). Distal renal tubular acidosis in mice lacking the AE1 (band3) Cl-/HCO3- exchanger (slc4a1). *Journal of the American Society of Nephrology, 18*, 1408–1418.

Sterling, D., & Casey, J. R. (1999). Transport activity of AE3 chloride/bicarbonate anion-exchange proteins and their regulation by intracellular pH. *The Biochemical Journal, 344*(Pt 1), 221–229.

Sterling, D., Reithmeier, R. A., & Casey, J. R. (2001). A transport metabolon. Functional interaction of carbonic anhydrase II and chloride/bicarbonate exchangers. *The Journal of Biological Chemistry, 276*, 47886–47894.

Stewart, A. K., Chernova, M. N., Kunes, Y. Z., & Alper, S. L. (2001). Regulation of AE2 anion exchanger by intracellular pH: Critical regions of the NH(2)-terminal cytoplasmic domain. *American Journal of Physiology. Cell Physiology, 281*, C1344–C1354.

Stewart, A. K., Kerr, N., Chernova, M. N., Alper, S. L., & Vaughan-Jones, R. D. (2004). Acute pH-dependent regulation of AE2-mediated anion exchange involves discrete local surfaces of the NH2-terminal cytoplasmic domain. *The Journal of Biological Chemistry, 279*, 52664–52676.

Stewart, A. K., Kurschat, C. K., Vaughan-Jones, R. D., & Alper, S. L. (2009). Putative re-entrant loop 1 of Ae2 transmembrane domain has a major role in acute regulation of anion exchange by pH. *The Journal of Biological Chemistry, 284*, 6126–6139.

Strapazon, E., & Steck, T. L. (1976). Binding of rabbit muscle aldolase to band 3, the predominant polypeptide of the human erythrocyte membrane. *Biochemistry, 15*, 1421–1424.

Su, Y., Blake-Palmer, K. G., Fry, A. C., Best, A., Brown, A. C., Hiemstra, T. F., et al. (2011). Glyceraldehyde 3-phosphate dehydrogenase is required for band 3 (anion exchanger 1) membrane residency in the mammalian kidney. *The American Journal of Renal Physiology, 300*, F157–F166.

Su, Y., Blake-Palmer, K. G., Sorrell, S., Javid, B., Bowers, K., Zhou, A., et al. (2008). Human H+ATPase a4 subunit mutations causing renal tubular acidosis reveal a role for interaction with phosphofructokinase-1. *American Journal of Physiology. Renal Physiology, 295*, F950–F958.

Suzuki, M., Vaisbich, M. H., Yamada, H., Horita, S., Li, Y., Sekine, T., et al. (2008). Functional analysis of a novel missense NBC1 mutation and of other mutations causing proximal renal tubular acidosis. *Pflügers Archiv, 455*, 583–593.

Tam, L. Y., Landolt-Marticorena, C., & Reithmeier, R. A. (1996). Glycosylation of multiple extracytosolic loops in Band 3, a model polytopic membrane protein. *The Biochemical Journal, 318*(Pt 2), 645–648.

Tam, L. Y., Loo, T. W., Clarke, D. M., & Reithmeier, R. A. F. (1994). Identification of an internal topogenic signal sequence in human band-3, the erythrocyte anion-exchanger. *Journal of Biological Chemistry, 269*, 32542–32550.

Tanner, M. J. (2002). Band 3 anion exchanger and its involvement in erythrocyte and kidney disorders. *Current Opinion in Hematology, 9*, 133–139.

Tanner, M. J., & Anstee, D. J. (1976). The membrane change in En(a-) human erythrocytes. Absence of the major erythrocyte sialoglycoprotein. *The Biochemical Journal, 153*, 271–277.

Tanner, M. J., Jenkins, R. E., Anstee, D. J., & Clamp, J. R. (1976). Abnormal carbohydrate composition of the major penetrating membrane protein of En(a-) human erythrocytes. *The Biochemical Journal, 155*, 701–703.

Tanner, M. J., Martin, P. G., & High, S. (1988). The complete amino acid sequence of the human erythrocyte membrane anion-transport protein deduced from the cDNA sequence. *The Biochemical Journal, 256*, 703–712.

Tanphaichitr, V. S., Sumboonnanonda, A., Ideguchi, H., Shayakul, C., Brugnara, C., Takao, M., et al. (1998). Novel AE1 mutations in recessive distal renal tubular acidosis. Loss-of-function is rescued by glycophorin A. *Journal of Clinical Investigation, 102*, 2173–2179.

Thevenin, B. J., & Low, P. S. (1990). Kinetics and regulation of the ankyrin-band 3 interaction of the human red blood cell membrane. *The Journal of Biological Chemistry, 265*, 16166–16172.

Thevenin, B. J., Willardson, B. M., & Low, P. S. (1989). The redox state of cysteines 201 and 317 of the erythrocyte anion exchanger is critical for ankyrin binding. *The Journal of Biological Chemistry, 264*, 15886–15892.

Toure, A., Lhuillier, P., Gossen, J. A., Kuil, C. W., Lhote, D., Jegou, B., et al. (2007). The testis anion transporter 1 (Slc26a8) is required for sperm terminal differentiation and male fertility in the mouse. *Human Molecular Genetics*, *16*, 1783–1793.

Toure, A., Morin, L., Pineau, C., Becq, F., Dorseuil, O., & Gacon, G. (2001). Tat1, a novel sulfate transporter specifically expressed in human male germ cells and potentially linked to rhogtpase signaling. *The Journal of Biological Chemistry*, *276*, 20309–20315.

Toye, A. M., Banting, G., & Tanner, M. J. (2004). Regions of human kidney anion exchanger 1 (kAE1) required for basolateral targeting of kAE1 in polarised kidney cells: Mis-targeting explains dominant renal tubular acidosis (dRTA). *Journal of Cell Science*, *117*, 1399–1410.

Toye, A. M., Bruce, L. J., Unwin, R. J., Wrong, O., & Tanner, M. J. (2002a). Band 3 Walton, a C-terminal deletion associated with distal renal tubular acidosis, is expressed in the red cell membrane but retained internally in kidney cells. *Blood*, *99*, 342–347.

Toye, A. M., Bruce, L. J., Unwin, R. J., Wrong, O., & Tanner, M. J. (2002b). Band 3 Walton, a C-terminal deletion associated with distal renal tubular acidosis, is expressed in the red cell membrane but retained internally in kidney cells. *Blood*, *99*, 342–347.

Toye, A. M., Ghosh, S., Young, M. T., Jones, G. K., Sessions, R. B., Ramauge, M., et al. (2005). Protein-4.2 association with band 3 (AE1, SLCA4) in Xenopus oocytes: Effects of three natural protein-4.2 mutations associated with hemolytic anemia. *Blood*, *105*, 4088–4095.

Toye, A. M., Parker, M. D., Daly, C. M., Lu, J., Virkki, L. V., Pelletier, M. F., et al. (2006). The human NBCe1-A mutant R881C, associated with proximal renal tubular acidosis, retains function but is mistargeted in polarized renal epithelia. *American Journal of Physiology. Cell Physiology*, *291*, C788–C801.

Toye, A. M., Williamson, R. C., Khanfar, M., Bader-Meunier, B., Cynober, T., Thibault, M., et al. (2008). Band 3 Courcouronnes (Ser667Phe): A trafficking mutant differentially rescued by wild-type band 3 and glycophorin A. *Blood*, *111*, 5380–5389.

Tsuji, T., Irimura, T., & Osawa, T. (1980). The carbohydrate moiety of band-3 glycoprotein of human erythrocyte membranes. *The Biochemical Journal*, *187*, 677–686.

Tsuji, T., Irimura, T., & Osawa, T. (1981). The carbohydrate moiety of band 3 glycoprotein of human erythrocyte membranes. Structures of lower molecular weight oligosaccharides. *The Journal of Biological Chemistry*, *256*, 10497–10502.

Tsukamoto, K., Suzuki, H., Harada, D., Namba, A., Abe, S., & Usami, S. (2003). Distribution and frequencies of PDS (SLC26A4) mutations in Pendred syndrome and nonsyndromic hearing loss associated with enlarged vestibular aqueduct: A unique spectrum of mutations in Japanese. *European Journal of Human Genetics*, *11*, 916–922.

Ungsupravate, D., Sawasdee, N., Khositseth, S., Udomchaiprasertkul, W., Khoprasert, S., Li, J., et al. (2010). Impaired trafficking and intracellular retention of mutant kidney anion exchanger 1 proteins (G701D and A858D) associated with distal renal tubular acidosis. *Molecular Membrane Biology*, *27*, 92–103.

Van Dort, H. M., Moriyama, R., & Low, P. S. (1998). Effect of band 3 subunit equilibrium on the kinetics and affinity of ankyrin binding to erythrocyte membrane vesicles. *The Journal of Biological Chemistry*, *273*, 14819–14826.

Van Hauwe, P., Everett, L. A., Coucke, P., Scott, D. A., Kraft, M. L., Ris-Stalpers, C., et al. (1998). Two frequent missense mutations in Pendred syndrome. *Human Molecular Genetics*, *7*, 1099–1104.

Van Zwieten, R., Francois, J. J., Van Leeuwen, K., Van Wesel, A. C., Van Bruggen, R., Van Solinge, W. W., et al. (2013). Hereditary spherocytosis due to band 3 deficiency: 15 novel mutations in SLC4A1. *American Journal of Hematology*, *88*, 159–160.

Vilas, G. L., Loganathan, S. K., Liu, J., Riau, A. K., Young, J. D., Mehta, J. S., et al. (2013). Transmembrane water flux through SLC4A11: A route defective in genetic corneal diseases. *Human Molecular Genetics, 22*, 4579–4590.

Vilas, G. L., Loganathan, S. K., Quon, A., Sundaresan, P., Vithana, E. N., & Casey, J. (2011). Oligomerization of SLC4A11 protein and the severity of FECD and CHED2 corneal dystrophies caused by SLC4A11 mutations. *Human Mutation, 33*, 419–428.

Vilas, G. L., Morgan, P. E., Loganathan, S. K., Quon, A., & Casey, J. R. (2011). A biochemical framework for SLC4A11, the plasma membrane protein defective in corneal dystrophies. *Biochemistry, 50*, 2157–2169.

Vince, J. W., Carlsson, U., & Reithmeier, R. A. F. (2000). Localization of the Cl-/HCO3- anion exchanger binding site to the amino-terminal region of carbonic anhydrase II. *Biochemistry, 39*, 13344–13349.

Vince, J. W., & Reithmeier, R. A. (1998). Carbonic anhydrase II binds to the carboxyl terminus of human band 3, the erythrocyte C1-/HCO3- exchanger. *The Journal of Biological Chemistry, 273*, 28430–28437.

Vince, J. W., & Reithmeier, R. A. (2000). Identification of the carbonic anhydrase II binding site in the Cl(-)/HCO(3)(-) anion exchanger AE1. *Biochemistry, 39*, 5527–5533.

Vithana, E. N., Morgan, P. E., Ramprasad, V., Tan, D. T., Yong, V. H., Venkataraman, D., et al. (2008). SLC4A11 mutations in Fuchs endothelial corneal dystrophy. *Human Molecular Genetics, 17*, 656–666.

Vithana, E. N., Morgan, P., Sundaresan, P., Ebenezer, N. D., Tan, D. T., Mohamed, M. D., et al. (2006). Mutations in sodium-borate cotransporter SLC4A11 cause recessive congenital hereditary endothelial dystrophy (CHED2). *Nature Genetics, 38*, 755–757.

Wada, H., Kanzaki, A., Yawata, A., Inoue, T., Kaku, M., Takezono, M., et al. (1999). Late expression of red cell membrane protein 4.2 in normal human erythroid maturation with seven isoforms of the protein 4.2 gene. *Experimental Hematology, 27*, 54–62.

Walder, J. A., Chatterjee, R., Steck, T. L., Low, P. S., Musso, G. F., Kaiser, E. T., et al. (1984). The interaction of hemoglobin with the cytoplasmic domain of band 3 of the human erythrocyte membrane. *The Journal of Biological Chemistry, 259*, 10238–10246.

Wang, L., Groves, J. D., Mawby, W. J., & Tanner, M. J. (1997). Complementation studies with Co-expressed fragments of the human red cell anion transporter (Band 3; AE1). The role of some exofacial loops in anion transport. *The Journal of Biological Chemistry, 272*, 10631–10638.

Wang, D. N., Kuhlbrandt, W., Sarabia, V. E., & Reithmeier, R. A. F. (1993). 2-Dimensional structure of the membrane domain of human band-3, the anion transport protein of the erythrocyte-membrane. *EMBO Journal, 12*, 2233–2239.

Wang, D. N., Sarabia, V. E., Reithmeier, R. A. F., & Kuhlbrandt, W. (1994). 3-Dimensional map of the dimeric membrane domain of the human erythrocyte anion-exchanger, band-3. *EMBO Journal, 13*, 3230–3235.

Wang, C. Z., Yano, H., Nagashima, K., & Seino, S. (2000). The Na+-driven Cl-/HCO3- exchanger. Cloning, tissue distribution, and functional characterization. *The Journal of Biological Chemistry, 275*, 35486–35490.

Wedenoja, S., Pekansaari, E., Hoglund, P., Makela, S., Holmberg, C., & Kere, J. (2011). Update on SLC26A3 mutations in congenital chloride diarrhea. *Human Mutation, 32*, 715–722.

Wickrema, A., Koury, S. T., Dai, C. H., & Krantz, S. B. (1994). Changes in cytoskeletal proteins and their mRNAs during maturation of human erythroid progenitor cells. *Journal of Cellular Physiology, 160*, 417–426.

Williamson, R. C., Brown, A. C., Mawby, W. J., & Toye, A. M. (2008). Human kidney anion exchanger 1 localisation in MDCK cells is controlled by the phosphorylation status of two critical tyrosines. *Journal of Cell Science, 121*, 3422–3432.

Williamson, R. C., & Toye, A. M. (2008). Glycophorin A: Band 3 aid. *Blood Cells, Molecules & Diseases*, *41*, 35–43.

Woods, C. M., Boyer, B., Vogt, P. K., & Lazarides, E. (1986). Control of erythroid differentiation: Asynchronous expression of the anion transporter and the peripheral components of the membrane skeleton in AEV- and S13-transformed cells. *The Journal of Cell Biology*, *103*, 1789–1798.

Wrong, O., Bruce, L. J., Unwin, R. J., Toye, A. M., & Tanner, M. J. (2002). Band 3 mutations, distal renal tubular acidosis, and Southeast Asian ovalocytosis. *Kidney International*, *62*, 10–19.

Wu, F., Saleem, M. A., Kampik, N. B., Satchwell, T. J., Williamson, R. C., Blattner, S. M., et al. (2010). Anion exchanger 1 interacts with nephrin in podocytes. *Journal of the American Society of Nephrology*, *21*, 1456–1467.

Wu, F., Satchwell, T. J., & Toye, A. M. (2011). Anion exchanger 1 in red blood cells and kidney: Band 3's in a pod. *Biochemistry and Cell Biology*, *89*, 106–114.

Xie, Q., Welch, R., Mercado, A., Romero, M. F., & Mount, D. B. (2002). Molecular characterization of the murine Slc26a6 anion exchanger: Functional comparison with Slc26a1. *American Journal of Physiology. Renal Physiology*, *283*, F826–F838.

Yamaguchi, T., Ikeda, Y., Abe, Y., Kuma, H., Kang, D., Hamasaki, N., et al. (2010). Structure of the membrane domain of human erythrocyte anion exchanger 1 revealed by electron crystallography. *Journal of Molecular Biology*, *397*, 179–189.

Yamazaki, O., Yamada, H., Suzuki, M., Horita, S., Shirai, A., Nakamura, M., et al. (2011). Functional characterization of nonsynonymous single nucleotide polymorphisms in the electrogenic Na+-HCO3- cotransporter NBCe1A. *Pflügers Archiv*, *461*, 249–259.

Yannoukakos, D., Stuart-Tilley, A., Fernandez, H. A., Fey, P., Duyk, G., & Alper, S. L. (1994). Molecular cloning, expression, and chromosomal localization of two isoforms of the AE3 anion exchanger from human heart. *Circulation Research*, *75*, 603–614.

Yenchitsomanus, P. T., Kittanakom, S., Rungroj, N., Cordat, E., & Reithmeier, R. A. (2005). Molecular mechanisms of autosomal dominant and recessive distal renal tubular acidosis caused by SLC4A1 (AE1) mutations. *Journal of Molecular and Genetic Medicine: An International Journal of Biomedical Research*, *1*, 49–62.

Yenchitsomanus, P. T., Sawasdee, N., Paemanee, A., Keskanokwong, T., Vasuvattakul, S., Bejrachandra, S., et al. (2003). Anion exchanger 1 mutations associated with distal renal tubular acidosis in the Thai population. *Journal of Human Genetics*, *48*, 451–456.

Yoon, J. S., Park, H. J., Yoo, S. Y., Namkung, W., Jo, M. J., Koo, S. K., et al. (2008). Heterogeneity in the processing defect of SLC26A4 mutants. *Journal of Medical Genetics*, *45*, 411–419.

Young, M. T., Beckmann, R., Toye, A. M., & Tanner, M. J. (2000). Red-cell glycophorin A-band 3 interactions associated with the movement of band 3 to the cell surface. *The Biochemical Journal*, *350*(Pt 1), 53–60.

Zhang, Y., Chernova, M. N., Stuart-Tilley, A. K., Jiang, L., & Alper, S. L. (1996). The cytoplasmic and transmembrane domains of AE2 both contribute to regulation of anion exchange by pH. *The Journal of Biological Chemistry*, *271*, 5741–5749.

Zhang, D., Kiyatkin, A., Bolin, J. T., & Low, P. S. (2000). Crystallographic structure and functional interpretation of the cytoplasmic domain of erythrocyte membrane band 3. *Blood*, *96*, 2925–2933.

Zhang, Z., Liu, K. X., He, J. W., Fu, W. Z., Yue, H., Zhang, H., et al. (2012). Identification of two novel mutations in the SLC4A1 gene in two unrelated Chinese families with distal renal tubular acidosis. *Archives of Medical Research*, *43*, 298–304.

Zhao, R. M., & Reithmeier, R. A. F. (2001). Expression and characterization of the anion transporter homologue YNL275w in Saccharomyces cerevisiae. *American Journal of Physiology. Cell Physiology*, *281*, C33–C45.

Zheng, J., Shen, W., He, D. Z., Long, K. B., Madison, L. D., & Dallos, P. (2000). Prestin is the motor protein of cochlear outer hair cells. *Nature*, *405*, 149–155.

Zhou, Z., DeSensi, S. C., Stein, R. A., Brandon, S., Song, L., Cobb, C. E., et al. (2007). Structure of the cytoplasmic domain of erythrocyte band 3 hereditary spherocytosis variant P327R: Band 3 Tuscaloosa. *Biochemistry*, *46*, 10248–10257.

Zhu, Q., & Casey, J. R. (2004). The substrate anion selectivity filter in the human erythrocyte Cl-/HCO3- exchange protein, AE1. *The Journal of Biological Chemistry*, *279*, 23565–23573.

CHAPTER TWO

Structural Dynamics and Regulation of the Mammalian SLC9A Family of Na^+/H^+ Exchangers

Ruth Hendus-Altenburger[*,†], **Birthe B. Kragelund**[*], **Stine Falsig Pedersen**[†,1]

[*]Section for Biomolecular Sciences, Department of Biology, University of Copenhagen, Copenhagen, Denmark
[†]Section for Cell and Developmental Biology, Department of Biology, University of Copenhagen, Copenhagen, Denmark
[1]Corresponding author: e-mail address: sfpedersen@bio.ku.dk

Contents

Current Topics in Membranes, Volume 73
ISSN 1063-5823
http://dx.doi.org/10.1016/B978-0-12-800223-0.00002-5

Abstract

Mammalian Na^+/H^+ exchangers of the SLC9A family are widely expressed and involved in numerous essential physiological processes. Their primary function is to mediate the 1:1 exchange of Na^+ for H^+ across the membrane in which they reside, and they play central roles in regulation of body, cellular, and organellar pH. Their function is tightly regulated through mechanisms involving interactions with multiple protein and lipid-binding partners, phosphorylations, and other posttranslational modifications. Biochemical and mutational analyses indicate that the SLC9As have a short intracellular N-terminus, 12 transmembrane (TM) helices necessary and sufficient for ion transport, and a C-terminal cytoplasmic tail region with essential regulatory roles. No high-resolution structures of the SLC9As exist; however, models based on crystal structures of the bacterial NhaAs support the 12 TM organization and suggest that TMIV and XI may form a central part of the ion-translocation pathway, whereas pH sensing may involve TMII, TMIX, and several intracellular loops. Similar to most ion transporters studied, SLC9As likely exist as coupled dimers in the membrane, and this appears to be important for the well-studied cooperativity of H^+ binding.

The aim of this work is to summarize and critically discuss the currently available evidence on the structural dynamics, regulation, and binding partner interactions of SLC9As, focusing in particular on the most widely studied isoform, SLC9A1/NHE1. Further, novel bioinformatic and structural analyses are provided that to some extent challenge the existing paradigm on how ions are transported by mammalian SLC9As.

ABBREVIATIONS

at *Amphiuma tridactylum*
CaM calmodulin
CaMKII calmodulin kinase II
CaN calcineurin
CHP calcineurin homologous protein
CK casein kinase
DAG diacylglycerol
EL extracellular loop
EM electron microscopy
ER endoplasmic reticulum
ERM ezrin, radixin, and moesin
FLIM fluorescence-lifetime imaging microscopy
FRET fluorescence resonance energy transfer

ID intrinsic disorder
IDPs intrinsically disordered proteins
IDR intrinsically disordered region
IFD interfacial domain
IL intracellular loop
IP immunoprecipitation
ITC ion-translocation center
ITD ion-translocation domain
LC liquid chromatography
LID lipid interaction domain
LPA lysophosphatitic acid
MAPK mitogen-activated protein kinase
Mj *M. jannaschii*
MS mass spectrometry
NHE Na^+/H^+ exchanger
NIK Ste-20 like Nck-interacting kinase
NMR nuclear magnetic resonance
NOE nuclear Overhauser effect
PDZ PSD-95/discs large/ZO-1
PE phorbol esters
PI(4,5)P$_2$ phosphatidylinositol-4,5-bisphosphate
PKA protein kinase A
PKB protein kinase B
PKC protein kinase C
POT proton-dependent oliogopeptide transporter
PP1 protein phosphatase 1
RACK1 receptor for activated C kinase-1
SAXS small-angle X-ray scattering
SGK1 serum- and glucocorticoid-inducible kinase 1
SLC9A solute carrier 9A
TM transmembrane
Y2H yeast 2 hybrid

1. INTRODUCTION

The solute carrier 9A (SLC9A) family of mammalian Na^+/H^+ exchangers (*NHEs*) play essential roles in cellular, organellar, and systemic pH regulation, in cell volume regulation, and in vectorial ion transport, and consequently, in a wide array of physiological events depending on these processes (see Alexander & Grinstein, 2006; Boedtkjer, Bunch, & Pedersen, 2012; Burckhardt, Di Sole, & Helmle-Kolb, 2002; Donowitz, Ming, & Fuster, 2013; Orlowski & Grinstein, 2004, 2011). *NHE* dysregulation is associated with a rapidly growing list of pathophysiological conditions, of which some of the best studied are ischemia/reperfusion damage, cancer, diabetes, hypertension, epilepsy, and several gastrointestinal disorders (see Donowitz et al., 2013; Orlowski & Grinstein, 2011; Pedersen, O'Donnell, Anderson, & Cala, 2006). All the *NHEs* exhibit a membrane topology of a short intracellular N-terminus, 12 predicted transmembrane (TM) helices mediating ion translocation, and a C-terminal cytoplasmic tail region with essential roles in *NHE* posttranslational modification and interactions with binding partners (Boedtkjer et al., 2012; Donowitz et al., 2013; Orlowski & Grinstein, 2011; Shrode, Gan, D'Souza, Orlowski, & Grinstein, 1998; Wakabayashi, Pang, Su, & Shigekawa, 2000), Fig. 2.1.

In the solute carrier (SLC) nomenclature (http://slc.bioparadigms.org/), the mammalian *NHEs* were recently divided into SLC9A1–9 (NHE1–9), SLC9B1–B2 (NHA1–2), and SLC9C1 (Sperm-NHE), reflecting major sequence divergence between these three groups (see also Donowitz et al., 2013). While the physiology of many of the mammalian *NHEs* is relatively well understood, a comprehensive understanding of their structural and biophysical properties is lacking. Here, we summarize and discuss the current insight into the structural dynamics, regulation, and binding partner interactions of the mammalian *NHEs*, with particular focus on *SLC9A1*/NHE1 (which will be referred to as NHE1 throughout this work). Furthermore, we provide novel bioinformatic and structural analyses that to some extent challenge the existing paradigm on the mechanism of ion transport by mammalian *NHEs*, and point to new open questions in the understanding of their regulation.

2. BASIC FUNCTIONAL PROPERTIES: SUBSTRATES, DRIVING FORCES AND KINETICS OF SLC9As

SLC9A activity is electroneutral with a stoichiometry of 1:1 (H^+:Na^+) (or 2:2, see Section 4). The transport is secondary active, that is, does not require ATP hydrolysis directly, but is driven by the combined driving forces for the substrates across the membrane. Under normal steady-state conditions, there is a substantial driving force for net cellular Na^+ uptake and H^+ efflux via *NHEs*, which can be described as: $RT\cdot\ln([Na^+]_i/[Na^+]_o) - RT\cdot\ln([H^+]_i/[H^+]_o)$, where R is the gas constant and T is the temperature in K. The reported affinities of NHE1, *-2*, and *-3* for extracellular Na^+ are in the range of 5–50 mM (Aronson, 1985; Montrose & Murer,

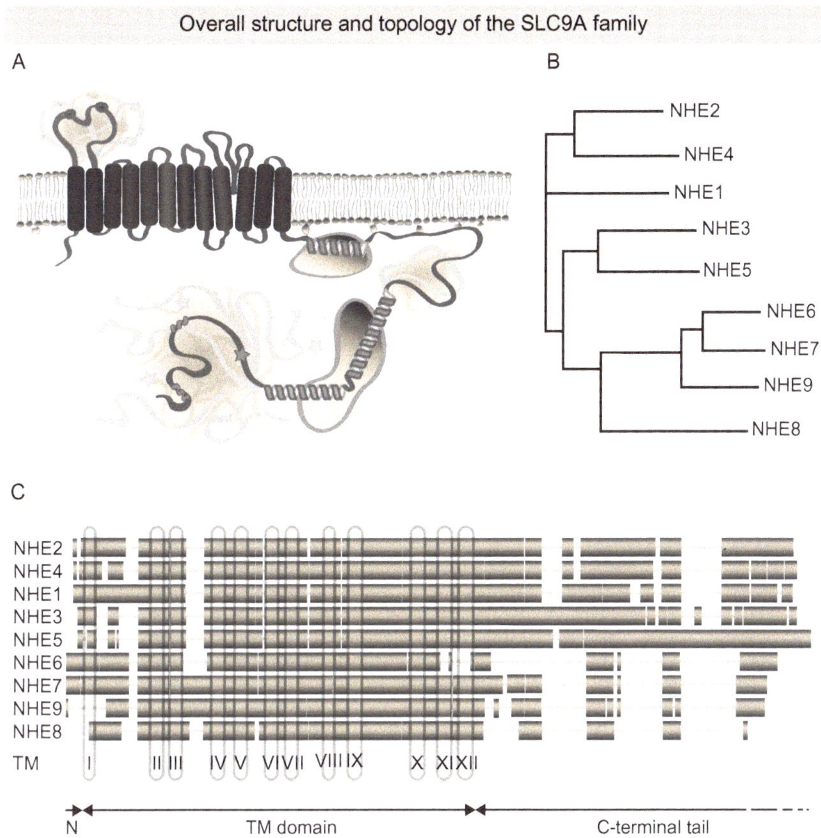

Figure 2.1 Overall structure and topology of the SLC9A family. (A) Presumed overall topology of NHE1 as representative for the SL9A family. The model shows the TM domain consisting of 12 TM α-helices and loops of varying length, as well as the short N-terminal and long C-terminal tails. Known structural elements of the tail as well as predicted and confirmed intrinsic disorder in the extracellular loop 1 (EL1) and tail are indicated. (B) Phylogram for the human SLC9A family showing the relative phylogenic distance between members. (C) Multiple sequence alignment of the SLC9A members. Horizontal bars represent aligning sequences and gaps in the alignment are depicted as horizontal lines. The positions of the TM α-helices from the Nygaard model of NHE1 are indicated. Note the long extracellular loops EL1 and extracellular loop 5 (EL5), as well as major gaps in the C-terminal tail leading to different lengths of the individual members. The multiple sequence alignment and the phylogram were made with the online software ClustalW2 (Goujon et al., 2010). Human sequences were retrieved from NCBI, and the accession numbers are as follows: NHE1 NP_003038.2, *NHE2* Q9UBY0.1, *NHE3* NP_004165.2, *NHE4* NP_001011552.2, *NHE5* AAC98696.1, *NHE6* Q92581.2, *NHE7* Q96T83.1, *NHE8* AAI12214.1, and *NHE9* NP_775924.1. (See the color plate.)

1988; Wakabayashi, Shigekawa, & Pouyssegur, 1997). The dependence of NHE1, *-2*, and *-3* on $[Na^+]_o$ shows simple Michaelis–Menten kinetics (Orlowski & Grinstein, 2011). For *NHE4*, both Michaelis–Menten-type $[Na^+]_o$ kinetics (Chambrey et al., 2001) and sigmoidal $[Na^+]_o$ dependence are reported (Bookstein et al., 1996). In addition to Na^+ and H^+, Li^+, K^+, and NH_4^+ are also substrates for *NHEs* (Blanchard et al., 1998; see Donowitz et al., 2013; Orlowski & Grinstein, 2011; Wakabayashi, Ikeda, Iwamoto, Pouyssegur, & Shigekawa, 1997; Wakabayashi, Shigekawa, et al., 1997). K^+ transport has been proposed to be a major function of the organellar NHE isoforms, *NHE6*, *-7*, and *-9* (see Donowitz et al., 2013).

NHE1–5 are generally found to be activated by internal H^+ with Hill coefficients well above 1, indicating cooperativity in intracellular H^+ binding (Aronson, Nee, & Suhm, 1982; Attaphitaya, Nehrke, & Melvin, 2001; Chambrey et al., 2001; Kapus, Grinstein, Wasan, Kandasamy, & Orlowski, 1994; Orlowski, 1993; Yu, Shull, & Orlowski, 1993; see Aronson, 1985; Wakabayashi, Ikeda, et al., 1997; Wakabayashi, Shigekawa, et al., 1997). This sigmoidal dependence on H^+_i ensures efficient activation at acidic pH_i and inactivation at alkaline pH_i, making these exchangers ideally suited for maintaining pH_i within a narrow range. The modulation of the *NHEs* by pH_i has been the subject of controversy in the field. A widely accepted mechanism originally proposed by Aronson et al. (1982) is that the sigmoidal activation by intracellular protons reflects the presence of an allosteric H^+-binding site. For NHE1, it was proposed that the stimulating effect of growth factors and other stimuli on NHE1 can be explained by these agents causing an increase in the apparent H^+ affinity of the allosteric H^+-binding site (Paris & Pouyssegur, 1984).

2.1. Pharmacological inhibitors of *NHEs*

Given the major clinical relevance of NHE1, development of pharmacological inhibitors has focused mostly on this isoform. The first *NHE* inhibitor developed was the diuretic drug amiloride, with K_i values of ~1 μM for NHE1 and *NHE2*, 21 μM for *NHE5*, and 100 μM for *NHE3* (Masereel, Pochet, & Laeckmann, 2003), but which also inhibits other Na^+ transporters including ENaC/Degenerin family Na^+ channels and the Na^+/Ca^{2+} exchanger (NCX) (Kellenberger & Schild, 2002; Qadri, Song, Fuller, & Benos, 2010). Later, *NHE* inhibitors include pyrazine derivatives such as 5′ *N*-ethylisopropylamiloride (K_i ~ 0.01–0.02 μM for NHE1) and dimethylamiloride ($K_i \sim 0.023$ μM) with negligible effects on NCX and ENaC (Masereel et al., 2003). Substitution of the pyrazine moiety gave rise to HOE-642/cariporide and eniporide, zoniporide, and sabiporide (Masereel et al., 2003; Touret et al., 2003) as well as KR-32570 and KR-33028 (Kim et al., 2007). More recently, several new NHE1 inhibitors have been reported including the aminoguanidine derivative T-162559 (Kusumoto et al., 2002) and the imidazolylpiperadine SL 59.1227 (Lorrain et al., 2000). A *NHE3*-specific inhibitor, S3226 ($K_i = 0.02$ μM for *NHE3*, 3.6 μM for NHE1) also exists (Schwark et al., 1998). A number of mutagenesis studies have addressed the mechanisms by which these inhibitors interact with NHE1, in particular, residues in TMIV and TMIX–TMXI appear to be important for the interaction, consistent with the partially competitive nature of *NHE* inhibition by these compounds (Counillon, Franchi, & Pouyssegur, 1993; Khadilkar, Iannuzzi, & Orlowski, 2001; Orlowski & Kandasamy, 1996; Pedersen, King, Nygaard, Rigor, & Cala, 2007; Touret, Poujeol, & Counillon, 2001; Yun et al., 1993).

3. STRUCTURAL ORGANIZATION AND DYNAMICS OF THE TM REGION OF SLC9As

3.1. Prokaryotic templates for understanding mammalian *NHEs*

Very early it was recognized that *NHEs* encompass a large membrane embedded part with a suggested topology of 10–13 TM segments of helical character (Orlowski, Kandasamy, & Shull, 1992; Rothman, Padan, & Schuldiner, 1996; Sardet, Franchi, & Pouyssegur, 1989; see Wakabayashi, Ikeda, et al., 1997; Wakabayashi, Shigekawa, et al., 1997). Although recognized more than 20 years ago, no experimentally determined three-dimensional structure yet exists of any mammalian *NHE*. Therefore, the current understanding of *SLC9A* structural dynamics relies strongly on biochemical data. In its place, structures of prokaryotic transporters have served as molecular scaffolds. A number of projection structures obtained from cryo-EM 2D crystals, first of *E. coli* (*Ec*) NhaA (Williams, 2000; Williams, Geldmacher-Kaufer, Padan, Schuldiner, & Kuhlbrandt, 1999), and later of *M. jannaschii* (*Mj*) NhaP1 (Goswami et al., 2011; Vinothkumar, Smits, & Kuhlbrandt, 2005), as well as the 3D crystal structures of monomeric (Hunte et al., 2005) and dimeric (4AU5) *Ec*NhaA have formed the basis for interpretation of biochemical data on the mammalian transporters.

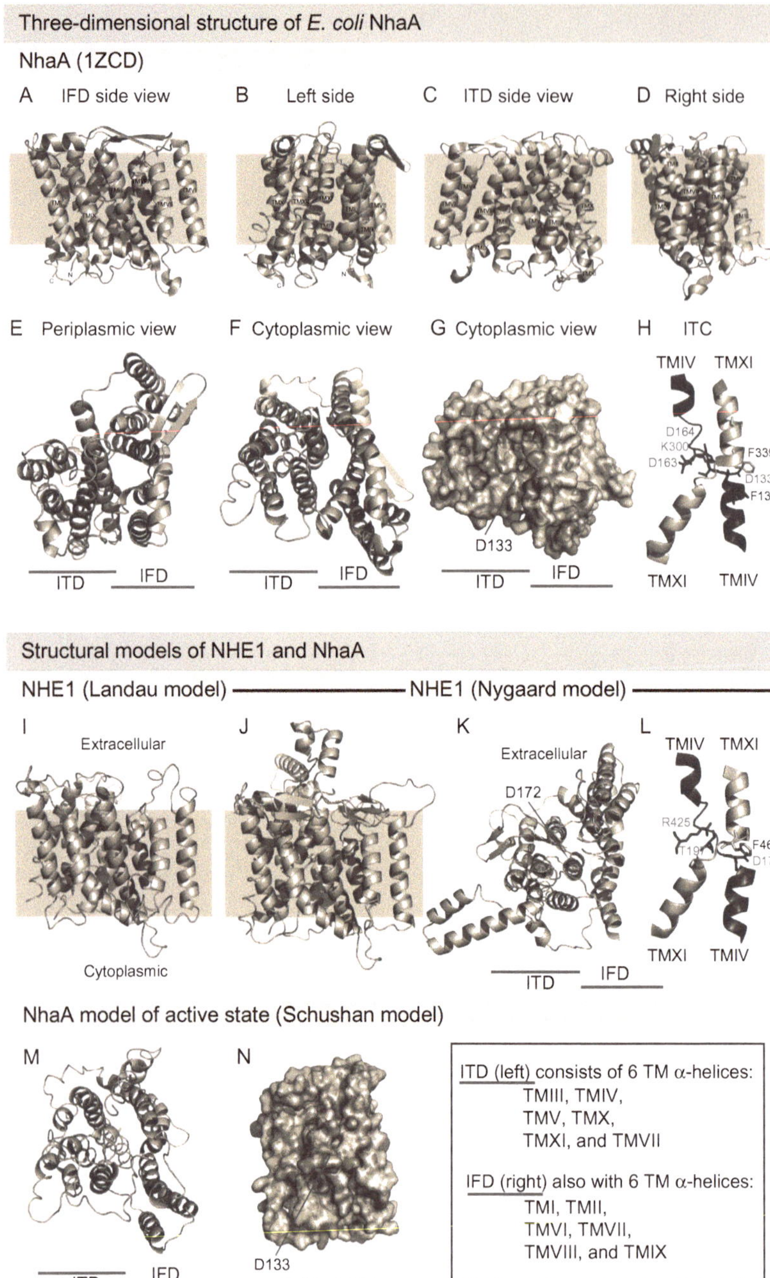

Figure 2.2 Three-dimensional structure of *E. coli* NhaA and structural models of human NHE1 and NhaA. Top: The three-dimensional crystal structure of *E. coli* NhaA (PDB code 1ZCD; Hunte et al., 2005) solved at pH 4.0 shown from all angles. (A) The IFD with view

The first structural insight was obtained from a 4 Å cryo-EM projection map of *Ec*NhaA that supported the notion that NhaA was a membrane-inserted dimer (Williams et al., 1999). This was later confirmed by its 6 Å crystal structure revealing 12 TM helices whose molecular orientation at the time defined a completely new structural fold (Williams, 2000). The details of the structure of the monomer became apparent from the 3.34 Å crystal structure of the inactive, low-pH structure of *Ec*NhaA, Fig. 2.2A–D (Hunte et al., 2005). The monomer is composed of 12 TM helices of various lengths with tilt angles of 10–27° relative to the membrane, and with diverse crossing angles. The helices are connected by 12 intra- and extracellular loops (ILs and ELs, respectively) of varying length, some with specific secondary structures, Fig. 2.2A–D. The monomer divides in two layers termed the interfacial domain (IFD) and the ion-translocation domain (ITD), each built by a 6 α-helix bundle, Fig. 2.2E and F. Three of the helices are discontinuous, and in particular, TMIV and TMXI of the ITD stretch at their crossing point accommodating several charged and polar residues, Fig. 2.2G and H. These charged and polar residues have been suggested to play a role in neutralizing the helix dipoles within the bilayer where the positive dipole ends of TMIVa and TMXIa are facing one another

into the dimer interface, (B) rotation to the right showing the structure from the left side, (C) the ITD side view, and (D) further rotation to the right. The TM helices are numbered and the gray horizontal bar represents the membrane layer, which has been drawn to guide the eye. The periplasmic side is shown at the top and the cytoplasmic side on the bottom. (E) The structure shown from the periplasmic view highlighting the two domains; the ITD to the left consists of 6 TM α-helices (TMIII, TMIV, TMV, TMX, TMXI, and TMVII) and the IFD to the right also with 6 TM α-helices (TMI, TMII, TMVI, TMVII, TMVIII, and TMIX). (F) Cytoplasmic view of *Ec*NhaA in ribbon representation and in (G) surface representation with $D133^{TMIV}$ highlighted in yellow indicated by an arrow showing access from the cytoplasmic side. (H) The anticipated ion-translocation center (ITC) at the interface between TMIV and TMXI. The helices are TMIV (red) and TMXI (blue) and important residues are labeled. K300 is located in TMX. Bottom: Structural models of NHE1 and *Ec*NhaA. (I) The Landau model (Landau, Herz, Padan, & Ben-Tal, 2007) and (J) the Nygaard model (Nygaard et al., 2011), orientations as in (A). The structures indicated in the loop regions are the result of force fields and not supported by any experimental data. (K) The Nygaard model seen from the extracellular side with $D172^{TMIV}$ highlighted in yellow. (L) The anticipated ion-translocation center (ITC) in NHE1. The helices are TMIV (red) and TMXI (blue) and residues are labeled; K425 is located in TMX. (M) Model of a periplasmically open NhaA simulated at high pH representing the active state of NhaA (Schushan et al., 2012). The helices belonging to each domain are indicated to the right. (N): Same orientation as in M shown in space-filling atoms indicating the position of D133. (See the color plate.)

and likewise the negative dipole ends of TMIVb and TMXIb meet (Hunte et al., 2005). They may, however, also be important for fulfilling the unpaired hydrogen bond donors and acceptors of the unwound helical backbone, and most likely play both roles. Although Na^+ could not be detected in the crystal structure, a number of studies have suggested this site to harbor the ion-translocation center (ITC) (Hunte et al., 2005; Screpanti & Hunte, 2007).

3.2. Mammalian SLC9A structures and models

Historically, the structure of the SLC9As has been a mystery, and several suggestions based on bioinformatical predictions have been presented to describe the overall topology of either 12 or 13 TM helices together with a large C-terminal soluble region, the length of which varies between isoforms. A number of nuclear magnetic resonance (NMR) analyses have been conducted on isolated TM- and loop segments of NHE1 (for a review, see Lee, Sykes, & Fliegel, 2011), but broad insight into the overall structure of the transporter has not yet been obtained. In 2000, an experimentally supported model of human NHE1 was described that redefined the topology of NHE1. This model was based on a systematic and very thorough cysteine-scanning mutagenesis and accessibility study involving 83 different sites in putative loop regions within human NHE1 (Wakabayashi et al., 2000). In cellular systems permeabilized or not for access to exposed cysteine residues, it was firmly established that both termini of NHE1 are cytoplasmic. Moreover, the data highlighted distinct differences to earlier structural interpretations, and in particular, the region from TMX to TMXII was extensively modified placing an extended hydrophobic patch between TMIX and TMX (EL5, corresponding to TMX in the original model) within the lipid bilayer. This is consistent with later studies in cell-free systems where this EL5 loop was accessible to the luminal space in endoplasmic reticulum (ER) but at the same time protected two of the three predicted N-glycosylation sites against enzymatic glycosylation (Sardet et al., 1989), thereby directing topogenesis (Sato, Ariyoshi, Mihara, & Sakaguchi, 2004). This inaccessibility and the observation of a sole N-glycosylation site, as well as an O-glycosylation site positioned in EL1, have been further experimentally supported (Counillon, Pouyssegur, & Reithmeier, 1994; Malapert, Pellissier, & Borgese, 1998). A portion of the loop between TMIV and TMV (IL2) was by similar arguments suggested to be positioned within the lipid bilayer (Sato et al., 2004). It is interesting to note that the glycosylation patterns among the different SLC9As differ considerably (Counillon et al., 1994; Malapert et al., 1998; Tse, Levine, Yun, Khurana, & Donowitz, 1994), and at present, the functionality of these glycosylation sites is not entirely known. A few residues predicted to locate to ILs were still accessible to external regents, in particular Arg180, Gln181 ($TMIV^{C\text{-term}}$) and Thr323, Ser324, and His325 ($TMIX^{C\text{-term}}$) (Wakabayashi et al., 2000). As will be apparent from the discussion below, their accessibility may relate directly to the conformational changes by NHE1 in response to pH sensing and ion exchange.

Although compared to NhaA, NHE1 has a different stoichiometry for transport and a reverse activity under physiological conditions, NhaA has served as a template for model building of human NHE1, thereby suggesting highly similar topology and mechanism of action (Landau et al., 2007; Nygaard et al., 2011). This is supported by several lines of experimental evidence. First, electron microscopy (EM) studies at a resolution of 22 Å confirmed NHE1 to be dimeric in the membrane with a size and shape that corresponds to those from the crystal structures of *Ec*NhaA (Moncoq, Kemp, Li, Fliegel, & Young, 2008). Second, structures of *Mj*NhaP1 (which is more similar in sequence as well as in function to the mammalian SLC9As than EcNhaA) have shown open and closed states similar to those of *Ec*NhaA (see below), yet also considerable differences in terms of overall dimensions (Vinothkumar et al., 2005). Likewise, in *Mj*NhaP1, the helices are more perpendicular and ion-induced conformational changes are more pronounced than those seen for *Ec*NhaA (Vinothkumar et al., 2005).

To date, three models of the TM domain of human NHE1 exist with distinct differences, Table 2.1. In addition to the experimentally built model mentioned above (Wakabayashi et al., 2000), two later models were proposed, Fig. 2.2. The first of these used the crystal structure of *Ec*NhaA as a template, together with conservation analyses and empirical data (Landau et al., 2007) (here termed the Landau model). The other was based on the NhaA model and the earlier experimentally based model of NHE1 (Wakabayashi et al., 2000) in conjunction with single electron paramagnetic resonance measurements exploiting distances between $Ala173^{TMIV}$ and $Ile461^{TMXI}$ (Nygaard et al., 2011) (here termed the Nygaard model). Although all models suggest 12 TM helices and a hydrophobic loop insertion between TMIX and TMX (EL5), their topology, helix assignments and loop extensions differ considerably, Table 2.1. In fact, the two latter models only agree to the extent and orientation of the last three helices. Importantly, neither of the later models can fully accommodate all available experimental

Table 2.1 Overview of NHE1 models and predictions

TMHMM[a,b]	HMMTOP[c]	SPLIT[d]	SOSUI[e]	TM	Waka/Cys[f]	Nygaard[g]	Landau[h]	NhaA[i]	Residues with an assigned role and other traits
15–37	15–32	17–38	18–40	I	15–35	16–34	130–146	10–31	
124–146	105–129	106–126	103–125	II	107–127	102–127	158–183	59–85	Phe-gate (F71[NhaA], F72[NhaA]; F103[NHE1])
161–179	160–177	133–153	130–152	III	133–149	132–150	189–207	94–117	
186–208	190–214	160–181	158–180	IV	155–175	161–182	227–246	121–143	Noncontinuous, Ion transport (D133[NhaA]; D172[NHE1])
228–247	229–246	187–218	192–214	V	190–210	191–208	256–277	151–175	Ion transport (D163[NhaA], D164[NhaA]; T197[NHE1])
254–276	257–274	225–251	227–249	VI	227–248	230–249	296–314	181–200	
291–313	291–315	256–280	258–280	VII	253–273	257–269	332–345	205–219	Dimer (R204[NhaA], L210[NhaA]; E262[NHE1], D267[NHE1])
326–343	328–345	289–322	294–316	VIII	295–315	303–315	349–363	222–237	Phe-gate (F299[NHE1])
347–366, 386–405	354–372, 385–404	331–366, 493–408	347–369, 388–409	IX	339–359	326–351	374–397	246–272	Dimerization (V254[NhaA], W258[NhaA]), pH sensor[NhaA]; Phe-gate (F267[NhaA])
415–437	413–437	415–441	415–437	X	411–431	419–438	416–437	291–313	K300[NhaA]; R425[NHE1]
450–469	450–469	452–472	453–473	XI	450–470	450–473	448–472	324–350	Noncontinuous, Phe-gate (F344[NhaA]; F463[NHE1], F474[NHE1])
479–501	480–497	480–505	484–506	XII	480–500	480–505	481–502	357–381	
				S					
38–123	33–104	39–105	41–102	EL1	36–106	35–101	147–157	32–58	N-glycosylation[NHE1,N75], disordered, dimerization

180–185	178–189	154–159	153–157	EL2	150–154	151–160	208–226	118–120	
228–253	247–256	219–224	215–226	EL3	211–226	209–229	278–295	176–180	
314–325	316–327	281–288	291–293	EL4	274–294	270–302	346–348	220–221	
406–414	405–412	409–414	410–414	EL5	360–410	352–418	397–415	273–290	Long, disordered
470–478	470–479	472–479	474–483	EL6	471–479	473–479	473–480	351–356	
1–14	1–14	1–16	1–17	N-ter	1–14	1–15	1–129	1–10	
147–160	130–159	127–132	126–129	IL1	128–132	128–131	183–188	86–93	
209–227	215–228	182–185	181–191	IL2	176–189	183–190	247–255	144–150	Access to extracellular reagents (182^{NHE1}, 183^{NHE1})
277–290	275–290	252–254	250–257	IL3	249–252	250–256	315–331	201–204	
344···	346···	323···	317···	IL4	316–338	316–325	363–373	238–245	
438–449	438–449	442–451	438–452	IL5	432–449	439–449	438–447	324–323	
502–	498–	504–	507–	C-ter	501–	506–	503–	382–	

[a]IL/EL suggested reversed with N-terminal out.
[b]Krogh et al. (2001).
[c]Abrol, Kim, Bray, Griffith, and Goddard (2011).
[d]Juretic, Zoranic, and Zucic (2002).
[e]Hirokawa, Boon-Chieng, and Mitaku (1998).
[f]model according to (Wakabayashi, Pang, Su, & Shigekawa, 2000)
[g]model according to (Nygaard et al., 2011)
[h]model according to (Landau, Herz, Padan, & Ben-Tal, 2007)
[i]model according to (Williams et al., 2000)
"..." indicates uncertainty in loop lengths, as helix predictions for this region suggested two helices instead of one.
All NHE1 numbering in this table refers to human NHE1.
Reversed in this table to confer to Wakabayashi et al. (2000); TM, transmembrane; S, soluble. Position and extent of structural units in the TM domain of NHE1, indicated by abbreviations in the middle, fifth column. To the left are bioinformatics predictions, and to the right are the three models of NHE1 (the Wakabayashi model, the Nygaard model, and the Landau model). For comparison, the same units from the crystal structure of NhaA are indicated. Residues of particular relevance are listed.

data, nor do they confer fully to the cysteine accessibility data (Wakabayashi et al., 2000) or incorporate the EL5 or IL2 into the bilayer plane. However, given the knowledge of glycosylation sites and the position of both termini intracellularly, in conjunction with the order–disorder predictions provided (Fig. 2.6), the plausibility of a long EL1 is strong, in conflict with the Landau model. Furthermore, consensus predictions of TM helices using a number of predictors (Abrol et al., 2011; Hirokawa et al., 1998; Juretic et al., 2002; Krogh, Larsson, von Heijne, & Sonnhammer, 2001), suggest a similar topology provided from the experimental model (Wakabayashi et al., 2000) and the Nygaard model (Nygaard et al., 2011), Table 2.1. The largest discrepancies and uncertainties between the models reside in the residues comprising TMIX.

4. SLC9As EXIST AS ASYMMETRICAL HOMODIMERS IN THE MEMBRANE

Using various approaches such as SDS-PAGE and disuccinimidyl suberate cross-linking in intact cells, NHE1 and *NHE3* specifically form homodimers (Fafournoux, Noel, & Pouyssegur, 1994; Fliegel, Haworth, & Dyck, 1993; Hisamitsu, Pang, Shigekawa, & Wakabayashi, 2004). Dimer formation is also a prerequisite for a complete description of the exchange kinetics (see Section 6), and dimer formation with dominant negative mutants of NHE1 strongly inhibit exchange (Hisamitsu, Ammar, Nakamura, & Wakabayashi, 2006). From both EM studies (Moncoq et al., 2008) and later higher resolution structures (PDB code 4AU5), a relatively open, irregular structure of the *Ec*NhaA dimer was revealed with a remarkable asymmetry involving many of the helices in the interface, Fig. 2.3A. The dimer appears to have the following three sites of interaction between monomers: one involving extracellular loops, one within the lipid bilayer with residues from TMI, TMII, TMVI, TMVII, and TMIX (Fig. 2.3A), and one within the cytoplasmic tail (Hisamitsu et al., 2004). Human NHE1 deletion mutations showed that lack of the entire C-terminal tail does not affect physical dimer formation within the membrane (Fafournoux et al., 1994). Instead, loss of the C-terminal tail removes cooperativity between monomers (Hisamitsu et al., 2004). In EL1 of *Ec*NhaA, new hydrogen bonds are formed across the interface stabilizing a short β-sheet, which subsequently runs consecutively across the monomers ($Gln47^{NhaA}$, $Arg49^{NhaA}$), Fig. 2.3A. The corresponding residues cannot

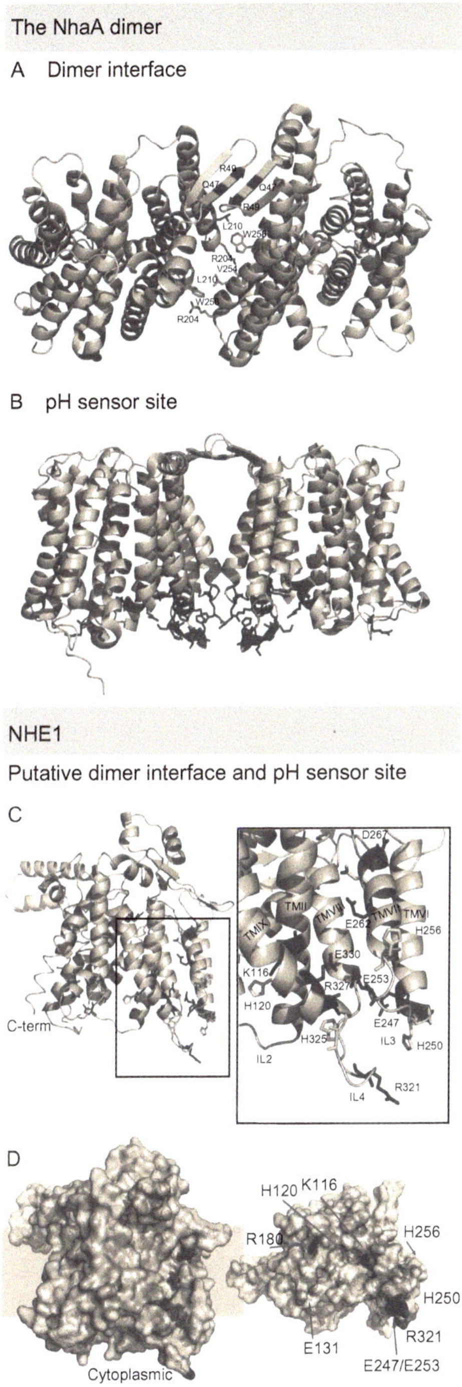

Figure 2.3 Dimer interfaces and pH sensor sites. Top: *Ec*NhaA dimer and pH sensor. (A) The two interaction sites within the asymmetric dimer of *Ec*NhaA (PDB code 4AU5) are highlighted. In blue is shown the backbone of Gln45^{EL1} and Arg47^{EL1} that form *(Continued)*

easily be defined from the NHE1 models, as the structure of EL1 is unresolved, but several possibilities exist.

There are a few NMR studies available on the potential EL structures of NHE1, particularly EL2 (G150–F155) (Lee, Li, Liu, Sykes, & Fliegel, 2009a) and EL4 (Y274–I291) (Lee, Liu, Li, Sykes, & Fliegel, 2012). EL2 leads into TMIV, which is suggested to be part of the ion-transport site (see below) and is rich in proline residues. NMR data revealed a number of different possible conformations for this loop that could be of significance to the function of the transporter (Lee et al., 2009a). In contrast, EL4 was mostly unstructured and thus likely acts as a flexible linker between TMVII and TMVIII (Lee et al., 2012). Interestingly, based on chemical cross-linking data, EL2, EL4, and EL5 have been suggested to participate in dimer contacts either directly or indirectly (Hisamitsu et al., 2006).

Cysteine cross-linking by introduction of a Ser375Cys mutation in EL5 of NHE1 markedly affected the exchanger function without affecting the affinity for ligands (Hisamitsu et al., 2006), suggesting a role for this loop at least in prohibiting allosteric cooperative structural changes across the dimer. Whether this loop directly forms inter-monomer contacts is not revealed from the study. Such an arrangement may, however, be questioned due the rather distant position to the expected dimer interface.

In *Ec*NhaA, dimer interactions within the bilayer are primarily formed by hydrophobic contacts and π–cation interactions between TMVII and TMIX, involving in particular, the residues $\text{Arg204}^{\text{NhaA/TMVII}}$, $\text{Val254}^{\text{NhaA/TMIX}}$, $\text{Leu210}^{\text{NhaA/TMVII}}$, and $\text{Trp258}^{\text{NhaA/TMIX}}$ (Fig. 2.3).

Figure 2.3—Cont'd hydrogen bonds in a consecutive β-sheet and in sticks colored by the individual atoms, residues $\text{Arg204}^{\text{TMVII}}$, $\text{Leu210}^{\text{TMVII}}$, $\text{Val254}^{\text{TMIX}}$, and $\text{Trp258}^{\text{TMIX}}$ forming a hydrophobic interaction surface of approximately 600 Å^2. (B) Residues of *Ec*NhaA suggested being responsible for the pH sensor site (residues at the C-terminal of TMII and the N-terminal of TMIX, as well as residues in IL2, -3 and -4) colored red and shown as sticks. Gln45 and Arg47 are shown in blue ribbon. Bottom: NHE1 putative dimer and pH sensor sites. (C) Charged residues at the C-terminal of TMII and the N-terminal of TMIX as well as in IL3 and -4 are shown as sticks and colored according to charge (red: negative charge, blue: positive charge). Histidines are colored yellow, and the residues are labeled in the zoom shown to the right. (D) Surface representation of the putative dimer interface highlighting its charged nature. The Nygaard model shown from the cytoplasmic side with the potential interaction site for the autoinhibitory region of the C-tail highlighted. Coloring as in (C). (See the color plate.)

This interface is not densely packed. However, the interface is directly in contact with a suggested pH sensor site from the *Ec*NhaA structure that involves mostly residues at the C-terminus of TMII and the N-terminus of TMIX, as well as residues in IL2, -3, and -4, and with $His226^{NhaA/TMVIII}$ and $Glu252^{NhaA/TMIX}$ as key residues (Gerchman et al., 1993; Olkhova, Hunte, Screpanti, Padan, & Michel, 2006; Padan et al., 2004; Tzubery, Rimon, & Padan, 2004). Proton binding to *Ec*NhaA causes an increase in the mean distance between helices suggesting an expanded pH-activated state (Appel, Hizlan, Vinothkumar, Ziegler, & Kuhlbrandt, 2009), and indicating that conformational changes occurring as a result of pH sensing can be directly sensed within the dimer interface and/or even in the ITC, see below. In support of this notion, $Lys249^{NhaA/TMIX}$ was accessible to trypsin cleavage only in the activated state at physiological pH (Rothman, Gerchman, Padan, & Schuldiner, 1997).

Although the position/range of TMIX is a matter of debate, inspection of the corresponding putative dimer interface in the Nygaard model of NHE1 reveals distinct differences and highlights a number of charged and polar residues ($Asp238^{TMVI}$, $Glu262^{TMVII}$, $Asn266^{TMVII}$, and $Asp267^{TMVII}$) within the assumed dimer site (Fig. 2.3C) whose protonation states may be linked to activation. Interestingly, in a mutational screen of 17 different sites, their negative charge, at least for $Glu262^{TMVII}$ and $Asp267^{TMVII}$ was shown to be obligate for NHE1 function (Murtazina, Booth, Bullis, Singh, & Fliegel, 2001). Importantly, Asp172 was not included in the screen. The mutation Glu262Leu is transport deficient (Hisamitsu et al., 2006) and inhibited heterodimers in a pH-dependent way (Fafournoux et al., 1994), suggesting that at neutral pH, dimerization is a prerequisite for activity. In contrast, if the proton concentration is high enough, then allosteric changes are not necessary for exchange and the monomer is capable of operating independently. Likewise mutation of the conserved $Glu262^{TMVII}$ and $Asp267^{TMVII}$ to their neutral counterparts, Gln and Asn effectively eliminated transporter activity (Murtazina et al., 2001). Furthermore, in rat NHE1, the Glu266Ile mutation, corresponding to $Glu262^{TMVII}$ in human, does not regulate H^+ efflux and is transport deficient (Denker, Huang, Orlowski, Furthmayr, & Barber, 2000). Finally, a Glu262Asp mutation changed Li^+ versus Na^+ affinity of NHE1 (Murtazina et al., 2001). Hence, charged residues within the proposed dimer interface are critical for activity, either by stabilization of the dimer or by providing an optimized chemical environment, perhaps even acting as key protonation sites.

4.1. pH-sensing site interrelated with the dimer interface

Inspection of the analogous pH sensor site in the NHE1 Nygaard model, which in *Ec*NhaA involves residues in IL2, -3, and -4, as well as residues in the C-terminus of TMII and the N-terminus of TMIX, shows that this site similarly hosts many charged and polar residues, Fig. 2.3D. In particular, four histidines (His120, His250, His256, and His325) are involved. However, none of these have been directly implicated in pH sensing when investigated by mutations, indicating that they may have other roles, for example, in binding partner interactions or microenvironment stabilization (Ding, Rainey, Xu, Sykes, & Fliegel, 2006; Hisamitsu, Yamada, Nakamura, & Wakabayashi, 2007; Wang, Balkovetz, & Warnock, 1995; Wiebe, Dibattista, & Fliegel, 2001). There are also other distinct differences to the bacterial pH sensor. For instance, whereas residues in IL2 have been implicated in pH sensing in *Ec*NhaA, this loop is hydrophobic and predicted to be membrane embedded in NHE1. Instead, residues in IL1 and TMIV, including Glu131$^{\text{IL1}}$, seem critical for proton sensing in NHE1 (Hisamitsu et al., 2007; Lacroix, Poet, Maehrel, & Counillon, 2004; Wakabayashi, Hisamitsu, Pang, & Shigekawa, 2003a). Thus, it appears that similar regions in the bacterial and mammalian transporters are responsible for functionality both for dimerization and for pH sensing, where the latter also involves the region TMII$^{\text{C-term}}$ and TMIX$^{\text{N-term}}$ and IL3 and -4. For NHE1, a surface representation of the Nygaard model reveals the charged hot spots both within the dimer interface (Fig. 2.3D, left) and at the cytosolic side of the transporter (Fig. 2.3D, right).

4.2. Peptide models of NHE1 TMs

A simple tactic for the study of membrane protein structure is to use a *divide-and-conquer* approach by dissecting single TM helices and loops to study them in isolation. While attractive in their simplicity, these studies suffer from loss of native interactions, as well as the generation of artificial interaction surfaces and structurally nonrelevant interactions with the membrane mimic (i.e. the lipid(s) and/or detergent(s) added to the preparation to mimic the membrane in which the protein is embedded in its native state). Nonetheless, the stabilities of α-helices and turn structures in the lipid environment are especially determined by local, short-range interactions and less by long-range tertiary ones, and *proof-of-principle* has been made on peptides from bacteriorhodopsin (Katragadda, Alderfer, & Yeagle, 2001), for example. A similar approach has been undertaken for parts of NHE1, and NMR structures of peptides have been solved in detergents corresponding to parts of TMIV (CD_3OH:$CDCl_3$:H_2O) (Slepkov et al., 2005), TMVI (Tzeng, Lee, Sykes, & Fliegel, 2010), TMVII (Ding et al., 2006), TMIX (Reddy et al., 2008), and TMXI (Lee, Li, Liu, Sykes, & Fliegel, 2009b).

Using NMR and membrane mimics to solve membrane protein structures is not a trivial task, and overlap of signals and weak nuclear Overhauser effects (NOEs) in the center of the helices are known obstacles (Rosner & Kragelund, 2012). Hence, chemical shifts of the backbone atoms as well as other NMR parameters such as residual dipolar couplings become important parameters in evaluating the presence of discontinuous helices. For the peptide studies on NHE1, both TMIV and TMXI were confirmed to be noncontinuous by C^{α} and/or H^{α} chemical shifts. For the other peptides (TMVI, TMVII, and TMIX), chemical shifts suggested continuous helices, and the lack of both, NOEs and peptide–lipid constraints, made the interpretation of possible kinks or breaks complicated. Similarly, the use of dodecylphosphocholine as membrane mimic may result in nonnative helical bending, as also noted from the paramagnetic relaxation enhancement measurements (Tzeng et al., 2010). Still, the structures of TMIV and TMXI of NHE1 are highly similar to the structures of these helices in NhaA, confirming their discontinuous character. In these studies, several accessibility measures were conducted that suggested a number of pore-lining residues. These are discussed below.

4.3. Structure-based hypotheses for the mechanism of ion transport by *NHEs*

It still remains an enigma how exactly ions are transported through the lipid bilayer in passages between TM helices of *NHEs*. The different structures and models available have, together with a larger number of mutational studies, suggested that residues in TMIV and TMXI play key roles in this process. In particular, conformational changes have been revealed from difference maps of cryo-EM 2D crystals both of *Ec*NhaA and of *Mj*NhaP1 grown at pH 4.0 (Goswami et al., 2011; Vinothkumar et al., 2005). In both cases, the crystals were titrated with ions and were modulated by changes in pH. Inspection of the difference maps showed the protein scaffolds to be highly flexible with changes in specific regions of the protein, Fig. 2.4A and B. A two-step mechanism was suggested to account for transport. Initially, for *Ec*NhaA, a pH-controlled priming of the structure occurred, which was independent of ions, initiated at low pH, and completed at

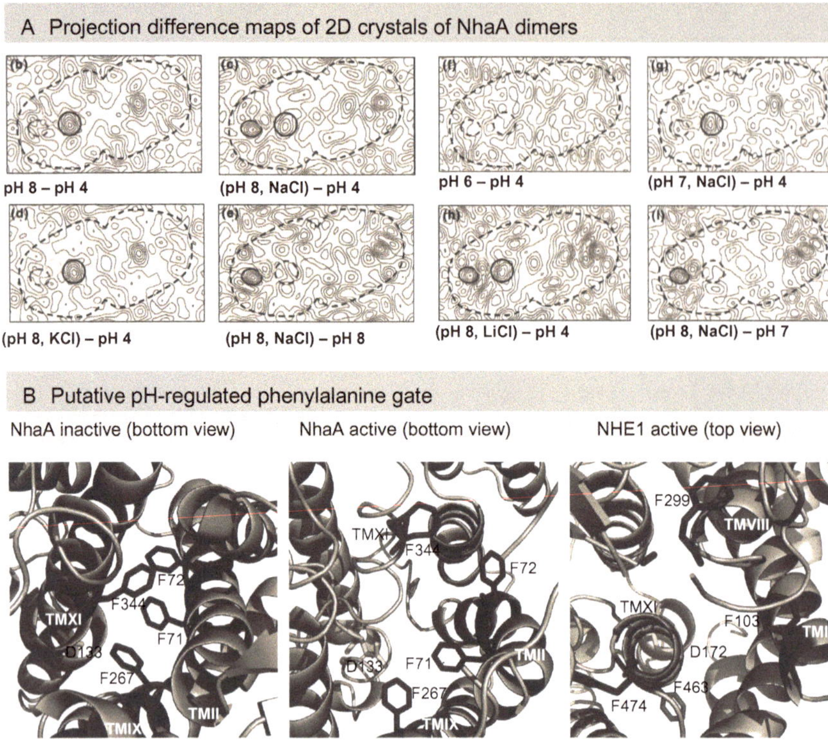

Figure 2.4 Dynamics of Na^+/H^+ exchangers. (A) Projection difference maps of 2D crystals of NhaA dimers incubated on the EM grid in buffers of pH 4.0, 6.0, 7.0, and 8.0 with and without substrate ions, or KCl as control. Positive contours (blue) in the maps indicate features that are stronger in map 1 than in the subtracted map 2. Negative contours are correspondingly weaker features and are colored red. The pH and ionic conditions compared are indicated below each square. The blue open circle indicates features that get stronger at higher pH, but only at pH > 6. The red open circle indicates features that get weaker only in the presence of Na^+, and not in the presence of K^+, and the yellow open circle is a feature getting simultaneously stronger. In (G), only changes within the blue circle are seen suggesting that the pH-induced conformational change is located within the monomers and between the ITD and the IFD domains. In (I), compensating changes in the ITD are seen and marked by red and yellow circles suggesting the movement of a helix. This change is induced by ions and not by changes in pH. (B) Putative pH-regulated phenylalanine gate. The change in density in the projection maps in the regions constituting the inter-domain space between the ITD and the IFD domains may involve changes around helices TMII, TMIX, and TMXI. Left: Zoom of the Phe-gate of the *Ec*NhaA inactive state (low pH, PDB code 1ZCD). Here access to the suggested ion-translocation site is closed. Middle: Zoom of the *Ec*NhaA active state model (high-pH, Schushan model), where the Phe-gate is open as a result of a slight rotation of helices. Right: Zoom of the Phe-gate of NHE1 in the active state (low pH, Nygaard model) where the gate is open. In *Ec*NhaA, residues $F71^{TMII}$, $F72^{TMII}$, $F267^{TMIX}$, and $F344^{TMXI}$ constitute the gate. In NHE1, it is not entirely clear which side chains constitute the Gate, but $F103^{TMII}$, $F299^{TMIX}$, $F463^{TMXI}$, and $F474^{TMXI}$ may be likely candidates. The phenylalanine side chains are shown as red sticks and the catalytic aspartic acids ($D133^{NhaA}$ and $D172^{NHE1}$) as yellow sticks. *(A) Figure reproduced with permission from Appel et al. (2009), Copyright 2009 Elsevier.* (See the color plate.)

pH 7. Subsequently, when the ion-transport step occurred, differences in specific helix orientations were observed.

The origin and molecular identity of the specific proton acceptor site is still not revealed, but likely involves two distinct sets of residues that are protonated and deprotonated in response to pH (Hellmer, Teubner, & Zeilinger, 2003). It is generally accepted that any proton-sensing site must be located on the intracellular side of the membrane, and that *NHEs* become quiescent at an optimized set point to preclude deleterious alkalinization. For *Ec*NhaA, deprotonation of the N-terminal tail histidines as a response to lower intracellular proton concentration has been suggested, causing the disordered tail to insert into the area between TMIV and TMXI. However, proton sensing in NhaA clearly also involves the above-mentioned sensor site as well as residues in TMV (Schushan et al., 2012). It is suggested that the N-terminal insertion fosters the allosteric change leading to an increase in the mean distance between helices, and an increase in the packing density of lipids surrounding the dimer, and that protonation of so far unidentified residues leads to rotation of TMII and TMIX (Schushan et al., 2012). Of particular importance is the observation that the pH-induced changes map to the interface between the IFD and the ITD. For mammalian SLC9As, this step is unresolved, but kinetics suggest that allostery plays a significant role (Lacroix et al., 2008) (for an in-depth discussion of this, please refer to Section 6). Many mutations of residues within the sensor site described above clearly and distinctly affect pH sensing (Section 6), whereas cooperativity seems to be linked to the C-terminal tail.

Subsequent to pH priming, the interface of the discontinuous helices TMIV and TMIX is anticipated to act as a passageway for ions in a manner depending on conformational change. The presence of open and closed states has been postulated for *Ec*NhaA (Padan et al., 2004) and human *NHE3* (Hayashi et al., 2002) and demonstrated for *Mj*NhaP1 in 2005 (Vinothkumar et al., 2005) and for *Ec*NhaA in 2008 (Appel et al., 2009). Difference cryo-EM maps as a result of a change in ion concentration are similar for *Ec*NhaA and *Mj*NhaP1, despite differences in activity and stoichiometry. Ion binding to *Ec*NhaA seems to involve significant motions of helices (entire helices or just parts), that bring the periplasmic part of TMIV closer to TMXI, thereby opening the periplasmic exit channel for ions (Fig. 2.4C). Although these motions are directly related to the translocation process, it is not understood how the ions enter or exit the transporter, or indeed whether this occurs by exploitation of the same sites. Recent computational analyses of *Ec*NhaA have been used to generate a periplasmically open image of *Ec*NhaA in the activated high-pH state (Schushan et al., 2012). Here, access to the exit tunnel from the periplasmic side, leading into the TMIV–TMXI crossing point, can be visualized (Fig. 2.4B, middle). The closure at the cytoplasmic side and opening on the periplasmic side shows alteration in the spatial closure between the IFD and ITD domains. It is currently not known if this process of transport in NHE1 confers to the alternating access mechanism described for a number of transporters such as for the glycerol-3-phosphate transporter (Huang, Lemieux, Song, Auer, & Wang, 2003), the sodium-hydantoin transporter Mhp1 (Shimamura et al., 2010), the calcium-proton exchanger (Waight et al., 2013), and the proton-dependent oliogopeptide transporter (POT) (Solcan et al., 2012). By this mechanism, the transporters exist in the following three different states: the outward-facing state, the occluded state, and the inward-facing state. POT is highly similar to *NHE* in topology, but senses protons extracellularly. Here, extracellular gating is controlled by charged arginines and glutamates, whereas the gating intracellularly is related to helical hinges. Given the known model structures of NHE1, the similar topology of NHE1 and POT, and the accessibility in NHE1 of both ILs (Wakabayashi et al., 2000) and pore-lining residues (Ding et al., 2006; Lee et al., 2009b; Reddy et al., 2008; Slepkov et al., 2005; Tzeng et al., 2010), it is not unlikely that a similar mechanism may operate in NHEs.

Particular residues of the two discontinuous helices TMIV and TMXI that are positioned at their crossing point are generally thought to engage actively in the actual ion-translocation step, aided by closely positioned charged and polar residues. The presence of charged/polar side chains (Asp133^{NhaA}/Asp172^{NHE1}/Asp132^{NhaP1}, Asp163^{NhaA}/Asp164^{NhaA}/Thr197^{NHE1}/Asp162^{NhaP1}, and Lys300^{NhaA}/Arg425^{NHE1}/Arg320^{NhaP1}) and prolines central in the crossing point has made this site attractive as the ITC. In fact, structure-derived pK_a values (calculated by propKa, http://propka.ki.ku.dk) for Asp133^{NhaA} (pK_a=5.68), Asp163^{NhaA} (pK_a=8.95), and Asp164^{NhaA} (pK_a=9.95) suggest that both Asp163 and Asp164 are uncharged at this interface, whereas Asp133^{NhaA} is responsive within the pH range of activation and can therefore switch between charged and uncharged states. This corresponds nicely with the Nygaard model of NHE1 where Asp172^{NHE1} has a pK_a of 6.51 and Thr197^{NHE1} substitutes for the two other Asp residues. The involvement of these residues has found support from several mutation studies, which show clear effects on translocation rates (Nygaard et al., 2011; Padan, Kozachkov, Herz, & Rimon, 2009; Slepkov et al., 2005). Whether the effects are caused by interference with the chemical step or conformational changes is unknown.

Deactivation (NhaA) and activation (NHE1) has been suggested to occur by deprotonation of Arg/Lys residues (Lys300^{NhaA}/Arg425^{NHE1}/Arg320^{NhaP1}), which are shown to participate in ion translocation directly (Hellmer et al., 2003; Nygaard et al., 2011; Padan et al., 2004). The pK_a of this amino acid side chain within the structure of *Ec*NhaA (Lys300) is 6.42 (calculated by propKa, http://propka.ki.ku.dk/), that makes it reactive within the pH range of activation.

In *NHE3*, the pH-dependent reaction (proton sensing) has been suggested to be a non-rate-determining step, whereas the subsequent allosteric conformational changes may be rate limiting depending on the changes needed (Alexander et al., 2007). Hysteresis, which results in a lag phase in response, can suggest either the presence of more than one initial state or the presence of a slow rate-limiting kinetic step early in the ion transport mechanism. Such hysteresis has been observed for *NHE3* and *NHE5*, but not for NHE1 or *NHE2* (Hayashi et al., 2002). This suggests either that there are different coupling schemes within the family or (and perhaps more likely) that the early steps in transport have different kinetic constants.

4.4. The Phe-nomenon: A potential phenylalanine gate for ion transport

A thorough inspection of the *Ec*NhaA structures in its active (cytoplasmic open) and inactive (periplasmic open) states (Hunte et al., 2005; Schushan et al., 2012), as well as of the Nygaard model of NHE1 at low pH (active state) (Nygaard et al., 2011), discloses a potential and highly interesting structural gating mechanism provided by a number of phenylalanine side chains, Fig. 2.4B. Whereas their hydrophobic side chains close the gate in the inactive low-pH structure of *Ec*NhaA, only a slight rotation of helices in the high-pH, open-state model of *Ec*NhaA (suggested earlier by Schushan et al., 2012), as well as in the active, low-pH model of NHE1, is needed to open the passage to the ITC. This phenylalanine gate is made up of at least Phe71$^{NhaA/TMII}$, Phe72$^{NhaA/TMII}$, Phe267$^{NhaA/TMIX}$, and Phe344$^{NhaA/TMXI}$ in *Ec*NhaA and, due to the lack of an experimentally determined structure of NHE1, at least some of the following residues in NHE1 extracted from the Nygaard model: Phe103^{TMII}, Phe192^{TMV}, Phe202^{TMV}, Phe203^{TMV}, Phe299^{EL4}, Phe335^{TMIX}, Phe463^{TMXI}, and Phe474^{TMXI}. Because tryptophan in some cases acts as a gate-keeper, Trp199^{TMV} may also be a candidate. To the best of our knowledge, the presence and the activity of this gate has not been described before. Of particular interest is the notion that these side chains provide an important link between the pH sensor site and the ITC. Specifically, this path goes from $TMII^{C\text{-term}}$ and $TMIX^{N\text{-term}}$ and IL1, -3, and -4, through the TMII ($Phe71/Phe72^{NhaA/TMII}$, $Phe103^{NHE1/TMII}$) and the TMIX ($Phe267^{NhaA/TMIX}$, $Phe335^{NHE1/TMIX}$) via TMXI ($Phe344^{NhaA/TMXI}$, $Ph463/Phe474^{NHE1/TMXI}$)-all TMs of the ITC. This is in full accordance with the observation that pH-induced conformational changes observed in 2D difference crystals pinpoint the region between ITD and IFD that is precisely the position of the Phe-gate, Fig. 2.4. Importantly, Phe463 has been shown previously to be highly important for activity (Lee et al., 2009b). Phe161, which has previously been suggested to be pore lining (Slepkov et al., 2005), is solvent exposed in the Nygaard model and distant to the Phe-gate.

A link between the pH-sensing site of *Ec*NhaA and two of the phenylalanines of this newly defined gate, $Phe267^{NhaA/TMIX}$ and $Phe344^{NhaA/TMXI}$ and the ITC was indirectly suggested because the substitution of these residues to cysteines inhibited the antiporter activity and strongly affected pH sensitivity (Tzubery, Rimon, & Padan, 2008). Interestingly, both TMII and TMIX have a slight kink in their helical structures that may be of importance for this allosteric reaction. This kink was not revealed by NMR studies (Reddy et al., 2008), which instead suggested a kink at Ser351. In the Nygaard model, this residue is at the border between TMIX and EL5, but we note that consensus regarding the range of TMIX has not been reached. Thus, we hypothesize that the mechanism of ion transport in NHE1 involves a pH-priming step where protonation within the pH-sensing site leads to allosteric changes that transform the phenylalanine gatekeepers via rotation of TMII, TMIX, and TMXI to allow ion access to the ITC. Subsequently, ion translocation involving movement of helices TMIV and TMXI can proceed. Because these two events may be decoupled (see Section 6), it is likely that priming of the gate occurs by mechanisms different from the pH-related conformational changes. One could imagine that changes in membrane pressure, interactions with a binding partner, or phosphorylation of the exchanger can inflict similar conformational changes irrespective of pH changes. We note that not all these aromatic groups listed here are fully conserved within the SLC9As, that is, some phenylalanines may be more important in the process than others. This of course also depends on their precise positions, which remain to be determined experimentally.

5. THE SLC9A C-TERMINAL TAIL: STRUCTURE/DISORDER, INTERACTION PARTNERS, AND PHOSPHORYLATION

In contrast to the highly sequence-conserved TM region described above, the C-terminal cytoplasmic tail regions are relatively poorly conserved across the SLC9As, especially in the distal half (see e.g., Pedersen, 2005). Early studies, mostly on NHE1, established that this part of the SLC9As is dispensable for ion transport, but essential for regulation by pH_i and growth factor signaling (see Wakabayashi, Ikeda, et al., 1997; Wakabayashi, Shigekawa, et al., 1997). Structural information on the NHE1 tail remains limited. For the proximal part of the tail, a few structures of very short regions of NHE1 are available—all in complex with binding partners (see Section 5, Fig. 2.5). For the distal part of the tail, no structural information was until recently available, when we demonstrated using a wide array of structural and biochemical approaches that this part of the NHE1 tail is intrinsically disordered in both human and *P. americanus* (Nørholm et al., 2011). Intrinsic disorder (ID) structurally in proteins refers to regions of low sequence complexity, that is, lack of hydrophobic residues that result in lack of a globular fold (see below). ID is widespread in the intracellular domains of TM proteins (Sigalov & Uversky, 2011), and notably, our *in silico* analyses predict that ID in the C-terminal tail is conserved among all the SLC9As. In fact, this is a much stronger conserved trait than sequence within this part of the SLC9A tail (Fig. 2.6).

In the following sections, we focus on binding partners, ID and phosphorylation motifs of NHE1, the isoform with—by far—most information

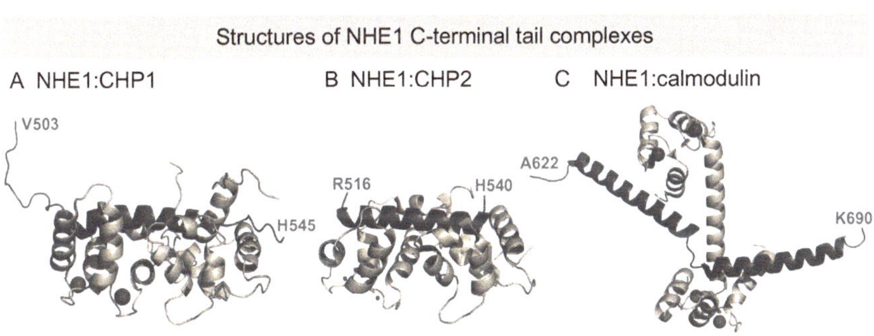

Figure 2.5 Structures of NHE1 C-terminal tail complexes. (A) NMR solution structure of calcineurin B homologous protein (CHP) 1 in complex with an NHE1 peptide Val503–His545 (PDB code 2E30; Mishima, Wakabayashi, & Kojima, 2007). (B) Crystal structure of CHP2 in complex with the NHE1 peptide Val503–His545 at 2.7 Å resolution (PDB code 2BEC; Ammar, Takeda, Hisamitsu, Mori, & Wakabayashi, 2006). (C) Crystal structure of calmodulin in complex with the NHE1 peptide Ala622–Lys690 at 2.23 Å resolution (PDB code 2YGG; Köster, Pavkov-Keller, Kuhlbrandt, & Yildiz, 2011). NHE1 peptides are shown in red, and the interaction partners are shown in gray. Ca^{2+} ions are shown as large magenta spheres and Li^{+} ions as small magenta spheres. (See the color plate.)

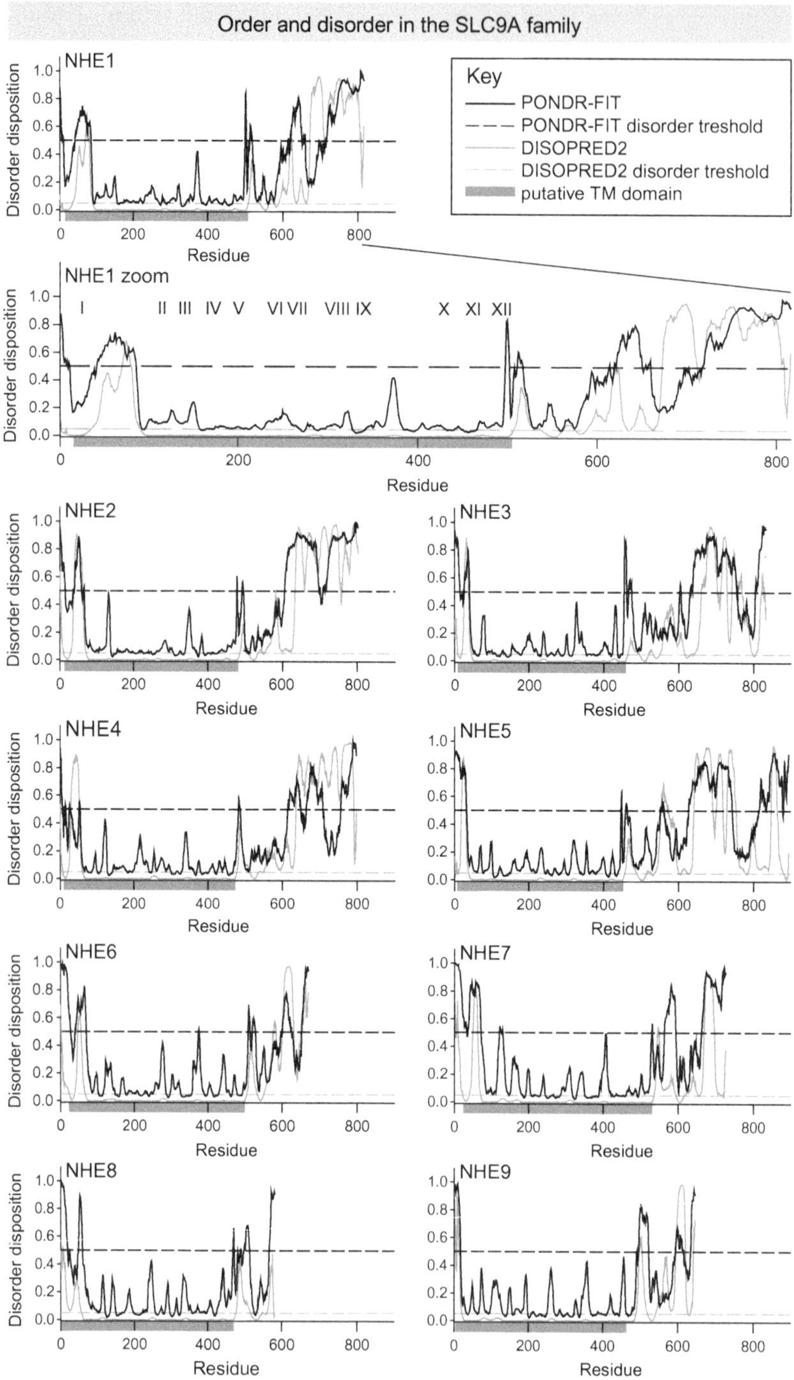

Order and disorder in the SLC9A family
Key
PONDR-FIT
PONDR-FIT disorder treshold
DISOPRED2
DISOPRED2 disorder treshold
putative TM domain
NHE1
NHE1 zoom
I
II III IV V
VI VII
VIII IX
X XI XII
NHE2
NHE3
NHE4
NHE5
NHE6
NHE7
NHE8
NHE9
Disorder disposition
Residue

available. Interacting partners of other isoforms are mostly understood in terms of roles in exchanger trafficking, and discussed in Section 7. In addition, Table 2.2 provides a comprehensive listing of reported binding partners for the SLC9As.

5.1. NHE1-binding partners

5.1.1 Interaction partners: Known structures of complexes

Detailed structural data involving the SLC9A regulatory tails come from three studies of NHE1 peptides, which were solved in complex with binding partners. These are an NMR solution structure of the Ca^{2+}-binding protein calcineurin B homologous protein 1 (CHP1; Fig. 2.5A) (Mishima et al., 2007) as well as two crystal structures, one of another calcineurin homologous protein (CHP) isoform, CHP2 (Fig. 2.5B) and one of another Ca^{2+}-binding protein, calmodulin (CaM; Fig. 2.5C) (Ammar et al., 2006; Köster et al., 2011). CHP1 was identified to be an NHE1-interacting partner in 1996 (Lin & Barber, 1996), followed by CHP2 and CHP3 (Inoue et al., 2003; Mailander et al., 2001). Several studies on the functional relation of CHP1–3 to NHE1 followed (Pang, Hisamitsu, Mori, Shigekawa, & Wakabayashi, 2004; Pang, Su, Wakabayashi, & Shigekawa, 2001; Zaun et al., 2008), leading to the two structures of CHP1 and CHP2 in complex with an NHE1-derived peptide (Ammar et al., 2006; Mishima et al., 2007). The overall structures of CHP1 and -2 are highly similar and are in complex with the same NHE1 peptide consisting of residues Val503–His545 (Fig. 2.5A and B). CHP1 and -2 are homologous to CaM and calcineurin B. They contain four EF hands, where only EF-3 and EF-4 can bind calcium with high affinity. CHP1 is also N-myristoylated, but binds NHE1 independently of its myristoylation state (Pang et al., 2001). The role of myristoylation for NHE1 activation is contradicting (Pang et al., 2001; Zaun et al., 2008).

Figure 2.6 Order and disorder in the SLC9A family. *In silico* disorder predictions for all human members of the entire SLC9A family using two independent predictors, PONDR-FIT in black (Xue, Dunbrack, Williams, Dunker, & Uversky, 2010) and DISOPRED2 in light gray (Ward, McGuffin, Bryson, Buxton, & Jones, 2004). The position of the TM domains based on the Nygaard model and the sequence alignment used in Fig. 2.1 are shown by dark gray bars. In the NHE1 zoom, the individual positions of TM helices are indicated in dark gray. For the PONDR-FIT analyses, a threshold is applied where disorder is assigned to values >0.5 as indicated by the black dashed line. For DISOPRED2, a 5% ID threshold is indicated by a light gray dashed line. (See the color plate.)

Table 2.2 Current knowledge on binding partners for all SLC9A family members

	Binding partner	Species, isoform	Method	Binding region/site on NHE	Reference(s)	Comments
(A) Three-dimensional structures of complexes						
Ca^{2+}-binding proteins	CHP1	Human NHE1	NMR	Ile518–Ile537	Mishima et al. (2007)	
	CHP2	Human NHE1	X-ray crystallography	Ser517–Gly539	Ammar et al. (2006)	
	Calmodulin	Human NHE1	X-ray crystallography, SAXS	Pro685–Val686 (site 1) Pro658–Gln671 (site 2)	Köster et al. (2011)	
(B) No structural data of complexes available: Binding site identified						
Ca^{2+}-binding proteins	CHP3/tescalcin	Human NHE1	Y2H, IF		Mailander, Muller-Esterl, and Dedio (2001)	
		Rat NHE1	Pull-down, co-IP	Leu505–Pro571	Zaun, Shrier, and Orlowski (2008)	
Cytoskeletal anchor proteins	Ezrin–radixin–moesin (ERM), phosphorylated ezrin	Human NHE1	*In vitro* binding assay, co-IP	$_{508}$VKKKQ$_{512}$ and $_{552}$RFNKKYVKK$_{560}$	Denker et al. (2000)	
	4.1R	Rat NHE1	*In vitro* pull-down, mutations	$_{519}$KR and $_{556}$RFNKKYVKK$_{564}$	Nunomura, Denker, Barber, Takakuwa, and Gascard (2012)	$K_D = 100$–200 nM, interaction requires both sites

	Ezrin/cytoskeleton via NHERF1/2	*NHE3*	*In vitro* binding assay	$_{656}$RKRL$_{659}$	Yun, Lamprecht, Forster, and Sidor (1998) and Kurashima et al. (1999)	Indirect association
Death effector domain protein	Daxx	Human NHE1	Y2H, *in vitro* pull-down, co-IP	Arg567–Asn637	Jung et al. (2008)	Competes with ezrin binding
Lipids, analogues, and nucleotides	PI(4,5)P_2	Rat NHE1	*In vitro* binding assay, mutations	Site 1: $_{513}$KKKQETKR$_{520}$, Site 2: $_{556}$RFNKKYVKK$_{564}$	Aharonovitz et al. (2000)	Corresponding sites in hNHE1: $_{509}$KKKQETKR$_{516}$ and $_{552}$RFNKKYVKK$_{560}$
	PI(4,5)P_2, other acidic lipids, and phorbol esters/DAG	Human NHE1	Confocal studies, *in vitro* liposomal pull-down, FRET, mutations	Gly542–Pro598	Wakabayashi, Nakamura, Kobayashi, and Hisamitsu (2010)	LID
	Phosphoinositides and other acidic lipids	Rat NHE1	Membrane overlay assay, surface plasmon resonance	–	Abu Jawdeh et al. (2011)	$PIP_3 > PIP_2 > PIP = PA > PS$
	PI(4,5)P_2 and PI(3,4,5)P_3	Rabbit *NHE3*	Liposomal pull-down, mutations	$_{501}$YLRDKWANFDRR$_{512}$ $_{520}$RQSAQKSRDRILNVFH-ELNLKDAISYVTEGERR-$_{552}$	Mohan et al. (2010)	Phe509 is required

Continued

Table 2.2 Current knowledge on binding partners for all SLC9A family members—cont'd

	Binding partner	Species, isoform	Method	Binding region/site on NHE	Reference(s)	Comments
	Anionic membranes	Rat *NHE3*	Liposome and cell-based assays, FRET, mutations	1. Leu456–Ala480 2. Lys503–Arg527 3. Lys645–Arg688	Alexander et al. (2011)	Electrostatic switch model
	Adenosine-5′-triphosphate (ATP)	Human NHE1	Photoaffinity label transfer, limited proteolysis, mutants, equilibrium dialysis	Gly542–Pro598	Shimada-Shimizu, Hisamitsu, Nakamura, and Wakabayashi (2013)	Also binds other nucleotides with lower affinity; identified binding site equals LID
Scaffold protein	14-3-3β	Human NHE1	Co-IP, tested by mutations	$_{700}$RIG<u>S</u>^{P}D<u>P</u>$_{705}$	Lehoux, Abe, Florian, and Berk (2001)	requires S703 phosphorylation
Enzyme	Carbonic anhydrase II	Rabbit NHE1	*In vitro* binding assay, affinity blotting, co-IP	–	Li, Alvarez, Casey, Reithmeier, and Fliegel (2002)	
			Co-IP, *in vitro* binding assays	$_{790}$RIQRCL<u>SD</u>PGPHP$_{802}$	Li, Liu, Alvarez, Casey, and Fliegel (2006)	Phosphorylation enhances binding
	Phospholipase gamma	Rabbit *NHE3*	Y2H, co-IP	586–605	Zachos et al. (2009)	
Signal protein	β Arrestin2	Human *NHE5*	*In vitro* pull-down, co-IP	$_{697}$ILTVESEEEEES-DSSETEKEDDEGII$_{723}$	Lukashova et al. (2011)	Dependent on *NHE5* phosphorylation by CK2

(C) No structural data of complexes available: Binding site unknown						
Peptide/hormone	Angiotensin	Human NHE1	*In vivo* activity assays, mutations	–	Eguti et al. (2010)	Effect on CaM/NHE1 interaction
Chaperones	Chaperone stress 70 protein (STCH)	Human NHE1	Co-IP	–	Bae et al. 2013)	
	dnaK/bovine heat shock protein 70 (hsp70)	Rabbit NHE1	Copurification, co-IP, immunofluorescence	Last 178 residues	Silva et al. (1995)	Bacterial analogue to Hsp70
		Mouse NHE1	Signal transduction antibody array, co-IP	–	Xue et al. (2007)	
Receptors	Integrin $\alpha_{IIb}\ \beta_3$	Human NHE1	FLIM–FRET	–	Yi et al. (2009)	Couples NHE1 with the Na^+/Ca^{2+} exchanger NCX1
						Lipid raft dependent
	CD44	Human NHE1	Co-IP	–	Bourguignon, Singleton, Diedrich, Stern, and Gilad (2004)	HA promotes colocalization in lipid rafts
	Megalin	Rabbit *NHE3*	Co-IP	–	Biemesderfer, Nagy, DeGray, and Aronson (1999)	
	Receptor for activated C kinase-1 (RACK1)	Human *NHE5*	Y2H, pull-down, immunofluorescence, co-IP	Gly492–Leu896	Onishi, Lin, Diering, Williams, and Numata (2007)	

Continued

Table 2.2 Current knowledge on binding partners for all SLC9A family members—cont'd

	Binding partner	Species, isoform	Method	Binding region/site on NHE	Reference(s)	Comments
Enzyme	Lactate dehydrogenase B (LDHB)	Human NHE1	Y2H, co-IP, *in vitro* pull-down	–	Kim (2009)	
	Dipeptidyl peptidase IV	Rabbit *NHE3*	Co-IP		Girardi, Degray et al. (2001)	May be indirect
Scaffold/adapter proteins	Na^+/H^+ exchanger regulatory factor 1/2 (NHERF1/2)	Human *NHE3*	Co-IP	–	Lamprecht, Weinman et al. (1998)	
			Co-IP	–	Weinman, Minkoff, and Shenolikar (2000)	Links *NHE3* to ezrin; alternative names: NHERF1/EBP50 and NHERF2/TKA-1/E3KARP
	Shank2	Human *NHE3*	Y2H, co-IP, surface plasmon resonance		Han et al. (2006)	
	PDZK1	Mouse *NHE3*	Y2H, *in vitro* pull-down		Gisler et al. (2003) and Han et al. (2006)	
Ca^{2+}-sensing membrane protein	Synaptotagmin (SytI)	Human *NHE3*	Co-IP, confocal microscopy	–	Musch et al. (2007)	Requires S605 phosphorylation
Signal protein	β Arrestin1	Mouse NHE1	Co-IP, confocal microscopy	-	Simonin and Fuster (2010)	β Arrestin1 links NHE1 and Nedd4-1 by direct interaction with both
Curvature sensor	Bin-1	Mouse NHE1	Signal transduction antibody array, co-IP	-	Xue et al. (2007)	

(D) Kinases and phosphatases-binding sites						
Kinases	Ste-20 like Nck-interacting kinase (NIK)	Rat NHE1	Y2H, co-IP, pull-down	Cys538–Asn638	Yan, Nehrke, Choi, and Barber (2001)	
	B-Raf	Human NHE1	Kinase array, *in vitro* pull-down and binding assay, co-IP	C-terminal 182 residues of NHE1	Karki, Li, Schrama, and Fliegel (2011)	
	p38 MAPK	Rabbit NHE1	Co-IP	–	Khaled et al. (2001)	
	Casein kinase 2α (CK2)	Rabbit *NHE3*	Co-IP, *in vitro* pull-down	Glu550–Ala667	Sarker et al. (2008)	
	CaMKII	Rabbit *NHE3*	Co-IP, *in vitro* pull-down	Leu586–Arg605	Zizak et al. (2012)	
Phosphatases	Calcineurin A (CaNA)	Human NHE1	Co-IP	^{715}PVITID^{720}	Hisamitsu, Nakamura, and Wakabayashi (2012)	Alternative name: PP3
	Protein phosphatase 2A (PP2A)	Opossum and rat *NHE3*	Y2H, co-IP, *in vitro* pull-down	The last 45 residues and between 557 and 794	Bobulescu et al. (2010)	
	Tyrosin phosphatase SHP2	Mouse NHE1	Signal transduction antibody array, co-IP, *in vitro* pull-down	–	Xue et al. (2007)	
	Protein phosphatase 1	Human NHE1	*In vitro* pull-down, co-IP	–	Misik, Perreault, Holmes, and Fliegel (2005)	

Binding partners are grouped according to the state of knowledge and function.

In complex with CHP1, the NHE1 peptide forms a five-turn amphipathic α-helix composed of residues Ile518–Ile537 that binds in the hydrophobic cleft of CHP1 (Mishima et al., 2007) in a head-to-tail orientation, that is, the C-terminal half binds the N-terminus of CHP1 and vice versa. The parts of the peptide proceeding (Val503–Ser517) and following (Cys538–His545) the bound helix were unstructured in the complex with no direct contact to CHP1. The entire peptide is unstructured in the absence of CHP1. Thus, CHP1 is needed for the helix to form. The apolar side of the helix binds the hydrophobic cleft of CHP1, and the NHE1–CHP1 interface is dominated by extensive hydrophobic interactions. Key residues for interaction of both CHP1 and the NHE1 helix were confirmed by mutagenesis (Mishima et al., 2007).

In the crystal structure of CHP2 with the same NHE1 peptide, only residues Arg516–His540 of NHE1 were resolved with the helix formed by Ser517–Gly539 (Ammar et al., 2006). Also here the complex is dominated by hydrophobic interactions, as expected due to the high similarity of CHP1 and -2, and again the interaction was probed by mutagenesis (Ammar et al., 2006). Both CHP1 and CHP2 possess a relatively long central linker connecting the two lobes, that is called the CHP-unique region. Based on functional analysis of a deletion mutant that lacks the CHP-unique region, this region was suggested to be involved in pH_i sensing of NHE1 and signal transduction from the tail to the TM domain (Ammar et al., 2006).

Since these structures were solved, more data on the previously contradicting functional role of CHPs for NHE1 function has become available, indicating that binding of CHP1 stabilizes NHE1 and is required for exchanger activity (Matsushita, Tanaka, Mitsui, & Kanazawa, 2011). CHP3 binding was further probed and demonstrated to be located at the same region as CHP1/2 binding. Overexpression of CHP3 promotes biosynthetic maturation and increases half-life at the cell surface (Zaun et al., 2008), presumably due to Ca^{2+} binding and N-myristoylation (Zaun, Shrier, & Orlowski, 2012). Thus, it is possible that the presence of CHPs is linked to dimer formation, a suggestion which has not been experimentally addressed yet.

CaM is a very versatile protein with numerous binding partners and functions in relaying Ca^{2+} signals. Several structures of CaM complexes have been solved, revealing a wide spectrum of binding conformations. CaM consists of two lobes connected by a short linker region each with two EF hands able to bind Ca^{2+}. CaM binding to NHE1 was first described in 1994 (Bertrand, Wakabayashi, Ikeda, Pouyssegur, & Shigekawa, 1994), followed by successive studies (Köster et al., 2011; Wakabayashi, Bertrand, Ikeda, Pouyssegur, & Shigekawa, 1994; Wakabayashi, Ikeda, et al., 1997). In contrast to CHP1 binding, CaM binding to NHE1 is highly Ca^{2+} dependent. CaM binds to two sites on the NHE1 tail, a high affinity and an intermediate affinity site ($K_D \sim 20$ nM and ~ 350 nM). The crystal structure of CaM in complex with the NHE1-derived peptide Ala622–Lys690 was solved to 2.23 Å resolution (Köster et al., 2011). The entire complex is very elongated, a conformation that was confirmed by small-angle X-ray scattering (SAXS) data, which is a solution-state technique. Thus, the elongated shape is not due to crystal packing effects. The NHE1-peptide forms two α-helices in complex with CaM (α_A: Asp626–Thr653 and α_B: Pro658–Val686). The first CaM binding site of NHE1 (α_A) interacts with the C-terminal lobe of CaM, and forms an amphiphilic α-helix. The polar and basic residues on one site interact with CaM, the hydrophobic residues are located on the opposite helix surface. Ca^{2+}-induced binding of CaM to this helix is thought to release an autoinhibitory interaction with the H^+ modifier site of the NHE1-TM domain, thereby activating NHE1 (Wakabayashi, Bertrand, Ikeda, et al., 1994; Wakabayashi, Bertrand, Shigekawa, Fafournoux, & Pouyssegur, 1994). Key residues for autoinhibition are Leu639, Arg651, and His652 (Ikeda, Schmitt, Pouyssegur, Wakabayashi, & Shigekawa, 1997; Wakabayashi, Ikeda, et al., 1997; Wakabayashi, Shigekawa, et al., 1997). This is highly interesting as the putative pH sensor site in NHE1 described above, contains glutamates (Glu246, Glu247, Glu253, Glu267, and Glu330) and polar residues (Thr124, Ser126, Ser127 Asn252, and Asn266) as potential interaction partners for these residues. Thus, the possibility arises that the autoinhibitory domain does interact with the putative pH sensor described above, and that this interaction stabilizes the inactive conformation of the pH sensor and of the Phe-gate. It has been proposed based on studies in cardiomyocytes that phosphorylation of Ser648 of NHE1 by protein kinase B (PKB)/Akt inhibits complex formation between NHE1 and CaM and thus prevents NHE1 activation by Ca^{2+}/CaM (Snabaitis, Cuello, & Avkiran, 2008). On the other hand, activation of NHE1 as a result of Ser648 phosphorylation by Akt was reported in fibroblasts (Meima, Webb, Witkowska, & Barber, 2009); thus, although phosphorylation similarly would be able to inhibit binding to the pH sensor site, the exact role of this residue as well as of its phosphorylation, remains enigmatic and may furthermore be cell-type dependent. Ser648 is located within the first CaM-binding site of NHE1, and phosphorylation presumably disrupts the interface between the autoinhibitory NHE1 helix and

CaM. The second NHE1 helix (α_B) is not amphiphilic and the contact area with CaM limited to the C-terminal part of the α-helix Pro658–Gln671 (Köster et al., 2011).

Despite the valuable insight gained by these structural studies, it is important to note that they were all based on peptides isolated from the context of the entire tail in its position at the inner plasma membrane. Thus, it is of utmost importance for the understanding, not only of NHE1 but also of the entire SCL9A family, that research on more complete structural system is performed. This will allow us to unveil the orchestration of regulatory events within the tail and lead to the understanding of how structural changes affect the ensemble and how signals are transferred to the TM domain.

5.2. Bioinformatical analyses and definition of subdomains of NHE1: "Walking along the tail"

Currently, the molecular sciences are challenged by the discovery that a sizeable fraction of the proteomes lacks three-dimensional globular folds, but still preserve functionality (Chouard, 2011; Dunker, Brown, Lawson, Iakoucheva, & Obradovic, 2002; Dyson & Wright, 2002; Tompa, 2002). The occurrence of such ID, as a characteristic of an entire protein (intrinsically disordered protein, IDP) or of a region (intrinsically disordered region, IDR), results from a relatively high number of charged amino acid residues and a low hydrophobicity (Dunker et al., 2001) This leads to an extraordinary flexibility and also forms the basis for *in silico* predictions of ID (Ward et al., 2004; Xue et al., 2010). IDPs are compliant and best described as ensembles of highly flexible, interchangeable conformations, with regions of lowly populated, transient structures and/or with conserved sequence motifs. Such local features are central to binding interaction partners and have been described as molecular recognition feature/elements (MoRF/MoRE) or preformed structural elements (PSEs) (Fuxreiter, Simon, Friedrich, & Tompa, 2004; Vacic et al., 2007) or by the characteristics of the motifs, such as linear motifs/short linear motifs (Davey, Shields, & Edwards, 2006; Fuxreiter, Tompa, & Simon, 2007). IDRs exploit an array of mechanisms of interaction and do so via a folding-upon-binding process (Dyson & Wright, 2005), via PSEs or MoRFs present in the unbound state (Kumar, Ma, Tsai, Sinha, & Nussinov, 2000), through preserving disorder in the bound state (Kumar et al., 2000; Tompa & Fuxreiter, 2008), or by so far undescribed mechanisms.

In the present work, we have performed two independent disorder predictions of the human sequences of the entire SLC9A family (Fig. 2.6), discussed in detail below, as well as predicted ordered and disordered regions for the NHE1 C-tail, Fig. 2.7. Here, *in silico* predictions suggest the presence of two distinct ID regions (Fig. 2.7A, bottom panel) and six putative α-helices (I–VI Fig. 2.7A, top panel). Based on these predictions, the C-terminal tail of NHE1 was divided in four subdomains A–D: Subdomain A (Leu506–Glu590) is the most membrane-proximal region consisting of four putative α-helices (I–IV); this region is presumably highly structured. Subdomain B is a stretch of 35 residues (Ser591–Lys625) that lacks propensity for secondary structure elements in consensus with high prediction scores for ID. Subdomain B therefore is potentially very flexible and might function as a structural spacer or linker between two ordered regions, allowing for important regulatory structural rearrangements within the tail. Subdomain C consists of two α-helices (V and VI, Asp626–Thr685) and domain D is intrinsically disordered (Val686–Gln815). Indeed, the structures available of NHE1 peptides in complex with CHP1/2 and CaM (Ammar et al., 2006; Köster et al., 2011; Mishima et al., 2007) confirm the existence of α-helices II, V, and VI, at least in their bound state, and ID in the distal part was confirmed by our earlier analyses (Kjaergaard et al., 2010; Nørholm et al., 2011). The predicted α-helix II overlaps the CHP1–3 binding helix, which is unstructured in isolation (Ammar et al., 2006). The α-helices V and VI overlap with the resolved α-helices in the crystal structure of the complex with CaM. Those two α-helices also have very high intrinsic α-helix propensities of up to 82% (Lys625–Leu639) and 72% (Pro658–Asn682) (predicted by AGADIR at http://agadir.crg.es/), suggesting that they might also exist in the CaM-free and autoinhibitory conformation.

Comparison of the bioinformatics-based dissection of the C-terminal tail of NHE1 with an earlier function-based dissection reveals a highly interesting overlap (Fig. 2.7A and B). The functional subdomains were identified from a set of truncation mutants where the effect on pH_i sensitivity was evaluated (Ikeda et al., 1997; see also Section 6). Four functional regions, I–IV, were described, which significantly overlap with the structural subdomains A–D defined here by our *in silico* analysis. Truncation of regions II and IV had no effect on pH_i sensitivity. Truncation of regions III shifted the pK_a up, truncation of region I shifted it down. Therefore, region I, overlapping with the ordered subdomain A, is linked to pH_i maintenance. Region II is proposed to function as a flexible loop, which concurs with the disorder of

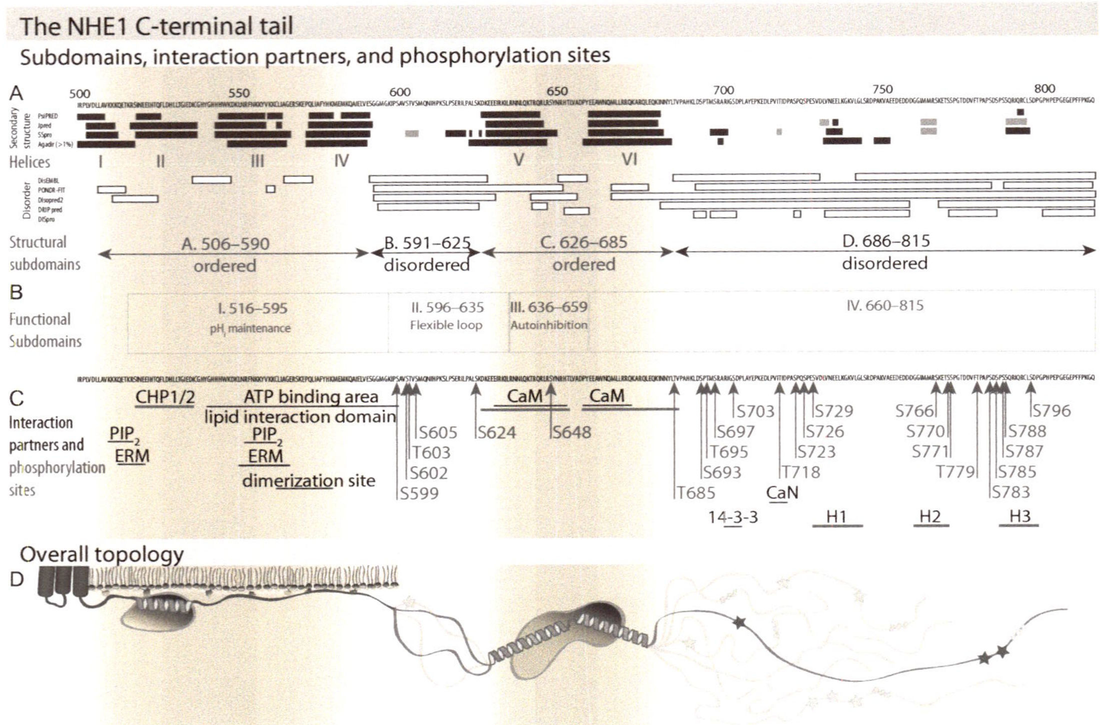

Figure 2.7 The NHE1 C-terminal tail: subdomains, interaction partners, and phosphorylation sites. (A) *In silico* secondary structure (top), disorder predictions (middle), and definition of structural subdomains (bottom) for the entire NHE1 C-terminal tail. Black bars indicate predictions of α-helices, gray bars β-strands, and white bars areas of intrinsic disorder. Consensus α-helices I–VI are indicated in roman numerals. Based on these predictions, the C-terminal tail was divided into four structural subdomains (A–D). *In silico* predictors used are PsiPRED (Jones, 1999), Jpred (Cole, Barber, & Barton, 2008), SSpro (Cheng, Randall, Sweredoski, & Baldi, 2005a), agadir (www.agadir.crg.es), DisEMBL (Linding et al., 2003), PONDR-FIT (Xue et al., 2010), Disopred2 (Ward et al., 2004), DRIPpred (http://www.sbc.su.se/~maccallr/disorder/), and Dispro (Cheng, Sweredoski, & Baldi, 2005b). (B) Functional regions of the tail as shown by Ikeda et al. (1997). (C) Interaction partners and phosphorylation sites. Binding sites are indicated by black lines and known structural elements by gray bold lines. Structural elements include the CHP binding helix II, the CaM-binding helices V and VI, and the transient helices H1–3 (MoRFs), phosphorylation sites are indicated by arrows and their numbers (see text and Table 2.3 for more information). (D) Average topology of the tail derived from consensus structural data and predictions, highlighting the four structural subdomains. (See the color plate.)

subdomain B. Region III is part of subdomain C and is concluded to be autoinhibitory. The distal region IV, consisting of helix VI and the disordered distal tail, has no effect on pH_i sensitivity (Ikeda et al., 1997).

5.3. ID as a conserved trait of the SLC9A family

The NHE1 distal tail (subdomain D) has a high score for ID and very limited secondary structure propensity. Indeed, as noted above, we recently showed experimentally by NMR spectroscopy that the distal tail is intrinsically disordered and this disorder is evolutionary conserved (Nørholm et al., 2011). Notably, a disordered distal tail is potentially present in SLC9A1–9. In the present work, we have performed *in silico* disorder predictions for the entire family using the predictors PONDR-FIT (Xue et al., 2010) and DISOPRED2 (Ward et al., 2004) (Fig. 2.6). Our work strongly indicates similar disorder profiles for all family members, SLC9A1–9, with a high score for disorder in the N-terminal region, within EL1, and along the C-terminal distal tails. Beyond residue 600, NHE1–5s have a long stretch of about 200 residues with high disorder scores. The disordered regions are shorter in *NHE6* and *NHE7*, but are still distinctly present. Interestingly, the disorder propensity is considerably lower for *NHE8* and *NHE9*, which are both evolutionarily most distant to NHE1 (Fig. 2.1B), with almost no disorder propensity for *NHE8*, which is also the shortest SLCA member.

5.4. Subdomain-related interaction partners

Subdomain A binds a number of reported binding partners for which structures of the complexes are missing (Fig. 2.7). However, information on the binding sequence on NHE1 is available for all of these. The putative α-helix I is located at the border between the TM domain and C-terminal tail and its exact boundaries have not been experimentally confirmed. Predictions and models define the end of the TM domain between Met497 and Leu506 (Table 2.1), that overlaps with the predicted helix I. The question remains whether α-helix I is simply the end of TM–XII or an independent secondary structural element. Following α-helix I, there is a lipid-interacting region (see below). Thus, the tail following the last TM-helix cannot stretch into the cytoplasm, but must instead bend toward the membrane bilayer. So even if α-helix I has a low propensity to form (AGADIR score of 5–13% for Val503–Lys511), the conformational constraints by the neighboring TM domain and the lipid-binding domain might prevent formation. The connection to the TM domain restricts the orientational freedom of the entire tail. Thus, the location of the NHE1 tail just proximal to the plasma membrane surface exposes it to an anisotropic environment, that is, restricted orientation and movement toward and along the membrane surface, which in case of the plasma membrane is highly negatively charged.

The sequences succeeding α-helix I and within α-helix III contain several basic residues identified to bind phosphatidylinositol-4,5-bisphosphate ($PI(4,5)P_2$); rat NHE1 site 1: 513–520, site 2: 556–564 (Aharonovitz et al., 2000); corresponding human NHE1 numbering; site 1: $_{509}$KKKQETKR$_{516}$, site 2: $_{552}$RFNKKYVKK$_{560}$. It was hypothesized that both sites can bind $PI(4,5)P_2$ independently, but simultaneous binding is required for optimal transport (Aharonovitz et al., 2000). A subsequent study extends the second $PI(4,5)P_2$-binding site to residues Gly542–Pro598 and refers to it as a lipid interaction domain (LID). This extension covers the putative α-helices III and IV. The N-terminus interacts with acidic lipids, whereas direct binding of LID to phorbol esters (PEs)/diacylglycerol (DAG) *in vitro* was reported to occur at the C-terminal part of the LID. It was suggested that the interaction with PEs/DAG increases the affinity of the LID for the plasma membrane with residues Leu573 and Ile574 crucial for this (Wakabayashi et al., 2010). Shortly after, a broad lipid-binding profile was described with the valence phosphatidylinositol(3,4,5) trisphosphate > phosphatidylinositol bisphosphates > phosphatidylinositol monophosphates = phosphatidic acid > phosphatidyl serine, and the driving force was suggested to be electrostatic (Abu Jawdeh et al., 2011). Also, *NHE3* has been shown to bind phosphoinositides as well as other acidic lipids in the corresponding region, and binding is required for activity (Alexander et al., 2011; Mohan et al., 2010). Many lipids of the intracellular leaflet of the membrane are negatively charged and the resulting membrane surface charge was indicated to dictate the structure and function of *NHE3*, leading to the proposal of an electrostatic switch model (Alexander et al., 2011). *NHE3* is stimulated by lysophosphatitic acid (LPA) (Murtazina et al., 2011), and lysophosphatidylcholine (LPC) stimulates Na^+–H^+ exchange. However, several lysophospholipids including LPC do not directly modulate Na^+–H^+ exchange (Goel, Ford, & Pierce, 2003).

Although ion exchange is secondary active, that is, no metabolic energy is required for ion translocation, early studies report inhibition of NHE1 by ATP depletion (Cassel, Katz, & Rotman, 1986). This inhibition was demonstrated for *NHE2* and *-3* as well (Cabado et al., 1996; Kapus et al., 1994). Thus, a modulating effect of ATP on SLC9A regulation was suggested (Cassel et al., 1986; Kapus et al., 1994). When the $PI(4,5)P_2$-binding sites within the cytosolic tail were identified, the effect of ATP depletion was, at least in part, attributed to the disturbed lipid metabolism as ATP depletion also leads to net dephosphorylation of PI(4,5)P_2 (Aharonovitz et al., 2000). Recently, however, evidence was provided that the cytosolic tail directly binds nucleotides with a preference for ATP within the region Gly542–Pro598, which overlaps with the previously defined LID (Shimada-Shimizu et al., 2013). Binding of PI(4,5)P_2 and ATP to LID is competitive (Shimada-Shimizu, Hisamitsu, Nakamura, Hirayama, & Wakabayashi, 2014). This is similar to the ATP-sensitive K^+ channels, where the inhibitory binding of nucleotides can be antagonized by PI(4,5)P_2 (MacGregor et al., 2002). Interestingly, mutations in the same positively charged residues in NHE1 proposed to mediate PI(4,5)P_2 binding were used to demonstrate a direct interaction with ezrin, radixin, and moesin (ERM) proteins. ERM proteins are major cytoskeletal-plasma membrane linker proteins with important roles in the regulation of cell morphology and motility (Denker et al., 2000). These data are consistent with the notion that ERM proteins can bind integral membrane proteins at clusters of positive charges within their juxta-membrane cytoplasmic domains (Yonemura et al., 1998). Furthermore, ERMs are themselves phosphoinositide binders (Jayasundar et al., 2012). Both of the positively charged clusters within the NHE1 tail are able to bind ERM proteins, where site 2 (α-helix III) is required and site 1 (between α-helix I and II) contributes, but is not sufficient on its own (Denker et al., 2000). Specifically, the interaction occurs between NHE1 and phosphorylated, active ERM proteins (Wu et al., 2004). In addition, the death effector domain protein Daxx and the key membrane skeletal protein in erythrocytes 4.1R (closely related to the ERM proteins) interact with NHE1 at the same basic clusters responsible for ERM protein binding (Jung et al., 2008; Nunomura et al., 2012). Analogously, *NHE3* is anchored by the cytoskeleton, with the putative-binding site $_{656}$RKRL$_{659}$. However, the association of *NHE3* with ERM proteins was suggested to be indirect, mediated by the adaptor proteins NHERF and E3KARP (Kurashima et al., 1999; Yun et al., 1998).

Thus, the same basic clusters within the NHE1 cytosolic tail mediate interaction with PI(4,5)P_2 and other negative-charged lipids, with ATP as well as with ERM proteins and related proteins. Additionally, cross-linking and coimmunoprecipitation experiments have shown that the proximal C-termini (Val503–Met580) of NHE1 monomers have a strong propensity to interact directly with each other in a dimer, specifically including residues Leu562–Lys579 (Hisamitsu et al., 2004). The putative dimerization site is located within the LID/ATP-binding site just after the second basic cluster. Thus, these many interactions as well as dimerization linked to the same regions of the transporter leave a number of urgent questions in regards to understanding SCL9A function. One of these is of course, how dimerization adds to the orchestration of binding partners within the proximal part of the tail. Secondly, for each of these interactions, only binary interaction data are available. The only exception is the competitive binding of PI(4,5)P_2 and ATP to LID. Thirdly, some of the binding partners are shown additionally to interact with each other, for example, ezrin also binds PI(4,5)P_2 (Tsukita & Yonemura, 1999). Thus, whether binding is mutually in- or exclusive and whether any interlinked effects such as cooperativity and antagonism exist are questions that so far have been left unaddressed both structurally and functionally.

No binding partners have yet been described that specifically bind to subdomain B, with the possible exception of Nck-interacting kinase, which binds somewhere in subdomains B or C (Yan et al., 2001). Subdomain C bears the autoinhibitory site and the Ca^{2+}/CaM-binding site, described in detail above.

5.5. Interactions and regulation by ID in the SLC9A family

Even though most of the interacting proteins bind NHE1 within the structured membrane-proximal part of the tail, a few binding partners have also been reported to localize to the disordered distal part. Many of these are regulatory proteins such as kinases and phosphatases, but also scaffolding proteins and proteins with other functions have been noted. An interesting example is carbonic anhydrase II, which was proposed to facilitate substrate availability for NHE1 (Li et al., 2002) similar to the contentious "metabolon" of an acid–base transporter and CAII proposed for several bicarbonate transporters (for a discussion, see Parker & Boron, 2013). Others include the scaffolding protein 14-3-3β (Lehoux et al., 2001), and the Ser/-Thr phosphatase calcineurin A (CaNA) (Hisamitsu et al., 2012). 14-3-3β binds the consensus motif $_{700}$RIGS^{P}DP$_{705}$ as demonstrated by immunoprecipitation and further confirmed by mutations. Binding is dependent on Ser703^{P} phosphorylation by p90^{RSK} and a proline in the +2 position (Pro705). Binding of 14-3-3β inhibits dephosphorylation of Ser703^{P} by phosphatase-1 and was suggested to participate in serum-stimulated antiporter activity (Lehoux et al., 2001). CaNA was identified to bind NHE1 directly at a consensus sequence motif $_{715}$PVITID$_{720}$. CaNA is the

CaM-binding catalytic subunit of calcineurin (CaN). Both subunits of calcineurin (CaNA and CaNB) coimmunoprecipitated with NHE1 but were dependent on the presence of CaNA. Thus, the entire CaN is suggested to interact with NHE1 via direct contact by CaNA (Hisamitsu et al., 2012).

Table 2.2 summarizes current knowledge on binding partners for all SLC9A family members. For a discussion of the roles of binding partners in SLC9A trafficking, please refer to Section 7.

5.6. Binding of protein kinases and phosphatases: Phosphorylation and dephosphorylation of NHE1

The occurrence of ID regions in the entire SCL9A family suggests the possibility of regulatory events that are linked to posttranslational modifications (Iakoucheva et al., 2004). For the NHE1 C-terminal tail region, subdomains B and D have many predicted phosphorylation sites (data not shown); several of those sites were reported to be phosphorylated (Fig. 2.7C and Table 2.3). Indeed, some kinases are shown to bind directly to the NHE1 C-terminal tail. A reciprocal regulation of NHE1 and mitogen-activated protein kinases (MAPKs) was previously recognized and reviewed (Pedersen, Darborg, Rentsch, & Rasmussen, 2007; Pedersen, King, et al., 2007). In 2010, NHE1 was proposed as a potential MAPK scaffold, that is, it is able to bind several kinases and their substrates simultaneously. In support of this proposal, NHE1 was recently found to interact with seven MAPK family kinases from all levels of the core MAPK hierarchy (MAPKKK–MAPKK–MAPK) (Bandyopadhyay et al., 2010). One of these is the mitogen-activated kinase 1 (MAPK1, *aka* ERK2). NHE1 was also suggested to be itself regulated by a MAP kinase (ERK), possibly through direct phosphorylation of the C-terminal tail (Wang et al., 1997). A successive study identified several *in vitro* ERK2 phosphorylation sites in the NHE1 tail, namely Ser693, Ser766, Ser770, Thr779, and Ser785 (Liu et al., 2004). Ser770 and Ser771 were hypothesized to mediate ERK-dependent NHE1 activation *in vivo* (Malo et al., 2007). It is noteworthy that only Ser693, Ser771, Thr779, and Ser785 are located in common MAPK substrate motifs, which require a proline in the +1 position [(S/T) P]. Subsequently, Ser771, Thr779, Ser785, and perhaps also Ser776 were identified to be involved in NHE1 phosphorylation after sustained intracellular acidosis in an ERK-dependent pathway (Odunewu & Fliegel, 2013). Moreover, activated p38 MAPK directly phosphorylates NHE1, and analysis by mass spectrometry (MS) identified the following four sequentially close *in vitro* phosphorylation sites in rabbit NHE1: Thr717, Ser722,

Table 2.3 Current knowledge on identified phosphorylation sites for all SLC9A family members

Corresponding site in human	Identified in species	Identified P-site	Method used for identification	Substrate motif for kinase	References
NHE1					
S599	Rat	S603	MS-based phosphoproteomics	CHK1, CSK3b, AMP	Lundby et al. (2012) and Rigor, Damoc, Phinney, and Cala (2011)
	At	S607	*In situ* phosphorylation, IP, LC-MS/MS		
S602	Rat	S606	MS-based phosphoproteomics	CK1, AMP	Lundby et al. (2012) and Rigor et al. (2011)
	At	S610	*In situ* phosphorylation, IP, LC-MS/MS		
T603	Rat	T607	MS-based phosphoproteomics	PKC, CamKII, CSK3	Lundby et al. (2012)
S605	Rat	S609	MS-based phosphoproteomics	CK1, AMP	Lundby et al. (2012) and Rigor et al. (2011)
	At	S613	*In situ* phosphorylation, IP, LC-MS/MS		
S624	Rat	S628	MS-based phosphoproteomics	CK2, Clk2, ATM, CDK1 motif 1	Lundby et al. (2012)
S648	Human	S648	*In vitro* kinase assay, LC-MS/MS	PKBα/Akt, Aurora-A/B, Clk2, PKC, PKA, CDK1 motif 1, Cdc2/5	Snabaitis et al. (2008) and Meima et al. (2009)
	Rat/rabbit	S648	*In vitro* kinase assay, LC-MS/MS		
T685	At	T693	*In situ* phosphorylation, IP, LC-MS/MS	Aurora-B, PKC, ATM	Rigor et al. (2011)

S693	Rat/human	S697/ S693	MS-based phosphoproteomics	GSK3/CSK3b, proline-dependent ERK, CDK1 motif 1/2, Cdc2/5	Lundby et al. (2012), Dephoure et al. (2008), Liu et al. (2004), and Malo, Li, and Fliegel (2007)
	Human	S693	MS-based phosphoproteomics		
			In vitro kinase assay		
			Mutants screen, IP and western blots		
	At	S701	*In situ* phosphorylation, IP, LC-MS/MS		Rigor et al. (2011)
T695	Rat	T699	MS-based phosphoproteomics	PKCd, CDK1 motif 1	Lundby et al. (2012)
S697	Rat	S701	MS-based phosphoproteomics	CK1, CSK3, AMP, Clk2, CDK1 motif 1/2	Lundby et al. (2012) and Dephoure et al. (2008)
	Human	S697	MS-based phosphoproteomics		
S703	Rat	S707	MS-based phosphoproteomics	CAMK2, PKA/ Akt, PKBα/Akt, Clk2, **p90^{RSK}**	Lundby et al. (2012), Dephoure et al. (2008), Takahashi et al. (1999), Snabaitis et al. (2008), and Rigor et al. (2011)
	Human	S703	MS-based phosphoproteomics		
			IP and 2D tryptic mapping, *in vitro* kinase assay		
			LC-MS/MS		
	At	S711	*In situ* phosphorylation, IP, LC-MS/MS		
(V716)	At	T727	*In situ* phosphorylation, IP, LC-MS/MS	–	Rigor et al. (2011)
T718	Rabbit	T717	IP and audioradioraphy, kinase assay	ATM	Khaled et al. (2001)

Continued

Table 2.3 Current knowledge on identified phosphorylation sites for all SLC9A family members—cont'd

Corresponding site in human	Identified in species	Identified P-site	Method used for identification	Substrate motif for kinase	References
S723	Rat	S727	MS-based phosphoproteomics	Proline-dependent ERK, p38 MAPK, GSK3, CDK1 motif 1/2, Cdc2/5	Lundby et al. (2012), Dephoure et al. (2008), and Dephoure et al. (2008)
	Human	S723	MS-based phosphoproteomics		
	Rabbit	S722	IP and audioradioraphy, kinase assay		
S726	Rat	S730	MS-based phosphoproteomics	CK1, proline-dependent ERK, p38 MAPK, DNA PK, CDK1 motif 1/2, Cdc2/5	Lundby et al. (2012), Dephoure et al. (2008), Grenier et al. (2008), and Khaled et al. (2001)
	Human	S726	MS-based phosphoproteomics		
			IP and immunoplot		
	Rabbit	S725	IP and audioradioraphy, kinase assay		
S729	Rat	S733	MS-based phosphoproteomics	CK1/2, CaMKII, ATM, DNA PK	Lundby et al. (2012), Dephoure et al. (2008), Grenier et al. (2008), and Khaled et al. (2001)
	Human	S729	MS-based phosphoproteomics		
			IP and immunoplot		
	Rabbit	S728	IP and audioradiography, kinase assay		
S766	Human	S766	*In vitro* kinase assay	CaMKII, CSK3	Liu et al. (2004) and Malo et al. (2007)
			Mutants screen, IP and western blots		
S770	Rat	S775	MS-based phosphoproteomics	GSK3/GSK3b, PLK1, Akt, Clk2	Lundby et al. (2012), Dephoure et al. (2008), Liu et al. (2004), and Malo et al. (2007)
	Human	S770	MS-based phosphoproteomics		
			In vitro kinase assay		
			Mutants screen, IP and western blots		

S771	Rat	S776	MS-based phosphoproteomics	Proline-dependent ERK, CDK1 motif 1/2, Cdc2/5	Lundby et al. (2012), Dephoure et al. (2008), Malo et al. (2007), and Odunewu and Fliegel (2013)
	Human	S771	MS-based phosphoproteomics		
			Mutants screen, IP and western blots IP and		
			audioradiography		
S776[a]	Human	S776	IP and audioradiography		Odunewu and Fliegel (2013)
T779	Human	T779	*In vitro* kinase assay	Proline-dependent ERK, GSK3/GSK3b, CK1/2, CDK1 motif 1/2, Cdc2/5	Liu et al. (2004), Malo et al. (2007), and Odunewu and Fliegel (2013)
			Mutants screen, IP and western blots		
			IP and audioradiography		
S783	Rat	S788	MS-based phosphoproteomics	GSK3/GSK3b, proline-dependent ERK	Lundby et al. (2012) and Dephoure et al. (2008)
	Human	S783	MS-based phosphoproteomics		
S785	Rat/human	S790/S785	MS-based phosphoproteomics	Proline-dependent ERK, CDK1 motif 1/2, Cdc2/5	Lundby et al. (2012), Dephoure et al. (2008), Liu et al. (2004), Malo et al. (2007), Odunewu and Fliegel (2013), and Rigor et al. (2011)
	Human	S785	MS-based phosphoproteomics		
			In vitro kinase assay		
			Mutants screen, IP and western blots IP and		
			audioradiography		
	At	S783	*In situ* phosphorylation, IP, LC-MS/MS		

Continued

Table 2.3 Current knowledge on identified phosphorylation sites for all SLC9A family members—cont'd

Corresponding site in human	Identified in species	Identified P-site	Method used for identification	Substrate motif for kinase	References
S787	Human	S787	MS-based phosphoproteomics	AMP, Cdc5	Dephoure et al. (2008)
S788	Rat	S793	MS-based phosphoproteomics	CK1, ATM, DNA PK, Cdc2	Lundby et al. (2012) and Dephoure et al. (2008)
	Human	S788	MS-based phosphoproteomics		
S796	Rat	S801	MS-based phosphoproteomics	CAMK2, CHK1, PKD, **PKBα/Akt**, AMP, Aurora-A/B, PKA, proline-dependent ERK	Lundby et al. (2012), Snabaitis et al. (2008), and Rigor et al. (2011)
	Human	S796	*In vitro* kinase assay, LC-MS/MS		
	At	S794	*In situ* phosphorylation, IP, LC-MS/MS		

Corresponding site in human	Identified in species	Identified P-site	Method used for identification	Consensus substrate motif for kinase	References
NHE3					
S555	Rat	S552	MS-based phosphoproteomics	CAMK2, Aurora-A/B, PKA, **PKC**	Lundby et al. (2012), Zhao et al. (1999), Kocinsky et al. (2005), and Liu and Jose (2013)
			In vivo and *in vitro* phosphorylation, autoradiography and immunoblot		
			Phospho-specific antibodies		
			Phospho-specific antibodies		

S563	Rat	S560	MS-based phosphoproteomics	CAMK2, Amp, Akt, Aurora-B, Clk2, PKA	Lundby et al. (2012)
–	Starfish	S590	*In vitro* kinase assay, audioradiography	–	Harada, Fukuda, Hirohashi, and Chiba (2010)
S607	Rat	S605	*In vivo* and *in vitro* phosphorylation, autoradiography and immunoblots	CAMK2, Akt, Aurora-A/B, Clk2, PKC, **PKA**	Zhao et al. (1999), Kurashima et al. (1997), and Kocinsky et al. (2005)
			IP and 2D tryptic mapping		
			Phospho-specific antibodies		
–	Starfish	S606	*In vitro* kinase assay, audioradiography	–	Harada et al. (2010)
S636	Rat	S634	IP and 2D tryptic mapping	Amp, Clk2	Kurashima et al. (1997)
S663	Rat	S661	MS-based phosphoproteomics	Akt, Aurora-A/B, CaMKII, Clk2, PKC, PKA, **SGK1**	Lundby et al. (2012) and Wang, Sun, Lang, and Yun (2005)
	Rabbit	S663	*In vitro* kinase assay		
–	Starfish	S673	*In vitro* kinase assay, audioradiography	–	Harada et al. (2010)
S718	Rabbit	S719	LC-MS/MS of IP, *in vitro* kinase assay	ATM, **CK2**, ERK, PLK1	Sarker et al. (2008)
S794	Rat	S791	MS-based phosphoproteomics	Proline-dependent ERK, PDK1 motif 1/2, Cdc2, Cdk5	Lundby et al. (2012)

Continued

Table 2.3 Current knowledge on identified phosphorylation sites for all SLC9A family members—cont'd

Corresponding site in human	Identified in species	Identified P-site	Method used for identification	Consensus substrate motif for kinase	References
(G796)	Rat	S793	MS-based phosphoproteomics		Lundby et al. (2012)
S807	Rat	S804	MS-based phosphoproteomics	AMP, Aurora-A/B, PKC, PKA	Lundby et al. (2012)
S810	Rat	S807	MS-based phosphoproteomics	Amp, ATM, Akt, CaMKII, CK1, Clk2, DNA PK, PKC, PKA	Lundby et al. (2012)
S813	Rat	S810	MS-based phosphoproteomics	Aurora-A, PKC, CK1	Lundby et al. (2012)

[a]There is no S776 in human NHE1; this may refer to S766.
For all identified phosphorylation sites the corresponding human positions are listed. Potential kinases were identified by searching the sites for kinase consensus motifs. Experimentally identified kinases are highlighted in bold letters.

Ser725, and Ser728 (corresponding positions in human NHE1: Thr718, Ser723, Ser726, Ser729; Table 2.3) (Khaled et al., 2001). Only Ser723 and Ser726 are located within common p38 MAPK substrate motifs. In addition, NHE1 activity is required for activation of p38 MAPK (Pedersen, Darborg, et al., 2007; Rentsch, Ossum, Hoffmann, & Pedersen, 2007). Phosphorylation of Ser726 and Ser729 was proposed to mediate apoptosis-induced alkalinization by NHE1 (Grenier et al., 2008).

In vivo Ser703 phosphorylation was identified by 2D tryptic peptide mapping of coimmunoprecipitated NHE1 and ERK-regulated p90 ribosomal S6 kinase ($p90^{RSK}$) was shown to be responsible (Cuello, Snabaitis, Cohen, Taunton, & Avkiran, 2007; Takahashi et al., 1999). Rho-associated kinase ($p160^{ROCK}$) mediates LPA-induced NHE1 phosphorylation *in vitro* and *in vivo*. Thus, NHE1 is a downstream, and possibly direct, target of $p160^{ROCK}$ (Park, Ryu, Yun, Kim, & Han, 2012; Tominaga, Ishizaki, Narumiya, & Barber, 1998). However, phosphorylation sites in *NHEs* by $p160^{ROCK}$ are still to be determined.

The Ser/Thr kinase B-Raf binds the NHE1 tail *in vitro* and *in vivo*, where it stimulates NHE1 activity. In a deletion study, binding was localized to the C-terminal 182 residues (Karki et al., 2011). An *in vitro* kinase assay using PKBα/Akt revealed three phosphorylation sites, which include substrate consensus motifs [RxRxx(S/T)] in the NHE1 distal tail, Ser703 and Ser648, as well as a nonconsensus motif involving Ser796 [RxxRxxS]. Mutational assays and kinetic analysis identified Ser648 as the major PKBα phosphorylation site (Snabaitis et al., 2008), and selective Ser648 phosphorylation by Akt was confirmed *in vitro* (Meima et al., 2009).

As described above, the Ste-20 like Nck-interacting kinase (NIK) (Yan et al., 2001) interacts with NHE1. NIK binds NHE1 between Cys538 and Asn638 and phosphorylates NHE1 distal to Asn638 *in vitro* and *in vivo*, although the exact sites for binding and phosphorylation are yet to be identified.

Vast phosphorylation screens exploiting phosphopeptide enrichment and MS analysis have identified many putative phosphorylation sites in rat and human NHE1 and *NHE3* (see Table 2.3) (Dephoure et al., 2008; Lundby et al., 2012). Several of those have not been identified previously, and for most of the phosphorylation sites the responsible kinases still await identification. In addition, for many of the NHE interacting kinases, the phosphorylation sites on NHE1 are likewise not yet determined. Importantly, their interactions may therefore also be indirect through other mediating proteins. In a study to identify phosphorylation events involved in cell cycle regulation, several sites in human NHE1 were identified. Human cells were arrested either in the G1 or mitotic phase prior to investigation (Dephoure et al., 2008). In another study, many rat NHE1 and *NHE3* phosphorylation sites were identified, most of which were organ and tissue specific, Table 2.3 (Lundby et al., 2012). Phosphorylation of NHE1 due to osmotic cell shrinkage was examined in the giant salamander, *Amphiuma tridactylum* (*at*). Osmotic cell shrinkage increased the *atNHE1* phosphorylation, and nine *in situ* phosphorylated Ser/Thr residues were identified by coimmunoprecipitation and liquid chromatography (LC) MS/MS (Rigor et al., 2011). The corresponding sites in human NHE1 are also listed in Table 2.3. Further proof is needed to confirm the *in vivo* relevance of most of the above-mentioned phosphorylation sites, as well as their interdependence and corresponding kinases.

Some phosphorylation data are also available for *NHE3*. Specific kinases involved in *NHE3* phosphorylation are protein kinase C (PKC) (Wiederkehr, Zhao, & Moe, 1999), protein kinase A (PKA) (Kocinsky et al., 2005), serum- and glucocorticoid-inducible kinase 1 (SGK1) (Wang et al., 2005), casein kinase 2 (CK2) (Sarker et al., 2008), CaM kinase II (CaMKII) (Zizak et al., 2012), and $p90^{RSK}$ (Harada et al., 2010). *NHE5* is also phosphorylated by CK2 (Lukashova et al., 2011). Identified phosphorylation sites for nonhuman SLC9As are listed in the end of Table 2.3. Phosphatases that are proposed to bind and/or dephosphorylate *NHE* include protein phosphatase 1 (Dynia, Steinmetz, & Kocinsky, 2010; Misik et al., 2005), CaNA (Hisamitsu et al., 2012), and serine/threonine protein phosphatase 2A (PP2A) (Bobulescu et al., 2010).

Evaluation of the relative positions of putative phosphorylation sites to each other reveals accumulation of several sites to distinct clusters. Interestingly, clusters of phosphorylation sites have recently drawn attention. Such clusters were suggested to orchestrate the time course of signaling events due to interdependent multiphosphorylation events (Moses, Heriche, & Durbin, 2007; Olsen et al., 2006; Schweiger & Linial, 2010). Hence, we suggest that the clusters of multiple phosphorylation sites within the NHE1 tail may contribute to the temporal control of signaling events involving NHE1.

6. BIOPHYSICS OF SLC9A FUNCTION

6.1. Proton sensing by *NHEs*: Allosteric H^+-binding site versus high- and low-affinity dimer

Extensive studies from the Wakabayashi group have implicated specific residues in the ILs between the TM helices, as well as in several regions in the

C-terminal tail, in H^+ sensing by NHE1 (see description above). Starting from the proximal part of the C-tail, it was found that preventing the interaction of NHE1 with its binding partner CHP1/2/3 (Section 5), which interacts with NHE1$^{Ile518-Ile537}$, not only reduces the V_{max} of the exchanger but also causes an acidic shift in H^+ sensitivity (Pang et al., 2004, 2001). Moving further along the C-tail, deletion of residues Arg567–Leu635 shifted pH_i dependence of NHE1 transport activity to more acidic values (Wakabayashi, Bertrand, Ikeda, et al., 1994; Wakabayashi, Bertrand, Shigekawa, et al., 1994), whereas deletion of Arg636–Ala656 caused the opposite shift to more alkaline values (Ikeda et al., 1997). Arg636–Ala656 contains the high-affinity CaM-binding site, which was proposed to act as an autoinhibitory domain that interacts with the putative allosteric H^+-binding site and inhibits its protonation (Wakabayashi, Bertrand, Ikeda, et al., 1994), see also Sections 3 and 5. Charge reversing mutations of four positively charged positions (Lys641, Arg643, Arg645, and Arg647) in the CaM-binding domain rendered NHE1 constitutively active by increasing its pH_i sensitivity (Wakabayashi, Bertrand, Ikeda, et al., 1994; Wakabayashi, Bertrand, Shigekawa, et al., 1994). These studies were further refined in later studies from the group, in which the cytoplasmic domain was divided into four subdomains according to their roles in H^+ sensing, with NHE1$^{Lys595-Leu635}$ proposed to serve as a linker region with no effect on H^+ sensitivity *per se*, and the section NHE1$^{Glu660-Gln815}$ also being inert in terms of H^+ sensing (Fig. 2.7B) (Ikeda et al., 1997). Consistent with (but not unique to [see below]) the interpretation that H^+_i sensitivity involves interactions between the C-tail and an allosteric H^+-binding site in the TM domain, Wakabayashi and coworkers also identified charged residues in the ILs important for H^+_i sensing, most notably Glu131^{IL1} (Hisamitsu et al., 2007) and Arg440^{IL5} and Gly455^{TMXI}/Gly456^{TMXI} (Wakabayashi, Hisamitsu, Pang, & Shigekawa, 2003b). It should be noted that these sites are different than the proposed pH-sensing site of NHE1 (see above). Yet, IL connectivity and especially effects imposed by glycine substitutions, with glycine being the most entropic amino acid residue, may clearly affect the ability of regions in NHE1 to transmit conformational changes.

While the allosteric site model gained widespread acceptance, other early kinetic studies on *NHE3* suggested that cooperativity might be most compatible with the transporter functioning as a dimer (Otsu, Kinsella, Heller, & Froehlich, 1993; Otsu, Kinsella, Sacktor, & Froehlich, 1989). This idea was followed up in elegant studies by Lacroix et al. (2004), who took advantage of a mutation in TMIX of NHE1 that converted the exchanger to a nonallosteric form (Arg327Glu). Interestingly, this site is linked via TMIX

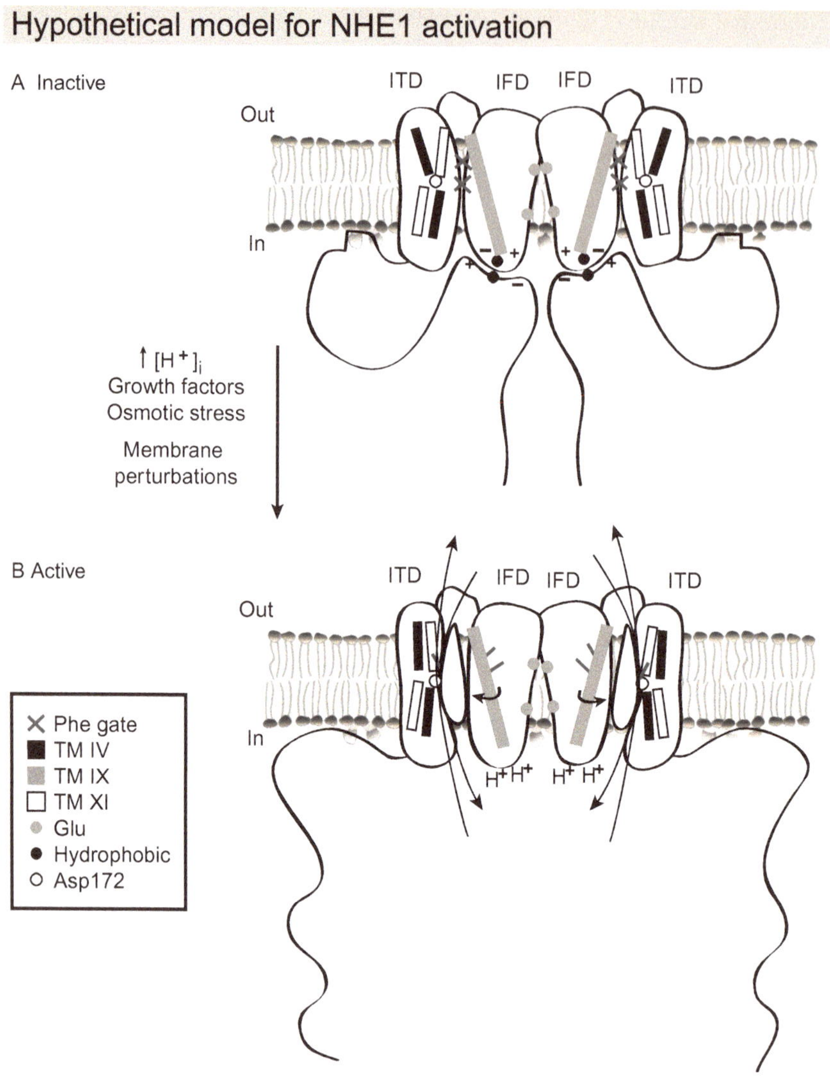

Figure 2.8 Proposed working model for H^+ sensing by NHE1. (A) NHE1 quiescent state and (B) NHE1-activated state. In the quiescent state, ion transport activity is prevented by the Phe-gate, which is closed due in particular to the conformation of TMIX and TMII and autoinhibitory interactions with a region of the C-terminal tail (Arg636–Ala656). Upon an increase in cytoplasmic H^+ concentration (or other stimuli that alter sensitivity toward the H^+ ions and hence increase NHE1 activity at steady-state pH_i), interactions of H^+ with Asp residues in the autoinhibitory site, and possibly in the dimer interface, result in disassembly of at least the interactions between the C-terminal tails and regions in the TM domain (likely involving IL3 and -4 and cytoplasmic regions of TMII and TMIX), rotation of TMII and TM IX and likely several other helices (not all shown for clarity), leading to opening of the Phe-gate (brown crosses). The allosteric change may

to the Phe-gate (see above). Based on a comparison of this mutant to wild-type NHE1, they used mathematical modeling to demonstrate that NHE1 activation by both acidification and growth factors is best described by a Monod–Wyman–Changeux concerted mechanism in which NHE1 exists as a dimer, the H^+ transport sites of which oscillate between a low-H^+ affinity and a high-H^+ affinity conformation. The former is the most prevalent at physiological pH, causing NHE1 to be essentially silent under these conditions; the latter is stabilized by intracellular H^+ as well as by growth factor stimulation, leading to exchanger activation and accounting for the observed cooperativity (Lacroix et al., 2004). Later modeling efforts by Fuster and collaborators based on NHE1 activity measurements using patch clamp and H^+-sensitive microelectrodes (Fuster, Moe, & Hilgemann, 2008) led to the suggestion that allostery in NHE1 may be explained by one of two different models—one in which NHE1 functions as a monomer at low pH_i, but as a coupled dimer at more alkaline pH_i, and another, in which NHE1 functions as a coupled monomer, denoted the "serial" model, in which each monomer transports two ions and translocation by one monomer allosterically favors translocation in the opposite direction by the other (Fuster et al., 2008). Such differentially regulated transport is not mutually exclusive and may relate to the possibility described above of the opening of the Phe-gate by multiple mechanisms. Based on the NhaA structures, the structural analyses performed in the present work (Fig. 2.4), the existing knowledge of NHE1 dimerization, the Lacroix et al. cooperativity studies, the strong evidence pointing to the existence of an autoinhibitory site, and the NHE1 mutations known to affect ion transport, we propose a working model (Fig. 2.8) in which H^+ sensing involves interactions of H^+ with (i) negatively charged residues in the cytoplasmic ends of TMII and TMIX, (ii) the dimer interface, and (iii) the autoinhibitory site of the C-tail. This results in disassembly of the interactions between the C-terminal tails and regions in the TM domain (likely also involving IL3 and -4), inducing an allosteric rotation of TM IX and TMII and likely several other helices, and finally opening the Phe-gate. It is also possible that interaction between

---

also affect dimerization between C-terminal tails. The dimer model is based on the NhaA structures, the structural analyses performed in the present work (Fig. 2.4), the existing knowledge of NHE1 dimerization, the Lacroix cooperativity studies, the strong evidence pointing to the existence of an autoinhibitory site, and the NHE1 mutations known to affect ion transport. "+" and "−" denote positive and negative charges, respectively. Other symbols are defined in the figure. (See the color plate.)

the two C-terminal tails across the dimer is integrated in this process, but this is unclear from available data.

Regardless of which model best represents NHE1, precisely how and where in NHE1 the structural changes underlying activation by various mechanisms are induced remains to be understood, and more structural insight is needed to distinguish between the proposed models, including the one proposed here. The finding that point mutations of charged residues in TMIV and TMIX (Arg180Lys, Arg327Glu, and Glu330Met) lead to loss of allostery (Lacroix et al., 2004), similarly to the effect observed by deletion of the C-terminal tail, might point to a role for this region to interact with the autoinhibitory domain of the C-tail. Whether this occurs intra-monomerically or inter-monomerically cannot be directly deduced from current data.

It is important to note that the cooperativity of NHE1 differs fundamentally from that of NhaA, which as noted above has a stoichiometry of 2 H^+:1 Na^+, and which shows strong negative cooperativity with a Hill coefficient of ~ -3 for the dimeric, and ~ -2 for the monomeric forms (Rimon, Tzubery, & Padan, 2007). As discussed by Lacroix et al. (2008), this by necessity requires several H^+-binding sites within each monomer. In contrast, the cooperativity of NHE1 can be explained by one binding site within one monomer that exhibits two different affinities for intracellular H^+, the transition between the high- and low-affinity form being associated with activation. We propose that the allosteric changes inflicted by proton binding involve opening of the Phe-gate. On this note, we observe aberrant pK_a values for Glu131^{IL2} (pK_a 5.96), Glu253^{IL3} (pK_a 6.39), Glu262^{TMVII} (pK_a 7.63), and Glu330^{TMIX} (pK_a 8.56) (propka http://propka.ki.ku.dk), which are all connected to the Phe-gate. Whether these numbers provide any meaning in relation to a link between proton sensing and a potential Phe-gate will have to await the solution of the atomic structure of NHE1.

The fact that *NHE2* and *NHE3*, similar to NHE1, exhibit cooperativity of H^+ binding (Aronson et al., 1982; Kapus et al., 1994; Orlowski, 1993; Yu et al., 1993) suggests the existence of a similar mechanism for acid sensing in the three isoforms, and indeed, *NHE3* has also been found to exist as a dimer in the membrane (Fafournoux et al., 1994). Additionally, however, acid sensing by *NHE3* additionally appears to involve a mechanism extrinsic to the exchanger *per se*. Thus, in kidney proximal tubule epithelial cells, the tyrosine kinase Pyk2 was identified as a cellular acid sensor that transmits the signal of intracellular acidity to *NHE3* in a manner involving Pyk2-mediated activation of Src (Li, Sato, Yang, Preisig, & Alpern, 2004).

6.2. Volume sensing by *NHEs*

Osmotic stress and concomitant cell shrinkage or swelling regulate a wide range of membrane transport proteins by mechanisms that are proposed to include altered membrane thickness, cytoskeletal reorganization, macromolecular crowding, ionic strength, specific ions, and numerous signaling pathways that are sensitive to osmotic perturbations or cellular volume changes (for a review, see Hoffmann, Lambert, & Pedersen, 2009). The different *NHE* isoforms respond differently to osmotic stress. Again, most available data on the mechanisms concern the shrinkage-activation of NHE1, and this is discussed in detail below. Intriguingly, for *NHE2*, shrinkage-activation (Kapus et al., 1994), shrinkage-inhibition (Nath et al., 1996), and no effect of shrinkage (Su, Pang, Wakabayashi, & Shigekawa, 2003) have been reported in various cell types, and may point to the importance of non-*NHE*-intrinsic components regulating its volume sensitivity, some of which may be membrane constituents binding to the C-tail. In PS120 fibroblasts, replacing the first extracellular loop in *NHE2* with that of NHE1 caused *NHE2* to become shrinkage-activated (which in these cells it is normally not). This was interpreted to reflect the existence of an inhibitory role of an *NHE2*-specific part of this loop in shrinkage-activation (Su et al., 2003). In contrast, there is agreement that *NHE3* is inhibited by osmotic shrinkage (Good, Di Mari, & Watts, 2000; Kapus et al., 1994; Nath et al., 1996; Soleimani et al., 1994; Watts & Good, 1999), and, conversely, it is found to be activated by cell swelling in most cell types studied (Alexander et al., 2007; Good et al., 2000; Watts & Good, 1999). Interestingly, although *NHE3* is generally regulated by insertion in/removal from the plasma membrane (Section 7), studies in MDCK cells showed that swelling-induced *NHE3* activation was dependent on membrane deformation but not on insertion of additional *NHE3* proteins in the plasma membrane, and, surprisingly, also not on altered interaction of the transporter with the cytoskeleton (Alexander et al., 2007). *NHE4* is efficiently shrinkage activated in a manner dependent on the actin cytoskeleton, and has been suggested to play a special role in regulatory volume increase in tissues exposed to very high osmolarity, for example, the kidney medulla (Bookstein et al., 1996; Bookstein et al., 1994). *NHE5*, which is most closely related to *NHE3*, is also inhibited by cell shrinkage (Attaphitaya et al., 2001), although to our knowledge, the mechanism remains to be elucidated.

Similar to the activation of NHE1 by growth factors, its activation by cell shrinkage has been shown to involve an alkaline shift in the H^+ sensitivity,

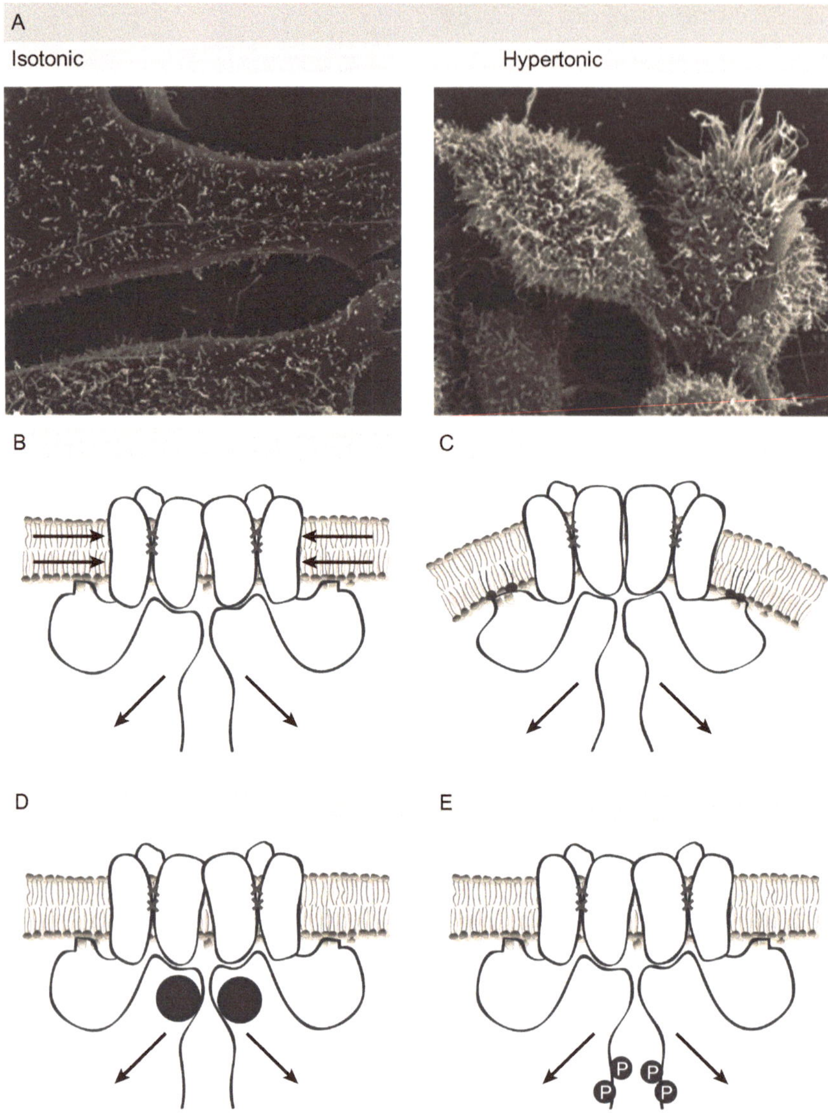

Figure 2.9 Possible mechanisms of activation of NHE1 by cell volume perturbations. (A) Scanning electron micrograph showing Ehrlich Lettre Ascites cells under isotonic conditions (left), and after 5 min of hypertonic shrinkage induced by doubling of extracellular osmolarity (right). (B–E) Working models for activation of NHE1 by cell shrinkage. In all cases, cell shrinkage is transmitted into allosteric changes in the NHE1 dimer, that lead to opening of the Phe-gate (brown crosses) and changes in the dimerization state involving the C-terminal tails. (B) Altered membrane tension may exert direct mechanical forces on NHE1 to elicit allosteric changes in the dimer and Phe-gate opening. (C) Altered membrane curvature induced mechanically or by changes in the membrane

causing the exchanger to be activated at physiological pH_i (Grinstein, Rothstein, & Cohen, 1985). In contrast, osmotic activation of NHE1 is not associated with a change in V_{max} (e.g., Lacroix et al., 2008; Nath et al., 1996), in good agreement with the Phe-gate model proposed above. Lacroix and coworkers also investigated the Monod–Wyman–Changeux model in the context of activation of NHE1 by osmotic cell shrinkage (Lacroix et al., 2008). They demonstrated that a hyperosmotic challenge modifies the *L0* parameter (the allosteric affinity constant) in favor of the activated state. Notably, stimulation of the exchanger with serum shifted the basal equilibrium of NHE1 toward the higher affinity state, but did not prevent osmotic activation of the exchanger (Lacroix et al., 2008).

Numerous molecular mechanisms that might mediate or contribute to the activating effect of cell shrinkage on NHE1 have been studied. Possible models are illustrated in Fig. 2.9. Firstly, as osmotic stress profoundly alters plasma membrane organization and mechanical parameters (e.g., Rasmussen et al., 2008; Fig. 2.9A), the roles of membrane reorganization in NHE1 regulation have been studied. Several mechanisms involving changes in the membrane tension, curvature, and/or composition have been proposed (Fig. 2.9B and C). In CHO cells, NHE1 was found to be sensitive to the lipid composition of the plasma membrane such that cholesterol, and other agents thickening the membrane activated NHE1 (Fuster, Moe, & Hilgemann, 2004). Studies in other cell types (Lacroix et al., 2008; Tekpli et al., 2008) have confirmed that NHE1 is highly sensitive to the biophysical properties of the membrane. However, in AP1 cells, in which NHE1 is also volume sensitive, its activity was not sensitive to the plasma membrane-altering agents that activated NHE1 in CHO cells (Fuster et al., 2004), arguing against a general and primary role of simple plasma membrane physics in the volume sensitivity of NHE1. The plasma

---

lipid composition, may exert direct mechanical forces on NHE1 to elicit allosteric changes in the dimer and Phe-gate opening. Altered membrane lipid composition may also affect NHE1 through specific lipid–NHE1 interactions via the NHE1 LID. (D) Cell shrinkage may induce NHE1 activation through cell volume-sensitive changes in binding partner interactions, in turn leading to Phe-gate opening. (E) Cell shrinkage-induced phosphorylation events in the distal part of the NHE1 *cytosolic* tail activation may alter NHE1 dimerization state and/or binding partner interactions, eventually leading to Phe-gate opening. See Fig. 2.8 for model details, and the text for details of each mechanism described above. *(A) Reproduced with permission from Rasmussen et al. (2008).* (See the color plate.)

membrane level of $PI(4,5)P_2$, which, as noted above, interacts with NHE1 (Abu Jawdeh et al., 2011; Aharonovitz et al., 2000; Fuster et al., 2004; Wakabayashi et al., 2010), has been shown to increase upon osmotic cell shrinkage (Nasuhoglu et al., 2002; Nielsen, Jensen, Harbak, Christensen, & Simonsen, 2007; Yamamoto et al., 2006). On the other hand, in PS120 fibroblasts exogenously overexpressing NHE1, the association of NHE1 with $PI(4,5)P_2$ was reported not to be involved in its activation by shrinkage (Lacroix et al., 2008). Although the organization of the actin-based cytoskeleton is dramatically altered by osmotic cell volume perturbations (Di Ciano et al., 2002; Hallows, Law, Packman, & Knauf, 1996; Hallows, Packman, & Knauf, 1991; Pedersen et al., 2002; Pedersen, Mills, & Hoffmann, 1999; Rasmussen et al., 2008; Rizoli, Rotstein, Parodo, Phillips, & Kapus, 2000), disruption of the actin cytoskeleton is generally found not to mediate the activation of NHE1 by cell shrinkage (Bookstein et al., 1996; Fuster et al., 2004; Lacroix et al., 2008). Similarly, knockdown of the NHE1 binding partner ezrin, which links NHE1 to the actin cytoskeleton, did not prevent shrinkage-activation of NHE1 (Rasmussen et al., 2008). Thus the mechanisms behind regulation of NHE1 by osmotic cell shrinkage remain incompletely understood. Possible, not mutually exclusive, mechanisms include the following scenarios: (i) NHE1 is in itself mechanosensitive—for example, regulated by membrane curvature changes or altered membrane tension (Fig. 2.9B and C); (ii) NHE1 is regulated by osmo- or mechanosensitive signaling events (e.g., protein kinase activation followed by NHE1 phosphorylation; Rigor et al., 2011), Fig. 2.9D; (iii) or NHE1 is activated because osmotic stress alters its interaction with specific binding partners (e.g., CaM; Garnovskaya, Mukhin, Vlasova, & Raymond, 2003) (Fig. 2.9E). It is possible that several mechanisms exist, the relative contributions of which may vary between cell types. It is also interesting to note that high-throughput screens identified the Bin-Amphiphysin-Rvsp (BAR) proteins Amphiphysin I and -II (Bin-1) as interaction partners for NHE1 (see also Table 2.1). As BAR proteins have been implicated in both, sensing and formation of membrane curvature, and because their membrane recruitment has been shown to be regulated by osmotic stress (Galic et al., 2012), it is possible that this interaction contributes to osmosensing by NHE1. However, to our knowledge, this possibility has never been experimentally addressed.

In terms of the Phe-gate mechanism which we suggest in the present work, hyperosmotic changes may lead to changes in membrane fluidity. Such a change may result in altered entropy of the lipid phase resulting in an altered packing of NHE1 helices. Thus, in relation to the changes observed in the lipid phase surrounding the dimer upon pH activation (Appel et al., 2009), changes in pressure within the bilayer may therefore mechanically lead to TMII and TMIX rotations, that could either open or close the Phe-gate. It is highly possible that the Phe-open state may be approximated to the high-affinity state of the transporter. Clearly, future studies are needed to address this possibility, as well as delineate membrane-NHE1 TM domain interactions, both functionally as well as structurally.

7. REGULATION OF SLC9A TURNOVER AND TRAFFICKING

The biosynthesis, trafficking, and degradation of the SLC9As remain relatively incompletely understood. The mature form of NHE1 (apparent molecular mass of 110 kDa) contains N-linked and O-linked oligosaccharides and is expressed at the cell surface, and a core glycosylated form (85 kDa) is present in the ER (Counillon et al., 1994). NHE1 contains a diacidic ER export motif ($_{770}$DEEED$_{774}$) in the distal, disordered part of its C-terminal tail, the mutation of whose corresponding helical MoRF (see Section 5) caused accumulation of the core glycosylated form in the ER (Nørholm et al., 2011). NHE1 has been shown to have a relatively long half-life (24 h) in the plasma membrane (Cavet et al., 2001). As described above, interaction with its binding partners CHP1, -2, and -3, appears to regulate NHE1 trafficking and stability (Section 5). NHE1 removal from the plasma membrane involves its interaction with β arrestin and Nedd4-1, followed by ubiquitinylation and endocytosis (Simonin & Fuster, 2010). Moreover, it was recently proposed that removal of NHE1 from the plasma membrane is dependent on the physical interaction of NHE1 with dynamin (Li et al., 2013).

As discussed in Section 6, most stimuli that activate NHE1 do so by increasing the apparent affinity for intracellular protons, with no effect on V_{max}. Based on such kinetic analyses, early investigators proposed that NHE1 regulation did not involve recruitment of new transporters, neither after acidification, osmotic shrinkage nor treatment with PEs (Bianchini et al., 1995; Dixon, Cohen, Cragoe, & Grinstein, 1987; Grinstein, Cohen, Goetz, & Rothstein, 1985). On the other hand, several recent papers clearly indicate that in response to certain stimuli, NHE1 activity may be increased by NHE1 recruitment to the plasma membrane, or to specific regions within the membrane. Thus, in cardiomyocytes, insulin and other stimuli activating NHE1 elicited the translocation of NHE1 to the plasma membrane in parallel with Glut-4 (Lawrence, Holman, & Koumanov, 2010). Furthermore, in CHO cells, FLIM–FRET experiments showed that stimulation of αIIb-β3 integrins elicited the recruited of NHE1 to αIIb-β3 integrins, where it physically interacted with the integrins and with the NCX (possibly via the integrins) (Yi et al., 2009). Another interesting issue in this regard is how NHE1 traffics to the leading edge lamellipodia upon migratory stimuli, as shown in migrating fibroblasts and cancer cells (Lagana et al., 2000; Martin, Pedersen, Schwab, & Stock, 2011; Schneider et al., 2009). While the mechanism of this trafficking is to our knowledge unresolved, we recently showed that NHE1 localization to the leading edge is dependent on Akt and ERK signaling (Clement et al., 2013).

For *NHE2* and *NHE3*, it has long been clear that their activation frequently involves an increase in V_{max} (Gens et al., 2007; Hensley, Bradley, & Mircheff, 1989; Janecki et al., 1998; Kurashima, Szabo, Lukacs, Orlowski, & Grinstein, 1998; Levine, Montrose, Tse, & Donowitz, 1993), suggesting either an increased number of exchangers in the plasma membrane or an increased turnover rate of exchangers resident in the membrane. Other lines of evidence also indicate that vesicle-mediated recruitment and retrieval is involved in short-term control of *NHE2* and *NHE3* activity. Interestingly, *NHE2* has a very short membrane half-life (3 h) compared to both NHE1 and *NHE3* (Cavet et al., 2001). For *NHE2*, the molecular mechanisms of trafficking are relatively poorly understood. It has, however, been shown that two proline-rich motifs ($_{743}$PPSVTPAP$_{750}$ and $_{786}$VPPKPPP$_{794}$ located in the disordered region; Fig. 2.7) that interact with SH3 domain-containing proteins are important for the apical membrane targeting of *NHE2* (Chow et al., 1999). Furthermore, similar to NHE1, both *NHE2* and *NHE3* interact with CHP (see Donowitz et al., 2013). Other *NHE3*-interacting proteins include NHERF1 and -2 (Kim et al., 2002; Weinman et al., 2000), ezrin (Kurashima et al., 1999; Yun et al., 1998), CaM kinase II (Zizak et al., 2012), PP2A (Bobulescu et al., 2010), phosphoinositides (Mohan et al., 2010), megalin (Biemesderfer et al., 1999), PDZK1 (Gisler et al., 2003), Shank2 (Han et al., 2006), phospholipase C gamma (Zachos et al., 2009), and synaptotagmin I (Musch et al., 2007). Several of these interaction partners have been shown to regulate *NHE3* trafficking, including NHERF1 and -2 (Kim et al., 2002; Weinman et al., 2000) and synaptotagmin I, which are all involved in cAMP-(NHERF1) and Ca^{2+}-(NHERF2) induced *NHE3* endocytosis (Musch et al., 2007). Shank2 is a PSD-95/discs large/ZO-1 (PDZ) domain protein, which similarly to NHERF1 and -2, interacts with *NHE3* through its C-terminal PDZ-binding domain. Shank2 favors the apical localization of *NHE3* and attenuates its endocytic retrieval, thus increasing net *NHE3* activity (Han et al., 2006). Finally, it has recently been proposed that myosin IV (Myo6) plays a role in *NHE3* retrieval from the apical membrane (Hegan, Giral, Levi, & Mooseker, 2012).

Several trafficking-related binding partners have also been identified for the brain-enriched SLC9A isoform, *NHE5*. A highly acidic, serine/threonine-rich, di-isoleucine motif (residues Ile697–Ile723) in the disordered region of the *NHE5* C-terminal tail (Fig. 2.6) is recognized by β arrestin2, and this interaction is important for *NHE5* internalization (Szabo, Numata, Lukashova, Iannuzzi, & Orlowski, 2005), whereas interaction with the secretory carrier membrane protein-2 appears to favor *NHE5* localization to the cell surface (Diering, Church, & Numata, 2009). Interestingly, receptor for activated C kinase-1 (RACK1) is a binding partner not only for *NHE5* (in which the binding site is localized within the disordered tail at residues Gly492–Leu896) (Onishi et al., 2007) but also for the organellar *NHE* isoforms *NHE6*, *-7*, and *-9*, the cell surface levels of which are regulated by RACK1 (Ohgaki, Fukura, Matsushita, Mitsui, & Kanazawa, 2008). Finally, two membrane-proximal motifs in the disordered region of the C-terminal tail of *NHE7* (residues His553–Val559 and Pro563–Pro568) were found to be required for *NHE7* localization to the Trans-Golgi network (Fukura et al., 2010). To our knowledge, little is known regarding the molecular mechanisms of trafficking of *NHE8*; however, it has recently been shown to exhibit regulated trafficking to the plasma membrane of rat kidney cells upon acid stress (Joseph et al., 2012).

8. CONCLUSIONS, PERSPECTIVES, AND OPEN QUESTIONS

In the present work, we have summarized and critically discussed the literature linking the still-limited understanding of the structure of the SLC9As to their possible modes of ion transport and regulation. Further, we have outlined the current landscape of known binding partners and phosphorylations of the SLC9As, and have addressed the possible mechanisms through which these modifications can regulate SLC9A function. Finally, we report here for the first time several structural and bioinformatics analyses that point to modes of SLC9A regulation that have to our knowledge not previously been proposed, and that should, in our view, be experimentally addressed. One major such feature is the possible presence of an allosteric gate formed by a group of phenylalanines from several TM helices, located at the border between the two helical bundles. Such a gate could be central to activation–inactivation of the SLC9As not just by the cytoplasmic H^+ concentration, but also by mechanical stimuli, interaction partners, and posttranslational modifications, for example mechanical stimuli, interaction partners, and posttranslational modifications.

Numerous interesting questions remain open. A central and urgent task is to obtain high-resolution structural information on the SLC9As. All existing models are based on bacterial NhaAs, and while there is some experimental support for the notion that there are important structural similarities, the NhaAs and SLC9As are clearly also fundamentally different, and, as pointed out above, exhibit distinctly different stoichiometry and cooperativity. The fact that the SLC9As, as opposed to the prokaryotic transporters, all exhibit a conserved region of ID in the distal part of the C-terminal tail suggests possible widespread roles for ID, MoRFs, and order–disorder transitions in the regulation of SLC9A function. The elucidation of the possible roles for ID in SLC9A regulation will, however, require numerous further studies, and will in part rely on novel experimental and bioinformatics tools that are only now becoming available.

Other important questions have arisen with the increasing knowledge of SLC9A interaction partners and posttranslational modifications. Essentially, all existing studies of interactions of SLC9As with their binding partners are binary, that is, addressing only the effects of a single binding partner at a time. Clearly, the *in vivo* situation is infinitely more complex with the likely scenario that the pattern of binding partner interactions is undergoing constant dynamic changes, and the precise composition of the complex at any given time will determine the outcome. Systems biology approaches will be necessary to begin to understand how the different binding partners and posttranslational modifications affect each other, as will structural studies on the interactions between specific binding partner effects. For instance, the binding of CaM to NHE1 engages polar and basic residues on one side of the helix, placing the hydrophobic residues on the opposite helix surface in a position that would allow interactions with other binding partners. Another example is that many binding sites on NHE1 overlap, and it has just recently been addressed whether binding of $PI(4,5)P_2$ and ATP to LID is mutually in- or exclusive. Yet, how this plays together with the binding of ERM proteins within the same small juxta-membrane region needs to be addressed. Furthermore, we hypothesize that the dimerization of the SLC9As orchestrates the binding partner pattern, a topic that to our knowledge is also completely open.

With respect to the extensive phosphorylation of the SLC9As, there are also numerous open questions. Only a few sites have really been exposed to a stringent analysis of their roles as *bona fide*, *in vivo* phosphorylated sites that serve as regulators of SLC9A function. Furthermore, for the most part, phosphorylations have been investigated in isolation. For example, the temporal sequence of SLC9A phosphorylations is completely unknown and could be of utmost importance for understanding their functions. Finally, the mechanisms of trafficking, membrane insertion, localization, stabilization, retrieval, and degradation of the SLC9As are surprisingly poorly understood, especially for some isoforms.

Currently, much of the basic knowledge essential for understanding how the structural dynamics of the SLC9As link to their regulation and function is available. However, the field is in desperate need for a high-resolution structure of the TM domain in order to delineate the intricate details of transport and regulation at an atomic level. With the large portfolio of interactions and regulations and with ID-linked functionality, issues highlighted in the current review will hopefully inspire researchers in the field of SLC9A to address the fundamental open questions regarding these widely expressed, physiologically and pathophysiologically important transporters.

ACKNOWLEDGMENTS

We thank the Novo Nordic Foundation (SFP/BBK), The Danish Research Councils (SFP/BBK), and the Carlsberg Foundation (BBK) for financial support for work in our laboratories related to this publication.

REFERENCES

Abrol, R., Kim, S. K., Bray, J. K., Griffith, A. R., & Goddard, W. A., III (2011). Characterizing and predicting the functional and conformational diversity of seven-transmembrane proteins. *Methods*, *55*, 405–414.

Abu Jawdeh, B. G., Khan, S., Deschenes, I., Hoshi, M., Goel, M., Lock, J. T., et al. (2011). Phosphoinositide binding differentially regulates NHE1 Na^+/H^+ exchanger-dependent proximal tubule cell survival. *Journal of Biological Chemistry*, *286*, 42435–42445.

Aharonovitz, O., Zaun, H. C., Balla, T., York, J. D., Orlowski, J., & Grinstein, S. (2000). Intracellular pH regulation by Na^+/H^+ exchange requires phosphatidylinositol 4,5-bisphosphate. *The Journal of Cell Biology*, *150*, 213–224.

Alexander, R. T., & Grinstein, S. (2006). Na^+/H^+ exchangers and the regulation of volume. *Acta Physiologica*, *187*, 159–167.

Alexander, R. T., Jaumouille, V., Yeung, T., Furuya, W., Peltekova, I., Boucher, A., et al. (2011). Membrane surface charge dictates the structure and function of the epithelial Na^+/H^+ exchanger. *The EMBO Journal, 30*, 679–691.

Alexander, R. T., Malevanets, A., Durkan, A. M., Kocinsky, H. S., Aronson, P. S., Orlowski, J., et al. (2007). Membrane curvature alters the activation kinetics of the epithelial Na^+/H^+ exchanger, NHE3. *The Journal of Biological Chemistry, 282*, 7376–7384.

Ammar, Y. B., Takeda, S., Hisamitsu, T., Mori, H., & Wakabayashi, S. (2006). Crystal structure of CHP2 complexed with NHE1-cytosolic region and an implication for pH regulation. *The EMBO Journal, 25*, 2315–2325.

Appel, M., Hizlan, D., Vinothkumar, K. R., Ziegler, C., & Kuhlbrandt, W. (2009). Conformations of NhaA, the Na^+/H^+ exchanger from Escherichia coli, in the pH-activated and ion-translocating states. *Journal of Molecular Biology, 388*, 659–672.

Aronson, P. S. (1985). Kinetic properties of the plasma membrane Na^+/H^+ exchanger. *Annual Review of Physiology, 47*, 545–560.

Aronson, P. S., Nee, J., & Suhm, M. A. (1982). Modifier role of internal H^+ in activating the Na^+-H^+ exchanger in renal microvillus membrane vesicles. *Nature, 299*, 161–163.

Attaphitaya, S., Nehrke, K., & Melvin, J. E. (2001). Acute inhibition of brain-specific Na^+/H^+ exchanger isoform 5 by protein kinases A and C and cell shrinkage. *AJP - Cell Physiology, 281*, C1146–C1157.

Bae, J. S., Koo, N. Y., Namkoong, E., Davies, A. J., Choi, S. K., Shin, Y., et al. (2013). Chaperone stress 70 protein (STCH) binds and regulates two acid/base transporters NBCe1-B and NHE1. *Journal of Biological Chemistry, 288*, 6295–6305.

Bandyopadhyay, S., Chiang, C. Y., Srivastava, J., Gersten, M., White, S., Bell, R., et al. (2010). A human MAP kinase interactome. *Nature Methods, 7*, 801–805.

Bertrand, B., Wakabayashi, S., Ikeda, T., Pouyssegur, J., & Shigekawa, M. (1994). The Na^+/H^+ exchanger isoform 1 (NHE1) is a novel member of the calmodulin-binding proteins. Identification and characterization of calmodulin-binding sites. *Journal of Biological Chemistry, 269*, 13703–13709.

Bianchini, L., Kapus, A., Lukacs, G., Wasan, S., Wakabayashi, S., Pouyssegur, J., et al. (1995). Responsiveness of mutants of NHE1 isoform of Na^+/H^+ antiport to osmotic stress. *The American Journal of Physiology, 269*, C998–C1007.

Biemesderfer, D., Nagy, T., DeGray, B., & Aronson, P. S. (1999). Specific association of megalin and the Na^+/H^+ exchanger isoform NHE3 in the proximal tubule. *Journal of Biological Chemistry, 274*, 17518–17524.

Blanchard, A., Eladari, D., Leviel, F., Tsimaratos, M., Paillard, M., & Podevin, R. A. (1998). NH4+ as a substrate for apical and basolateral Na^+/H^+ exchangers of thick ascending limbs of rat kidney: Evidence from isolated membranes. *The Journal of Physiology, 506*(Pt 3), 689–698.

Bobulescu, I. A., Quinones, H., Gisler, S. M., Di, S. F., Hu, M. C., Shi, M., et al. (2010). Acute regulation of renal Na+/H+exchanger NHE3 by dopamine: Role of protein phosphatase 2A. *American Journal of Physiology Renal Physiology, 298*, F1205–F1213.

Boedtkjer, E., Bunch, L., & Pedersen, S. F. (2012). Physiology, pharmacology and pathophysiology of the pH regulatory transport proteins NHE1 and NBCn1: Similarities, differences, and implications for cancer therapy. *Current Pharmaceutical Design, 18*, 1345–1371.

Bookstein, C., Musch, M. W., DePaoli, A., Xie, Y., Rabenau, K., Villereal, M., et al. (1996). Characterization of the rat Na^+/H^+ exchanger isoform NHE4 and localization in rat hippocampus. *The American Journal of Physiology, 271*, C1629–C1638.

Bookstein, C., Musch, M. W., DePaoli, A., Xie, Y., Villereal, M., Rao, M. C., et al. (1994). A unique sodium-hydrogen exchange isoform (NHE-4) of the inner medulla of the rat kidney is induced by hyperosmolarity. *The Journal of Biological Chemistry, 269*, 29704–29709.

Bourguignon, L. Y., Singleton, P. A., Diedrich, F., Stern, R., & Gilad, E. (2004). CD44 interaction with Na + -H + exchanger (NHE1) creates acidic microenvironments leading to hyaluronidase-2 and cathepsin B activation and breast tumor cell invasion. *The Journal of Biological Chemistry, 279*, 26991–27007.

Burckhardt, G., Di Sole, F., & Helmle-Kolb, C. (2002). The Na^+/H^+ exchanger gene family. *Journal of Nephrology, 15*(Suppl. 5), S3–S21.

Cabado, A. G., Yu, F. H., Kapus, A., Lukacs, G., Grinstein, S., & Orlowski, J. (1996). Distinct structural domains confer cAMP sensitivity and ATP dependence to the Na^+/H^+ exchanger NHE3 isoform. *The Journal of Biological Chemistry, 271*, 3590–3599.

Cassel, D., Katz, M., & Rotman, M. (1986). Depletion of cellular ATP inhibits Na +/H + antiport in cultured human cells. Modulation of the regulatory effect of intracellular protons on the antiporter activity. *The Journal of Biological Chemistry, 261*, 5460–5466.

Cavet, M. E., Akhter, S., Murtazina, R., Sanchez de Medina, F., Tse, C. M., & Donowitz, M. (2001). Half-lives of plasma membrane Na +/H + exchangers NHE1–3: Plasma membrane NHE2 has a rapid rate of degradation. *American Journal of Physiology. Cell Physiology, 281*, C2039–C2048.

Chambrey, R., St John, P. L., Eladari, D., Quentin, F., Warnock, D. G., Abrahamson, D. R., et al. (2001). Localization and functional characterization of Na +/H + exchanger isoform NHE4 in rat thick ascending limbs. *American Journal of Physiology. Renal Physiology, 281*, F707–F717.

Cheng, J., Randall, A. Z., Sweredoski, M. J., & Baldi, P. (2005). SCRATCH: A protein structure and structural feature prediction server. *Nucleic Acids Research, 33*, W72–W76.

Cheng, J., Sweredoski, M. J., & Baldi, P. (2005). Accurate prediction of protein disordered regions by mining protein structure data. *Data Mining and Knowledge Discovery, 11*, 213–222.

Chouard, T. (2011). Structural biology: Breaking the protein rules. *Nature, 471*, 151–153.

Chow, C. W., Woodside, M., Demaurex, N., Yu, F. H., Plant, P., Rotin, D., et al. (1999). Proline-rich motifs of the Na^+/H^+ exchanger 2 isoform. Binding of Src homology domain 3 and role in apical targeting in epithelia. *Journal of Biological Chemistry, 274*, 10481–10488.

Clement, D. L., Mally, S., Stock, C., Lethan, M., Satir, P., Schwab, A., et al. (2013). PDGFRalpha signaling in the primary cilium regulates NHE1-dependent fibroblast migration via coordinated differential activity of MEK1/2-ERK1/2-p90RSK and AKT signaling pathways. *Journal of Cell Science, 126*, 953–965.

Cole, C., Barber, J. D., & Barton, G. J. (2008). The Jpred 3 secondary structure prediction server. *Nucleic Acids Research, 36*, W197–W201.

Counillon, L., Franchi, A., & Pouyssegur, J. (1993). A point mutation of the Na^+/H^+ exchanger gene (NHE1) and amplification of the mutated allele confer amiloride resistance upon chronic acidosis. *Proceedings of the National Academy of Sciences of the United States of America, 90*, 4508–4512.

Counillon, L., Pouyssegur, J., & Reithmeier, R. A. (1994). The Na^+/H^+ exchanger NHE-1 possesses N- and O-linked glycosylation restricted to the first N-terminal extracellular domain. *Biochemistry, 33*, 10463–10469.

Cuello, F., Snabaitis, A. K., Cohen, M. S., Taunton, J., & Avkiran, M. (2007). Evidence for direct regulation of myocardial Na^+/H^+ exchanger isoform 1 phosphorylation and activity by 90-kDa ribosomal S6 kinase (RSK): Effects of the novel and specific RSK inhibitor fmk on responses to alpha1-adrenergic stimulation. *Molecular Pharmacology, 71*, 799–806.

Davey, N. E., Shields, D. C., & Edwards, R. J. (2006). SLiMDisc: Short, linear motif discovery, correcting for common evolutionary descent. *Nucleic Acids Research, 34*, 3546–3554.

Denker, S. P., Huang, D. C., Orlowski, J., Furthmayr, H., & Barber, D. L. (2000). Direct binding of the Na–H exchanger NHE1 to ERM proteins regulates the cortical cytoskeleton and cell shape independently of H+translocation. *Molecular Cell*, *6*, 1425–1436.

Dephoure, N., Zhou, C., Villen, J., Beausoleil, S. A., Bakalarski, C. E., Elledge, S. J., et al. (2008). A quantitative atlas of mitotic phosphorylation. *Proceedings of the National Academy of Sciences of the United States of America*, *105*, 10762–10767.

Di Ciano, C., Nie, Z., Szaszi, K., Lewis, A., Uruno, T., Zhan, X., et al. (2002). Osmotic stress-induced remodeling of the cortical cytoskeleton. *American Journal of Physiology. Cell Physiology*, *283*, C850–C865.

Diering, G. H., Church, J., & Numata, M. (2009). Secretory carrier membrane protein 2 regulates cell-surface targeting of brain-enriched Na^+/H^+ exchanger NHE5. *Journal of Biological Chemistry*, *284*, 13892–13903.

Ding, J., Rainey, J. K., Xu, C., Sykes, B. D., & Fliegel, L. (2006). Structural and functional characterization of transmembrane segment VII of the Na+/H+exchanger isoform 1. *Journal of Biological Chemistry*, *281*, 29817–29829.

Dixon, S. J., Cohen, S., Cragoe, E. J., Jr., & Grinstein, S. (1987). Estimation of the number and turnover rate of Na^+/H^+ exchangers in lymphocytes. Effect of phorbol ester and osmotic shrinking. *The Journal of Biological Chemistry*, *262*, 3626–3632.

Donowitz, M., Ming, T. C., & Fuster, D. (2013). SLC9/NHE gene family, a plasma membrane and organellar family of Na^+/H^+ exchangers. *Molecular Aspects of Medicine*, *34*, 236–251.

Dunker, A. K., Brown, C. J., Lawson, J. D., Iakoucheva, L. M., & Obradovic, Z. (2002). Intrinsic disorder and protein function. *Biochemistry*, *41*, 6573–6582.

Dunker, A. K., Lawson, J. D., Brown, C. J., Williams, R. M., Romero, P., Oh, J. S., et al. (2001). Intrinsically disordered protein. *Journal of Molecular Graphics & Modelling*, *19*, 26–59.

Dynia, D. W., Steinmetz, A. G., & Kocinsky, H. S. (2010). NHE3 function and phosphorylation are regulated by a calyculin A-sensitive phosphatase. *American Journal of Physiology. Renal Physiology*, *298*, F745–F753.

Dyson, H. J., & Wright, P. E. (2002). Coupling of folding and binding for unstructured proteins. *Current Opinion in Structural Biology*, *12*, 54–60.

Dyson, H. J., & Wright, P. E. (2005). Intrinsically unstructured proteins and their functions. *Nature Reviews. Molecular Cell Biology*, *6*, 197–208.

Eguti, D. M., Thieme, K., Leung, G. P., Mello-Aires, M., & Oliveira-Souza, M. (2010). Regulation of Na+/H+ exchanger isoform 1 (NHE1) by calmodulin-binding sites: Role of angiotensin II. *Cellular Physiology and Biochemistry*, *26*, 541–552.

Fafournoux, P., Noel, J., & Pouyssegur, J. (1994). Evidence that Na^+/H^+ exchanger isoforms NHE1 and NHE3 exist as stable dimers in membranes with a high degree of specificity for homodimers. *The Journal of Biological Chemistry*, *269*, 2589–2596.

Fliegel, L., Haworth, R. S., & Dyck, J. R. (1993). Characterization of the placental brush border membrane Na^+/H^+ exchanger: Identification of thiol-dependent transitions in apparent molecular size. *The Biochemical Journal*, *289*(Pt 1), 101–107.

Fukura, N., Ohgaki, R., Matsushita, M., Nakamura, N., Mitsui, K., & Kanazawa, H. (2010). A membrane-proximal region in the C-terminal tail of NHE7 is required for its distribution in the trans-Golgi network, distinct from NHE6 localization at endosomes. *The Journal of Membrane Biology*, *234*, 149–158.

Fuster, D., Moe, O. W., & Hilgemann, D. W. (2004). Lipid- and mechanosensitivities of sodium/hydrogen exchangers analyzed by electrical methods. *Proceedings of the National Academy of Sciences of the United States of America*, *101*, 10482–10487.

Fuster, D., Moe, O. W., & Hilgemann, D. W. (2008). Steady-state function of the ubiquitous mammalian Na/H exchanger (NHE1) in relation to dimer coupling models with 2Na/2H stoichiometry. *The Journal of General Physiology*, *132*, 465–480.

Fuxreiter, M., Simon, I., Friedrich, P., & Tompa, P. (2004). Preformed structural elements feature in partner recognition by intrinsically unstructured proteins. *Journal of Molecular Biology, 338*, 1015–1026.
Fuxreiter, M., Tompa, P., & Simon, I. (2007). Local structural disorder imparts plasticity on linear motifs. *Bioinformatics, 23*, 950–956.
Galic, M., Jeong, S., Tsai, F. C., Joubert, L. M., Wu, Y. I., Hahn, K. M., et al. (2012). External push and internal pull forces recruit curvature-sensing N-BAR domain proteins to the plasma membrane. *Nature Cell Biology, 14*, 874–881.
Garnovskaya, M. N., Mukhin, Y. V., Vlasova, T. M., & Raymond, J. R. (2003). Hypertonicity activates Na^+/H^+ exchange through Janus kinase 2 and calmodulin. *The Journal of Biological Chemistry, 278*, 16908–16915.
Gens, J. S., Du, H., Tackett, L., Kong, S. S., Chu, S., & Montrose, M. H. (2007). Different ionic conditions prompt NHE2 and NHE3 translocation to the plasma membrane. *Biochimica et Biophysica Acta, 1768*, 1023–1035.
Gerchman, Y., Olami, Y., Rimon, A., Taglicht, D., Schuldiner, S., & Padan, E. (1993). Histidine-226 is part of the pH sensor of NhaA, a Na^+/H^+ antiporter in Escherichia coli. *Proceedings of the National Academy of Sciences of the United States of America, 90*, 1212–1216.
Girardi, A. C., Degray, B. C., Nagy, T., Biemesderfer, D., & Aronson, P. S. (2001). Association of Na(+)-H(+) exchanger isoform NHE3 and dipeptidyl peptidase IV in the renal proximal tubule. *Journal of Biological Chemistry, 276*, 46671–46677.
Gisler, S. M., Pribanic, S., Bacic, D., Forrer, P., Gantenbein, A., Sabourin, L. A., et al. (2003). PDZK1: I. A major scaffolder in brush borders of proximal tubular cells. *Kidney International, 64*, 1733–1745.
Goel, D. P., Ford, L. D., & Pierce, G. N. (2003). Lysophospholipids do not directly modulate Na^+/H^+ exchange. *Molecular and Cellular Biochemistry, 251*, 3–7.
Good, D. W., Di Mari, J. F., & Watts, B. A., III. (2000). Hyposmolality stimulates Na^+/H^+ exchange and $HCO3^-$ absorption in thick ascending limb via PI 3-kinase. *AJP - Cell Physiology, 279*, C1443–C1454.
Goswami, P., Paulino, C., Hizlan, D., Vonck, J., Yildiz, O., & Kuhlbrandt, W. (2011). Structure of the archaeal Na+/H+antiporter NhaP1 and functional role of transmembrane helix 1. *The EMBO Journal, 30*, 439–449.
Goujon, M., McWilliam, H., Li, W., Valentin, F., Squizzato, S., Paern, J., et al. (2010). A new bioinformatics analysis tools framework at EMBL-EBI. *Nucleic Acids Research, 38*, W695–W699.
Grenier, A. L., bu-Ihweij, K., Zhang, G., Ruppert, S. M., Boohaker, R., Slepkov, E. R., et al. (2008). Apoptosis-induced alkalinization by the Na+/H+exchanger isoform 1 is mediated through phosphorylation of amino acids Ser726 and Ser729. *AJP - Cell Physiology, 295*, C883–C896.
Grinstein, S., Cohen, S., Goetz, J. D., & Rothstein, A. (1985). Na^+/H^+ exchange in volume regulation and cytoplasmic pH homeostasis in lymphocytes. *Federation Proceedings, 44*, 2508–2512.
Grinstein, S., Rothstein, A., & Cohen, S. (1985). Mechanism of osmotic activation of Na^+/H^+ exchange in rat thymic lymphocytes. *The Journal of General Physiology, 85*, 765–787.
Hallows, K. R., Law, F. Y., Packman, C. H., & Knauf, P. A. (1996). Changes in cytoskeletal actin content, F-actin distribution, and surface morphology during HL-60 cell volume regulation. *Journal of Cellular Physiology, 167*, 60–71.
Hallows, K. R., Packman, C. H., & Knauf, P. A. (1991). Acute cell volume changes in anisotonic media affect F-actin content of HL-60 cells. *The American Journal of Physiology, 261*, C1154–C1161.
Han, W., Kim, K. H., Jo, M. J., Lee, J. H., Yang, J., Doctor, R. B., et al. (2006). Shank2 associates with and regulates Na^+/H^+ exchanger 3. *Journal of Biological Chemistry, 281*, 1461–1469.

Harada, K., Fukuda, E., Hirohashi, N., & Chiba, K. (2010). Regulation of intracellular pH by p90Rsk-dependent activation of an Na^{+}/H^{+} exchanger in starfish oocytes. *Journal of Biological Chemistry*, *285*, 24044–24054.

Hayashi, H., Szaszi, K., Coady-Osberg, N., Orlowski, J., Kinsella, J. L., & Grinstein, S. (2002). A slow pH-dependent conformational transition underlies a novel mode of activation of the epithelial Na+/H+exchanger-3 isoform. *Journal of Biological Chemistry*, *277*, 11090–11096.

Hegan, P. S., Giral, H., Levi, M., & Mooseker, M. S. (2012). Myosin VI is required for maintenance of brush border structure, composition, and membrane trafficking functions in the intestinal epithelial cell. *Cytoskeleton (Hoboken N.J.)*, *69*, 235–251.

Hellmer, J., Teubner, A., & Zeilinger, C. (2003). Conserved arginine and aspartate residues are critical for function of MjNhaP1, a Na^{+}/H^{+} antiporter of M. jannaschii. *FEBS Letters*, *542*, 32–36.

Hensley, C. B., Bradley, M. E., & Mircheff, A. K. (1989). Parathyroid hormone-induced translocation of Na-H antiporters in rat proximal tubules. *The American Journal of Physiology*, *257*, C637–C645.

Hirokawa, T., Boon-Chieng, S., & Mitaku, S. (1998). SOSUI: Classification and secondary structure prediction system for membrane proteins. *Bioinformatics*, *14*, 378–379.

Hisamitsu, T., Ammar, Y. B., Nakamura, T. Y., & Wakabayashi, S. (2006). Dimerization is crucial for the function of the Na^{+}/H^{+} exchanger NHE1. *Biochemistry*, *45*, 13346–13355.

Hisamitsu, T., Nakamura, T. Y., & Wakabayashi, S. (2012). Na^{+}/H^{+} exchanger 1 directly binds to calcineurin A and activates downstream NFAT signaling, leading to cardiomyocyte hypertrophy. *Molecular and Cellular Biology*, *32*, 3265–3280.

Hisamitsu, T., Pang, T., Shigekawa, M., & Wakabayashi, S. (2004). Dimeric interaction between the cytoplasmic domains of the Na^{+}/H^{+} exchanger NHE1 revealed by symmetrical intermolecular cross-linking and selective co-immunoprecipitation. *Biochemistry*, *43*, 11135–11143.

Hisamitsu, T., Yamada, K., Nakamura, T. Y., & Wakabayashi, S. (2007). Functional importance of charged residues within the putative intracellular loops in pH regulation by Na^{+}/H^{+} exchanger NHE1. *The FEBS Journal*, *274*, 4326–4335.

Hoffmann, E. K., Lambert, I. H., & Pedersen, S. F. (2009). Physiology of cell volume regulation in vertebrates. *Physiological Reviews*, *89*, 193–277.

Huang, Y., Lemieux, M. J., Song, J., Auer, M., & Wang, D. N. (2003). Structure and mechanism of the glycerol-3-phosphate transporter from Escherichia coli. *Science*, *301*, 616–620.

Hunte, C., Screpanti, E., Venturi, M., Rimon, A., Padan, E., & Michel, H. (2005). Structure of a Na^{+}/H^{+} antiporter and insights into mechanism of action and regulation by pH. *Nature*, *435*, 1197–1202.

Iakoucheva, L. M., Radivojac, P., Brown, C. J., O'Connor, T. R., Sikes, J. G., Obradovic, Z., et al. (2004). The importance of intrinsic disorder for protein phosphorylation. *Nucleic Acids Research*, *32*, 1037–1049.

Ikeda, T., Schmitt, B., Pouyssegur, J., Wakabayashi, S., & Shigekawa, M. (1997). Identification of cytoplasmic subdomains that control pH-sensing of the Na^{+}/H^{+} exchanger (NHE1): pH-maintenance, ATP-sensitive, and flexible loop domains. *Journal of Biochemistry*, *121*, 295–303.

Inoue, H., Nakamura, Y., Nagita, M., Takai, T., Masuda, M., Nakamura, N., et al. (2003). Calcineurin homologous protein isoform 2 (CHP2), Na^{+}/H^{+} exchangers-binding protein, is expressed in intestinal epithelium. *Biological & Pharmaceutical Bulletin*, *26*, 148–155.

Janecki, A. J., Montrose, M. H., Zimniak, P., Zweibaum, A., Tse, C. M., Khurana, S., et al. (1998). Subcellular redistribution is involved in acute regulation of the brush border Na^{+}/H^{+} exchanger isoform 3 in human colon adenocarcinoma cell line Caco-2. Protein kinase C-mediated inhibition of the exchanger. *Journal of Biological Chemistry*, *273*, 8790–8798.

Jayasundar, J. J., Ju, J. H., He, L., Liu, D., Meilleur, F., Zhao, J., et al. (2012). Open conformation of ezrin bound to phosphatidylinositol 4,5-bisphosphate and to F-actin revealed by neutron scattering. *Journal of Biological Chemistry, 287*, 37119–37133.

Jones, D. T. (1999). Protein secondary structure prediction based on position-specific scoring matrices. *Journal of Molecular Biology, 292*, 195–202.

Joseph, C., Twombley, K., Gattineni, J., Zhang, Q., Dwarakanath, V., & Baum, M. (2012). Acid increases NHE8 surface expression and activity in NRK cells. *American Journal of Physiology. Renal Physiology, 302*, F495–F503.

Jung, Y. S., Kim, H. Y., Kim, J., Lee, M. G., Pouyssegur, J., & Kim, E. (2008). Physical interactions and functional coupling between Daxx and sodium hydrogen exchanger 1 in ischemic cell death. *The Journal of Biological Chemistry, 283*, 1018–1025.

Juretic, D., Zoranic, L., & Zucic, D. (2002). Basic charge clusters and predictions of membrane protein topology. *Journal of Chemical Information and Computer Sciences, 42*, 620–632.

Kapus, A., Grinstein, S., Wasan, S., Kandasamy, R., & Orlowski, J. (1994). Functional characterization of three isoforms of the Na^+/H^+ exchanger stably expressed in Chinese hamster ovary cells. ATP dependence, osmotic sensitivity, and role in cell proliferation. *The Journal of Biological Chemistry, 269*, 23544–23552.

Karki, P., Li, X., Schrama, D., & Fliegel, L. (2011). B-Raf associates with and activates the NHE1 isoform of the Na^+/H^+ exchanger. *The Journal of Biological Chemistry, 286*, 13096–13105.

Katragadda, M., Alderfer, J. L., & Yeagle, P. L. (2001). Assembly of a polytopic membrane protein structure from the solution structures of overlapping peptide fragments of bacteriorhodopsin. *Biophysical Journal, 81*, 1029–1036.

Kellenberger, S., & Schild, L. (2002). Epithelial sodium channel/degenerin family of ion channels: A variety of functions for a shared structure. *Physiological Reviews, 82*, 735–767.

Khadilkar, A., Iannuzzi, P., & Orlowski, J. (2001). Identification of sites in the second exomembrane loop and ninth transmembrane helix of the mammalian Na^+/H^+ exchanger important for drug recognition and cation translocation. *Journal of Biological Chemistry, 276*, 43792–43800.

Khaled, A. R., Moor, A. N., Li, A., Kim, K., Ferris, D. K., Muegge, K., et al. (2001). Trophic factor withdrawal: p38 mitogen-activated protein kinase activates NHE1, which induces intracellular alkalinization. *Molecular and Cellular Biology, 21*, 7545–7557.

Kim, E. (2009). Functional and physical interaction between human lactate dehydrogenase B and Na+/H+ exchanger isoform 1. *Animal Cells and Systems, 13*, 283–288.

Kim, J., Jung, Y. S., Han, W., Kim, M. Y., Namkung, W., Lee, B. H., et al. (2007). Pharmacodynamic characteristics and cardioprotective effects of new NHE1 inhibitors. *European Journal of Pharmacology, 567*, 131–138.

Kim, J. H., Lee-Kwon, W., Park, J. B., Ryu, S. H., Yun, C. H., & Donowitz, M. (2002). Ca(2+)-dependent inhibition of Na^+/H^+ exchanger 3 (NHE3) requires an NHE3-E3KARP-alpha-actinin-4 complex for oligomerization and endocytosis. *Journal of Biological Chemistry, 277*, 23714–23724.

Kjaergaard, M., Nørholm, A. B., Hendus-Altenburger, R., Pedersen, S. F., Poulsen, F. M., & Kragelund, B. B. (2010). Temperature dependent structural changes in intrinsically disordered proteins: Formation of alpha-helices or loss of polyproline II? *Protein Science, 19*, 1555–1564.

Kocinsky, H. S., Girardi, A. C., Biemesderfer, D., Nguyen, T., Mentone, S., Orlowski, J., et al. (2005). Use of phospho-specific antibodies to determine the phosphorylation of endogenous Na+/H+exchanger NHE3 at PKA consensus sites. *American Journal of Physiology. Renal Physiology, 289*, F249–F258.

Köster, S., Pavkov-Keller, T., Kuhlbrandt, W., & Yildiz, O. (2011). Structure of human Na+/H+exchanger NHE1 regulatory region in complex with calmodulin and Ca^{2+}. *Journal of Biological Chemistry, 286*, 40954–40961.

Krogh, A., Larsson, B., von Heijne, H., & Sonnhammer, E. L. (2001). Predicting transmembrane protein topology with a hidden Markov model: Application to complete genomes. *Journal of Molecular Biology, 305*, 567–580.

Kumar, S., Ma, B., Tsai, C. J., Sinha, N., & Nussinov, R. (2000). Folding and binding cascades: Dynamic landscapes and population shifts. *Protein Science, 9*, 10–19.

Kurashima, K., D'Souza, S., Szaszi, K., Ramjeesingh, R., Orlowski, J., & Grinstein, S. (1999). The apical Na^+/H^+ exchanger isoform NHE3 is regulated by the actin cytoskeleton. *The Journal of Biological Chemistry, 274*, 29843–29849.

Kurashima, K., Szabo, E. Z., Lukacs, G., Orlowski, J., & Grinstein, S. (1998). Endosomal recycling of the Na^+/H^+ exchanger NHE3 isoform is regulated by the phosphatidylinositol 3-kinase pathway. *Journal of Biological Chemistry, 273*, 20828–20836.

Kurashima, K., Yu, F. H., Cabado, A. G., Szabo, E. Z., Grinstein, S., & Orlowski, J. (1997). Identification of sites required for down-regulation of Na^+/H^+ exchanger NHE3 activity by cAMP-dependent protein kinase. Phosphorylation-dependent and -independent mechanisms. *Journal of Biological Chemistry, 272*, 28672–28679.

Kusumoto, K., Igata, H., Abe, A., Ikeda, S., Tsuboi, A., Imamiya, E., et al. (2002). In vitro and in vivo pharmacology of a structurally novel Na^+/H^+ exchange inhibitor, T-162559. *British Journal of Pharmacology, 135*, 1995–2003.

Lacroix, J., Poet, M., Huc, L., Morello, V., Djerbi, N., Ragno, M., et al. (2008). Kinetic analysis of the regulation of the Na^+/H^+ exchanger NHE-1 by osmotic shocks. *Biochemistry, 47*, 13674–13685.

Lacroix, J., Poet, M., Maehrel, C., & Counillon, L. (2004). A mechanism for the activation of the Na/H exchanger NHE-1 by cytoplasmic acidification and mitogens. *EMBO Reports, 5*, 91–96.

Lagana, A., Vadnais, J., Le, P. U., Nguyen, T. N., Laprade, R., Nabi, I. R., et al. (2000). Regulation of the formation of tumor cell pseudopodia by the Na(+)/H(+) exchanger NHE1. *Journal of Cell Science, 113*(Pt 20), 3649–3662.

Lamprecht, G., Weinman, E. J., & Yun, C. H. (1998). The role of NHERF and E3KARP in the cAMP-mediated inhibition of NHE3. *Journal of Biological Chemistry, 273*, 29972–29978.

Landau, M., Herz, K., Padan, E., & Ben-Tal, N. (2007). Model structure of the Na^+/H^+ exchanger 1 (NHE1): Functional and clinical implications. *The Journal of Biological Chemistry, 282*, 37854–37863.

Lawrence, S. P., Holman, G. D., & Koumanov, F. (2010). Translocation of the Na^+/H^+ exchanger 1 (NHE1) in cardiomyocyte responses to insulin and energy-status signalling. *The Biochemical Journal, 432*, 515–523.

Lee, B. L., Li, X., Liu, Y., Sykes, B. D., & Fliegel, L. (2009a). Structural and functional analysis of extracellular loop 2 of the Na^+/H^+ exchanger. *Biochimica et Biophysica Acta, 1788*, 2481–2488.

Lee, B. L., Li, X., Liu, Y., Sykes, B. D., & Fliegel, L. (2009b). Structural and functional analysis of transmembrane XI of the NHE1 isoform of the Na^+/H^+ exchanger. *The Journal of Biological Chemistry, 284*, 11546–11556.

Lee, B. L., Liu, Y., Li, X., Sykes, B. D., & Fliegel, L. (2012). Structural and functional analysis of extracellular loop 4 of the Nhe1 isoform of the Na^+/H^+ exchanger. *Biochimica et Biophysica Acta, 1818*, 2783–2790.

Lee, B. L., Sykes, B. D., & Fliegel, L. (2011). Structural analysis of the Na^+/H^+ exchanger isoform 1 (NHE1) using the divide and conquer approach. *Biochemistry and Cell Biology, 89*, 189–199.

Lehoux, S., Abe, J., Florian, J. A., & Berk, B. C. (2001). 14-3-3 Binding to Na^+/H^+ exchanger isoform-1 is associated with serum-dependent activation of Na^+/H^+ exchange. *Journal of Biological Chemistry, 276*, 15794–15800.

Levine, S. A., Montrose, M. H., Tse, C. M., & Donowitz, M. (1993). Kinetics and regulation of three cloned mammalian Na^+/H^+ exchangers stably expressed in a fibroblast cell line. *Journal of Biological Chemistry, 268*, 25527–25535.
Li, X., Alvarez, B., Casey, J. R., Reithmeier, R. A., & Fliegel, L. (2002). Carbonic anhydrase II binds to and enhances activity of the Na^+/H^+ exchanger. *Journal of Biological Chemistry, 277*, 36085–36091.
Li, X., Liu, Y., Alvarez, B. V., Casey, J. R., & Fliegel, L. (2006). A novel carbonic anhydrase II binding site regulates NHE1 activity. *Biochemistry, 45*, 2414–2424.
Li, S., Sato, S., Yang, X., Preisig, P. A., & Alpern, R. J. (2004). Pyk2 activation is integral to acid stimulation of sodium/hydrogen exchanger 3. *The Journal of Clinical Investigation, 114*, 1782–1789.
Li, J., Xu, L., Ye, J., Li, X., Zhang, D., Liang, D., et al. (2013). Aberrant dynamin 2-dependent Na/H exchanger-1 trafficking contributes to cardiomyocyte apoptosis. *Journal of Cellular and Molecular Medicine, 17*, 1119–1127.
Lin, X., & Barber, D. L. (1996). A calcineurin homologous protein inhibits GTPase-stimulated Na-H exchange. *Proceedings of the National Academy of Sciences of the United States of America, 93*, 12631–12636.
Linding, R., Jensen, L. J., Diella, F., Bork, P., Gibson, T. J., & Russell, R. B. (2003). Protein disorder prediction: Implications for structural proteomics. *Structure, 11*, 1453–1459.
Liu, H., Stupak, J., Zheng, J., Keller, B. O., Brix, B. J., Fliegel, L., et al. (2004). Open tubular immobilized metal ion affinity chromatography combined with MALDI MS and MS/MS for identification of protein phosphorylation sites. *Analytical Chemistry, 76*, 4223–4232.
Liu, T., & Jose, P. A. (2013). Gastrin induces sodium-hydrogen exchanger 3 phosphorylation and mTOR activation via a phosphoinositide 3-kinase-/protein kinase C-dependent but AKT-independent pathway in renal proximal tubule cells derived from a normotensive male human. *Endocrinology, 154*, 865–875.
Lorrain, J., Briand, V., Favennec, E., Duval, N., Grosset, A., Janiak, P., et al. (2000). Pharmacological profile of SL 59.1227, a novel inhibitor of the sodium/hydrogen exchanger. *British Journal of Pharmacology, 131*, 1188–1194.
Lukashova, V., Szabo, E. Z., Jinadasa, T., Mokhov, A., Litchfield, D. W., & Orlowski, J. (2011). CK2 phosphorylation of an acidic Ser/Thr di-isoleucine motif in the Na^+/H^+ exchanger NHE5 isoform promotes association with beta-arrestin2 and endocytosis. *Journal of Biological Chemistry, 286*, 11456–11468.
Lundby, A., Secher, A., Lage, K., Nordsborg, N. B., Dmytriyev, A., Lundby, C., et al. (2012). Quantitative maps of protein phosphorylation sites across 14 different rat organs and tissues. *Nature Communications, 3*, 876.
MacGregor, G. G., Dong, K., Vanoye, C. G., Tang, L., Giebisch, G., & Hebert, S. C. (2002). Nucleotides and phospholipids compete for binding to the C terminus of KATP channels. *Proceedings of the National Academy of Sciences of the United States of America, 99*, 2726–2731.
Mailander, J., Muller-Esterl, W., & Dedio, J. (2001). Human homolog of mouse tescalcin associates with Na^+/H^+ exchanger type-1. *FEBS Letters, 507*, 331–335.
Malapert, M., Pellissier, B., & Borgese, F. (1998). Asn49 is the unique glycosylation site of the trout red blood cell Na^+/H^+ exchanger. *European Journal of Biochemistry, 257*, 228–235.
Malo, M. E., Li, L., & Fliegel, L. (2007). Mitogen-activated protein kinase-dependent activation of the Na^+/H^+ exchanger is mediated through phosphorylation of amino acids Ser770 and Ser771. *The Journal of Biological Chemistry, 282*, 6292–6299.
Martin, C., Pedersen, S. F., Schwab, A., & Stock, C. (2011). Intracellular pH gradients in migrating cells. *American Journal of Physiology. Cell Physiology, 300*, C490–C495.
Masereel, B., Pochet, L., & Laeckmann, D. (2003). An overview of inhibitors of Na^+/H^+ exchanger. *European Journal of Medicinal Chemistry, 38*, 547–554.

Matsushita, M., Tanaka, H., Mitsui, K., & Kanazawa, H. (2011). Dual functional significance of calcineurin homologous protein 1 binding to Na+/H+exchanger isoform 1. *American Journal of Physiology. Cell Physiology, 301*, C280–C288.

Meima, M. E., Webb, B. A., Witkowska, H. E., & Barber, D. (2009). The Na-H exchanger NHE1 is an Akt substrate necessary for actin filament reorganization by growth factors. *The Journal of Biological Chemistry, 284*, 26666–26675.

Mishima, M., Wakabayashi, S., & Kojima, C. (2007). Solution structure of the cytoplasmic region of Na^+/H^+ exchanger 1 complexed with essential cofactor calcineurin B homologous protein 1. *The Journal of Biological Chemistry, 282*, 2741–2751.

Misik, A. J., Perreault, K., Holmes, C. F., & Fliegel, L. (2005). Protein phosphatase regulation of Na^+/H^+ exchanger isoform I. *Biochemistry, 44*, 5842–5852.

Mohan, S., Tse, C. M., Gabelli, S. B., Sarker, R., Cha, B., Fahie, K., et al. (2010). NHE3 activity is dependent on direct phosphoinositide binding at the N terminus of its intracellular cytosolic region. *Journal of Biological Chemistry, 285*, 34566–34578.

Moncoq, K., Kemp, G., Li, X., Fliegel, L., & Young, H. S. (2008). Dimeric structure of human Na^+/H^+ exchanger isoform 1 overproduced in Saccharomyces cerevisiae. *The Journal of Biological Chemistry, 283*, 4145–4154.

Montrose, M. H., & Murer, H. (1988). Kinetics of Na+/H+exchange. In *Na+/H+exchange* (pp. 58–75). Boca Raton, FL: CRC Press.

Moses, A. M., Heriche, J. K., & Durbin, R. (2007). Clustering of phosphorylation site recognition motifs can be exploited to predict the targets of cyclin-dependent kinase. *Genome Biology, 8*, R23.

Murtazina, R., Booth, B. J., Bullis, B. L., Singh, D. N., & Fliegel, L. (2001). Functional analysis of polar amino-acid residues in membrane associated regions of the NHE1 isoform of the mammalian Na^+/H^+ exchanger. *European Journal of Biochemistry, 268*, 4674–4685.

Murtazina, R., Kovbasnjuk, O., Chen, T. E., Zachos, N. C., Chen, Y., Kocinsky, H. S., et al. (2011). NHERF2 is necessary for basal activity, second messenger inhibition, and LPA stimulation of NHE3 in mouse distal ileum. *American Journal of Physiology. Cell Physiology, 301*, C126–C136.

Musch, M. W., Arvans, D. L., Walsh-Reitz, M. M., Uchiyama, K., Fukuda, M., & Chang, E. B. (2007). Synaptotagmin I binds intestinal epithelial NHE3 and mediates cAMP- and Ca^{2+}-induced endocytosis by recruitment of AP2 and clathrin. *The American Journal of Physiology, 292*, G1549–G1558.

Nasuhoglu, C., Feng, S., Mao, Y., Shammat, I., Yamamato, M., Earnest, S., et al. (2002). Modulation of cardiac PIP2 by cardioactive hormones and other physiologically relevant interventions. *AJP - Cell Physiology, 283*, C223–C234.

Nath, S. K., Hang, C. Y., Levine, S. A., Yun, C. H., Montrose, M. H., Donowitz, M., et al. (1996). Hyperosmolarity inhibits the Na^+/H^+ exchanger isoforms NHE2 and NHE3: An effect opposite to that on NHE1. *The American Journal of Physiology, 270*, G431–G441.

Nielsen, D. K., Jensen, A. K., Harbak, H., Christensen, S. C., & Simonsen, L. O. (2007). Cell content of phosphatidylinositol (4,5)bisphosphate in Ehrlich mouse ascites tumour cells in response to cell volume perturbations in anisotonic and in isosmotic media. *The Journal of Physiology, 582*, 1027–1036.

Nørholm, A. B., Hendus-Altenburger, R., Bjerre, G., Kjaergaard, M., Pedersen, S. F., & Kragelund, B. B. (2011). The intracellular distal tail of the Na^+/H^+ exchanger NHE1 is intrinsically disordered: Implications for NHE1 trafficking. *Biochemistry, 50*, 3469–3480.

Nunomura, W., Denker, S. P., Barber, D. L., Takakuwa, Y., & Gascard, P. (2012). Characterization of cytoskeletal protein 4.1R interaction with NHE1 (Na+/H+exchanger isoform 1). *The Biochemical Journal, 446*, 427–435.

Nygaard, E. B., Lagerstedt, J. O., Bjerre, G., Shi, B., Budamagunta, M., Poulsen, K. A., et al. (2011). Structural modeling and electron paramagnetic resonance spectroscopy of the human Na^+/H^+ exchanger isoform 1, NHE1. *The Journal of Biological Chemistry, 286*, 634–648.

Odunewu, A., & Fliegel, L. (2013). Acidosis-mediated regulation of the NHE1 isoform of the Na+/H+exchanger in renal cells. *American Journal of Physiology. Renal Physiology, 305*, F370–F381.

Ohgaki, R., Fukura, N., Matsushita, M., Mitsui, K., & Kanazawa, H. (2008). Cell surface levels of organellar Na^+/H^+ exchanger isoform 6 are regulated by interaction with RACK1. *Journal of Biological Chemistry, 283*, 4417–4429.

Olkhova, E., Hunte, C., Screpanti, E., Padan, E., & Michel, H. (2006). Multiconformation continuum electrostatics analysis of the NhaA Na^+/H^+ antiporter of Escherichia coli with functional implications. *Proceedings of the National Academy of Sciences of the United States of America, 103*, 2629–2634.

Olsen, J. V., Blagoev, B., Gnad, F., Macek, B., Kumar, C., Mortensen, P., et al. (2006). Global, in vivo, and site-specific phosphorylation dynamics in signaling networks. *Cell, 127*, 635–648.

Onishi, I., Lin, P. J., Diering, G. H., Williams, W. P., & Numata, M. (2007). RACK1 associates with NHE5 in focal adhesions and positively regulates the transporter activity. *Cellular Signalling, 19*, 194–203.

Orlowski, J. (1993). Heterologous expression and functional properties of amiloride high affinity (NHE-1) and low affinity (NHE-3) isoforms of the rat Na/H exchanger. *Journal of Biological Chemistry, 268*, 16369–16377.

Orlowski, J., & Grinstein, S. (2004). Diversity of the mammalian sodium/proton exchanger SLC9 gene family. *Pflügers Archiv, 447*, 549–565.

Orlowski, J., & Grinstein, S. (2011). Na+/H+exchangers. *Comprehensive Physiology, 1*, 2083–2100.

Orlowski, J., & Kandasamy, R. A. (1996). Delineation of transmembrane domains of the Na^+/H^+ exchanger that confer sensitivity to pharmacological antagonists. *Journal of Biological Chemistry, 271*, 19922–19927.

Orlowski, J., Kandasamy, R. A., & Shull, G. E. (1992). Molecular cloning of putative members of the Na/H exchanger gene family. cDNA cloning, deduced amino acid sequence, and mRNA tissue expression of the rat Na/H exchanger NHE-1 and two structurally related proteins. *Journal of Biological Chemistry, 267*, 9331–9339.

Otsu, K., Kinsella, J. L., Heller, P., & Froehlich, J. P. (1993). Sodium dependence of the Na^+/H^+ exchanger in the pre-steady state. Implications for the exchange mechanism. *Journal of Biological Chemistry, 268*, 3184–3193.

Otsu, K., Kinsella, J., Sacktor, B., & Froehlich, J. P. (1989). Transient state kinetic evidence for an oligomer in the mechanism of Na^+/H^+ exchange. *Proceedings of the National Academy of Sciences of the United States of America, 86*, 4818–4822.

Padan, E., Kozachkov, L., Herz, K., & Rimon, A. (2009). NhaA crystal structure: Functional-structural insights. *The Journal of Experimental Biology, 212*, 1593–1603.

Padan, E., Tzubery, T., Herz, K., Kozachkov, L., Rimon, A., & Galili, L. (2004). NhaA of Escherichia coli, as a model of a pH-regulated Na^+/H^+ antiporter. *Biochimica et Biophysica Acta, 1658*, 2–13.

Pang, T., Hisamitsu, T., Mori, H., Shigekawa, M., & Wakabayashi, S. (2004). Role of calcineurin B homologous protein in pH regulation by the Na^+/H^+ exchanger 1: Tightly bound Ca^{2+} ions as important structural elements. *Biochemistry, 43*, 3628–3636.

Pang, T., Su, X., Wakabayashi, S., & Shigekawa, M. (2001). Calcineurin homologous protein as an essential cofactor for Na^+/H^+ exchangers. *Journal of Biological Chemistry, 276*, 17367–17372.

Paris, S., & Pouyssegur, J. (1984). Growth factors activate the Na^+/H^+ antiporter in quiescent fibroblasts by increasing its affinity for intracellular H+. *Journal of Biological Chemistry, 259*, 10989–10994.

Park, J. H., Ryu, J. M., Yun, S. P., Kim, M. O., & Han, H. J. (2012). Fibronectin stimulates migration through lipid raft dependent NHE-1 activation in mouse embryonic stem cells: Involvement of RhoA, Ca^{2+}/CaM, and ERK. *Biochimica et Biophysica Acta, 1820*, 1618–1627.

Parker, M. D., & Boron, W. F. (2013). The divergence, actions, roles, and relatives of sodium-coupled bicarbonate transporters. *Physiological Reviews*, *93*, 803–959.

Pedersen, S. F. (2005). A novel NHE1 from red blood cells of the winter flounder: Regulation by multiple signaling pathways. In P. K. Lauf & N. Adragna (Eds.), *Cell volume and signaling. Advances in experimental medicine and biology* (pp. 89–99). New York, NY: Springer Science+Business Media, Inc.

Pedersen, S. F., Beisner, K. H., Hougaard, C., Willumsen, B. M., Lambert, I. H., & Hoffmann, E. K. (2002). Rho family GTP binding proteins are involved in the regulatory volume decrease process in NIH3T3 mouse fibroblasts. *The Journal of Physiology*, *541*, 779–796.

Pedersen, S. F., Darborg, B. V., Rentsch, M. L., & Rasmussen, M. (2007). Regulation of mitogen-activated protein kinase pathways by the plasma membrane Na^+/H^+ exchanger, NHE1. *Archives of Biochemistry and Biophysics*, *462*, 195–201.

Pedersen, S. F., King, S. A., Nygaard, E. B., Rigor, R. R., & Cala, P. M. (2007). NHE1 inhibition by amiloride- and benzoylguanidine-type compounds. Inhibitor binding loci deduced from chimeras of NHE1 homologues with endogenous differences in inhibitor sensitivity. *The Journal of Biological Chemistry*, *282*, 19716–19727.

Pedersen, S. F., Mills, J. W., & Hoffmann, E. K. (1999). Role of the F-actin cytoskeleton in the RVD and RVI processes in Ehrlich ascites tumor cells. *Experimental Cell Research*, *252*, 63–74.

Pedersen, S. F., O'Donnell, M. E., Anderson, S. E., & Cala, P. M. (2006). Physiology and pathophysiology of Na+/H+exchange and Na+-K+-2Cl- cotransport in the heart, brain, and blood. *American Journal of Physiology. Regulatory, Integrative and Comparative Physiology*, *291*, R1–R25.

Qadri, Y. J., Song, Y., Fuller, C. M., & Benos, D. J. (2010). Amiloride docking to acid-sensing ion channel-1. *The Journal of Biological Chemistry*, *285*, 9627–9635.

Rasmussen, M., Alexander, R. T., Darborg, B. V., Mobjerg, N., Hoffmann, E. K., Kapus, A., et al. (2008). Osmotic cell shrinkage activates ezrin/radixin/moesin (ERM) proteins: Activation mechanisms and physiological implications. *AJP - Cell Physiology*, *294*, C197–C212.

Reddy, T., Ding, J., Li, X., Sykes, B. D., Rainey, J. K., & Fliegel, L. (2008). Structural and functional characterization of TM IX of the NHE1 isoform of the Na^+/H^+ exchanger. *The Journal of Biological Chemistry*, *283*, 22018–22030.

Rentsch, M. L., Ossum, C. G., Hoffmann, E. K., & Pedersen, S. F. (2007). Roles of Na^+/H^+ exchange in regulation of p38 mitogen-activated protein kinase activity and cell death after chemical anoxia in NIH3T3 fibroblasts. *Pflügers Archiv*, *454*, 649–662.

Rigor, R. R., Damoc, C., Phinney, B. S., & Cala, P. M. (2011). Phosphorylation and activation of the plasma membrane Na^+/H^+ exchanger (NHE1) during osmotic cell shrinkage. *PLoS One*, *6*, e29210.

Rimon, A., Tzubery, T., & Padan, E. (2007). Monomers of the NhaA Na^+/H^+ antiporter of Escherichia coli are fully functional yet dimers are beneficial under extreme stress conditions at alkaline pH in the presence of Na^+ or Li^+. *Journal of Biological Chemistry*, *282*, 26810–26821.

Rizoli, S. B., Rotstein, O. D., Parodo, J., Phillips, M. J., & Kapus, A. (2000). Hypertonic inhibition of exocytosis in neutrophils: Central role for osmotic actin skeleton remodeling. *AJP - Cell Physiology*, *279*, C619–C633.

Rosner, H. I., & Kragelund, B. B. (2012). Structure and dynamic properties of membrane proteins using NMR. *Comprehensive Physiology*, *2*, 1491–1539.

Rothman, A., Gerchman, Y., Padan, E., & Schuldiner, S. (1997). Probing the conformation of NhaA, a Na^+/H^+ antiporter from Escherichia coli, with trypsin. *Biochemistry*, *36*, 14572–14576.

Rothman, A., Padan, E., & Schuldiner, S. (1996). Topological analysis of NhaA, a Na^+/H^+ antiporter from Escherichia coli. *Journal of Biological Chemistry*, *271*, 32288–32292.

Sardet, C., Franchi, A., & Pouyssegur, J. (1989). Molecular cloning, primary structure, and expression of the human growth factor-activatable Na^+/H^+ antiporter. *Cell, 56*, 271–280.
Sarker, R., Gronborg, M., Cha, B., Mohan, S., Chen, Y., Pandey, A., et al. (2008). Casein kinase 2 binds to the C terminus of Na+/H+exchanger 3 (NHE3) and stimulates NHE3 basal activity by phosphorylating a separate site in NHE3. *Molecular Biology of the Cell, 19*, 3859–3870.
Sato, Y., Ariyoshi, N., Mihara, K., & Sakaguchi, M. (2004). Topogenesis of NHE1: Direct insertion of the membrane loop and sequestration of cryptic glycosylation and processing sites just after TM9. *Biochemical and Biophysical Research Communications, 324*, 281–287.
Schneider, L., Stock, C. M., Dieterich, P., Jensen, B. H., Pedersen, L. B., Satir, P., et al. (2009). The Na+/H+ exchanger NHE1 is required for directional migration stimulated via PDGFR-alpha in the primary cilium. *The Journal of Cell Biology, 185*, 163–176.
Schushan, M., Rimon, A., Haliloglu, T., Forrest, L. R., Padan, E., & Ben-Tal, N. (2012). A model-structure of a periplasm-facing state of the NhaA antiporter suggests the molecular underpinnings of pH-induced conformational changes. *Journal of Biological Chemistry, 287*, 18249–18261.
Schwark, J. R., Jansen, H. W., Lang, H. J., Krick, W., Burckhardt, G., & Hropot, M. (1998). S3226, a novel inhibitor of Na^+/H^+ exchanger subtype 3 in various cell types. *Pflügers Archiv, 436*, 797–800.
Schweiger, R., & Linial, M. (2010). Cooperativity within proximal phosphorylation sites is revealed from large-scale proteomics data. *Biology Direct, 5*, 6.
Screpanti, E., & Hunte, C. (2007). Discontinuous membrane helices in transport proteins and their correlation with function. *Journal of Structural Biology, 159*, 261–267.
Shimada-Shimizu, N., Hisamitsu, T., Nakamura, T. Y., & Wakabayashi, S. (2013). Evidence that Na^+/H^+ exchanger 1 is an ATP-binding protein. *The FEBS Journal, 280*, 1430–1442.
Shimada-Shimizu, N., Hisamitsu, T., Nakamura, T. Y., Hirayama, N., & Wakabayashi, S. (2014). Na^+/H^+ exchanger 1 is regulated via its lipid-interacting domain which functions as a molecular switch: A pharmacological approach using indolocarbazole compounds. *Molecular Pharmacology, 85*, 18–28.
Shimamura, T., Weyand, S., Beckstein, O., Rutherford, N. G., Hadden, J. M., Sharples, D., et al. (2010). Molecular basis of alternating access membrane transport by the sodium-hydantoin transporter Mhp1. *Science, 328*, 470–473.
Silva, N. L., Haworth, R. S., Singh, D., & Fliegel, L. (1995). The carboxyl-terminal region of the Na^+/H^+ exchanger interacts with mammalian heat shock protein. *Biochemistry, 34*, 10412–10420.
Shrode, L. D., Gan, B. S., D'Souza, S. J., Orlowski, J., & Grinstein, S. (1998). Topological analysis of NHE1, the ubiquitous Na^+/H^+ exchanger using chymotryptic cleavage. *The American Journal of Physiology, 275*, C431–C439.
Sigalov, A. B., & Uversky, V. N. (2011). Differential occurrence of protein intrinsic disorder in the cytoplasmic signaling domains of cell receptors. *Self Nonself, 2*, 55–72.
Simonin, A., & Fuster, D. (2010). Nedd4-1 and beta-arrestin-1 are key regulators of Na^+/H^+ exchanger 1 ubiquitylation, endocytosis, and function. *Journal of Biological Chemistry, 285*, 38293–38303.
Slepkov, E. R., Rainey, J. K., Li, X., Liu, Y., Cheng, F. J., Lindhout, D. A., et al. (2005). Structural and functional characterization of transmembrane segment IV of the NHE1 isoform of the Na^+/H^+ exchanger. *The Journal of Biological Chemistry, 280*, 17863–17872.
Snabaitis, A. K., Cuello, F., & Avkiran, M. (2008). Protein kinase B/Akt phosphorylates and inhibits the cardiac Na^+/H^+ exchanger NHE1. *Circulation Research, 103*, 881–890.
Solcan, N., Kwok, J., Fowler, P. W., Cameron, A. D., Drew, D., Iwata, S., et al. (2012). Alternating access mechanism in the POT family of oligopeptide transporters. *The EMBO Journal, 31*, 3411–3421.

Soleimani, M., Bookstein, C., McAteer, J. A., Hattabaugh, Y. J., Bizal, G. L., Musch, M. W., et al. (1994). Effect of high osmolality on Na^+/H^+ exchange in renal proximal tubule cells. *Journal of Biological Chemistry, 269*, 15613–15618.
Su, X., Pang, T., Wakabayashi, S., & Shigekawa, M. (2003). Evidence for involvement of the putative first extracellular loop in differential volume sensitivity of the Na^+/H^+ exchangers NHE1 and NHE2. *Biochemistry, 42*, 1086–1094.
Szabo, E. Z., Numata, M., Lukashova, V., Iannuzzi, P., & Orlowski, J. (2005). beta-Arrestins bind and decrease cell-surface abundance of the Na+/H+exchanger NHE5 isoform. *Proceedings of the National Academy of Sciences of the United States of America, 102*, 2790–2795.
Takahashi, E., Abe, J., Gallis, B., Aebersold, R., Spring, D. J., Krebs, E. G., et al. (1999). p90 (RSK) is a serum-stimulated Na+/H+exchanger isoform-1 kinase. Regulatory phosphorylation of serine 703 of Na+/H+exchanger isoform-1. *Journal of Biological Chemistry, 274*, 20206–20214.
Tekpli, X., Huc, L., Lacroix, J., Rissel, M., Poet, M., Noel, J., et al. (2008). Regulation of Na+/H+ exchanger 1 allosteric balance by its localization in cholesterol- and caveolin-rich membrane microdomains. *Journal of Cellular Physiology, 216*, 207–220.
Tominaga, T., Ishizaki, T., Narumiya, S., & Barber, D. L. (1998). p160ROCK mediates RhoA activation of Na-H exchange. *The EMBO Journal, 17*, 4712–4722.
Tompa, P. (2002). Intrinsically unstructured proteins. *Trends in Biochemical Sciences, 27*, 527–533.
Tompa, P., & Fuxreiter, M. (2008). Fuzzy complexes: Polymorphism and structural disorder in protein-protein interactions. *Trends in Biochemical Sciences, 33*, 2–8.
Touret, N., Poujeol, P., & Counillon, L. (2001). Second-site revertants of a low-sodium-affinity mutant of the Na^+/H^+ exchanger reveal the participation of TM4 into a highly constrained sodium-binding site. *Biochemistry, 40*, 5095–5101.
Touret, N., Tanneur, V., Godart, H., Seidler, R., Taki, N., Burger, E., et al. (2003). Characterization of sabiporide, a new specific NHE-1 inhibitor exhibiting slow dissociation kinetics and cardioprotective effects. *European Journal of Pharmacology, 459*, 151–158.
Tse, C. M., Levine, S. A., Yun, C. H., Khurana, S., & Donowitz, M. (1994). Na^+/H^+ exchanger-2 is an O-linked but not an N-linked sialoglycoprotein. *Biochemistry, 33*, 12954–12961.
Tsukita, S., & Yonemura, S. (1999). Cortical actin organization: Lessons from ERM (ezrin/radixin/moesin) proteins. *The Journal of Biological Chemistry, 274*, 34507–34510.
Tzeng, J., Lee, B. L., Sykes, B. D., & Fliegel, L. (2010). Structural and functional analysis of transmembrane segment VI of the NHE1 isoform of the Na^+/H^+ exchanger. *Journal of Biological Chemistry, 285*, 36656–36665.
Tzubery, T., Rimon, A., & Padan, E. (2004). Mutation E252C increases drastically the Km value for Na+and causes an alkaline shift of the pH dependence of NhaA Na^+/H^+ antiporter of Escherichia coli. *Journal of Biological Chemistry, 279*, 3265–3272.
Tzubery, T., Rimon, A., & Padan, E. (2008). Structure-based functional study reveals multiple roles of transmembrane segment IX and loop VIII-IX in NhaA Na^+/H^+ antiporter of Escherichia coli at physiological pH. *Journal of Biological Chemistry, 283*, 15975–15987.
Vacic, V., Oldfield, C. J., Mohan, A., Radivojac, P., Cortese, M. S., Uversky, V. N., et al. (2007). Characterization of molecular recognition features, MoRFs, and their binding partners. *Journal of Proteome Research, 6*, 2351–2366.
Vinothkumar, K. R., Smits, S. H., & Kuhlbrandt, W. (2005). pH-induced structural change in a sodium/proton antiporter from Methanococcus jannaschii. *The EMBO Journal, 24*, 2720–2729.
Waight, A. B., Pedersen, B. P., Schlessinger, A., Bonomi, M., Chau, B. H., Roe-Zurz, Z., et al. (2013). Structural basis for alternating access of a eukaryotic calcium/proton exchanger. *Nature, 499*, 107–110.

Wakabayashi, S., Bertrand, B., Ikeda, T., Pouyssegur, J., & Shigekawa, M. (1994). Mutation of calmodulin-binding site renders the Na^+/H^+ exchanger (NHE1) highly H^+-sensitive and Ca^{2+} regulation-defective. *Journal of Biological Chemistry, 269*, 13710–13715.

Wakabayashi, S., Bertrand, B., Shigekawa, M., Fafournoux, P., & Pouyssegur, J. (1994). Growth factor activation and "H^+-sensing" of the Na^+/H^+ exchanger isoform 1 (NHE1). Evidence for an additional mechanism not requiring direct phosphorylation. *Journal of Biological Chemistry, 269*, 5583–5588.

Wakabayashi, S., Hisamitsu, T., Pang, T., & Shigekawa, M. (2003a). Kinetic dissection of two distinct proton binding sites in Na^+/H^+ exchangers by measurement of reverse mode reaction. *Journal of Biological Chemistry, 278*, 43580–43585.

Wakabayashi, S., Hisamitsu, T., Pang, T., & Shigekawa, M. (2003b). Mutations of Arg440 and Gly455/Gly456 oppositely change pH sensing of Na^+/H^+ exchanger 1. *Journal of Biological Chemistry, 278*, 11828–11835.

Wakabayashi, S., Ikeda, T., Iwamoto, T., Pouyssegur, J., & Shigekawa, M. (1997). Calmodulin-binding autoinhibitory domain controls "pH-sensing" in the Na^+/H^+ exchanger NHE1 through sequence-specific interaction. *Biochemistry, 36*, 12854–12861.

Wakabayashi, S., Nakamura, T. Y., Kobayashi, S., & Hisamitsu, T. (2010). Novel phorbol ester-binding motif mediates hormonal activation of Na^+/H^+ exchanger. *The Journal of Biological Chemistry, 285*, 26652–26661.

Wakabayashi, S., Pang, T., Su, X., & Shigekawa, M. (2000). A novel topology model of the human Na^+/H^+ exchanger isoform 1. *Journal of Biological Chemistry, 275*, 7942–7949.

Wakabayashi, S., Shigekawa, M., & Pouyssegur, J. (1997). Molecular physiology of vertebrate Na^+/H^+ exchangers. *Physiological Reviews, 77*, 51–74.

Wang, D., Balkovetz, D. F., & Warnock, D. G. (1995). Mutational analysis of transmembrane histidines in the amiloride-sensitive Na^+/H^+ exchanger. *The American Journal of Physiology, 269*, C392–C402.

Wang, H., Silva, N. L., Lucchesi, P. A., Haworth, R., Wang, K., Michalak, M., et al. (1997). Phosphorylation and regulation of the Na^+/H^+ exchanger through mitogen-activated protein kinase. *Biochemistry, 36*, 9151–9158.

Wang, D., Sun, H., Lang, F., & Yun, C. C. (2005). Activation of NHE3 by dexamethasone requires phosphorylation of NHE3 at Ser663 by SGK1. *American Journal of Physiology. Cell Physiology, 289*, C802–C810.

Ward, J. J., McGuffin, L. J., Bryson, K., Buxton, B. F., & Jones, D. T. (2004). The DISOPRED server for the prediction of protein disorder. *Bioinformatics, 20*, 2138–2139.

Watts, B. A., III, & Good, D. W. (1999). Hyposmolality stimulates apical membrane Na^+/H^+ exchange and HCO_3^- absorption in renal thick ascending limb. *The Journal of Clinical Investigation, 104*, 1593–1602.

Weinman, E. J., Minkoff, C., & Shenolikar, S. (2000). Signal complex regulation of renal transport proteins: NHERF and regulation of NHE3 by PKA. *American Journal of Physiology. Renal Physiology, 279*, F393–F399.

Wiebe, C. A., Dibattista, E. R., & Fliegel, L. (2001). Functional role of polar amino acid residues in Na^+/H^+ exchangers. *The Biochemical Journal, 357*, 1–10.

Wiederkehr, M. R., Zhao, H., & Moe, O. W. (1999). Acute regulation of Na/H exchanger NHE3 activity by protein kinase C: Role of NHE3 phosphorylation. *The American Journal of Physiology, 276*, C1205–C1217.

Williams, K. A. (2000). Three-dimensional structure of the ion-coupled transport protein NhaA. *Nature, 403*, 112–115.

Williams, K. A., Geldmacher-Kaufer, U., Padan, E., Schuldiner, S., & Kuhlbrandt, W. (1999). Projection structure of NhaA, a secondary transporter from Escherichia coli, at 4.0 A resolution. *The EMBO Journal, 18*, 3558–3563.

Wu, K. L., Khan, S., Lakhe-Reddy, S., Jarad, G., Mukherjee, A., Obejero-Paz, C. A., et al. (2004). The NHE1 Na^+/H^+ exchanger recruits ezrin/radixin/moesin proteins to regulate Akt-dependent cell survival. *The Journal of Biological Chemistry, 279*, 26280–26286.

Xue, B., Dunbrack, R. L., Williams, R. W., Dunker, A. K., & Uversky, V. N. (2010). PONDR-FIT: A meta-predictor of intrinsically disordered amino acids. *Biochimica et Biophysica Acta*, *1804*, 996–1010.

Xue, J., Zhou, D., Yao, H., Gavrialov, O., McConnell, M. J., Gelb, B. D., et al. (2007). Novel functional interaction between Na+/H+exchanger 1 and tyrosine phosphatase SHP-2. *American Journal of Physiology Regulatory, Integrative and Comparative Physiology*, *292*, R2406–R2416.

Yamamoto, M., Chen, M. Z., Wang, Y. J., Sun, H. Q., Wei, Y., Martinez, M., et al. (2006). Hypertonic stress increases PI(4,5)P2 levels by activating PIP5KIbeta. *The Journal of Biological Chemistry*, *281*, 32630–32638.

Yan, W., Nehrke, K., Choi, J., & Barber, D. L. (2001). The Nck-interacting kinase (NIK) phosphorylates the Na^+/H^+ exchanger NHE1 and regulates NHE1 activation by platelet-derived growth factor. *Journal of Biological Chemistry*, *276*, 31349–31356.

Yi, Y. H., Ho, P. Y., Chen, T. W., Lin, W. J., Gukassyan, V., Tsai, T. H., et al. (2009). Membrane targeting and coupling of NHE1-integrinalphaIIbβ3-NCX1 by lipid rafts following integrin-ligand interactions trigger Ca^{2+} oscillations. *The Journal of Biological Chemistry*, *284*, 3855–3864.

Yonemura, S., Hirao, M., Doi, Y., Takahashi, N., Kondo, T., Tsukita, S., et al. (1998). Ezrin/radixin/moesin (ERM) proteins bind to a positively charged amino acid cluster in the juxta-membrane cytoplasmic domain of CD44, CD43, and ICAM-2. *The Journal of Cell Biology*, *140*, 885–895.

Yu, F. H., Shull, G. E., & Orlowski, J. (1993). Functional properties of the rat Na/H exchanger NHE-2 isoform expressed in Na/H exchanger-deficient Chinese hamster ovary cells. *Journal of Biological Chemistry*, *268*, 25536–25541.

Yun, C. H., Lamprecht, G., Forster, D. V., & Sidor, A. (1998). NHE3 kinase A regulatory protein E3KARP binds the epithelial brush border Na^+/H^+ exchanger NHE3 and the cytoskeletal protein ezrin. *Journal of Biological Chemistry*, *273*, 25856–25863.

Yun, C. H., Little, P. J., Nath, S. K., Levine, S. A., Pouyssegur, J., Tse, C. M., et al. (1993). Leu143 in the putative fourth membrane spanning domain is critical for amiloride inhibition of an epithelial Na^+/H^+ exchanger isoform (NHE-2). *Biochemical and Biophysical Research Communications*, *193*, 532–539.

Zachos, N. C., van Rossum, D. B., Li, X., Caraveo, G., Sarker, R., Cha, B., et al. (2009). Phospholipase C-gamma binds directly to the Na^+/H^+ exchanger 3 and is required for calcium regulation of exchange activity. *Journal of Biological Chemistry*, *284*, 19437–19444.

Zaun, H. C., Shrier, A., & Orlowski, J. (2008). Calcineurin B homologous protein 3 promotes the biosynthetic maturation, cell surface stability, and optimal transport of the Na^+/H^+ exchanger NHE1 isoform. *The Journal of Biological Chemistry*, *283*, 12456–12467.

Zaun, H. C., Shrier, A., & Orlowski, J. (2012). N-myristoylation and Ca^{2+} binding of calcineurin B homologous protein CHP3 are required to enhance Na^+/H^+ exchanger NHE1 half-life and activity at the plasma membrane. *Journal of Biological Chemistry*, *287*, 36883–36895.

Zhao, H., Wiederkehr, M. R., Fan, L., Collazo, R. L., Crowder, L. A., & Moe, O. W. (1999). Acute inhibition of Na/H exchanger NHE-3 by cAMP. Role of protein kinase a and NHE-3 phosphoserines 552 and 605. *Journal of Biological Chemistry*, *274*, 3978-3987.

Zizak, M., Chen, T., Bartonicek, D., Sarker, R., Zachos, N. C., Cha, B., et al. (2012). Calmodulin kinase II constitutively binds, phosphorylates, and inhibits brush border Na^+/H^+ exchanger 3 (NHE3) by a NHERF2 protein-dependent process. *Journal of Biological Chemistry*, *287*, 13442–13456.

CHAPTER THREE

Vesicular Neurotransmitter Transporters: Mechanistic Aspects

Christine Anne, Bruno Gasnier[1]
Université Paris Descartes, Sorbonne Paris Cité, Centre National de la Recherche Scientifique, Unité Mixte de Recherche 8192, Centre Universitaire des Saints-Pères, Paris, France
[1]Corresponding author: e-mail address: bruno.gasnier@parisdescartes.fr

Contents

Abstract

Secondary transporters driven by a V-type H^+-ATPase accumulate nonpeptide neurotransmitters into synaptic vesicles. Distinct transporter families are involved depending on the neurotransmitter. Monoamines and acetylcholine on the one hand, and glutamate and ATP on the other hand, are accumulated by SLC18 and SLC17 transporters, respectively, which belong to the major facilitator superfamily (MFS). GABA and glycine accumulate through a common SLC32 transporter from the amino acid/polyamine/organocation (APC) superfamily. Although crystallographic structures are not yet available for any vesicular transporter, homology modeling studies of MFS-type vesicular transporters based on distantly related bacterial structures recently provided significant advances, such as the characterization of substrate-binding pockets or the identification of spatial clusters acting as hinge points during the alternating-access cycle. However, several basic issues, such as the ion stoichiometry of vesicular amino acid transporters, remain unsettled.

1. INTRODUCTION

Neurotransmission requires the controlled release of neurotransmitters at synapses by exocytosis of synaptic vesicles. Released transmitters then diffuse across the synaptic cleft and bind to cognate receptors on the

Current Topics in Membranes, Volume 73
ISSN 1063-5823
http://dx.doi.org/10.1016/B978-0-12-800223-0.00003-7

postsynaptic neuron. They are eventually cleared from the extracellular medium by neuronal and glial plasma membrane transporters. In the case of classical (nonpeptide) transmitters, the presynaptic steps include the biosynthesis of transmitter molecules in the cytosol, or their reuptake from the extracellular medium, followed by their storage into synaptic vesicles. This chapter focuses on the transporters responsible for the vesicular uptake of transmitters.

Vesicular neurotransmitter transporters are fueled by a V-type H^+-ATPase, which builds up an electric potential (positive inside) across the vesicular membrane and acidifies the vesicle lumen to a pH value of ~5.5. Active vesicular uptake generally involves the exchange of cytosolic transmitters for luminal protons (Fig. 3.1), but this mechanism is only established for monoamine and acetylcholine transmitters. The bioenergetics of vesicular uptake remains debated for glutamate and inhibitory amino acids.

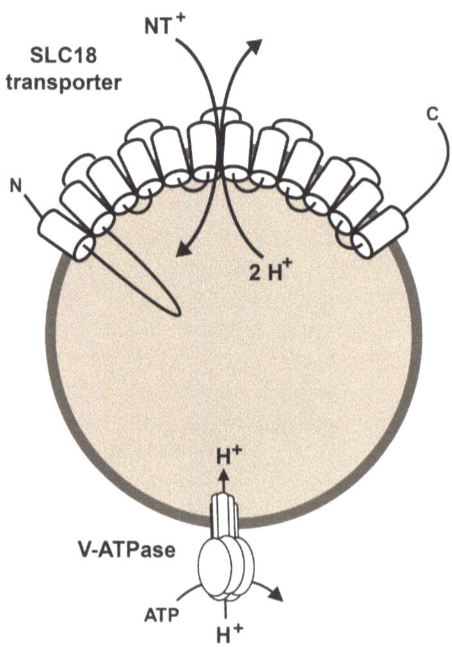

Figure 3.1 Nonpeptide neurotransmitters are accumulated into synaptic vesicles by secondary transporters that use the H^+ gradient established by a V-type H^+-ATPase. In the case of SLC18 proteins, which are responsible for the vesicular uptake of monoamines or acetylcholine, the cationic transmitter is exchanged for two luminal protons. The ion-coupling mechanism is still debated for glutamate and inhibitory amino acid transporters. (See the color plate.)

The biology of vesicular neurotransmitter transporters and their role in synaptic transmission have been covered in several recent reviews (Edwards, 2007; El Mestikawy, Wallen-Mackenzie, Fortin, Descarries, & Trudeau, 2011; Hnasko & Edwards, 2012). We focus here on transport mechanisms, in agreement with the scope of this volume. Functional studies of vesicular transporters are hindered by technical difficulties (e.g., poor accessibility and low substrate affinity) and the lack of any crystallographied homologous protein. Therefore, the mechanistic knowledge of vesicular transporters is scarce in comparison with that for other neurotransmitter transporters such as SLC1 and SLC6 (Boudker & Verdon, 2010; Krishnamurthy, Piscitelli, & Gouaux, 2009). However, some vesicular transporters have been subjected to extensive site-directed mutagenesis, and the emergence of experimentally validated 3D homology models offers a way to translate these structure–function data into specific structural hypotheses.

2. CLASSIFICATION OF VESICULAR NEUROTRANSMITTER TRANSPORTERS

Three transporter families, named SLC17, SLC18, and SLC32 according to the human genome nomenclature, are involved in vesicular uptake (Table 3.1). In mammals, the SLC17 family comprises the following three vesicular glutamate transporters (VGLUTs): VGLUT1, -2, and -3, coded by genes *SLC17A7*, *SLC17A6*, and *SLC17A8*, respectively (Bellocchio, Reimer, Fremeau, & Edwards, 2000; Edwards, 2007; El Mestikawy et al., 2011; Takamori, Rhee, Rosenmund, & Jahn, 2000). VGLUT1 and VGLUT2 ensure the vesicular uptake of glutamate at most excitatory synapses, whereas VGLUT3 is expressed in very limited number of neurons that usually show dual neurotransmitter specificity. A homologous vesicular nucleotide transporter (VNUT, human gene *SLC17A9*) responsible for the release of ATP at purinergic synapses was recently identified (Sawada et al., 2008). The SLC17 family also comprises four transporters of organic anions, including one for urate that is expressed in peripheral tissues (Reimer, 2013) and one for sialic acids (named sialin, human gene *SLC17A5*) expressed in lysosomes (Morin, Sagné, & Gasnier, 2004; Verheijen et al., 1999). Sialin has also been reported to act as a nitrate transporter at the plasma membrane (Qin et al., 2012), and as a transporter for aspartate and the dipeptide *n*-acetyl-aspartyl-glutamate into synaptic vesicles (Lodder-Gadaczek, Gieselmann, & Eckhardt, 2013;

Table 3.1 Phylogenetic classification and functional properties of vesicular neurotransmitter transporters

Transporter	Superfamily	Human gene[a]	Transporter classification database family[b]	Substrates	Inhibitors	Ion coupling
VMAT1	Major facilitator (MFS)	*SLC18A1*	2.A.1.2 Drug:H^+ antiporter-1 (DHA1)	Monoamines	Reserpine	$2H^+$/amine$^+$ antiport
VMAT2	MFS	*SLC18A2*	2.A.1.2	Monoamines, histamine	Reserpine, tetrabenazine	$2H^+$/amine$^+$ antiport
VAChT	MFS	*SLC18A3*	2.A.1.2	Acetylcholine (ACh)	Vesamicol	$2H^+$/ACh$^+$ antiport
VGLUT1	MFS	*SLC17A7*	2.A.1.14 Anion:cation symporter (ACS)	Glutamate (Glu)	Azo dyes, Rose Bengal	Unclear
VGLUT2	MFS	*SLC17A6*	2.A.1.14	Glu	Azo dyes, Rose Bengal	Unclear
VGLUT3	MFS	*SLC17A8*	2.A.1.14	Glu	Azo dyes, Rose Bengal	Unclear
Sialin	MFS	*SLC17A5*	2.A.1.14	Sialic acid nitrate, Glu, aspartate *N*-acetyl-Asp-Glu		Substrate dependent
VNUT	MFS	*SLC17A9*	2.A.1.14	ATP	Unknown	Unknown
VIAAT	Amino acid–polyamine–organocation (APC)	*SLC32A1*	2.A.18.5	GABA, glycine	Unknown	Unclear

[a]See http://slc.bioparadigms.org/.
[b]See http://www.tcdb.org/.

Miyaji et al., 2008). However, see Morland et al. (2012) for a diverging view on the aspartatergic function.

The SLC18 family includes two vesicular monoamine transporters (VMAT1 and VMAT2, encoded by genes *SLC18A1* and *SLC18A2*, respectively) (Erickson, Eiden, & Hoffman, 1992; Liu et al., 1992) and a vesicular acetylcholine transporter (VAChT; *SLC18A3* gene) (Alfonso, Grundahl, Duerr, Han, & Rand, 1993). Biogenic monoamines include serotonin, dopamine, noradrenalin, adrenalin, histamine, and, in invertebrates, tyramine and octopamine. Invertebrates have a single VMAT gene (Duerr et al., 1999; Greer et al., 2005). Interestingly, the VAChT gene is nested in an intron of the gene for choline acetyltransferase, the synthetic enzyme of acetylcholine. This nesting feature is conserved in vertebrates and invertebrates (Alfonso et al., 1993; Bejanin, Cervini, Mallet, & Berrard, 1994; Roghani et al., 1994). An additional putative neurotransmitter transporter, SLC18A4 or portabella, has been discovered in Drosophila, but its substrate is currently unknown (Brooks et al., 2011).

Finally, the SLC32 family comprises a single member, the vesicular inhibitory amino acid transporter (VIAAT, also named VGAT for vesicular GABA transporter), which packages GABA and/or glycine into inhibitory synaptic vesicles (McIntire, Reimer, Schuske, Edwards, & Jorgensen, 1997; Sagné, El Mestikawy, et al., 1997).

Notably, the uptake of D-serine by an unknown protein has been recently described in vesicles purified from cultured astrocytes (Martineau et al., 2013). Therefore, the list of vesicular transporters for signaling molecules may extend in the future.

SLC17 and SLC18 transporters belong to the major facilitator superfamily (MFS), the largest group of secondary active transporters in prokaryotic and eukaryotic species (Reddy, Shlykov, Castillo, Sun, & Saier, 2012). MFS proteins are responsible for the passive or active transport across cellular and intracellular membranes of a wide diversity of compounds, including inorganic ions, sugars, amino acids, and xenobiotics. Several prokaryotic members, including *Escherichia coli* lactose permease, have been intensively studied (Guan & Kaback, 2006; Law, Maloney, & Wang, 2008; Yan, 2013). VIAAT belongs to a distinct superfamily, the amino acid/polyamine/organocation (APC) superfamily, which also includes SLC6 transporters and their prokaryotic homologue LeuT (Krishnamurthy et al., 2009; Wong et al., 2012). The structure and mechanism of VIAAT should thus greatly differ from those of VMATs, VAChT, and VGLUTs.

3. NEUROTRANSMITTER TRANSPORTERS FROM THE MFS

3.1. Monoamine and acetylcholine transporters (SLC18)

3.1.1 Putative structure

VMATs and VAChT are ~500-amino acid polypeptides consisting of 12 transmembrane helices (TMs), with N and C termini in the cytosol, and a large *N*-glycosylated loop between the first and second TMs. VMAT1 and VMAT2 share ~60% amino acid identity, and they are ~40% identical to VAChT. No crystallographic structure is available yet for SLC18 proteins, but their distant homology (~20% identity) to crystallographied MFS proteins has allowed investigators to generate 3D homology models (Vardy, Arkin, Gottschalk, Kaback, & Schuldiner, 2004; Yaffe, Radestock, Shuster, Forrest, & Schuldiner, 2013).

All reported MFS structures share a common architecture in which the first six and last six TMs form two compact bundles surrounding a single central binding site (Law et al., 2008; Yan, 2013). This site opens to one side of the membrane through a funnel-shaped aqueous cavity, whereas its access is blocked on the other side by contacts between the N and C domains. Rocker-switch movement of these domains alternately exposes the substrate-binding site to either side of the membrane with, at least in some transporters, an intermediate occluded state in which the site is insulated from both cytosolic and extracytosolic compartments (Law et al., 2008; Yan, 2013).

Interestingly, MFS structures display symmetry at two levels (Huang, Lemieux, Song, Auer, & Wang, 2003; Radestock & Forrest, 2011). First, the N and C domains are related by a pseudo twofold symmetry with an axis running normal to the membrane through the center of the transporter. Second, within each domain, the first three TMs are structurally related to the next three TMs by a twofold symmetry axis running parallel to the membrane through the center of the domain. The substrate translocation pathway is lined by the first two TMs of each 3-TM repeat (TMs 1, 2, 4, 5, 7, 8, 10, and 11), while remaining TMs 3, 6, 9, and 12 are located on the peripheral side of the N and C halves. Swapping the conformations of the two 3-TM bundles within each domain provides an elegant way to predict an outward-open conformation from a known inward-open structure, and vice versa (Radestock & Forrest, 2011).

Two 3D homology models of VMAT2 based on known prokaryotic MFS structures have been published (Vardy et al., 2004; Yaffe et al.,

2013). In the most recent study, multiple MFS sequences and conserved MFS motifs were used to improve sequence alignment between VMAT2 and the *E. coli* lactose permease template (Abramson et al., 2003). This altered the position of several TMs relative to that in the previous 3D model, but in good agreement with earlier biochemical data (see below). Interestingly, this study also revealed the existence of two clusters of interactions between the N and C domains, involving hydrogen bonds between K139 and Q143 in TM2, and D427 in TM11 on one hand, and hydrophobic interactions between V233 and L234 in TM5, and F335 and L336 in TM8 on the other hand (Fig. 3.2). The existence of these clusters is essential for monoamine transport (Yaffe et al., 2013). K139 and D427 had been previously suggested to form an ion pair (Merickel, Kaback, & Edwards, 1997);

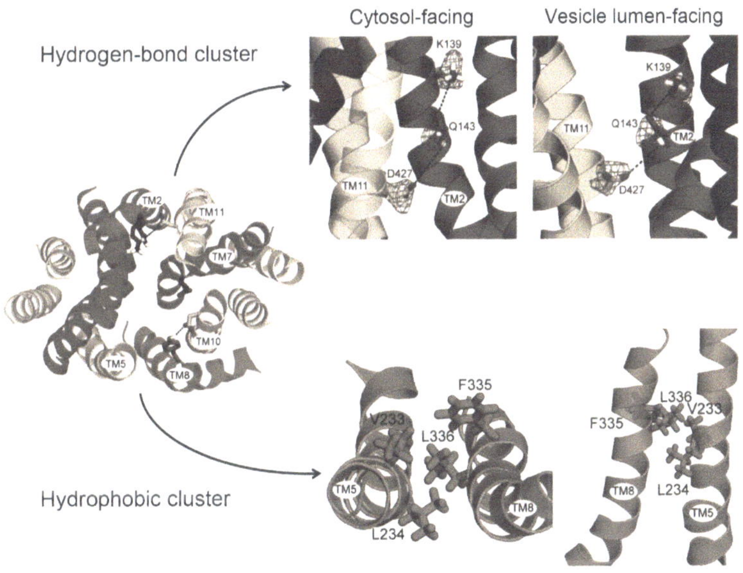

Figure 3.2 Identification of hinge points between the N and C domains of VMAT2 by homology modeling. A model of the cytosol-facing conformation of rat VMAT2 viewed from the cytosol is shown on the left. Predicted spatial clusters of interactions between the N and C domains are shown in more detail on the right. Upper panels: The hydrogen-bond cluster is showed in cytosol-facing (left) and lumen-facing (right) states. Pictures are views along the plane of the membrane with the cytosol toward the bottom. Alternating access preserves the cluster. Lower panels: The hydrophobic cluster is shown from the cytosol (left) and along the plane of the membrane (right). Side chains of key residues are shown as sticks. Predicted interactions are shown as dashed lines. *Reproduced with permission from Yaffe et al. (2013).* (See the color plate.)

however, novel mutagenesis experiments guided by the 3D model confirmed that their interaction is mediated by Q143 (Yaffe et al., 2013).

These investigators predicted the alternate conformation of VMAT using the aforementioned repeat-swapping technique. Remarkably, although TM2, TM5, TM8, and TM11 undergo large conformational changes between the two states, the two clusters remain essentially unchanged (Fig. 3.2). It was thus proposed that these clusters act as hinge points on which the N and C domains rotate during the rocker-switch transitions of the transport cycle (Yaffe et al., 2013).

Another contact between the N and C domains of VMAT revealed by earlier biochemical studies is a disulfide bond linking the large luminal loop (residue C118 upstream of TM2 [rat VMAT2 numbering]) to the loop connecting TM7 and TM8 (C325) (Thiriot, Sievert, & Ruoho, 2002). However, disrupting this bond reduced, but did not abolish, transport. The disulfide bond of VMAT, which is not conserved in VAChT, may thus have a stabilizing, rather than functional, role.

A 3D homology model of VAChT has also been reported (Khare, Mulakaluri, & Parsons, 2010), but it was based on the former VMAT model (Vardy et al., 2004). Therefore, VAChT mutagenesis data should be reinterpreted in light of the new VMAT model (Yaffe et al., 2013). For instance, the Q143 residue of rat VMAT2 is conserved in VAChT (Q135). Therefore, the K131–D425 ion pair (rat sequence numbering) previously reported for VAChT between TM2 and TM11 (Kim, Lu, Kelly, & Hersh, 2000) is more likely a K131–Q143–D425 hydrogen-bond cluster as described in VMAT.

3.1.2 Ion coupling

The ion-coupling stoichiometry of VMATs has been determined by measuring the concentration gradient of monoamines at equilibrium in resealed secretory granules from bovine adrenal medulla (Johnson, Carty, & Scarpa, 1981; Knoth, Zallakian, & Njus, 1981), a preparation in which the VMAT2 isoform predominates (Gasnier, Krejci, Botton, Massoulie, & Henry, 1994; Howell et al., 1994). Combined with measurement of the vesicular voltage ($\Delta\Psi$) and pH (ΔpH) gradients using radiolabeled thiocyanate and methylamine, respectively, these experiments showed that the logarithm of the intravesicular to extravesicular monoamine ratio depends linearly on $F/2.3RT \times \Delta\Psi$ with a slope of 1 but depends linearly on ΔpH with a slope of 2. It was thus concluded that one cationic monoamine is exchanged for two intravesicular protons (Johnson et al., 1981; Knoth et al., 1981; Schuldiner, Shirvan, & Linial, 1995).

In the case of VAChT, equilibrium gradients could not be achieved (see below). The ion-coupling stoichiometry was thus indirectly determined with resealed synaptic vesicles from the *Torpedo* electric organ by fitting one-proton and two-proton models to the kinetics of acetylcholine uptake at different pH_{in} and pH_{out} values (Nguyen, Cox, & Parsons, 1998). The two-proton model provided the best fit, suggesting that the exchange of one cationic transmitter for two protons is conserved within the SLC18 family. The electrogenic two-proton model is also consistent with the fact that a positive electric polarization of the vesicles stimulates acetylcholine uptake (Nguyen et al., 1998).

With pH_{in} and pH_{out} values of 5.5 and ~7.2, and a $\Delta\Psi$ of ~60 mV (positive inside), the two-proton antiport mechanism of SLC18 transporters can build up a vesicle:cytosol concentration ratio of cationic transmitter of $\sim 3 \times 10^4$. In the case of monoamines, which are synthesized at micromolar concentrations in the cytosol, this ratio is consistent with the quantal release of ~3000–5000 molecules of transmitter, and an intravesicular concentration of ~250 mM in monoaminergic synaptic vesicles (Bruns & Jahn, 1995; Pothos, Davila, & Sulzer, 1998). Therefore, VMATs accumulate monoamines at thermodynamic equilibrium.

In contrast, VAChT should accumulate acetylcholine at a much lower level than the thermodynamic equilibrium to preserve the osmotic balance of synaptic vesicles because the cytosolic concentration of acetylcholine is in the millimolar range. Consistently, a quantal size of ~10,000 acetylcholine molecules has been determined at the neuromuscular junction (Van der Kloot & Molgo, 1994). The physiological requirement for a two-proton-coupling mechanism thus remains unclear in the case of VAChT.

3.1.3 Substrate and inhibitor binding

SLC18 transporters show broad substrate promiscuity. VMATs recognize not only structurally diverse biogenic amines but also amphetamines, meta-iodobenzylguanidine, the neurotoxin 1-methyl-4-phenylpyridinium, and bulkier synthetic amines, including fluorescent substrates used to monitor their activity in live neurons or brain slices (Gros & Schuldiner, 2010; Gubernator et al., 2009; Henry, Sagné, Bedet, & Gasnier, 1998; Schuldiner et al., 1995). Similarly, VAChT transports not only acetylcholine but also bulkier analogues and unrelated synthetic cations such as tetraphenylphosphonium (Bravo, Kolmakova, & Parsons, 2005b; Clarkson, Rogers, & Parsons, 1992). Even choline is transported, albeit with lower affinity than acetylcholine (Bravo, Kolmakova, & Parsons, 2004a).

The most selective VMAT inhibitors are reserpine and tetrabenazine. Reserpine binds at subnanomolar concentrations and its binding is accelerated by a proton electrochemical gradient, suggesting that binding or translocation of a proton is required to expose the high-affinity site (Darchen, Scherman, & Henry, 1989; Deupree & Weaver, 1984; Rudnick, Steiner-Mordoch, Fishkes, Stern-Bach, & Schuldiner, 1990; Scherman & Henry, 1984). Tetrabenazine binds preferentially to the VMAT2 isoform with nanomolar affinity, apparently to a site distinct from those for substrates and reserpine because its binding is inhibited by millimolar rather than micromolar concentrations of biogenic amines (Scherman, Jaudon, & Henry, 1983).

3.1.3.1 VMAT-binding site

To our knowledge, no ligand molecular docking has been performed on VMAT homology models. Therefore, delineating substrate and inhibitor-binding sites is limited by the poor resolution of biochemical approaches and classical (not structure-guided) mutagenesis. Photoactivatable derivatives of tetrabenazine and ketanserin—another ligand of the tetrabenazine site—labeled VMAT2 regions encompassing TM10 and TM11, and the N-terminus and TM1, respectively (Sagné, Isambert, et al., 1997; Sievert & Ruoho, 1997). These findings are consistent with the canonical architecture of MFS proteins.

Several groups performed mutagenesis studies to identify determinants of monoamine and tetrabenazine binding based on the fact that the VMAT1 and VMAT2 isoforms differ in this respect. A prominent finding was the requirement for an aromatic residue in TM11 (Y434, rat VMAT2 numbering) for high-affinity tetrabenazine binding (Finn & Edwards, 1997). Interestingly, another study found that a nearby endogenous cysteine (C431 in rat VMAT2) is protected from thiol reagents by tetrabenazine (Thiriot & Ruoho, 2001)[1]—a finding consistent with Y434 directly interacting with the inhibitor. Mutagenesis studies also identified a residue required for tetrabenazine binding in TM12 (D461) (Finn & Edwards, 1997). However, this result is difficult to reconcile with the structural role of TM12 in the MFS fold, and more consistent with an indirect effect.

More recently, an unbiased genetic screen of VMAT2 in yeast identified another residue required for tetrabenazine binding—F136 in TM2. An

[1] Residue numbering is shifted in this reference. The reported "C439" residue of human VMAT2 should read as C430.

aromatic side chain is essential at this position (Gros & Schuldiner, 2010). Interestingly, F136 lies about one helical turn above K139, a residue implicated in the aforementioned polar hinge point between the N and C domains (Yaffe et al., 2013). It might be interesting to examine whether a cation–pi interaction occurs between these two residues. If so, then tetrabenazine binding might be indirectly affected by a shift in the alternate access equilibrium. In a very recent study, Schuldiner and coworkers identified novel determinants of tetrabenazine binding using the yeast genetic screen targeted to specific regions of VMAT2. Interestingly, tetrabenazine-resistant mutants resulted from replacements at or near conserved helix-breaking residues (G308 and P314 in TM7, and V41 and V132 adjacent to P42, and G133 in TM1 and TM2, respectively), strongly suggesting that VMAT2 undergoes conformational changes to bind tetrabenazine (Ugolev, Segal, Yaffe, Gros, & Schuldiner, 2013).

Mutagenesis studies also identified residues associated with monoamine recognition. Mutation of Y434 in TM11 affects histamine, but not serotonin, recognition (Finn & Edwards, 1997). The yeast genetic screen of VMAT2 identified two neighboring mutations (I425F, V428A) that shift substrate selectivity toward toxic compounds (Gros & Schuldiner, 2010). Neutralizing an aspartate in TM1 (D33N) abolished serotonin inhibition of reserpine binding (Merickel, Rosandich, Peter, & Edwards, 1995), suggesting a role for D33 in monoamine binding or monoamine/H^+ coupling.

In several structurally known MFS proteins, substrate binding predominantly involves a single 6-TM domain (see Yan, 2013 for review). However, it is unclear from the above mutagenesis data whether this rule applies to VMAT. Molecular docking and structure-guided mutagenesis and, better, future structural studies should help delineate the monoamine-binding site and unveil the basis of VMAT substrate promiscuity.

3.1.3.2 Substrate and inhibitor binding to VAChT

As already mentioned, VAChT transports structurally diverse substrates (Bravo et al., 2005b). Vesamicol is the best-characterized inhibitor and binds to VAChT with nanomolar affinity (Bahr & Parsons, 1986). Similar to the interaction between tetrabenazine and monoamines, acetylcholine inhibits vesamicol binding at concentrations much higher (~200-fold) than those required to saturate transport, indicating allosteric interaction (Bahr, Clarkson, Rogers, Noremberg, & Parsons, 1992). Intriguingly, although the proton electrochemical gradient has no effect on vesamicol binding, it weakens its inhibition by acetylcholine (Bravo, Kolmakova, & Parsons, 2004b). This observation has been interpreted by a kinetic model in which the default orientation of VAChT faces the cytosol. In this model, the proton electrochemical gradient reorients acetylcholine-loaded, but not empty, VAChT molecules toward a lumen-facing state in which the allosteric interaction between acetylcholine and vesamicol is weaker (Bahr et al., 1992; Bravo et al., 2004b).

Mutagenesis studies have implicated several residues in vesamicol or acetylcholine binding. A C391Y, but not C391A, mutation in TM10 abolishes vesamicol sensitivity while preserving acetylcholine transport, suggesting steric hindrance between this residue and the larger inhibitor molecule (Zhu et al., 2001). The conservative mutation D398E in the same TM has a similar effect (Bravo, Kolmakova, & Parsons, 2005a; Kim, Lu, Lim, Chai, & Hersh, 1999). Mutation F335A in TM8 reduces affinity for vesamicol threefold without altering acetylcholine transport (Ojeda, Kolmakova, & Parsons, 2004). Conversely, mutations F220A and A228V in TM5, S252F in TM6, W331A and Y343A in TM8, and Y428A in TM11 substantially increase the K_M for acetylcholine without altering vesamicol binding (Khare, Mulakaluri, et al., 2010; Ojeda et al., 2004; Zhu et al., 2001). Mutation A334F decreases affinity for both acetylcholine and vesamicol (Khare, Mulakaluri, et al., 2010). The clustering of several of these residues in the VAChT homology model led to speculation that the acetylcholine-binding site has a "deep" location close to the vesicle lumen in contrast with the central binding site observed in known MFS structures (Khare, Mulakaluri, et al., 2010). However, this proposal awaits direct experimental evidence. Moreover, W331 is equivalent in VMAT2 to F335, which participates in the central hydrophobic hinge between the N and C domains in the revised homology model of VMAT2 (Yaffe et al., 2013).

3.1.4 Transport mechanism

A kinetic model has been proposed for the transport cycle of VAChT based on the bell-shaped pH dependence of acetylcholine transport, its saturation kinetics, and the interactions between acetylcholine and vesamicol (Bravo & Parsons, 2002; Parsons, 2000). According to this model, the two protons would be translocated at distinct steps of the cycle. One luminal proton would be exchanged for one cytosolic acetylcholine molecule during reorientation of the acetylcholine-loaded binding site toward the vesicle lumen, whereas another proton—bound to a distinct site—would be translocated during reorientation of empty VAChT toward the cytosol. These authors suggested that a similar kinetic model applies to VMAT (Parsons, 2000). To account for the intriguing concerted 1:1 H^+/acetylcholine exchange in the first reorientation step, they speculated that VAChT might possess two hemichannels open to opposite sides of the membrane, separated by a central rotating domain. The central domain would harbor one proton-binding site and the acetylcholine-binding site on opposite sides, and upon rotation around an axis perpendicular to the membrane, it would alternately expose each site to the lumen-open and cytosol-open hemichannels (Bravo & Parsons, 2002). This speculative scheme, somehow reminiscent of rotating ATPases, was proposed 1 year before the beginning of the transporter structural era. It is obviously hard to reconcile with known MFS structures! Then, is the concerted H^+/substrate exchange of the kinetic model compatible with current structural knowledge? We believe such a mechanism could be possible if we hypothesize that protons reach their binding site along hydrogen bonds (proton wires) rather than through the aqueous cavity between the N and C domains. The presence of proton wires in secondary transporters is suggested by structural and functional evidence (Pedersen et al., 2013; Ruivo et al., 2012). Such proton wires would provide independent pathways for the proton and the organic substrate, thereby allowing concerted opposite movements. For instance, in the cytosol-facing conformation of the transporter, the cytosolic substrate might diffuse to its site through the central aqueous cavity while a luminal proton would be transferred in opposite direction across the barrier formed by the contacting N and C domains on the luminal side. This proposal, however, remains speculative.

Molecular insight into the mechanism of VMATs and VAChT has been provided by mutagenesis studies. Mutation of H419 in the TM10/TM11 intervening loop of rat VMAT1 abolished monoamine transport, as well as ATP-dependent, but not ATP-independent, reserpine binding. Thus, this conserved histidine appears involved in H^+ coupling (Shirvan, Laskar, Steiner-Mordoch, & Schuldiner, 1994). However, this role is apparently specific to VMAT because mutation of the equivalent residue in VAChT (H413) had no effect on acetylcholine transport (Kim et al., 2000). A similar discrepancy between the two types of transporters occurs with a conserved glutamate located in the middle of TM7 and exposed to the central aqueous cavity according to the revised homology model of VMAT2. This residue is irreplaceable as conservative (E313D) and isosteric (E313Q) mutations impaired, or abolished, monoamine transport and tetrabenazine binding (Yaffe et al., 2013). In contrast, equivalent mutations in VAChT (E309D, E309Q) did not significantly alter acetylcholine transport (Bravo et al., 2005a; Khare, Ojeda, Chandrasekaran, & Parsons, 2010; Kim et al., 1999). The pH profile of vesamicol binding was strongly altered in the E309 mutants, suggesting indirect interactions with the inhibitor-binding site (Khare, Ojeda, et al., 2010). These apparent discrepancies between VMAT and VAChT suggest that the two types of transporters display significant differences in mechanism in addition to the neurotransmitter specificity.

Structure/function studies have also yielded convergent findings in VMAT and VAChT. The conserved aspartate in TM11 (D427 in rat VMAT2, D425 in VAChT) is irreplaceable for neurotransmitter transport in the three vesicular SLC18 proteins (Bravo et al., 2005a; Kim et al., 1999; Merickel et al., 1997; Steiner-Mordoch, Shirvan, & Schuldiner, 1996). On the other hand, conservative or nonconservative mutations at this position preserved ATP-dependent binding of reserpine to VMAT and vesamicol binding to VAChT, implying that this aspartate is required to complete the transport cycle. Inhibition of vesamicol binding by acetylcholine, a partial reaction believed to involve reorientation of VAChT, was also altered (Bravo et al., 2005a; Kim et al., 1999). Therefore, the TM11 aspartate may be required for conformational changes. The participation of this residue in the K139–Q143–D427 hydrogen bond cluster acting as a hinge point in VMAT2 (Fig. 3.2) provides a good rationale for these effects. In strong support to the existence of this cluster, restoring a negative charge in the bridging glutamine (Q143E) fully rescued monoamine transport in the dead mutant D427N (Yaffe et al., 2013). In VAChT, D425 was suggested to form an ion pair with K131 in TM2 (Bravo et al., 2005a). However, the presence of an equivalent glutamine residue one helix turn downstream of K131 suggests that a similar cluster occurs in VAChT.

Another aspartate residue with convergent roles in VMATs and VAChT is found in TM10 (D400 in rat VMAT2, D398 in VAChT). A negative charge is required at this position as nonconservative mutations abolished neurotransmitter transport in both types of transporters (Kim et al., 1999; Merickel et al., 1997; Steiner-Mordoch et al., 1996; Yaffe et al., 2013). However, the effect of specific mutations at this site on inhibitor binding depended on the type of SLC18 protein and the ligand. ATP-dependent reserpine binding to VMAT1 was abolished by nonconservative mutations, but fully preserved when this aspartate was replaced with glutamate (Steiner-Mordoch et al., 1996). In contrast, D400 mutations preserved tetrabenazine binding in VMAT2 (Steiner-Mordoch et al., 1996; Yaffe et al., 2013). In VAChT, D398 mutations, including D398E, abolished vesamicol binding (Khare, Ojeda, et al., 2010; Kim et al., 1999). Interestingly, a charge reversal mutation (H338D/D398H) between D398 and H338 in TM8 restored vesamicol binding in the D398 mutant, suggesting the presence of an ion pair between these residues (Kim et al., 2000). Homology modeling suggested a similar TM8/TM10 ion pair between D400 and Y342 in VMAT2. In agreement with this prediction, there is a "synthetic lethal" interaction between these residues—the Y342H/D400E double mutant has no activity even though Y342H and D400E single mutants robustly transport monoamines (Yaffe et al., 2013). Therefore, the interaction of the TM10 aspartate with a protonatable residue in TM8 is a conserved feature of VMATs and VAChT.

Another interesting feature of this aspartate is its potential involvement in proton binding and/or translocation, and possibly, in proton-coupled conformational changes. The D404E mutation in VMAT1 shifts to the acid side and strongly sharpens the bell-shaped pH dependence of monoamine transport (Steiner-Mordoch et al., 1996). Similarly, D398 in VAChT might correspond to the residue that must deprotonate for vesamicol binding (Khare, Ojeda, et al., 2010). Protonation/deprotonation of the TM10 aspartate may thus break the TM8/TM10 hydrogen bond during the transport cycle, similar to a paradigm observed in bacterial MFS transporters (Guan & Kaback, 2006; Law et al., 2008; Yan, 2013). Interestingly, the hydrogen-bonded residue in TM8 (Y342 in VMAT2) is distant by only two helix turns from the hydrophobic cluster (F335 and L336 for TM8; Fig. 3.2). Because of this proximity and the presumed hinge point role of the hydrophobic cluster, protonation/deprotonation of the TM10 aspartate may propagate to this hydrophobic cluster and couple proton translocation to alternating access (Yaffe et al., 2013). Further studies are needed to test this attractive model of ion binding/substrate translocation coupling.

3.2. Vesicular glutamate transporters (SLC17)

VGLUT polypeptides consist of 12 TMs with cytosolic N and C termini, but in contrast to VMATs and VAChT, the intravesicular loop between TM1 and TM2 is much shorter (~20 amino acids). The three VGLUT isoforms are highly similar (75–80% amino acid identity in humans), and their closest homologue in the SLC17 family is sialin (~40% identity and ~60% similarity). Although the K_M of VGLUTs for glutamate (~2 mM) is 100–1000 times higher than that of EAAT transporters at the plasma membrane, this value is approximately the cytosolic concentration of glutamate. VGLUTs do not recognize aspartate or glutamine, and they have a marked preference for L-glutamate over D-glutamate. Several inhibitors have been identified, including azo dyes such as Evans Blue and Rose Bengal which binds with 90 and 25 nM affinity, respectively (Shigeri, Seal, & Shimamoto, 2004; Thompson et al., 2005). Another unique feature of VGLUTs is their biphasic dependence on chloride (Bellocchio et al., 2000; Naito & Ueda, 1985).

In expression or reconstitution assays, VGLUTs also transport inorganic phosphate in a sodium-dependent manner (Aihara et al., 2000; Juge, Yoshida, Yatsushiro, Omote, & Moriyama, 2006; Ni, Rosteck, Nadi, & Paul, 1994), similar to the first discovered SLC17 protein, NPT1 (Werner et al., 1991). In fact, this phosphate transport activity delayed the recognition of VGLUTs as glutamate transporters (Bellocchio et al., 2000; Takamori et al., 2000). However, its functional significance is unclear. For instance, the phosphate uptake activity associated with VGLUT2 is intriguingly resistant to inhibitors and mutations (Juge et al., 2006), raising the possibility that it might correspond to a nonspecific effect.

3.2.1 Putative structure and ligand binding

The 3D structure of VGLUTs remains unknown. However, homology models of VGLUT1 (Almqvist, Huang, Laaksonen, Wang, & Hovmoller, 2007) and VGLUT2 (Juge et al., 2006) have been generated based on the cytosol-open crystallographic structure of *E. coli* glycerol-3-phosphate transporter (GlpT)—a distantly related MFS protein (Huang et al., 2003). Homology models have also been generated for sialin (Courville, Quick, & Reimer, 2010; Pietrancosta et al., 2012). In the VGLUT2 model, three conserved charged residues required for glutamate transport (H128 in TM2 and R184 and E191 in TM4) were found at the closed end of the cavity between the N and C domains (Juge et al., 2006). The human VGLUT1 model, based on a distinct sequence alignment with GlpT, was validated by molecular dynamics simulations and characterized by docking studies (Almqvist et al., 2007). These studies suggested the existence of the following two glutamate-binding sites: a central site located about halfway across the membrane, and an "upper" binding site located at the closed end of the cavity. The upper site is closer to the vesicle lumen than the central site (Almqvist et al., 2007). Because the vast majority of residues facing the cavity are conserved among the three isoforms, this binding site prediction should also apply to VGLUT2 and VGLUT3. The upper binding site comprises the aforementioned essential histidine and arginine residues (H120 and R176 in VGLUT1), suggesting that it might be involved in glutamate transport. Docking studies with azo dye inhibitors suggested that Evans Blue and Chicago Sky Blue only reach the central binding site, whereas Trypan Blue goes deeper into the cavity and interacts with both central and upper binding sites (Almqvist et al., 2007). However, the predicted binding sites have not been validated by mutagenesis studies. Because these mammalian models are based on a weak homology with the bacterial template, it is essential to subject them to thorough experimental testing.

In this respect, sialin provides a useful alternative starting point for SLC17 modeling because the existence of an efficient functional assay allows intensive mutagenesis and pharmacological tests (Morin et al., 2004). Published sialin models (Courville et al., 2010; Pietrancosta et al., 2012) were based on either the GlpT template or a lumen-open crystallographic structure of *E. coli* fucose permease (Dang et al., 2010). However, only GlpT-based models are supported by strong experimental evidence. In one study, investigators focused their wet lab tests on TM4 and showed that the tolerance to mutations and accessibility to reagents in this helix are consistent with its cavity-lining orientation in the model (Courville et al., 2010). In another study, we focused on sialic-acid binding and showed the existence of a single central binding site by docking high-affinity sialic acid analogues to the GlpT-based model (Pietrancosta et al., 2012). This binding site was validated by two types of experiments. First, two residues contacting diametrically opposite sides of the docked ligand were identified and mutated to bulkier side chains to narrow the binding site (Fig. 3.3). In agreement with the model, these mutations selectively impaired analogue binding, with a cumulative effect of individual mutations (Pietrancosta et al., 2012). In a second approach, the binding site model was validated further by showing its capacity to identify structurally unrelated, high-affinity competitive inhibitors in virtual high-throughput screening (Pietrancosta et al., 2012). Therefore, docking studies located the substrate-binding site at distinct depths in VGLUT and sialin models (Almqvist et al., 2007; Pietrancosta et al., 2012). Further research is needed to determine whether these differences reflect prediction errors or a structural divergence between the two subtypes of SLC17 proteins.

3.2.2 Ion coupling and regulation by chloride

The ion-coupling mechanism of VGLUT remains unclear. In contrast to VMAT, VGLUT depends more strongly on $\Delta\Psi$ than ΔpH, suggesting

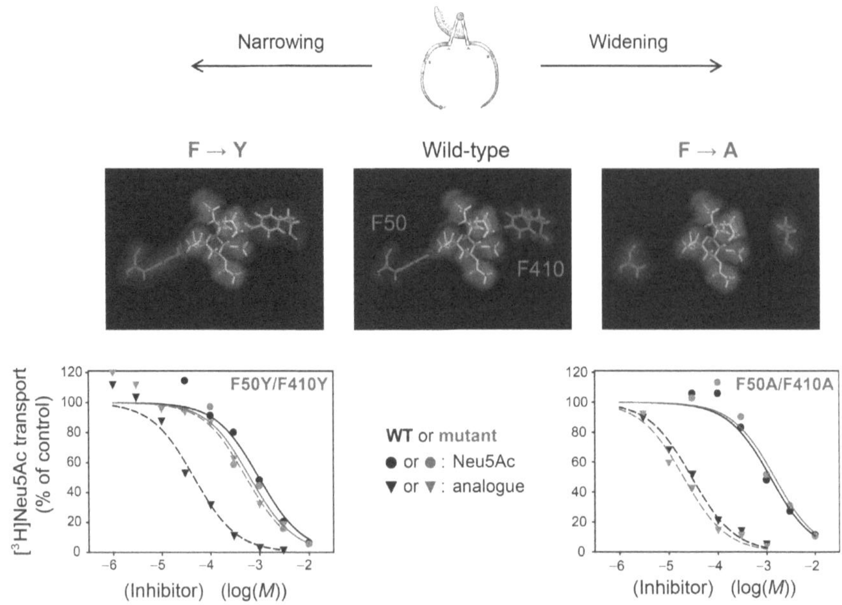

Figure 3.3 Mutagenesis validation of the substrate-binding site in a sialin homology model. Sialin is a VGLUT homologue that transports sialic acids such as neuraminic acid (Neu5Ac). The substrate-binding site was identified in the model by docking high-affinity sialic acid analogues. Upper panels: Two phenylalanine residues in TMs 1 and 10 (shown in green) making van der Waals contacts on diametrically opposite sides of the docked analogue (orange) were identified in the 3D model, thus providing a sort of "caliper" to test the accuracy of the docking site. Lower panels: In agreement with the model, mutating these phenylalanines to bulkier tyrosine residues impaired analogue, but not Neu5Ac, recognition. In contrast, ablation of the phenyl side chains by mutation to alanine had no effect, in agreement with the existence of numerous other interactions between the ligand and the binding site. *These data were originally published in the Journal of Biological Chemistry. Pietrancosta et al. (2012). © The American Society for Biochemistry and Molecular Biology.* (See the color plate.)

either a glutamate uniport or an antiport of one glumatate anion for one proton (Maycox, Deckwerth, Hell, & Jahn, 1988; Tabb, Kish, Van Dyke, & Ueda, 1992). The fact that ΔpH significantly promotes glutamate uptake in the absence of $\Delta\Psi$ favors the antiport mechanism (Edwards, 2007). However, the matter is complicated by the biphasic dependence of VGLUT on chloride (Bellocchio et al., 2000; Hartinger & Jahn, 1993; Maycox et al., 1988), and the interplay of this dependence with $\Delta\Psi$ and ΔpH (Tabb et al., 1992; Wolosker, de Souza, & de Meis, 1996). The inhibitory part of the biphasic curve at high chloride concentrations is well understood.

It merely reflects the fact that chloride influx into the vesicle shunts the electrogenic activity of the V-ATPase. Therefore, ΔpH increases at the expense of $\Delta\Psi$ and attenuates glutamate uptake. However, the mechanism underlying the activation of VGLUT at low chloride concentrations (<5 mM) is debated because conflicting data were reported in two studies on the purified protein reconstituted in liposomes. In one study, investigators identified a cytosolic-binding site for chloride that allosterically regulates VGLUT (Juge et al., 2010). Binding of chloride or bromide to this site strongly activated glutamate uptake with a Hill coefficient of 3. Interestingly, ketone bodies and pyruvate also bound to this site and competed with chloride, but they did not activate VGLUT, resulting in decreased glutamate uptake under standard ionic conditions (10 mM Cl^-). The cytosolic location of this allosteric site was supported by the lack of chloride permeation into VGLUT proteoliposomes, and the fact that 100 mM chloride in the proteoliposome lumen did not alter glutamate uptake (Juge et al., 2010).

In contrast, in another study, investigators identified a chloride conductance in VGLUT proteoliposomes and found that intracellular chloride strongly stimulated glutamate uptake (Schenck, Wojcik, Brose, & Takamori, 2009). When proteoliposomes were resealed in high chloride concentrations (100 mM), glutamate uptake still showed a biphasic dependence on extravesicular chloride, but the stimulation at low extravesicular chloride concentrations was much weaker than in the absence of luminal chloride. These authors suggested that when luminal chloride concentrations increase, VGLUT shifts from a H^+ antiport mode to a chloride antiport mode. The stimulation of glutamate uptake at low extravesicular chloride concentrations in the absence of luminal chloride would reflect a kinetic requirement for a minimal acidification of the vesicle lumen (Schenck et al., 2009). The origin of these discrepancies needs further investigation.

4. VESICULAR INHIBITORY AMINO ACID TRANSPORTER (SLC32)

VIAAT is the least well-characterized vesicular neurotransmitter transporter. When it was discovered, it was initially predicted to possess 10 TMs (McIntire et al., 1997; Sagné, El Mestikawy, et al., 1997). However, subsequent immunological and proteolysis-based evidence ruled out this topological model and supported a model with nine TMs, and N and C termini located in the cytosol and vesicle lumen, respectively (Martens et al., 2008).

In *C. elegans*, the orthologue of VIAAT (UNC-47) interacts, at least transiently, with UNC-46, a monotopic protein harboring a lysosomal-associated membrane protein (LAMP) repeat in the vesicle lumen (Schuske, Palfreyman, Watanabe, & Jorgensen, 2007). The UNC-46 protein is almost exclusively expressed at GABAergic nerve terminals; it is required for GABA exocytosis and functionally, and probably physically, interacts with UNC-47. These data could reflect either a participation of UNC-46 as an accessory subunit in a UNC-47/UNC-46 hetero-oligomer, or a specific role of UNC-46 in UNC-47 trafficking. These investigators favored the latter model because the residual pool of functional GABAergic vesicles in *unc-46* worms secreted a normal amount of GABA and UNC-47 overexpression rescued the loss of UNC-46, indicating that UNC-46 is dispensable for vesicular GABA transport (Schuske et al., 2007). The precise role of UNC-46 in UNC-47 trafficking needs further investigation. In mammals, a protein homologous to UNC-46, termed BAD-LAMP, has been described (David et al., 2007), but its role in inhibitory neurons has not yet been documented.

Another issue is the ion-coupling mechanism of VIAAT. Because GABA and glycine are zwitterionic, the fact that VIAAT activity equally depends on $\Delta\Psi$ and ΔpH in synaptic vesicles suggested a H^+ antiport mechanism (Burger et al., 1991). However, according to a recent study on purified VIAAT reconstituted in proteoliposomes, VIAAT activity exclusively depends on $\Delta\Psi$, and the protein may instead function as a chloride/neurotransmitter symporter (Juge, Muroyama, Hiasa, Omote, & Moriyama, 2009).

ACKNOWLEDGMENTS

We thank Lucy Forrest and Shimon Schuldiner for their kind permission to reproduce published data, and our collaborators Nicolas Pietrancosta and Francine Acher for their contribution to the experiment shown in Fig. 3.3. Our research is supported by the Paris Descartes University and the *Centre National de la Recherche Scientifique*. C. A. is a scientist from the *Institut National de la Santé et de la Recherche Médicale*.

REFERENCES

Abramson, J., Smirnova, I., Kasho, V., Verner, G., Kaback, H. R., & Iwata, S. (2003). Structure and mechanism of the lactose permease of Escherichia coli. *Science*, *301*(5633), 610–615.

Aihara, Y., Mashima, H., Onda, H., Hisano, S., Kasuya, H., Hori, T., et al. (2000). Molecular cloning of a novel brain-type Na(+)-dependent inorganic phosphate cotransporter. *Journal of Neurochemistry*, *74*, 2622–2625.

Alfonso, A., Grundahl, K., Duerr, J. S., Han, H. P., & Rand, J. B. (1993). The Caenorhabditis elegans unc-17 gene: A putative vesicular acetylcholine transporter. *Science*, *261*(5121), 617–619.

Almqvist, J., Huang, Y., Laaksonen, A., Wang, D. N., & Hovmoller, S. (2007). Docking and homology modeling explain inhibition of the human vesicular glutamate transporters. *Protein Science*, *16*, 1819–1829.

Bahr, B. A., Clarkson, E. D., Rogers, G. A., Noremberg, K., & Parsons, S. M. (1992). A kinetic and allosteric model for the acetylcholine transporter-vesamicol receptor in synaptic vesicles. *Biochemistry*, *31*(25), 5752–5762.

Bahr, B. A., & Parsons, S. M. (1986). Demonstration of a receptor in Torpedo synaptic vesicles for the acetylcholine storage blocker L-trans-2-(4-phenyl[3,4-3H]-piperidino) cyclohexanol. *Proceedings of the National Academy of Sciences of the United States of America*, *83*(7), 2267–2270.

Bejanin, S., Cervini, R., Mallet, J., & Berrard, S. (1994). A unique gene organization for two cholinergic markers, choline acetyltransferase and a putative vesicular transporter of acetylcholine. *Journal of Biological Chemistry*, *269*(35), 21944–21947.

Bellocchio, E. E., Reimer, R. J., Fremeau, R. T., Jr., & Edwards, R. H. (2000). Uptake of glutamate into synaptic vesicles by an inorganic phosphate transporter. *Science*, *289*(5481), 957–960.

Boudker, O., & Verdon, G. (2010). Structural perspectives on secondary active transporters. *Trends in Pharmacological Sciences*, *31*(9), 418–426.

Bravo, D. T., Kolmakova, N. G., & Parsons, S. M. (2004a). Choline is transported by vesicular acetylcholine transporter. *Journal of Neurochemistry*, *91*(3), 766–768.

Bravo, D. T., Kolmakova, N. G., & Parsons, S. M. (2004b). Transmembrane reorientation of the substrate-binding site in vesicular acetylcholine transporter. *Biochemistry*, *43*(27), 8787–8793.

Bravo, D. T., Kolmakova, N. G., & Parsons, S. M. (2005a). Mutational and pH analysis of ionic residues in transmembrane domains of vesicular acetylcholine transporter. *Biochemistry*, *44*, 7955–7966.

Bravo, D. T., Kolmakova, N. G., & Parsons, S. M. (2005b). New transport assay demonstrates vesicular acetylcholine transporter has many alternative substrates. *Neurochemistry International*, *47*(4), 243–247.

Bravo, D., & Parsons, S. M. (2002). Microscopic kinetics and structure-function analysis in the vesicular acetylcholine transporter. *Neurochemistry International*, *41*, 285–289.

Brooks, E. S., Greer, C. L., Romero-Calderon, R., Serway, C. N., Grygoruk, A., Haimovitz, J. M., et al. (2011). A putative vesicular transporter expressed in Drosophila mushroom bodies that mediates sexual behavior may define a neurotransmitter system. *Neuron*, *72*(2), 316–329.

Bruns, D., & Jahn, R. (1995). Real-time measurement of transmitter release from single synaptic vesicles. *Nature*, *377*(6544), 62–65.

Burger, P. M., Hell, J., Mehl, E., Krasel, C., Lottspeich, F., & Jahn, R. (1991). GABA and glycine in synaptic vesicles: Storage and transport characteristics. *Neuron*, *7*, 287–293.

Clarkson, E. D., Rogers, G. A., & Parsons, S. M. (1992). Binding and active transport of large analogues of acetylcholine by cholinergic synaptic vesicles in vitro. *Journal of Neurochemistry*, *59*(2), 695–700.

Courville, P., Quick, M., & Reimer, R. J. (2010). Structure-function studies of the SLC17 transporter sialin identify crucial residues and substrate-induced conformational changes. *Journal of Biological Chemistry*, *285*, 19316–19323.

Dang, S., Sun, L., Huang, Y., Lu, F., Liu, Y., Gong, H., et al. (2010). Structure of a fucose transporter in an outward-open conformation. *Nature*, *467*(7316), 734–738.

Darchen, F., Scherman, D., & Henry, J. P. (1989). Reserpine binding to chromaffin granules suggests the existence of two conformations of the monoamine transporter. *Biochemistry*, *28*(4), 1692–1697.

David, A., Tiveron, M. C., Defays, A., Beclin, C., Camosseto, V., Gatti, E., et al. (2007). BAD-LAMP defines a subset of early endocytic organelles in subpopulations of cortical projection neurons. *Journal of Cell Science*, *120*, 353–365.

Deupree, J. D., & Weaver, J. A. (1984). Identification and characterization of the catecholamine transporter in bovine chromaffin granules using [3H]reserpine. *Journal of Biological Chemistry*, *259*(17), 10907–10912.

Duerr, J. S., Frisby, D. L., Gaskin, J., Duke, A., Asermely, K., Huddleston, D., et al. (1999). The cat-1 gene of Caenorhabditis elegans encodes a vesicular monoamine transporter required for specific monoamine-dependent behaviors. *Journal of Neuroscience*, *19*(1), 72–84.

Edwards, R. H. (2007). The neurotransmitter cycle and quantal size. *Neuron*, *55*, 835–858.

El Mestikawy, S., Wallen-Mackenzie, A., Fortin, G. M., Descarries, L., & Trudeau, L. E. (2011). From glutamate co-release to vesicular synergy: Vesicular glutamate transporters. *Nature Reviews. Neuroscience*, *12*, 204–216.

Erickson, J. D., Eiden, L. E., & Hoffman, B. J. (1992). Expression cloning of a reserpine-sensitive vesicular monoamine transporter. *Proceedings of the National Academy of Sciences of the United States of America*, *89*(22), 10993–10997.

Finn, J. P., 3rd., & Edwards, R. H. (1997). Individual residues contribute to multiple differences in ligand recognition between vesicular monoamine transporters 1 and 2. *Journal of Biological Chemistry*, *272*(26), 16301–16307.

Gasnier, B., Krejci, E., Botton, D., Massoulie, J., & Henry, J. P. (1994). Expression of a bovine vesicular monoamine transporter in COS cells. *FEBS Letters*, *342*(3), 225–229.

Greer, C. L., Grygoruk, A., Patton, D. E., Ley, B., Romero-Calderon, R., Chang, H. Y., et al. (2005). A splice variant of the Drosophila vesicular monoamine transporter contains a conserved trafficking domain and functions in the storage of dopamine, serotonin, and octopamine. *Journal of Neurobiology*, *64*(3), 239–258.

Gros, Y., & Schuldiner, S. (2010). Directed evolution reveals hidden properties of VMAT, a neurotransmitter transporter. *Journal of Biological Chemistry*, *285*(7), 5076–5084.

Guan, L., & Kaback, H. R. (2006). Lessons from lactose permease. *Annual Review of Biophysics and Biomolecular Structure*, *35*, 67–91.

Gubernator, N. G., Zhang, H., Staal, R. G., Mosharov, E. V., Pereira, D. B., Yue, M., et al. (2009). Fluorescent false neurotransmitters visualize dopamine release from individual presynaptic terminals. *Science*, *324*(5933), 1441–1444.

Hartinger, J., & Jahn, R. (1993). An anion binding site that regulates the glutamate transporter of synaptic vesicles. *Journal of Biological Chemistry*, *268*, 23122–23127.

Henry, J. P., Sagné, C., Bedet, C., & Gasnier, B. (1998). The vesicular monoamine transporter: From chromaffin granule to brain. *Neurochemistry International*, *32*(3), 227–246.

Hnasko, T. S., & Edwards, R. H. (2012). Neurotransmitter corelease: Mechanism and physiological role. *Annual Review of Physiology*, *74*, 225–243.

Howell, M., Shirvan, A., Stern-Bach, Y., Steiner-Mordoch, S., Strasser, J. E., Dean, G. E., et al. (1994). Cloning and functional expression of a tetrabenazine sensitive vesicular monoamine transporter from bovine chromaffin granules. *FEBS Letters*, *338*(1), 16–22.

Huang, Y., Lemieux, M. J., Song, J., Auer, M., & Wang, D. N. (2003). Structure and mechanism of the glycerol-3-phosphate transporter from Escherichia coli. *Science*, *301*(5633), 616–620.

Johnson, R. G., Carty, S. E., & Scarpa, A. (1981). Proton: Substrate stoichiometries during active transport of biogenic amines in chromaffin ghosts. *Journal of Biological Chemistry*, *256*(11), 5773–5780.

Juge, N., Gray, J. A., Omote, H., Miyaji, T., Inoue, T., Hara, C., et al. (2010). Metabolic control of vesicular glutamate transport and release. *Neuron*, *68*, 99–112.

Juge, N., Muroyama, A., Hiasa, M., Omote, H., & Moriyama, Y. (2009). Vesicular inhibitory amino acid transporter is a Cl-/gamma-aminobutyrate co-transporter. *Journal of Biological Chemistry*, *284*, 35073–35078.

Juge, N., Yoshida, Y., Yatsushiro, S., Omote, H., & Moriyama, Y. (2006). Vesicular glutamate transporter contains two independent transport machineries. *Journal of Biological Chemistry*, *281*, 39499–39506.

Khare, P., Mulakaluri, A., & Parsons, S. M. (2010). Search for the acetylcholine and vesamicol binding sites in vesicular acetylcholine transporter: The region around the lumenal end of the transport channel. *Journal of Neurochemistry, 115*(4), 984–993.

Khare, P., Ojeda, A. M., Chandrasekaran, A., & Parsons, S. M. (2010). Possible important pair of acidic residues in vesicular acetylcholine transporter. *Biochemistry, 49*, 3049–3059.

Kim, M. H., Lu, M., Kelly, M., & Hersh, L. B. (2000). Mutational analysis of basic residues in the rat vesicular acetylcholine transporter. Identification of a transmembrane ion pair and evidence that histidine is not involved in proton translocation. *Journal of Biological Chemistry, 275*, 6175–6180.

Kim, M. H., Lu, M., Lim, E. J., Chai, Y. G., & Hersh, L. B. (1999). Mutational analysis of aspartate residues in the transmembrane regions and cytoplasmic loops of rat vesicular acetylcholine transporter. *Journal of Biological Chemistry, 274*(2), 673–680.

Knoth, J., Zallakian, M., & Njus, D. (1981). Stoichiometry of H+-linked dopamine transport in chromaffin granule ghosts. *Biochemistry, 20*(23), 6625–6629.

Krishnamurthy, H., Piscitelli, C. L., & Gouaux, E. (2009). Unlocking the molecular secrets of sodium-coupled transporters. *Nature, 459*(7245), 347–355.

Law, C. J., Maloney, P. C., & Wang, D. N. (2008). Ins and outs of major facilitator superfamily antiporters. *Annual Review of Microbiology, 62*, 289–305.

Liu, Y., Peter, D., Roghani, A., Schuldiner, S., Prive, G. G., Eisenberg, D., et al. (1992). A cDNA that suppresses MPP+ toxicity encodes a vesicular amine transporter. *Cell, 70*(4), 539–551.

Lodder-Gadaczek, J., Gieselmann, V., & Eckhardt, M. (2013). Vesicular uptake of N-acetylaspartylglutamate is catalysed by sialin (SLC17A5). *The Biochemical Journal, 454*(1), 31–38.

Martens, H., Weston, M. C., Boulland, J. L., Gronborg, M., Grosche, J., Kacza, J., et al. (2008). Unique luminal localization of VGAT-C terminus allows for selective labeling of active cortical GABAergic synapses. *Journal of Neuroscience, 28*, 13125–13131.

Martineau, M., Shi, T., Puyal, J., Knolhoff, A. M., Dulong, J., Gasnier, B., et al. (2013). Storage and uptake of D-serine into astrocytic synaptic-like vesicles specify gliotransmission. *Journal of Neuroscience, 33*(8), 3413–3423.

Maycox, P. R., Deckwerth, T., Hell, J. W., & Jahn, R. (1988). Glutamate uptake by brain synaptic vesicles. Energy dependence of transport and functional reconstitution in proteoliposomes. *Journal of Biological Chemistry, 263*, 15423–15428.

McIntire, S. L., Reimer, R. J., Schuske, K., Edwards, R. H., & Jorgensen, E. M. (1997). Identification and characterization of the vesicular GABA transporter. *Nature, 389*(6653), 870–876.

Merickel, A., Kaback, H. R., & Edwards, R. H. (1997). Charged residues in transmembrane domains II and XI of a vesicular monoamine transporter form a charge pair that promotes high affinity substrate recognition. *Journal of Biological Chemistry, 272*(9), 5403–5408.

Merickel, A., Rosandich, P., Peter, D., & Edwards, R. H. (1995). Identification of residues involved in substrate recognition by a vesicular monoamine transporter. *Journal of Biological Chemistry, 270*(43), 25798–25804.

Miyaji, T., Echigo, N., Hiasa, M., Senoh, S., Omote, H., & Moriyama, Y. (2008). Identification of a vesicular aspartate transporter. *Proceedings of the National Academy of Sciences of the United States of America, 105*(33), 11720–11724.

Morin, P., Sagné, C., & Gasnier, B. (2004). Functional characterization of wild-type and mutant human sialin. *EMBO Journal, 23*, 4560–4570.

Morland, C., Nordengen, K., Larsson, M., Prolo, L. M., Farzampour, Z., Reimer, R. J., et al. (2012). Vesicular uptake and exocytosis of L-aspartate is independent of sialin. *The FASEB Journal, 27*(3), 1264–1274.

Naito, S., & Ueda, T. (1985). Characterization of glutamate uptake into synaptic vesicles. *Journal of Neurochemistry, 44*, 99–109.

Nguyen, M. L., Cox, G. D., & Parsons, S. M. (1998). Kinetic parameters for the vesicular acetylcholine transporter: Two protons are exchanged for one acetylcholine. *Biochemistry*, *37*(38), 13400–13410.

Ni, B., Rosteck, P. R., Jr., Nadi, N. S., & Paul, S. M. (1994). Cloning and expression of a cDNA encoding a brain-specific Na(+)-dependent inorganic phosphate cotransporter. *Proceedings of the National Academy of Sciences of the United States of America*, *91*, 5607–5611.

Ojeda, A. M., Kolmakova, N. G., & Parsons, S. M. (2004). Acetylcholine binding site in the vesicular acetylcholine transporter. *Biochemistry*, *43*(35), 11163–11174.

Parsons, S. M. (2000). Transport mechanisms in acetylcholine and monoamine storage. *FASEB Journal: Official Publication of the Federation of American Societies for Experimental Biology*, *14*, 2423–2434.

Pedersen, B. P., Kumar, H., Waight, A. B., Risenmay, A. J., Roe-Zurz, Z., Chau, B. H., et al. (2013). Crystal structure of a eukaryotic phosphate transporter. *Nature*, *496*(7446), 533–536.

Pietrancosta, N., Anne, C., Prescher, H., Ruivo, R., Sagné, C., Debacker, C., et al. (2012). Successful prediction of substrate-binding pocket in SLC17 transporter sialin. *Journal of Biological Chemistry*, *287*, 11489–11497.

Pothos, E. N., Davila, V., & Sulzer, D. (1998). Presynaptic recording of quanta from midbrain dopamine neurons and modulation of the quantal size. *The Journal of Neuroscience*, *18*(11), 4106–4118.

Qin, L., Liu, X., Sun, Q., Fan, Z., Xia, D., Ding, G., et al. (2012). Sialin (SLC17A5) functions as a nitrate transporter in the plasma membrane. *Proceedings of the National Academy of Sciences of the United States of America*, *109*(33), 13434–13439.

Radestock, S., & Forrest, L. R. (2011). The alternating-access mechanism of MFS transporters arises from inverted-topology repeats. *Journal of Molecular Biology*, *407*(5), 698–715.

Reddy, V. S., Shlykov, M. A., Castillo, R., Sun, E. I., & Saier, M. H., Jr. (2012). The major facilitator superfamily (MFS) revisited. *The FEBS Journal*, *279*(11), 2022–2035.

Reimer, R. J. (2013). SLC17: A functionally diverse family of organic anion transporters. *Molecular Aspects of Medicine*, *34*(2–3), 350–359.

Roghani, A., Feldman, J., Kohan, S. A., Shirzadi, A., Gundersen, C. B., Brecha, N., et al. (1994). Molecular cloning of a putative vesicular transporter for acetylcholine. *Proceedings of the National Academy of Sciences of the United States of America*, *91*(22), 10620–10624.

Rudnick, G., Steiner-Mordoch, S. S., Fishkes, H., Stern-Bach, Y., & Schuldiner, S. (1990). Energetics of reserpine binding and occlusion by the chromaffin granule biogenic amine transporter. *Biochemistry*, *29*(3), 603–608.

Ruivo, R., Bellenchi, G. C., Chen, X., Zifarelli, G., Sagné, C., Debacker, C., et al. (2012). Mechanism of proton/substrate coupling in the heptahelical lysosomal transporter cystinosin. *Proceedings of the National Academy of Sciences of the United States of America*, *109*, E210–E217.

Sagné, C., El Mestikawy, S., Isambert, M. F., Hamon, M., Henry, J. P., Giros, B., et al. (1997). Cloning of a functional vesicular GABA and glycine transporter by screening of genome databases. *FEBS Letters*, *417*(2), 177–183.

Sagné, C., Isambert, M. F., Vandekerckhove, J., Henry, J. P., & Gasnier, B. (1997). The photoactivatable inhibitor 7-azido-8-iodoketanserin labels the N terminus of the vesicular monoamine transporter from bovine chromaffin granules. *Biochemistry*, *36*(11), 3345–3352.

Sawada, K., Echigo, N., Juge, N., Miyaji, T., Otsuka, M., Omote, H., et al. (2008). Identification of a vesicular nucleotide transporter. *Proceedings of the National Academy of Sciences of the United States of America*, *105*(15), 5683–5686.

Schenck, S., Wojcik, S. M., Brose, N., & Takamori, S. (2009). A chloride conductance in VGLUT1 underlies maximal glutamate loading into synaptic vesicles. *Nature Neuroscience*, *12*, 156–163.

Scherman, D., & Henry, J. P. (1984). Reserpine binding to bovine chromaffin granule membranes. Characterization and comparison with dihydrotetrabenazine binding. *Molecular Pharmacology*, *25*(1), 113–122.
Scherman, D., Jaudon, P., & Henry, J. P. (1983). Characterization of the monoamine carrier of chromaffin granule membrane by binding of [2-3H]dihydrotetrabenazine. *Proceedings of the National Academy of Sciences of the United States of America*, *80*(2), 584–588.
Schuldiner, S., Shirvan, A., & Linial, M. (1995). Vesicular neurotransmitter transporters: From bacteria to humans. *Physiological Reviews*, *75*, 369–392.
Schuske, K., Palfreyman, M. T., Watanabe, S., & Jorgensen, E. M. (2007). UNC-46 is required for trafficking of the vesicular GABA transporter. *Nature Neuroscience*, *10*, 846–853.
Shigeri, Y., Seal, R. P., & Shimamoto, K. (2004). Molecular pharmacology of glutamate transporters, EAATs and VGLUTs. *Brain Research Reviews*, *45*, 250–265.
Shirvan, A., Laskar, O., Steiner-Mordoch, S., & Schuldiner, S. (1994). Histidine-419 plays a role in energy coupling in the vesicular monoamine transporter from rat. *FEBS Letters*, *356*, 145–150.
Sievert, M. K., & Ruoho, A. E. (1997). Peptide mapping of the [125I]Iodoazidoketanserin and [125I]2-N-[(3'-iodo-4'-azidophenyl)propionyl]tetrabenazine binding sites for the synaptic vesicle monoamine transporter. *Journal of Biological Chemistry*, *272*(41), 26049–26055.
Steiner-Mordoch, S., Shirvan, A., & Schuldiner, S. (1996). Modification of the pH profile and tetrabenazine sensitivity of rat VMAT1 by replacement of aspartate 404 with glutamate. *Journal of Biological Chemistry*, *271*, 13048–13054.
Tabb, J. S., Kish, P. E., Van Dyke, R., & Ueda, T. (1992). Glutamate transport into synaptic vesicles. Roles of membrane potential, pH gradient, and intravesicular pH. *Journal of Biological Chemistry*, *267*, 15412–15418.
Takamori, S., Rhee, J. S., Rosenmund, C., & Jahn, R. (2000). Identification of a vesicular glutamate transporter that defines a glutamatergic phenotype in neurons. *Nature*, *407*(6801), 189–194.
Thiriot, D. S., & Ruoho, A. E. (2001). Mutagenesis and derivatization of human vesicle monoamine transporter 2 (VMAT2) cysteines identifies transporter domains involved in tetrabenazine binding and substrate transport. *Journal of Biological Chemistry*, *276*(29), 27304–27315.
Thiriot, D. S., Sievert, M. K., & Ruoho, A. E. (2002). Identification of human vesicle monoamine transporter (VMAT2) lumenal cysteines that form an intramolecular disulfide bond. *Biochemistry*, *41*(20), 6346–6353.
Thompson, C. M., Davis, E., Carrigan, C. N., Cox, H. D., Bridges, R. J., & Gerdes, J. M. (2005). Inhibitor of the glutamate vesicular transporter (VGLUT). *Current Medicinal Chemistry*, *12*, 2041–2056.
Ugolev, Y., Segal, T., Yaffe, D., Gros, Y., & Schuldiner, S. (2013). Identification of conformationally sensitive residues essential for inhibition of vesicular monoamine transport by the noncompetitive inhibitor-tetrabenazine. *Journal of Biological Chemistry*, *288*, 32160–32171.
Van der Kloot, W., & Molgo, J. (1994). Quantal acetylcholine release at the vertebrate neuromuscular junction. *Physiological Reviews*, *74*(4), 899–991.
Vardy, E., Arkin, I. T., Gottschalk, K. E., Kaback, H. R., & Schuldiner, S. (2004). Structural conservation in the major facilitator superfamily as revealed by comparative modeling. *Protein Science*, *13*(7), 1832–1840.
Verheijen, F. W., Verbeek, E., Aula, N., Beerens, C. E., Havelaar, A. C., Joosse, M., et al. (1999). A new gene, encoding an anion transporter, is mutated in sialic acid storage diseases. *Nature Genetics*, *23*(4), 462–465.
Werner, A., Moore, M. L., Mantei, N., Biber, J., Semenza, G., & Murer, H. (1991). Cloning and expression of cDNA for a Na/Pi cotransport system of kidney cortex. *Proceedings of the National Academy of Sciences of the United States of America*, *88*, 9608–9612.

Wolosker, H., de Souza, D. O., & de Meis, L. (1996). Regulation of glutamate transport into synaptic vesicles by chloride and proton gradient. *Journal of Biological Chemistry*, *271*, 11726–11731.

Wong, F. H., Chen, J. S., Reddy, V., Day, J. L., Shlykov, M. A., Wakabayashi, S. T., et al. (2012). The amino acid-polyamine-organocation superfamily. *Journal of Molecular Microbiology and Biotechnology*, *22*(2), 105–113.

Yaffe, D., Radestock, S., Shuster, Y., Forrest, L. R., & Schuldiner, S. (2013). Identification of molecular hinge points mediating alternating access in the vesicular monoamine transporter VMAT2. *Proceedings of the National Academy of Sciences of the United States of America*, *110*, E1332–E1341.

Yan, N. (2013). Structural advances for the major facilitator superfamily (MFS) transporters. *Trends in Biochemical Sciences*, *38*(3), 151–159.

Zhu, H., Duerr, J. S., Varoqui, H., McManus, J. R., Rand, J. B., & Erickson, J. D. (2001). Analysis of point mutants in the Caenorhabditis elegans vesicular acetylcholine transporter reveals domains involved in substrate translocation. *Journal of Biological Chemistry*, *276*(45), 41580–41587.

CHAPTER FOUR

Biology of the Major Facilitative Folate Transporters SLC19A1 and SLC46A1

Zhanjun Hou[*,†,1], **Larry H. Matherly**[*,†,‡,1]
[*]Department of Oncology, Wayne State University School of Medicine, Detroit, Michigan, USA
[†]Molecular Therapeutics Program, Barbara Ann Karmanos Cancer Institute, Detroit, Michigan, USA
[‡]Department of Pharmacology, Wayne State University School of Medicine, Detroit, Michigan, USA
[1]Corresponding authors: e-mail address: houz@karmanos.org; matherly@karmanos.org

Contents

Abstract

This chapter focuses on the biology of the major facilitative membrane folate transporters, the reduced folate carrier (RFC), and the proton-coupled folate transporter (PCFT). Folates are essential vitamins, and folate deficiency contributes to a variety of heath disorders. RFC is ubiquitously expressed and is the major folate transporter in mammalian cells and tissues. PCFT mediates intestinal absorption of dietary folates. Clinically relevant antifolates such as methotrexate (MTX) are transported by RFC, and the loss of RFC transport is an important mechanism of MTX resistance. PCFT is abundantly expressed in human tumors and is active under pH conditions associated with the tumor microenvironment. Pemetrexed (PMX) is an excellent substrate for PCFT as well as for RFC. Novel tumor-targeted antifolates related to PMX with selective membrane transport by PCFT over RFC are being developed. The molecular picture of RFC and PCFT continues to evolve relating to membrane topology, N-glycosylation, energetics, and identification of structurally and functionally important domains and amino acids. The molecular bases for MTX resistance associated with loss of RFC function, and for the rare autosomal recessive condition, hereditary folate malabsorption (HFM),

Current Topics in Membranes, Volume 73
ISSN 1063-5823
http://dx.doi.org/10.1016/B978-0-12-800223-0.00004-9

attributable to mutant PCFT, have been established. From structural homologies to the bacterial transporters GlpT and LacY, homology models were developed for RFC and PCFT, enabling new mechanistic insights and experimentally testable hypotheses. RFC and PCFT exist as homo-oligomers, and evidence suggests that homo-oligomerization of RFC and PCFT monomeric proteins may be important for intracellular trafficking and/or transport function. Better understanding of the structure and function of RFC and PCFT should facilitate the rational development of new therapeutic strategies for cancer as well as for HFM.

1. INTRODUCTION

The folates are members of the B family of vitamins that require active membrane transport systems for cellular uptake (Matherly & Goldman, 2003; Zhao, Diop-Bove, Visentin, & Goldman, 2011). Studies of the membrane transport of folate cofactors over the past five decades have established its nutritional importance in providing cofactors for essential metabolic reactions required for cell proliferation and tissue repair. Transport is also required for uptake of antifolate drugs such as methotrexate (MTX) into tumor cells and is a major determinant of chemotherapeutic efficacy for these agents (Matherly, Hou, & Deng, 2007; Zhao & Goldman, 2003).

The reduced folate carrier (RFC; SLC19A1) belongs to the solute carrier (SLC) group of transporters and is the principal mechanism by which folates and clinically used antifolates are delivered to mammalian cells and tissues from the systemic circulation at neutral pH (Matherly et al., 2007). Although a transporter with unique properties from RFC was recognized for over 30 years to mediate intestinal absorption of folates, and more recently to drive uptake of the new generation antifolate pemetrexed (PMX) into tumor cells, it was not until 2006 that this system was cloned and characterized (Qiu et al., 2006). Thus, the proton-coupled folate transporter (PCFT; SLC46A1) was recognized as distinct from RFC or other known transporters capable of transporting folates such as the family of organic anion transporters (Matherly et al., 2007; Rizwan & Burckhardt, 2007). Identification of loss-of-function mutations in PCFT from human subjects with hereditary folate malabsorption (HFM) provided the molecular basis for this rare condition and unequivocally established PCFT as the principal mechanism for intestinal absorption of dietary folates (Qiu et al., 2006).

This review focuses on the biology of the major facilitative folate transporters, RFC and PCFT, including their physiology and structural and functional properties.

2. THE BIOLOGICAL ROLES OF FOLATES AND THERAPY WITH ANTIFOLATES

Folate is the generic term for water-soluble members of the B class of vitamins that are required for normal tissue growth and development in mammalian cells. The biological importance of reduced folates relates to their essential roles in one-carbon transfer reactions leading to thymidylate, purine nucleotides, serine, and methionine (Stokstad, 1990). Folates are also required for vitamin-B12-dependent synthesis of methionine, a precursor of *S*-adenosyl methionine, which is needed for methylation of DNA, histones, lipids, and neurotransmitters (Lu, 2000). Folic acid (Fig. 4.1) is the synthetic form of the metabolically important folates found in cells that differ in the level of oxidation of the pteridine ring, the nature of the one-carbon substituent at the N^5 and N^{10} positions, and the extent of γ-glutamate conjugation (Stokstad, 1990). The major circulating folate is 5-methyl tetrahydrofolate (THF) (Fig. 4.1). Following cellular uptake, 5-methyl THF is converted to various THF polyglutamates required for folate-dependent metabolic reactions (Stokstad, 1990).

Mammalian cells, unlike bacteria, cannot synthesize folates *de novo*. Hence, folate requirements must be met entirely from dietary sources. Traditionally, folates are derived from foods such as liver and dark green leafy vegetables. In the United States and Canada, the fortification of cereals, grains, and bread with folic acid now represents an important source of dietary folate and has resulted in a rise in folate levels in tissues and blood (Jacques, Selhub, Bostom, Wilson, & Rosenberg, 1999).

Folates are hydrophilic molecules that are anions at physiological pH and thus cross biological membranes poorly by diffusion. Reflecting this, mammalian cells have evolved sophisticated uptake systems for facilitating cellular uptake of folate cofactors (Kugel Desmoulin, Hou, Gangjee, & Matherly, 2012; Matherly & Goldman, 2003; Zhao et al., 2011). Major facilitative folate transporters include: (i) the RFC, an anionic exchanger and the major route of delivery of folates to systemic tissues at physiological pH (Matherly & Goldman, 2003; Matherly et al., 2007); and (ii) PCFT, like RFC, a member of the superfamily of solute carriers, but with an acidic pH optimum and distinct transport specificities (Kugel Desmoulin et al., 2012; Zhao & Goldman, 2007). In certain epithelial tissues such as renal cells and macrophages, folate receptors (FRs) α and β are embedded in the plasma membrane by a glycosylphosphatidylinositol anchor and

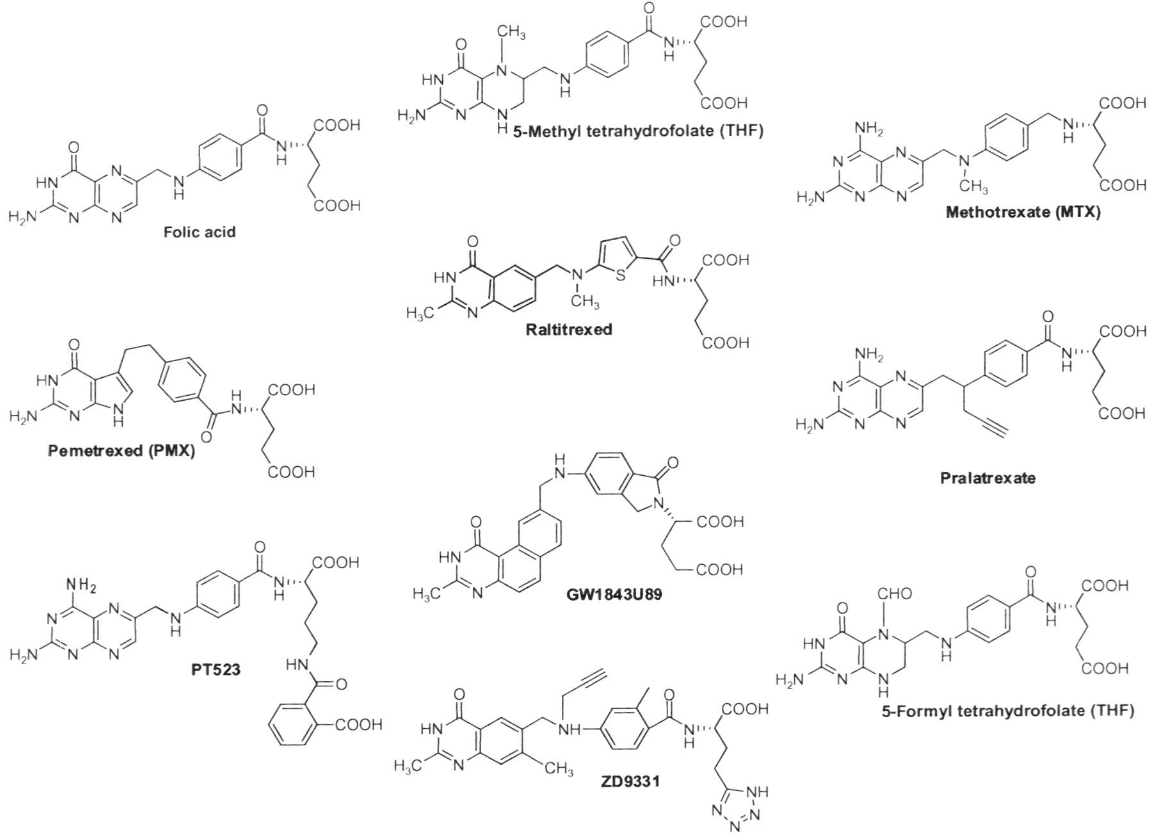

Figure 4.1 Structures of folates and clinically relevant antifolates. RFC substrates including MTX, PMX, ratitrexed, pralatrexate, 5-methyl THF, and 5-formyl THF are all excellent PCFT substrates. However, the antifolates PT523 and GW1843U89 do not appear to be transported by PCFT. RFC has a low affinity but PCFT has a high affinity for folic acid. GW1843U89 is an excellent substrate for human RFC, but a poor substrate for murine RFC.

mediate uptake of folates into cells via receptor-mediated endocytosis at neutral to mildly acidic pHs (Elnakat & Ratnam, 2004; Kamen & Smith, 2004). Whereas RFC is ubiquitously expressed in human tissues (Whetstine, Flatley, & Matherly, 2002), PCFT is more narrowly expressed, although substantial levels of PCFT were detected in the duodenum and jejunum, kidney, and liver, as well as choroid plexus (Kugel Desmoulin et al., 2012). FRα is expressed in the proximal renal tubules, choroid plexus, uterus, and placenta, whereas FRβ is expressed in placenta and hematopoietic cells, and in activated macrophages (Elnakat & Ratnam, 2004).

Proliferating tumor cells have unique metabolic requirements characterized by enhanced nutrient uptake and metabolic pathways to support the biosynthesis of macromolecules needed for cell growth and division. These pathways include the folate-dependent *de novo* synthesis of purine nucleotides and thymidylate. Structural analogs of folate compounds (i.e., antifolates) have been developed and continue to be important drugs for treating a variety of cancers (Gonen & Assaraf, 2012a; Kugel Desmoulin et al., 2012; Visentin, Zhao, & Goldman, 2012a) and nonmalignant diseases such as psoriasis and rheumatoid arthritis (Chladek et al., 1998; Wessels, Huizinga, & Guchelaar, 2008). Clinically relevant antifolates include MTX, PMX, raltitrexed, and pralatrexate (Fig. 4.1). These classic antifolates use the same transporters as physiological folates to enter tumor cells, particularly RFC (Kugel Desmoulin et al., 2012; Matherly et al., 2007). Membrane transport of antifolates is an important determinant of clinical efficacy for cancer, and impaired transport is a common mechanism of drug resistance (Zhao & Goldman, 2003). In recent years, there has been emphasis on developing tumor-targeted folate-based therapies reflecting selective cellular uptake of folate-conjugated cytotoxins or cytotoxic antifolates by FRs or the PCFT by tumor cells (Deng, Wang, et al., 2008; Deng et al., 2009; Gibbs et al., 2005; Kugel Desmoulin et al., 2012, 2011a, 2010; Wang et al., 2012, 2010, 2011; Xia & Low, 2010; Yang, Vlashi, & Low, 2012). This has fostered a new paradigm, namely the rational development of tumor-targeted therapies based on tumor-specific expression and/or function of the major folate transporters.

In the following sections, we focus on biological properties of RFC and PCFT, including structure, function, and mechanisms of facilitative transport.

3. REDUCED FOLATE CARRIER (SLC19A1)

3.1. Properties of RFC

As noted above, RFC is expressed ubiquitously and is widely considered to be the major transport system for folates in mammalian cells and tissues (Matherly et al., 2007). Very high levels of RFC transcripts are detected in liver and placenta, with appreciable levels in other tissues, including kidney, lung, bone marrow, intestine, brain, and portions of the central nervous system (Whetstine et al., 2002). By immunohistochemistry in mouse tissues, RFC was detected at the basolateral membrane of the renal tubule epithelium, the apical brush border membrane of the small intestine and colon, hepatocyte membranes, the apical surface of the choroid plexus, and the apical membrane of the cells lining the spinal canal (Wang, Zhao, Russell, & Goldman, 2001). Reflecting its physiological importance, low levels of RFC could result in a number of pathophysiological states associated with folate deficiency, ranging from cardiovascular disease, fetal abnormalities, neurological disorders, to possibly cancer (Matherly, 2004; Matherly et al., 2007). RFC is important for development since inactivating both RFC alleles by targeted homologous recombination in mice is embryonic lethal (Zhao, Russell, et al., 2001).

The properties of RFC-mediated transport in both human and rodent models have been comprehensively studied in numerous laboratories (Matherly et al., 2007). Investigators have used commercially available radioactive MTX as a surrogate substrate in functional studies on RFC because MTX is not metabolized over short intervals and because of the ease and accuracy of influx determinations and of distinguishing between free and protein-bound drug within cells (Goldman, Lichtenstein, & Oliverio, 1968). With minor exceptions, transport properties are remarkably conserved among species, in large part reflecting significant conservation of primary sequence (see below). The murine RFC was cloned in 1994 (Dixon, Lanpher, Chiu, Kelley, & Cowan, 1994), followed shortly thereafter by the human RFC by a number of laboratories (Moscow et al., 1995; Prasad, Ramamoorthy, Leibach, & Ganapathy, 1995; Williams & Flintoff, 1995; Wong, Proefke, Bhushan, & Matherly, 1995), thus paving the way to further detailed studies of RFC structure and function. The human RFC gene is localized to chromosome 21q22.3 (Moscow et al., 1995) and is subjected to complex regulation involving multiple promoters and noncoding exons that likely ensures its ubiquitous tissue expression (Matherly et al., 2007).

For most reduced folates and many antifolate substrates including clinically relevant drugs such as MTX and pralatrexate, transport by RFC is saturable at low micromolar concentrations ($K_t \sim 1$–$5\ \mu M$) (Matherly et al., 2007). The novel δ-hemiphthaloylorinithine antifolate PT523 (Fig. 4.1) is one of the best substrates for human RFC with a K_t value less than 1 μM (Rosowsky, Bader, Wright, Keyomarsi, & Matherly, 1994). Likewise, the benzoquinazoline antifolate GW1843U89 (Fig. 4.1) has a reported K_t of 0.33 μM and is an excellent substrate for human RFC ($V_{max}/K_t = 20.3$), although it is a surprisingly poor substrate for the murine RFC ($V_{max}/K_t = 0.25$) (Duch et al., 1993). The affinity of RFC for folic acid is orders of magnitude less than for reduced folates (Goldman et al., 1968; Westerhof et al., 1995). While RFC transport is not stereospecific for 5-methyl THF (Fig. 4.1) (Sirotnak & Donsbach, 1974; White, Bailey, & Goldman, 1978), the (6*R*) isomer of 5-formyl THF (Fig. 4.1) has a far lower affinity than the natural (6*S*) stereoisomer (Sirotnak et al., 1979). Thus, the (6*R*) stereoisomer of 5-formyl THF in leucovorin [(6*R*,*S*)5-formyl THF] should have no impact on cellular uptake of the (6*S*) stereoisomer of 5-formyl THF.

RFC is an anion transporter, and its activity is competitively inhibited by inorganic anions such as chloride, bicarbonate, and phosphate, or by structurally diverse organic anions such as adenine nucleotides and thiamine phosphate (Goldman, 1971a). Furthermore, replacing anionic buffers with nonanionic HEPES–sucrose buffers results in a dramatic stimulation of concentrative uptake of MTX (Henderson & Zevely, 1983b). When cells are incubated in anion-free buffers (without glucose) and preloaded with radio-labeled MTX, the rate of MTX efflux is inhibited. Efflux can be induced upon addition of both inorganic and organic anions (e.g., chloride, sulfate, phosphate, thiamine pyrophosphate, AMP, ADP, and 5-formyl THF) (Henderson & Zevely, 1980, 1981).

Folates include a glutamate moiety for which both α- and γ-carboxyl groups are ionized at physiological pH. Replacing the glutamate with valine or 2-aminosuberate in the novel antifolate substrate ICI-198,583 preserved binding to RFC (Westerhof et al., 1995). Modifications of the glutamate γ-carboxyl group appear to be well tolerated; replacing either the glutamate with a δ-hemiphthaloylornithine in PT523 (Fig. 4.1) or the γ-carboxyl with a tetrazole in ZD9331 (Fig. 4.1) preserved substrate activity (Jansen, 1999; Rosowsky et al., 1994). A series of diaminofuro[2,3-*d*]pyrimidine antifolates with hydrogen or methyl substitutions at the α- or γ-carboxyl groups of the terminal glutamate were used to systematically assess the importance of these groups for substrate binding to RFC (Deng, Hou, et al., 2008). Analogs with only a single α-carboxyl but no γ-carboxyl were potent inhibitors of MTX transport by RFC, whereas analogs with only a γ-carboxyl group but no α-carboxyl group were poor inhibitors. Thus, the α-carboxyl group is important for binding of (anti)folate substrates to RFC.

Transport by RFC is temperature-dependent and sodium-independent (Goldman, 1971a; Goldman et al., 1968) and shows a distinctly neutral pH optimum. Transport decreases dramatically below pH 7 (Sierra, Brigle, Spinella, & Goldman, 1997). Transport by RFC is concentrative, although the energy source for this uphill process is not directly linked to ATP hydrolysis (Goldman, 1971a; Henderson & Zevely, 1983a). Concentrative uptake by RFC is driven by gradients of organic anions that bind and exit cells via RFC and block folate export via this same mechanism (Goldman, 1971a). Consistent with this model, influx of radiolabeled MTX by RFC is enhanced (trans-stimulated) in cells "preloaded" with high concentrations of 5-formyl or 5-methyl THF (Goldman, 1971b). Similarly, in studies with membrane vesicles loaded with sulfate or phosphate anions, uptake of MTX in the "trans" compartment was dramatically stimulated via a counter-transport mechanism (Yang, Sirotnak, & Dembo, 1984). Finally, thiamine pyrophosphate and ZMP (5-aminoimidazole-4-carboxamide ribonucleotide) are *bona fide* RFC substrates, which, when present intracellularly, can trans-stimulate folate influx by RFC while inhibiting (anti)folate export (Visentin, Zhao, & Goldman, 2012b; Zhao, Gao, et al., 2001). Collectively, these studies support a model whereby concentrative folate uptake by RFC is driven by a counter-transport mechanism involving an anionic cellular metabolite, perhaps an organic phosphate. However, neither the identity of the actual physiologic counter-anion(s) nor the detailed mechanism by which bidirectional anion fluxes are coupled is firmly established.

3.2. Structure and function of RFC

RFC belongs to the major facilitator superfamily (MFS) of transporters. MFS proteins transport an assortment of substrates including amino acids, neurotransmitters, sugars, vitamins, nucleosides, and organic phosphate in a uniport, symport, or antiport fashion (Chang, Lin, Keith Studley, Tran, & Saier, 2004). While no structure of RFC is yet available, three-dimensional crystal structures are available for RFC structural homologs including the bacterial lactose/proton symporter (LacY) (Abramson et al., 2003) and the inorganic phosphate/glycerol-3-phosphate antiporter

(GlpT) (Huang, Lemieux, Song, Auer, & Wang, 2003) from which to draw mechanistic inferences and to generate predictive homology models (Hou, Ye, Haska, & Matherly, 2006) that can be experimentally tested. By analogy with these bacterial transporters, membrane transport of folates by RFC would be expected to involve a physical movement of the carrier within the plasma membrane accompanied by alternate accessibilities of the aqueous substrate-binding cavity between the intra- and extracellular compartments. For RFC, as noted above, transport is driven by extrusion of anions down a large (intra- to extracellular) concentration gradient.

While the functional properties of RFC in cultured murine leukemia cells were first documented over 40 years ago (Goldman et al., 1968), it was only following cloning of RFCs from multiple species (Dixon et al., 1994; Moscow et al., 1995; Prasad et al., 1995; Williams & Flintoff, 1995; Williams, Murray, Underhill, & Flintoff, 1994; Wong et al., 1995) that it became possible to explore structural determinants of RFC transport that impact its physiology and pharmacology. These include its membrane topology, N-glycosylation, functionally and structurally important domains and amino acids, and, more recently, tertiary and quaternary structures.

RFC is characterized by 12 transmembrane domains (TMDs) and cytoplasmically oriented N- and C-termini (Fig. 4.2) (Cao & Matherly, 2004; Ferguson & Flintoff, 1999; Liu & Matherly, 2002). Human RFC includes 591 amino acids and is N-glycosylated at Asn58 in the loop domain connecting TMD1 and TMD2 (Matherly, Czajkowski, & Angeles, 1991; Wong, Zhang, Proefke, & Matherly, 1998). RFCs from various species are conserved (64–66% between humans and rodents) with substantial conservation within the TMDs and the lowest homologies in the loop domain connecting TMD6 and TMD7, and in the N- and C-termini (Matherly & Hou, 2008). In structure–function studies of RFC, investigators have sought to identify domains and/or residues that contribute to binding and/or translocation of (anti)folate substrates, as well as TMD helix packing associations that facilitate folate substrate binding and translocation.

Deletions of the N- and C-termini exclusive of the TMDs from ectopically expressed human RFC did not impact surface targeting or transport activity in HEK-293 and HuTu-80 cells (Marchant, Subramanian, Parker, & Said, 2002). While the same findings were obtained with rodent RFCs for the N-terminus, the impact of deleting the C-terminal domain ranged from modest to a complete loss of expression (Sadlish, Williams, & Flintoff, 2002; Sharina, Zhao, Wang, Babani, & Goldman, 2002). Larger deletions of human RFC that included TMD segments abolished surface targeting (Marchant et al., 2002).

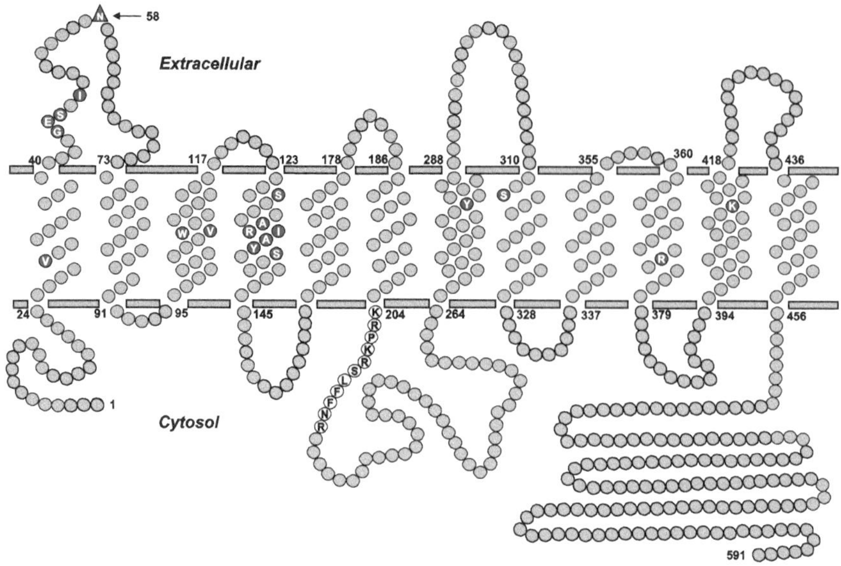

Figure 4.2 Human RFC topology model. A topology model is shown for human RFC, with 12 TMDs, internal N- and C-termini, and a loop domain connecting TMD6 and 7. The structurally and functionally important amino acids, as described in the text, are shown as red circles. A conserved stretch of amino acids (Lys204–Arg214) in the TMD6/TMD7 loop domain, which is important for transport activity, is shown as yellow circles. N-glycosylation occurs at Asn58, which is labeled as a red triangle. (See the color plate.)

The linker domain connecting TMD6 and TMD7 in RFC (Fig. 4.2) is poorly conserved with the exception of a Lys204–Arg214 segment. Deleting segments (49 or 60 amino acids; positions 215–263 and 204–263, respectively) of the TMD6/TMD7 linker from human RFC abolished transport activity (Liu, Witt, & Matherly, 2003). Interestingly, replacing the deleted segments with 73 or 84 amino acid segments of nonhomologous sequence from SLC19A2 restored transport, although there was an absolute requirement for the human RFC 204–214 peptide sequence. Finally, when human RFC was expressed as individual TMD1–6 and TMD7–12 half molecules, transport activity was restored (Witt, Stapels, & Matherly, 2004). These studies establish that neither the N- or C-termini nor the TMD6/TMD7 linker is critical for folate substrate binding and membrane translocation. The primary role of the TMD6-TMD7 loop domain is to provide appropriate spacing between the TMD1–6 and TMD7–12 segments for optimal carrier function.

Studies with RFC mutants have implicated conserved amino acids in TMD1, TMD2, TMD4, TMD8, TMD10, and TMD11 as important for transport (Fig. 4.2) (Matherly et al., 2007). Structurally or functionally important amino acids identified from mutant studies include (numbers based on human RFC sequence) Val29, Gly44, Glu45, Ser46, Ile48, Val106, Trp107, Ser127, Ala132, Arg133, Tyr281, Ser313, and Arg373 (Brigle, Spinella, Sierra, & Goldman, 1995; Drori, Jansen, Mauritz, Peters, & Assaraf, 2000; Jansen et al., 1998; Liu & Matherly, 2001; Roy, Tolner, Chiao, & Sirotnak, 1998; Sadlish et al., 2002; Sharina et al., 2002; Wong et al., 1999; Zhao, Assaraf, & Goldman, 1998; Zhao, Gao, Babani, & Goldman, 2000; Zhao, Gao, & Goldman, 1999). Mutation of Lys404 in TMD11 of murine RFC (Lys411 in human RFC) to leucine resulted in 11- and 4-fold increases in K_i values for 5-methyl THF and 5-formyl THF, respectively (Sharina, Zhao, Wang, Babani, & Goldman, 2001). Binding of the γ-carboxyl group of MTX was localized to Lys411 of human RFC based on *N*-hydroxysuccinimide [^{3}H]MTX radioaffinity-labeling results (Deng, Hou, et al., 2008).

To identify domains that form the putative substrate-binding pocket/translocation pathway, Hou and coworkers (Hou, Stapels, Haska, & Matherly, 2005; Hou et al., 2006) used scanning cysteine accessibility methods (SCAM) with a functional "Cys-less" human RFC in which the 11 cysteine residues were replaced with serines (Cao & Matherly, 2003). In these studies, 282 cysteine mutants (in the Cys-less background) were individually expressed in RFC-null HeLa cells and treated with 2-sulfonatoethyl methanethiosulfonate (MTSES) so as to identify positions that are aqueous accessible (in spite of their TMD localization) and that may participate in folate substrate binding and/or translocation (Hou et al., 2005, 2006). From the patterns of transport inhibition by MTSES and protection from MTSES inhibition by leucovorin, residues localized in TMD4, TMD5, TMD7, TMD8, TMD10, and TMD11 were implicated in forming the putative substrate-binding pocket (Hou et al., 2005, 2006). Of the 282 RFC cysteine mutants tested, only 10 were completely inactive. Interestingly, these included a stretch of amino acids in TMD4 (Arg133, Ile134, Ala135, Tyr136, Ser138), Tyr281 in TMD7, Ser313 in TMD8, and Arg373 in TMD10. As described above, Arg133, Ser313, and Arg373 were previously implicated as structurally or functionally important (Liu & Matherly, 2001; Sadlish et al., 2002; Sharina et al., 2002; Zhao et al., 1999).

From the solved structures of the LacY and GlpT proteins, a three-dimensional homology model for human RFC was generated that included a membrane translocation pathway formed from TMD1, TMD2, TMD4, TMD5, TMD7, TMD8, TMD10, and TMD11, and mechanistically important roles for Tyr281, Ser313, Arg373, and Lys411 (Fig. 4.3) (Hou et al., 2006). Based on this homology model, site-directed mutagenesis of Ser313 (localized to TMD8) and homo-bifunctional cross-linking between TMD8 and juxtaposed TMD5 were used to further examine the functional significance of Ser313 and TMD8 in RFC membrane transport (Hou, Wu, Ye, Cherian, & Matherly, 2010). The results showed that there are substantial differences between structurally diverse RFC (anti)folate substrates in their binding to RFCs with mutated Ser313, and in inducing conformational changes involving the proximal end of TMD8, determined by protein cross-linking. Thus, there is direct support for an essential role of TMD8 and Ser313 in binding and/or membrane translocation of (anti)folate substrates (Hou, Wu, et al., 2010).

Human RFC, like many MFS proteins, exists as a homo-oligomer (Hou & Matherly, 2009). RFC monomers appear to function independently (i.e., oligomerization is not a prerequisite for transport) (Hou, Cherian, Drews, Wu, & Matherly, 2010). Nonetheless, co-folding of RFC monomers to form oligomeric RFC seems to be functionally important for trafficking from the ER to the cell surface (Hou & Matherly, 2009). Co-expressing wild-type and inactive mutant S138C RFCs produces a dominant-negative phenotype due to decreased surface expression of both mutant and wild-type RFCs. Thus, it appears that RFC oligomerization may serve an important regulatory function. Clearly, further characterization of oligomeric RFC, including structural and regulatory determinants of oligomer formation, is essential, given its potential importance to understanding the broader physiologic and pharmacologic role of this transporter.

4. PROTON-COUPLED FOLATE TRANSPORTER (SLC46A1)

4.1. Properties of PCFT

In 2006, a new folate transporter was identified—PCFT (Qiu et al., 2006). While PCFT also belongs to the superfamily of solute carriers, it is functionally distinct from the RFC in that PCFT functions optimally at acidic rather than neutral pH, and it shows several differences in its specificities for particular (anti)folate substrates (Kugel Desmoulin et al., 2012; Zhao & Goldman, 2007).

The cloning of PCFT provided a molecular basis for the long-recognized but uncharacterized low pH folate transport activity in many normal and

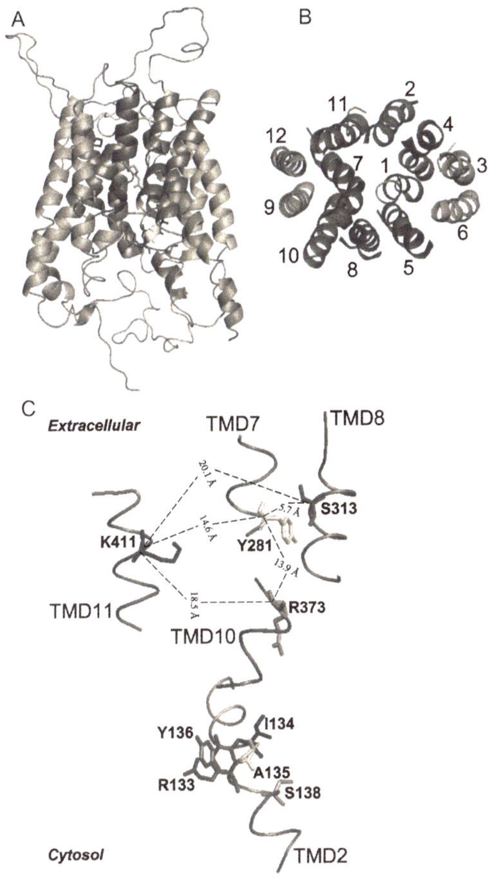

Figure 4.3 Three-dimensional (3D) homology models of human RFC. A 3D model for human RFC is presented, based on structure alignments between RFC and LacY/GlpT and experimental data. (A) A side view of the RFC for which the extended C-terminal segment is truncated at Lys479. TMD1, TMD2, TMD4, and TMD5 of the N-terminal region and TMD7, TMD8, TMD10, and TMD11 of the C-terminal region are involved in formation of the hydrophilic binding site for anionic folates (colored *sky blue*). TMD3, TMD6, TMD9, and TMD12 are buried in the lipid bilayer and do not directly participate in substrate binding (colored *green*). Panel (A) also depicts key amino acids (shown in assorted colors) that may contribute to the binding pocket for anionic folate substrates, as described in the text. (B) A cytosolic view of only the TMD segments of the human RFC molecule so that the order of helix packing can be seen easily. TMD coloring is the same as described in (A). (C) Enhanced view of the hypothetical substrate-binding site comprising the same key amino acids depicted in (A), including Lys411, Ser313, Tyr281, and Arg373, as described in the text. Other residues that may contribute to the substrate-binding pocket are also shown and include Arg133, Ile134, Ala135, Tyr136, and Ser138. The physical distances between the α carboxyl groups of Lys411, Ser313, Tyr281, and Arg373 are given in angstroms. *This figure was originally published in* The Journal of Biological Chemistry, *by Hou et al. (2006). © The American Society for Biochemistry and Molecular Biology.* (See the color plate.)

malignant cells, most notably in the small intestine (Assaraf, Babani, & Goldman, 1998; Henderson & Strauss, 1990; Kuhnel, Chiao, & Sirotnak, 2000; Sierra & Goldman, 1998; Zhao, Gao, Hanscom, & Goldman, 2004; Zhao & Goldman, 2007). The critical role of PCFT in intestinal absorption of dietary folates was confirmed by the demonstration of loss-of-function mutations in the human PCFT gene in patients with HFM (Atabay et al., 2010; Borzutzky et al., 2009; Diop-Bove, Jain, Scaglia, & Goldman, 2013; Lasry et al., 2008; Mahadeo et al., 2011, 2010; Meyer et al., 2010; Min et al., 2008; Qiu et al., 2006; Shin et al., 2011, 2010; Zhao et al., 2007). PCFT knockout mice were developed that largely recapitulate the HFM syndrome seen in humans (Salojin et al., 2011). These mice fail to absorb dietary folates, resulting in undetectable levels of serum folate and elevated plasma homocysteine.

Within the intestine, the highest PCFT levels are found at the apical brush border membrane of the proximal jejunum and duodenum, while levels decrease markedly in other segments of the intestine and colon (Inoue et al., 2008; Qiu et al., 2006, 2007; Urquhart et al., 2010). Other major sites of PCFT expression include the kidney, placenta, spleen, the sinusoidal membrane of the liver, the basolateral membrane of the choroid plexus, and the retinal pigment epithelium (Inoue et al., 2008; Kugel Desmoulin et al., 2012; Qiu et al., 2006; Umapathy et al., 2007; Wollack et al., 2008; Zhao, Min, et al., 2009). PCFT is abundantly expressed in human tumor cell lines (e.g., breast, lung, ovarian, and lung) and at very low levels in human leukemia cell lines (Gonen, Bram, & Assaraf, 2008; Kugel Desmoulin et al., 2012). While PCFT is likely to be active in the acidic microenvironment characterizing the upper GI and in solid tumors, it seems unlikely that PCFT represents an important mechanism of folate uptake in most normal tissues in which it is expressed. From results in HFM patients, PCFT is important for folate transport across the choroid plexus (Wollack et al., 2008; Zhao, Min, et al., 2009). Clearly, the physiologic role of PCFT in tissues not normally associated with a low pH microclimate needs to be clarified. It is conceivable that PCFT transports 5-methyl THF and related folates in normal tissues in areas of localized acidification or when expressed at high levels (Zhao, Matherly, & Goldman, 2009). However, at comparatively neutral pHs of most tissues, RFC is more efficient at delivering reduced folates than PCFT (Zhao et al., 2008).

Investigators have extensively characterized the transport properties of PCFT expressed in transfected cell lines and in *Xenopus* oocytes microinjected with PCFT cRNAs (Kugel Desmoulin et al., 2011b, 2010; Nakai et al., 2007; Qiu et al., 2006; Wang et al., 2011; Zhao & Goldman, 2007). A distinguishing characteristic of PCFT is its acidic optimum (pH 5–5.5) (Nakai et al., 2007; Qiu et al., 2006; Zhao & Goldman, 2007). As the pH increases from pH 5.5, transport decreases dramatically. Above pH 7, transport is almost undetectable, although the extent of residual transport varies for different substrates (Kugel Desmoulin et al., 2011a, 2010; Wang et al., 2010, 2011; Zhao & Goldman, 2007; Zhao et al., 2008). RFC substrates including MTX, PMX, raltitrexed, aminopterin, pralatrexate, 5-methyl THF, and 5-formyl THF are all PCFT substrates (Deng et al., 2009; Kugel Desmoulin et al., 2012; Menter et al., 2012; Zhao et al., 2008). However, the antifolates PT523 and GW1843U89 do not appear to be transported by PCFT (Deng et al., 2009; Zhao & Goldman, 2007). In contrast to RFC, PCFT has similar K_ts for transport of reduced folates (5-methyl THF, 5-formyl THF) and folic acid (Zhao & Goldman, 2007). PCFT is stereospecific for (6*S*)5-formyl THF, and for L- over D-aminopterin (Menter et al., 2012; Zhao & Goldman, 2007). Whereas the 5-substituted pyrrolo[2,3-*d*]pyrimidine antifolate PMX is among the best substrates for transport by PCFT (Zhao & Goldman, 2007), a novel series of 6-subsitituted pyrrolo[2,3-*d*]pyrimidine thienoyl antifolate analogs were recently reported with affinities and transport activities for PCFT comparable to those for PMX (Kugel Desmoulin et al., 2011a; Kugel Desmoulin, Wang, et al., 2012; Wang et al., 2010, 2011). For the lead compound of this series, PCFT transport by malignant mesothelioma xenografts in severe combined immunodeficient mice resulted in potent antitumor effects in early- and late-stage disease (Cherian et al., 2013). Notably, the binding affinities for both 5- and 6-substituted pyrrolo[2,3-*d*]pyrimidine analogs appear to be less affected by pH than for other PCFT substrates.

PCFT activity is not affected by removing extracellular Na^+, K^+, Ca^{2+}, Mg^{2+}, or Cl^- (Qiu et al., 2006) or by changing membrane potential (Inoue et al., 2008), data consistent with an electroneutral mode of PCFT transport. However, there is a requirement for an inwardly directed proton gradient for PCFT-mediated folate transport based on the following observations. (i) Dissipation of the transmembrane proton gradient by a proton ionophore in *Xenopus* oocytes (Qiu et al., 2006) and a K^+/H^+-exchanging ionophore in HEK293 cells (Inoue et al., 2008) reduced transport by PCFT. (ii) A transvesicular pH gradient was reported to result in increased unidirectional folate transport and substantial transmembrane folate concentration gradients in rabbit jejunum, from the low pH to the high pH compartment, consistent with a proton-coupled process (Schron, Washington, & Blitzer, 1985). (iii) Treatment of HeLa cells with nitrate or bisulfite inhibited PCFT transport by abolishing the pH gradient (Zhao, Visentin, Suadicani, & Goldman, 2013). Finally, (iv) cellular acidification accompanied folate transport into *Xenopus* oocytes, confirming proton coupling (Unal, Zhao, et al., 2009).

Since folates are bivalent anions, this means that more than two protons must be co-transported with each folate molecule to account for the net positive charge of the PCFT–folate–proton complex (Qiu et al., 2006, 2007). Thus, PCFT functions as a folate–proton symporter such that the downhill flow of protons is coupled to the uphill flow of folates into cells. Interestingly, PCFT can also function in the absence of a transmembrane pH gradient. Under these conditions, folate transport is driven by membrane potentials (Qiu et al., 2006; Umapathy et al., 2007), similar to the divalent metal-ion transporter DMT1 (Mackenzie, Ujwal, Chang, Romero, & Hediger, 2006). At low pH, PCFT exhibited channel-like activities (i.e., protons can flow independent of folates) (Mahadeo et al., 2010; Unal, Zhao, et al., 2009).

4.2. Structure and function of PCFT

Human PCFT includes 459 amino acids and has a predicted molecular mass of 49.8 kDa. PCFT is predicted to contain 12 TMDs comprised of hydrophobic stretches of amino acids with cytosolic-oriented N- and C-termini (Fig. 4.4). This structure has been experimentally confirmed by immunofluorescence staining of hemagglutinin (HA)-tagged PCFT constructs and by SCAM with MTSEA (2-aminoethyl methanethiosulfonate)-biotin (Unal, Zhao, Qiu, & Goldman, 2008; Zhao, Unal, Shin, & Goldman, 2010). PCFT has two N-glycosylation sites (Asn58 and Asn68) in the loop domain between TMD1 and TMD2 (Unal et al., 2008). Mutating either of the asparagine residues to glutamine had no effect on PCFT expression and function (Unal et al., 2008). However, mutating both Asn58 and Asn68 to glutamine decreased transport activity to 40% of that for wild-type PCFT. In contrast to RFC which is expressed at the basolateral membrane (see above), PCFT (with a C-terminal, yellow fluorescent protein tag) was localized to the apical membrane of Madin-Darby canine kidney and Caco-2 cells (Subramanian, Marchant, & Said, 2008). Truncating the C-terminus of human PCFT at position 449 did not impact apical membrane targeting or transporter activity (Subramanian et al., 2008).

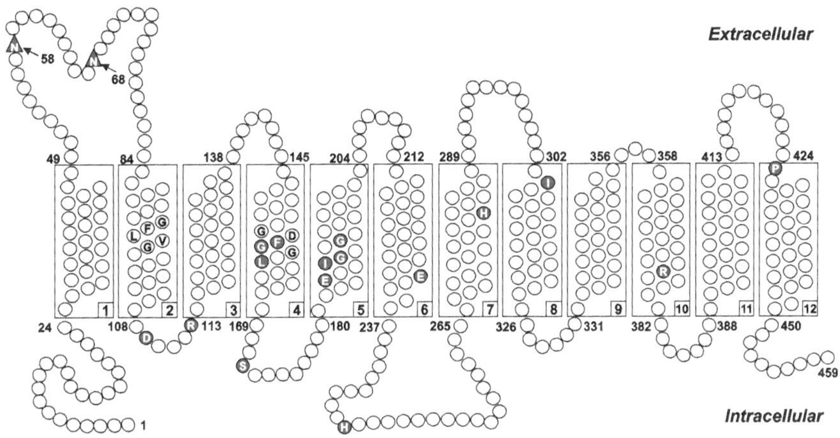

Figure 4.4 Schematic structure of human PCFT membrane topology. A topology model is shown for human PCFT, with 12 TMDs and internal N- and C-termini. Structurally or functionally important amino acids, as determined from published mutagenesis studies and in patients with hereditary folate malabsorption (HFM), are shown as red circles. GXXXG putative oligomerization motifs are shown as yellow circles (Phe157 and Gly158 in the $G^{155}XXXG^{159}$ motif are shown as red circles because they are also structurally and functionally important). N-glycosylation occurs at Asn58 and Asn68 (shown as red triangles). (See the color plate.)

A number of structurally and functionally important amino acids in human PCFT have been identified. This includes molecular characterization of human PCFT mutants that result in loss of function in HFM cases, and systematic mutagenesis of these HFM mutants. From considerations of species homologies and amino acid charge characteristics in relation to their TMD localization, other residues have been identified as important to PCFT transport. Thus, residues have been implicated as critical to proton coupling (i.e., Glu185 (TMD5); Unal, Zhao, & Goldman, 2009) or substrate binding (i.e., His281 (TMD7); Unal, Zhao, et al., 2009), or both (i.e., Arg376 (TMD10); Mahadeo et al., 2010) (Fig. 4.4).

His247, which is localized to the loop region separating TMD6 and TMD7, is predicted to lie in the cytoplasmic opening of the water-filled translocation pathway where it interacts with Ser172 to limit substrate access to the folate-binding pocket (thus determining substrate selectivity) (Unal, Zhao, et al., 2009). His281 faces the extracellular region in TMD7 (Fig. 4.4) and is believed to play an important role in PCFT protonation, which augments substrate binding to the carrier (Unal, Zhao, et al., 2009).

The conserved segment (DXXGRR) linking TMD2 and TMD3 (residues 109–114), including a β-turn, is of particular interest (Lasry et al., 2008; Shin et al., 2010; Subramanian et al., 2008; Zhao et al., 2007). Key residues in this stretch include Asp109 and Arg113 that are likely to play important roles in substrate binding and/or translocation since both conservative and nonconservative replacements completely abolish transport (Lasry, Berman, Glaser, Jansen, & Assaraf, 2009; Shin et al., 2010). According to a PCFT homology model based on the GlpT template, Arg113 protrudes into a hydrophobic cavity formed by TMD1, TMD3, TMD4, and TMD6 (Lasry et al., 2008). However, this has not been experimentally confirmed.

Other residues that may be functionally important in the human PCFT include Leu161 (TMD4), Glu232 (TMD6), and Ile304 (TMD8) (Fig. 4.4) (Zhao, Shin, Diop-Bove, Ovits, & Goldman, 2011). Loss of transport was associated with a reduced rate of carrier translocation (Glu232Gly mutant) or lower substrate affinities (Ile304Phe and Leu161Arg mutants). Mutating Pro425 to Arg in the loop junction flanking TMD12 (Fig. 4.4) eliminated binding of MTX and other substrates to PCFT, but preserved binding of PMX (Shin, Zhao, Yap, Fiser, & Goldman, 2012). Given its location, Pro425 is unlikely to directly participate in substrate binding. Rather, mutation at this position likely produces a conformational change that selectively inhibits binding of particular (anti)folate substrates. Finally, from mutant studies, Gly189 and Gly192 (TMD5) were implicated as functionally important (Zhao, Shin, Fiser, & Goldman, 2012).

In SCAM studies on human PCFT, membrane-impermeable methanethiosulfonate reagents reacted with Cys replacements at positions 157, 158, and 161 in TMD4 and at position 188 in TMD5 (Shin, Zhao, Fiser, & Goldman, 2013; Zhao et al., 2012). Reactions could be protected by PMX, placing these residues within or near the substrate-binding site(s) in human PCFT.

Based on the results with several complementary techniques, the human PCFT forms homo-oligomers both in detergent solution and *in situ* (Hou et al., 2012). Oligomerization of PCFT was demonstrated by protein cross-linking of the ectopically expressed carrier with 1,1-methanediyl bismethanethiosulfonate (MTS-1-MTS), and by fluorescence resonance energy transfer between co-expressed YPet- and ECFP*-tagged PCFT monomers (Hou et al., 2012). Physical associations between HA- and His_{10}-tagged PCFT monomers were established by co-expression in PCFT-null HeLa cells and co-binding to nickel affinity columns. Furthermore, the PCFT dimer was shown to be the dominant species by blue native gel electrophoresis (Hou et al., 2012). The functional significance of oligomerization was established by co-expression of wild-type and mutant Pro425Arg PCFT forms in PCFT-null HeLa cells. In these experiments, there was a "dominant-positive" functional phenotype, establishing positive cooperativity between PCFT monomers and a functional "rescue" of inactive mutant PCFT by wild-type protein (Hou et al., 2012). Based on these results, an "alternate access" model for PCFT transport, analogous to that suggested for LacY (Abramson et al., 2003) and adapted from that for monomeric PCFT (Unal et al., 2009), was proposed (Hou et al., 2012) that incorporates a role for PCFT oligomerization.

Duddempudi, Nakashe, Blanton, and Jansen (2013) suggested that PCFT may not oligomerize in plasma membranes prepared by polymerization with colloidal silica and polyacrylic acid from *Xenopus* oocytes or in Chinese hamster ovary cells. However, this conclusion was based on a limited number of methods. Furthermore, since different methods and metrics were employed in this report and the more extensive study of Hou et al. (2012), it is impossible to reconcile the conclusions of these reports.

The human PCFT primary sequence includes GXXXG motifs in TMD2 (amino acids 93–97) and TMD4 (amino acids 155–159) (Fig. 4.4) that are analogous to "dimerization motifs" in other amphipathic proteins (Duan, Wu, & You, 2011; Polgar et al., 2010). Mutating Gly93 and Gly97 to Ala did not inhibit transport activity or oligomer formation, based on thiol-reactive (MTS-1-MTS) protein cross-linking (Zhao et al., 2012). Unfortunately, parallel studies on the GXXXG motif in TMD4 were not performed. Interestingly, a TMD6 interface between PCFT monomers was suggested by the finding that MTS-1-MTS treatment generated cross-links between Cys229 (in TMD6) in individual PCFT monomers, but not when Cys229 was mutated to Ser (Zhao et al., 2012).

5. SUMMARY

This chapter focuses on the biology of the major facilitative membrane transport systems for folate cofactors, RFC and PCFT. Folates are essential vitamins, and folate deficiency has been implicated in a variety of heath disorders ranging from cardiovascular disease to neurodegenerative disease and cancer (Lucock, 2000; Matherly, 2004). Further, folate-based therapeutics (e.g., antifolates) continue to occupy an important niche for treating cancer as well as other pathological conditions (Gonen & Assaraf, 2012a; Kugel Desmoulin et al., 2012; Salojin et al., 2011; Visentin et al., 2012a). These include recent FDA-approved drugs such as PMX and pralatrexate, with many more folate-based therapies in the pipeline. Losses of RFC or PCFT transport can have potential physiological and developmental consequences, particularly in the context of folate deficiency (Matherly, 2004; Matherly et al., 2007; Zhao et al., 2009), or may result in drug resistance with antifolates used for cancer therapy (Gonen & Assaraf, 2012b; Matherly et al., 2007; Zhao & Goldman, 2003).

RFC was functionally characterized beginning in the late 1960s as a neutral pH membrane transport mechanism for clinically relevant antifolates such as MTX that when lost resulted in MTX resistance (Goldman et al., 1968; Sirotnak, Kurita, & Hutchison, 1968). RFC was cloned and its molecular features were exhaustively documented beginning in the mid-1990s (Matherly & Hou, 2008; Matherly et al., 2007). It was soon recognized that RFC was the major folate transport system in mammalian cells, consistent with its ubiquitous tissue expression (Whetstine et al., 2002) and its absolute requirement for fetal development (Zhao, Russell, et al., 2001). By contrast, although an acidic pH transporter had long been recognized, the molecular characteristics of PCFT were not demonstrated until 2005 when it was originally classified as a heme transporter (Shayeghi et al., 2005). By 2006, PCFT was redefined as the molecular entity responsible for the acidic folate transport activity involved in intestinal uptake of dietary folates (Qiu et al., 2006). Studies soon followed that established that a low pH transport activity detected in many human solid tumor cells was in fact PCFT, and that PCFT was abundantly expressed in a wide range of human tumors and was highly active at pH conditions approximating those found in the tumor microenvironment (Kugel Desmoulin et al., 2011a). This soon led to the development of novel tumor-targeted therapies typified by 6-substituted pyrrolo [2,3-*d*]pyrimidine antifolates with selective membrane transport by PCFT over RFC (Kugel Desmoulin et al., 2012).

An important prerequisite for effective clinical applications of RFC and PCFT transport associated with folate deficiency or in targeted therapy of cancer is understanding of the structure and function of these critical transport systems. Indeed, since their cloning, structure–function research has led to remarkably detailed molecular information, including the membrane topology, N-glycosylation, energetics, and roles of functionally and structurally important domains and amino acids for both RFC and PCFT. These insights have shed important new light on the causal basis for chemotherapy resistance to MTX associated with the loss of RFC function, and for the rare autosomal recessive condition, HFM, involving mutations in the PCFT gene and protein. Based on structural homologies to the bacterial MFS transporters GlpT and LacY, homology models for human RFC and human PCFT have been developed. The models provide new mechanistic insights and experimentally testable hypotheses that could further facilitate efforts to design a new generation of targeted therapeutics with selective membrane transport by PCFT over RFC. Unfortunately, detailed X-ray diffraction studies on membrane proteins such as RFC and PCFT remain a formidable challenge.

Structure and function considerations for human RFC and human PCFT must now include higher-order structures such as homo-oligomers that may impact transport mechanisms and regulate transporter expression and function. Indeed, evidence suggests that formation of homo-oligomers between monomeric RFC and PCFT proteins may be essential for intracellular trafficking and/or for carrier function. Formation of PCFT or RFC oligomers, including both wild-type and mutant forms, may also be important to clinical phenotypes involving mutant transporters in patients with HFM or in cancer patients being treated with cytotoxic antifolates. Clearly, better understanding of the broader significance and structural determinants of these higher-order oligomeric structures will be an important area for future study as these may lead to approaches for modulating the formation of oligomers for therapeutic impact.

ACKNOWLEDGMENT

This study was supported by Grant R01 CA53535 from the National Cancer Institute, National Institutes of Health.

REFERENCES

Abramson, J., Smirnova, I., Kasho, V., Verner, G., Kaback, H. R., & Iwata, S. (2003). Structure and mechanism of the lactose permease of Escherichia coli. *Science*, *301*(5633), 610–615.

Assaraf, Y. G., Babani, S., & Goldman, I. D. (1998). Increased activity of a novel low pH folate transporter associated with lipophilic antifolate resistance in Chinese hamster ovary cells. *The Journal of Biological Chemistry*, *273*(14), 8106–8111.

Atabay, B., Turker, M., Ozer, E. A., Mahadeo, K., Diop-Bove, N., & Goldman, I. D. (2010). Mutation of the proton-coupled folate transporter gene (PCFT-SLC46A1) in Turkish siblings with hereditary folate malabsorption. *Pediatric Hematology and Oncology*, *27*(8), 614–619.

Borzutzky, A., Crompton, B., Bergmann, A. K., Giliani, S., Baxi, S., Martin, M., et al. (2009). Reversible severe combined immunodeficiency phenotype secondary to a mutation of the proton-coupled folate transporter. *Clinical Immunology*, *133*(3), 287–294.

Brigle, K. E., Spinella, M. J., Sierra, E. E., & Goldman, I. D. (1995). Characterization of a mutation in the reduced folate carrier in a transport defective L1210 murine leukemia cell line. *The Journal of Biological Chemistry*, *270*(39), 22974–22979.

Cao, W., & Matherly, L. H. (2003). Characterization of a cysteine-less human reduced folate carrier: Localization of a substrate-binding domain by cysteine-scanning mutagenesis and cysteine accessibility methods. *The Biochemical Journal*, *374*(Pt 1), 27–36.

Cao, W., & Matherly, L. H. (2004). Analysis of the membrane topology for transmembrane domains 7-12 of the human reduced folate carrier by scanning cysteine accessibility methods. *The Biochemical Journal*, *378*(Pt 1), 201–206.

Chang, A. B., Lin, R., Keith Studley, W., Tran, C. V., & Saier, M. H., Jr. (2004). Phylogeny as a guide to structure and function of membrane transport proteins. *Molecular Membrane Biology*, *21*(3), 171–181.

Cherian, C., Kugel Desmoulin, S., Wang, L., Polin, L., White, K., Kushner, J., et al. (2013). Therapeutic targeting malignant mesothelioma with a novel 6-substituted pyrrolo[2,3-d] pyrimidine thienoyl antifolate via its selective uptake by the proton-coupled folate transporter. *Cancer Chemotherapy and Pharmacology*, *71*(4), 999–1011.

Chladek, J., Martinkova, J., Simkova, M., Vaneckova, J., Koudelkova, V., & Nozickova, M. (1998). Pharmacokinetics of low doses of methotrexate in patients with psoriasis over the early period of treatment. *European Journal of Clinical Pharmacology*, *53*(6), 437–444.

Deng, Y., Hou, Z., Wang, L., Cherian, C., Wu, J., Gangjee, A., et al. (2008). Role of lysine 411 in substrate carboxyl group binding to the human reduced folate carrier, as determined by site-directed mutagenesis and affinity inhibition. *Molecular Pharmacology*, *73*(4), 1274–1281.

Deng, Y., Wang, Y., Cherian, C., Hou, Z., Buck, S. A., Matherly, L. H., et al. (2008). Synthesis and discovery of high affinity folate receptor-specific glycinamide ribonucleotide formyltransferase inhibitors with antitumor activity. *Journal of Medicinal Chemistry*, *51*(16), 5052–5063.

Deng, Y., Zhou, X., Kugel Desmoulin, S., Wu, J., Cherian, C., Hou, Z., et al. (2009). Synthesis and biological activity of a novel series of 6-substituted thieno[2,3-d]pyrimidine antifolate inhibitors of purine biosynthesis with selectivity for high affinity folate receptors over the reduced folate carrier and proton-coupled folate transporter for cellular entry. *Journal of Medicinal Chemistry*, *52*(9), 2940–2951.

Diop-Bove, N., Jain, M., Scaglia, F., & Goldman, I. D. (2013). A novel deletion mutation in the proton-coupled folate transporter (PCFT; SLC46A1) in a Nicaraguan child with hereditary folate malabsorption. *Gene*, *527*(2), 673–674.

Dixon, K. H., Lanpher, B. C., Chiu, J., Kelley, K., & Cowan, K. H. (1994). A novel cDNA restores reduced folate carrier activity and methotrexate sensitivity to transport deficient cells. *The Journal of Biological Chemistry*, *269*(1), 17–20.

Drori, S., Jansen, G., Mauritz, R., Peters, G. J., & Assaraf, Y. G. (2000). Clustering of mutations in the first transmembrane domain of the human reduced folate carrier in GW1843U89-resistant leukemia cells with impaired antifolate transport and augmented folate uptake. *The Journal of Biological Chemistry*, *275*(40), 30855–30863.

Duan, P., Wu, J., & You, G. (2011). Mutational analysis of the role of GXXXG motif in the function of human organic anion transporter 1 (hOAT1). *International Journal of Biochemistry and Molecular Biology*, *2*(1), 1–7.

Duch, D. S., Banks, S., Dev, I. K., Dickerson, S. H., Ferone, R., Heath, L. S., et al. (1993). Biochemical and cellular pharmacology of 1843U89, a novel benzoquinazoline inhibitor of thymidylate synthase. *Cancer Research*, *53*(4), 810–818.

Duddempudi, P. K., Nakashe, P., Blanton, M. P., & Jansen, M. (2013). The monomeric state of the proton-coupled folate transporter represents the functional unit in the plasma membrane. *The FEBS Journal*, *280*(12), 2900–2915.

Elnakat, H., & Ratnam, M. (2004). Distribution, functionality and gene regulation of folate receptor isoforms: Implications in targeted therapy. *Advanced Drug Delivery Reviews*, *56*(8), 1067–1084.

Ferguson, P. L., & Flintoff, W. F. (1999). Topological and functional analysis of the human reduced folate carrier by hemagglutinin epitope insertion. *The Journal of Biological Chemistry, 274*(23), 16269–16278.

Gibbs, D. D., Theti, D. S., Wood, N., Green, M., Raynaud, F., Valenti, M., et al. (2005). BGC 945, a novel tumor-selective thymidylate synthase inhibitor targeted to alpha-folate receptor-overexpressing tumors. *Cancer Research, 65*(24), 11721–11728.

Goldman, I. D. (1971a). The characteristics of the membrane transport of amethopterin and the naturally occurring folates. *Annals of the New York Academy of Sciences, 186*, 400–422.

Goldman, I. D. (1971b). A model system for the study of heteroexchange diffusion: Methotrexate-folate interactions in L1210 leukemia and Ehrlich ascites tumor cells. *Biochimica et Biophysica Acta, 233*(3), 624–634.

Goldman, I. D., Lichtenstein, N. S., & Oliverio, V. T. (1968). Carrier-mediated transport of the folic acid analogue, methotrexate, in the L1210 leukemia cell. *The Journal of Biological Chemistry, 243*(19), 5007–5017.

Gonen, N., & Assaraf, Y. G. (2012a). Antifolates in cancer therapy: Structure, activity and mechanisms of drug resistance. *Drug Resistance Updates, 15*(4), 183–210.

Gonen, N., & Assaraf, Y. G. (2012b). Antifolates in cancer therapy: Structure, activity and mechanisms of drug resistance. *Drug Resistance Updates, 15*(4), 183–210.

Gonen, N., Bram, E. E., & Assaraf, Y. G. (2008). PCFT/SLC46A1 promoter methylation and restoration of gene expression in human leukemia cells. *Biochemical and Biophysical Research Communications, 376*(4), 787–792.

Henderson, G. B., & Strauss, B. P. (1990). Characteristics of a novel transport system for folate compounds in wild-type and methotrexate-resistant L1210 cells. *Cancer Research, 50*(6), 1709–1714.

Henderson, G. B., & Zevely, E. M. (1980). Transport of methotrexate in L1210 cells: Effect of ions on the rate and extent of uptake. *Archives of Biochemistry and Biophysics, 200*(1), 149–155.

Henderson, G. B., & Zevely, E. M. (1981). Anion exchange mechanism for transport of methotrexate in L1210 cells. *Biochemical and Biophysical Research Communications, 99*(1), 163–169.

Henderson, G. B., & Zevely, E. M. (1983a). Structural requirements for anion substrates of the methotrexate transport system in L1210 cells. *Archives of Biochemistry and Biophysics, 221*(2), 438–446.

Henderson, G. B., & Zevely, E. M. (1983b). Use of non-physiological buffer systems in the analysis of methotrexate transport in L1210 cells. *Biochemistry International, 6*(4), 507–515.

Hou, Z., Cherian, C., Drews, J., Wu, J., & Matherly, L. H. (2010). Identification of the minimal functional unit of the homo-oligomeric human reduced folate carrier. *The Journal of Biological Chemistry, 285*(7), 4732–4740.

Hou, Z., Kugel Desmoulin, S., Etnyre, E., Olive, M., Hsiung, B., Cherian, C., et al. (2012). Identification and functional impact of homo-oligomers of the human proton-coupled folate transporter. *The Journal of Biological Chemistry, 287*(7), 4982–4995.

Hou, Z., & Matherly, L. H. (2009). Oligomeric structure of the human reduced folate carrier: Identification of homo-oligomers and dominant-negative effects on carrier expression and function. *The Journal of Biological Chemistry, 284*(5), 3285–3293.

Hou, Z., Stapels, S. E., Haska, C. L., & Matherly, L. H. (2005). Localization of a substrate binding domain of the human reduced folate carrier to transmembrane domain 11 by radioaffinity labeling and cysteine-substituted accessibility methods. *The Journal of Biological Chemistry, 280*(43), 36206–36213.

Hou, Z., Wu, J., Ye, J., Cherian, C., & Matherly, L. H. (2010). Substrate-specific binding and conformational changes involving Ser313 and transmembrane domain 8 of the human reduced folate carrier, as determined by site-directed mutagenesis and protein cross-linking. *The Biochemical Journal, 430*(2), 265–274.

Hou, Z., Ye, J., Haska, C. L., & Matherly, L. H. (2006). Transmembrane domains 4, 5, 7, 8, and 10 of the human reduced folate carrier are important structural or functional components of the transmembrane channel for folate substrates. *The Journal of Biological Chemistry, 281*(44), 33588–33596.

Huang, Y., Lemieux, M. J., Song, J., Auer, M., & Wang, D. N. (2003). Structure and mechanism of the glycerol-3-phosphate transporter from Escherichia coli. *Science, 301*(5633), 616–620.

Inoue, K., Nakai, Y., Ueda, S., Kamigaso, S., Ohta, K. Y., Hatakeyama, M., et al. (2008). Functional characterization of PCFT/HCP1 as the molecular entity of the carrier-mediated intestinal folate transport system in the rat model. *American Journal of Physiology. Gastrointestinal and Liver Physiology, 294*(3), G660–G668.

Jacques, P. F., Selhub, J., Bostom, A. G., Wilson, P. W., & Rosenberg, I. H. (1999). The effect of folic acid fortification on plasma folate and total homocysteine concentrations. *The New England Journal of Medicine, 340*(19), 1449–1454.

Jansen, G. (1999). Receptor-and carrier-mediated transport systems for folates and antifolates. In A. L. Jackman (Ed.), *Antifolate drugs in cancer therapy* (pp. 293–321). Totowa, NJ: Humana Press Inc.

Jansen, G., Mauritz, R., Drori, S., Sprecher, H., Kathmann, I., Bunni, M., et al. (1998). A structurally altered human reduced folate carrier with increased folic acid transport mediates a novel mechanism of antifolate resistance. *The Journal of Biological Chemistry, 273*(46), 30189–30198.

Kamen, B. A., & Smith, A. K. (2004). A review of folate receptor alpha cycling and 5-methyltetrahydrofolate accumulation with an emphasis on cell models in vitro. *Advanced Drug Delivery Reviews, 56*(8), 1085–1097.

Kugel Desmoulin, S., Hou, Z., Gangjee, A., & Matherly, L. H. (2012). The human proton-coupled folate transporter: Biology and therapeutic applications to cancer. *Cancer Biology & Therapy, 13*(14), 1355–1373.

Kugel Desmoulin, S., Wang, L., Hales, E., Polin, L., White, K., Kushner, J., et al. (2011a). Therapeutic targeting of a novel 6-substituted pyrrolo [2,3-d]pyrimidine thienoyl antifolate to human solid tumors based on selective uptake by the proton-coupled folate transporter. *Molecular Pharmacology, 80*(6), 1096–1107.

Kugel Desmoulin, S., Wang, L., Hales, E., Polin, L., White, K., Kushner, J., et al. (2011b). Therapeutic targeting of a novel 6-substituted pyrrolo [2,3-d]pyrimidine thienoyl antifolate to human solid tumors based on selective uptake by the proton-coupled folate transporter. *Molecular Pharmacology, 80*(6), 1096–1107.

Kugel Desmoulin, S., Wang, L., Polin, L., White, K., Kushner, J., Stout, M., et al. (2012). Functional loss of the reduced folate carrier enhances the antitumor activities of novel antifolates with selective uptake by the proton-coupled folate transporter. *Molecular Pharmacology, 82*(4), 591–600.

Kugel Desmoulin, S., Wang, Y., Wu, J., Stout, M., Hou, Z., Fulterer, A., et al. (2010). Targeting the proton-coupled folate transporter for selective delivery of 6-substituted pyrrolo[2,3-d]pyrimidine antifolate inhibitors of de novo purine biosynthesis in the chemotherapy of solid tumors. *Molecular Pharmacology, 78*(4), 577–587.

Kuhnel, J. M., Chiao, J. H., & Sirotnak, F. M. (2000). Contrasting effects of oncogene expression on two carrier-mediated systems internalizing folate compounds in Fisher rat 3T3 cells. *Journal of Cellular Physiology, 184*(3), 364–372.

Lasry, I., Berman, B., Glaser, F., Jansen, G., & Assaraf, Y. G. (2009). Hereditary folate malabsorption: A positively charged amino acid at position 113 of the proton-coupled folate transporter (PCFT/SLC46A1) is required for folic acid binding. *Biochemical and Biophysical Research Communications, 386*(3), 426–431.

Lasry, I., Berman, B., Straussberg, R., Sofer, Y., Bessler, H., Sharkia, M., et al. (2008). A novel loss-of-function mutation in the proton-coupled folate transporter from a patient with hereditary folate malabsorption reveals that Arg 113 is crucial for function. *Blood, 112*(5), 2055–2061.

Liu, X. Y., & Matherly, L. H. (2001). Functional interactions between arginine-133 and aspartate-88 in the human reduced folate carrier: Evidence for a charge-pair association. *The Biochemical Journal, 358*(Pt 2), 511–516.

Liu, X. Y., & Matherly, L. H. (2002). Analysis of membrane topology of the human reduced folate carrier protein by hemagglutinin epitope insertion and scanning glycosylation insertion mutagenesis. *Biochimica et Biophysica Acta, 1564*(2), 333–342.

Liu, X. Y., Witt, T. L., & Matherly, L. H. (2003). Restoration of high-level transport activity by human reduced folate carrier/ThTr1 thiamine transporter chimaeras: Role of the transmembrane domain 6/7 linker region in reduced folate carrier function. *The Biochemical Journal, 369*(Pt 1), 31–37.

Lu, S. C. (2000). S-Adenosylmethionine. *The International Journal of Biochemistry & Cell Biology, 32*(4), 391–395.

Lucock, M. (2000). Folic acid: Nutritional biochemistry, molecular biology, and role in disease processes. *Molecular Genetics and Metabolism, 71*(1–2), 121–138.

Mackenzie, B., Ujwal, M. L., Chang, M. H., Romero, M. F., & Hediger, M. A. (2006). Divalent metal-ion transporter DMT1 mediates both H^+-coupled Fe^{2+} transport and uncoupled fluxes. *Pflügers Archiv, 451*(4), 544–558.

Mahadeo, K. M., Diop-Bove, N., Ramirez, S. I., Cadilla, C. L., Rivera, E., Martin, M., et al. (2011). Prevalence of a loss-of-function mutation in the proton-coupled folate transporter gene (PCFT-SLC46A1) causing hereditary folate malabsorption in Puerto Rico. *The Journal of Pediatrics, 159*(4), 623–627.

Mahadeo, K., Diop-Bove, N., Shin, D., Unal, E. S., Teo, J., Zhao, R., et al. (2010). Properties of the Arg376 residue of the proton-coupled folate transporter (PCFT-SLC46A1) and a glutamine mutant causing hereditary folate malabsorption. *American Journal of Physiology. Cell Physiology, 299*(5), C1153–C1161.

Marchant, J. S., Subramanian, V. S., Parker, I., & Said, H. M. (2002). Intracellular trafficking and membrane targeting mechanisms of the human reduced folate carrier in mammalian epithelial cells. *The Journal of Biological Chemistry, 277*(36), 33325–33333.

Matherly, L. H. (2004). Human reduced folate carrier gene and transcript variants: Functional, physiologic, and pharmacologic consequences. *Current Pharmacogenetics, 2*, 287–298.

Matherly, L. H., Czajkowski, C. A., & Angeles, S. M. (1991). Identification of a highly glycosylated methotrexate membrane carrier in K562 human erythroleukemia cells up-regulated for tetrahydrofolate cofactor and methotrexate transport. *Cancer Research, 51*(13), 3420–3426.

Matherly, L. H., & Goldman, D. I. (2003). Membrane transport of folates. *Vitamins and Hormones, 66*, 403–456.

Matherly, L. H., & Hou, Z. (2008). Structure and function of the reduced folate carrier a paradigm of a major facilitator superfamily mammalian nutrient transporter. *Vitamins and Hormones, 79*, 145–184.

Matherly, L. H., Hou, Z., & Deng, Y. (2007). Human reduced folate carrier: Translation of basic biology to cancer etiology and therapy. *Cancer Metastasis Reviews, 26*(1), 111–128.

Menter, A., Thrash, B., Cherian, C., Matherly, L. H., Wang, L., Gangjee, A., et al. (2012). Intestinal transport of aminopterin enantiomers in dogs and humans with psoriasis is stereoselective: Evidence for a mechanism involving the proton-coupled folate transporter. *The Journal of Pharmacology and Experimental Therapeutics, 342*(3), 696–708.

Meyer, E., Kurian, M. A., Pasha, S., Trembath, R. C., Cole, T., & Maher, E. R. (2010). A novel PCFT gene mutation (p.Cys66LeufsX99) causing hereditary folate malabsorption. *Molecular Genetics and Metabolism, 99*(3), 325–328.

Min, S. H., Oh, S. Y., Karp, G. I., Poncz, M., Zhao, R., & Goldman, I. D. (2008). The clinical course and genetic defect in the PCFT gene in a 27-year-old woman with hereditary folate malabsorption. *The Journal of Pediatrics*, *153*(3), 435–437.

Moscow, J. A., Gong, M., He, R., Sgagias, M. K., Dixon, K. H., Anzick, S. L., et al. (1995). Isolation of a gene encoding a human reduced folate carrier (RFC1) and analysis of its expression in transport-deficient, methotrexate-resistant human breast cancer cells. *Cancer Research*, *55*(17), 3790–3794.

Nakai, Y., Inoue, K., Abe, N., Hatakeyama, M., Ohta, K. Y., Otagiri, M., et al. (2007). Functional characterization of human proton-coupled folate transporter/heme carrier protein 1 heterologously expressed in mammalian cells as a folate transporter. *The Journal of Pharmacology and Experimental Therapeutics*, *322*(2), 469–476.

Polgar, O., Ierano, C., Tamaki, A., Stanley, B., Ward, Y., Xia, D., et al. (2010). Mutational analysis of threonine 402 adjacent to the GXXXG dimerization motif in transmembrane segment 1 of ABCG2. *Biochemistry (Mosc)*, *49*(10), 2235–2245.

Prasad, P. D., Ramamoorthy, S., Leibach, F. H., & Ganapathy, V. (1995). Molecular cloning of the human placental folate transporter. *Biochemical and Biophysical Research Communications*, *206*(2), 681–687.

Qiu, A., Jansen, M., Sakaris, A., Min, S. H., Chattopadhyay, S., Tsai, E., et al. (2006). Identification of an intestinal folate transporter and the molecular basis for hereditary folate malabsorption. *Cell*, *127*(5), 917–928.

Qiu, A., Min, S. H., Jansen, M., Malhotra, U., Tsai, E., Cabelof, D. C., et al. (2007). Rodent intestinal folate transporters (SLC46A1): Secondary structure, functional properties, and response to dietary folate restriction. *American Journal of Physiology. Cell Physiology*, *293*(5), C1669–C1678.

Rizwan, A. N., & Burckhardt, G. (2007). Organic anion transporters of the SLC22 family: Biopharmaceutical, physiological, and pathological roles. *Pharmaceutical Research*, *24*(3), 450–470.

Rosowsky, A., Bader, H., Wright, J. E., Keyomarsi, K., & Matherly, L. H. (1994). Synthesis and biological activity of N omega-hemiphthaloyl-alpha, omega-diaminoalkanoic acid analogues of aminopterin and 3',5-dichloroaminopterin. *Journal of Medicinal Chemistry*, *37*(14), 2167–2174.

Roy, K., Tolner, B., Chiao, J. H., & Sirotnak, F. M. (1998). A single amino acid difference within the folate transporter encoded by the murine RFC-1 gene selectively alters its interaction with folate analogues. Implications for intrinsic antifolate resistance and directional orientation of the transporter within the plasma membrane of tumor cells. *The Journal of Biological Chemistry*, *273*(5), 2526–2531.

Sadlish, H., Williams, F. M., & Flintoff, W. F. (2002). Cytoplasmic domains of the reduced folate carrier are essential for trafficking, but not function. *The Biochemical Journal*, *364*(Pt 3), 777–786.

Salojin, K. V., Cabrera, R. M., Sun, W., Chang, W. C., Lin, C., Duncan, L., et al. (2011). A mouse model of hereditary folate malabsorption: Deletion of the PCFT gene leads to systemic folate deficiency. *Blood*, *117*(18), 4895–4904.

Schron, C. M., Washington, C., Jr., & Blitzer, B. L. (1985). The transmembrane pH gradient drives uphill folate transport in rabbit jejunum. Direct evidence for folate/hydroxyl exchange in brush border membrane vesicles. *The Journal of Clinical Investigation*, *76*(5), 2030–2033.

Sharina, I. G., Zhao, R., Wang, Y., Babani, S., & Goldman, I. D. (2001). Mutational analysis of the functional role of conserved arginine and lysine residues in transmembrane domains of the murine reduced folate carrier. *Molecular Pharmacology*, *59*(5), 1022–1028.

Sharina, I. G., Zhao, R., Wang, Y., Babani, S., & Goldman, I. D. (2002). Role of the C-terminus and the long cytoplasmic loop in reduced folate carrier expression and function. *Biochemical Pharmacology*, *63*(9), 1717–1724.

Shayeghi, M., Latunde-Dada, G. O., Oakhill, J. S., Laftah, A. H., Takeuchi, K., Halliday, N., et al. (2005). Identification of an intestinal heme transporter. *Cell, 122*(5), 789–801.

Shin, D. S., Mahadeo, K., Min, S. H., Diop-Bove, N., Clayton, P., Zhao, R., et al. (2011). Identification of novel mutations in the proton-coupled folate transporter (PCFT-SLC46A1) associated with hereditary folate malabsorption. *Molecular Genetics and Metabolism, 103*(1), 33–37.

Shin, D. S., Min, S. H., Russell, L., Zhao, R., Fiser, A., & Goldman, I. D. (2010). Functional roles of aspartate residues of the proton-coupled folate transporter (PCFT-SLC46A1); a D156Y mutation causing hereditary folate malabsorption. *Blood, 116*(24), 5162–5169.

Shin, D. S., Zhao, R., Fiser, A., & Goldman, I. D. (2013). Role of the fourth transmembrane domain in proton-coupled folate transporter function as assessed by the substituted cysteine accessibility method. *American Journal of Physiology. Cell Physiology, 304*(12), C1159–C1167.

Shin, D. S., Zhao, R., Yap, E. H., Fiser, A., & Goldman, I. D. (2012). A P425R mutation of the proton-coupled folate transporter causing hereditary folate malabsorption produces a highly selective alteration in folate binding. *American Journal of Physiology. Cell Physiology, 302*(9), C1405–C1412.

Sierra, E. E., Brigle, K. E., Spinella, M. J., & Goldman, I. D. (1997). pH dependence of methotrexate transport by the reduced folate carrier and the folate receptor in L1210 leukemia cells. Further evidence for a third route mediated at low pH. *Biochemical Pharmacology, 53*(2), 223–231.

Sierra, E. E., & Goldman, I. D. (1998). Characterization of folate transport mediated by a low pH route in mouse L1210 leukemia cells with defective reduced folate carrier function. *Biochemical Pharmacology, 55*(9), 1505–1512.

Sirotnak, F. M., Chello, P. L., Moccio, D. M., Kisliuk, R. L., Combepine, G., Gaumont, Y., et al. (1979). Stereospecificity at carbon 6 of fomyltetrahydrofolate as a competitive inhibitor of transport and cytotoxicity of methotrexate in vitro. *Biochemical Pharmacology, 28*(19), 2993–2997.

Sirotnak, F. M., & Donsbach, R. C. (1974). Stereochemical characteristics of the folate-antifolate transport mechanism in L1210 Leukemia cells. *Cancer Research, 34*(2), 371–377.

Sirotnak, F. M., Kurita, S., & Hutchison, D. J. (1968). On the nature of a transport alteration determining resistance to amethopterin in the L1210 leukemia. *Cancer Research, 28*(1), 75–80.

Stokstad, E. L. R. (1990). Historical perspective on key advances in the biochemistry and physiology of folates. In M. F. Picciano, E. L. R. Stokstad, & J. F. Gregory (Eds.), *Contemporary issues in clinical nutrition: Folic acid metabolism in health and disease* (pp. 1–21). New York: Wiley-Liss.

Subramanian, V. S., Marchant, J. S., & Said, H. M. (2008). Apical membrane targeting and trafficking of the human proton-coupled transporter in polarized epithelia. *American Journal of Physiology. Cell Physiology, 294*(1), C233–C240.

Umapathy, N. S., Gnana-Prakasam, J. P., Martin, P. M., Mysona, B., Dun, Y., Smith, S. B., et al. (2007). Cloning and functional characterization of the proton-coupled electrogenic folate transporter and analysis of its expression in retinal cell types. *Investigative Ophthalmology & Visual Science, 48*(11), 5299–5305.

Unal, E. S., Zhao, R., Chang, M. H., Fiser, A., Romero, M. F., & Goldman, I. D. (2009). The functional roles of the His247 and His281 residues in folate and proton translocation mediated by the human proton-coupled folate transporter SLC46A1. *The Journal of Biological Chemistry, 284*(26), 17846–17857.

Unal, E. S., Zhao, R., & Goldman, I. D. (2009). Role of the glutamate 185 residue in proton translocation mediated by the proton-coupled folate transporter SLC46A1. *American Journal of Physiology. Cell Physiology, 297*(1), C66–C74.

Unal, E. S., Zhao, R., Qiu, A., & Goldman, I. D. (2008). N-linked glycosylation and its impact on the electrophoretic mobility and function of the human proton-coupled folate transporter (HsPCFT). *Biochimica et Biophysica Acta, 1778*(6), 1407–1414.

Urquhart, B. L., Gregor, J. C., Chande, N., Knauer, M. J., Tirona, R. G., & Kim, R. B. (2010). The human proton-coupled folate transporter (hPCFT): Modulation of intestinal expression and function by drugs. *American Journal of Physiology. Gastrointestinal and Liver Physiology, 298*(2), G248–G254.

Visentin, M., Zhao, R., & Goldman, I. D. (2012a). The antifolates. *Hematology/Oncology Clinics of North America, 26*(3), 629–648.

Visentin, M., Zhao, R., & Goldman, I. D. (2012b). Augmentation of reduced folate carrier-mediated folate/antifolate transport through an antiport mechanism with 5-aminoimidazole-4-carboxamide riboside monophosphate. *Molecular Pharmacology, 82*(2), 209–216.

Wang, L., Cherian, C., Kugel Desmoulin, S., Mitchell-Ryan, S., Hou, Z., Matherly, L. H., et al. (2012). Synthesis and biological activity of 6-substituted pyrrolo[2,3-d]pyrimidine thienoyl regioisomers as inhibitors of de novo purine biosynthesis with selectivity for cellular uptake by high affinity folate receptors and the proton-coupled folate transporter over the reduced folate carrier. *Journal of Medicinal Chemistry, 55*(4), 1758–1770.

Wang, L., Cherian, C., Kugel Desmoulin, S., Polin, L., Deng, Y., Wu, J., et al. (2010). Synthesis and antitumor activity of a novel series of 6-substituted pyrrolo[2,3-d]pyrimidine thienoyl antifolate inhibitors of purine biosynthesis with selectivity for high affinity folate receptors and the proton-coupled folate transporter over the reduced folate carrier for cellular entry. *Journal of Medicinal Chemistry, 53*(3), 1306–1318.

Wang, L., Kugel Desmoulin, S., Cherian, C., Polin, L., White, K., Kushner, J., et al. (2011). Synthesis, biological, and antitumor activity of a highly potent 6-substituted pyrrolo[2,3-d]pyrimidine thienoyl antifolate inhibitor with proton-coupled folate transporter and folate receptor selectivity over the reduced folate carrier that inhibits beta-glycinamide ribonucleotide formyltransferase. *Journal of Medicinal Chemistry, 54*(20), 7150–7164.

Wang, Y., Zhao, R., Russell, R. G., & Goldman, I. D. (2001). Localization of the murine reduced folate carrier as assessed by immunohistochemical analysis. *Biochimica et Biophysica Acta, 1513*(1), 49–54.

Wessels, J. A., Huizinga, T. W., & Guchelaar, H. J. (2008). Recent insights in the pharmacological actions of methotrexate in the treatment of rheumatoid arthritis. *Rheumatology (Oxford), 47*(3), 249–255.

Westerhof, G. R., Schornagel, J. H., Kathmann, I., Jackman, A. L., Rosowsky, A., Forsch, R. A., et al. (1995). Carrier- and receptor-mediated transport of folate antagonists targeting folate-dependent enzymes: Correlates of molecular-structure and biological activity. *Molecular Pharmacology, 48*(3), 459–471.

Whetstine, J. R., Flatley, R. M., & Matherly, L. H. (2002). The human reduced folate carrier gene is ubiquitously and differentially expressed in normal human tissues: Identification of seven non-coding exons and characterization of a novel promoter. *The Biochemical Journal, 367*(Pt 3), 629–640.

White, J. C., Bailey, B. D., & Goldman, I. D. (1978). Lack of stereospecificity at carbon 6 of methyltetrahydrofolate transport in Ehrlich ascites tumor cells. Carrier-mediated transport of both stereoisomers. *The Journal of Biological Chemistry, 253*(1), 242–245.

Williams, F. M., & Flintoff, W. F. (1995). Isolation of a human cDNA that complements a mutant hamster cell defective in methotrexate uptake. *The Journal of Biological Chemistry, 270*(7), 2987–2992.

Williams, F. M., Murray, R. C., Underhill, T. M., & Flintoff, W. F. (1994). Isolation of a hamster cDNA clone coding for a function involved in methotrexate uptake. *The Journal of Biological Chemistry, 269*(8), 5810–5816.

Witt, T. L., Stapels, S. E., & Matherly, L. H. (2004). Restoration of transport activity by co-expression of human reduced folate carrier half-molecules in transport-impaired K562 cells: Localization of a substrate binding domain to transmembrane domains 7-12. *The Journal of Biological Chemistry, 279*(45), 46755–46763.

Wollack, J. B., Makori, B., Ahlawat, S., Koneru, R., Picinich, S. C., Smith, A., et al. (2008). Characterization of folate uptake by choroid plexus epithelial cells in a rat primary culture model. *Journal of Neurochemistry, 104*(6), 1494–1503.

Wong, S. C., Proefke, S. A., Bhushan, A., & Matherly, L. H. (1995). Isolation of human cDNAs that restore methotrexate sensitivity and reduced folate carrier activity in methotrexate transport-defective Chinese hamster ovary cells. *The Journal of Biological Chemistry, 270*(29), 17468–17475.

Wong, S. C., Zhang, L., Proefke, S. A., & Matherly, L. H. (1998). Effects of the loss of capacity for N-glycosylation on the transport activity and cellular localization of the human reduced folate carrier. *Biochimica et Biophysica Acta, 1375*(1–2), 6–12.

Wong, S. C., Zhang, L., Witt, T. L., Proefke, S. A., Bhushan, A., & Matherly, L. H. (1999). Impaired membrane transport in methotrexate-resistant CCRF-CEM cells involves early translation termination and increased turnover of a mutant reduced folate carrier. *The Journal of Biological Chemistry, 274*(15), 10388–10394.

Xia, W., & Low, P. S. (2010). Folate-targeted therapies for cancer. *Journal of Medicinal Chemistry, 53*(19), 6811–6824.

Yang, C. H., Sirotnak, F. M., & Dembo, M. (1984). Interaction between anions and the reduced folate/methotrexate transport system in L1210 cell plasma membrane vesicles: Directional symmetry and anion specificity for differential mobility of loaded and unloaded carrier. *The Journal of Membrane Biology, 79*(3), 285–292.

Yang, J., Vlashi, E., & Low, P. (2012). Folate-linked drugs for the treatment of cancer and inflammatory diseases. *Sub-Cellular Biochemistry, 56*, 163–179.

Zhao, R., Assaraf, Y. G., & Goldman, I. D. (1998). A mutated murine reduced folate carrier (RFC1) with increased affinity for folic acid, decreased affinity for methotrexate, and an obligatory anion requirement for transport function. *The Journal of Biological Chemistry, 273*(30), 19065–19071.

Zhao, R., Diop-Bove, N., Visentin, M., & Goldman, I. D. (2011). Mechanisms of membrane transport of folates into cells and across epithelia. *Annual Review of Nutrition, 31*, 177–201.

Zhao, R., Gao, F., Babani, S., & Goldman, I. D. (2000). Sensitivity to 5,10-dideazatetrahydrofolate is fully conserved in a murine leukemia cell line highly resistant to methotrexate due to impaired transport mediated by the reduced folate carrier. *Clinical Cancer Research, 6*(8), 3304–3311.

Zhao, R., Gao, F., & Goldman, I. D. (1999). Discrimination among reduced folates and methotrexate as transport substrates by a phenylalanine substitution for serine within the predicted eighth transmembrane domain of the reduced folate carrier. *Biochemical Pharmacology, 58*(10), 1615–1624.

Zhao, R., Gao, F., Hanscom, M., & Goldman, I. D. (2004). A prominent low-pH methotrexate transport activity in human solid tumors: Contribution to the preservation of methotrexate pharmacologic activity in HeLa cells lacking the reduced folate carrier. *Clinical Cancer Research, 10*(2), 718–727.

Zhao, R., Gao, F., Wang, Y., Diaz, G. A., Gelb, B. D., & Goldman, I. D. (2001). Impact of the reduced folate carrier on the accumulation of active thiamin metabolites in murine leukemia cells. *The Journal of Biological Chemistry, 276*(2), 1114–1118.

Zhao, R., & Goldman, I. D. (2003). Resistance to antifolates. *Oncogene, 22*(47), 7431–7457.

Zhao, R., & Goldman, I. D. (2007). The molecular identity and characterization of a proton-coupled folate transporter—PCFT; biological ramifications and impact on the activity of pemetrexed. *Cancer Metastasis Reviews, 26*(1), 129–139.

Zhao, R., Matherly, L. H., & Goldman, I. D. (2009). Membrane transporters and folate homeostasis: Intestinal absorption and transport into systemic compartments and tissues. *Expert Reviews in Molecular Medicine*, *11*, e4.
Zhao, R., Min, S. H., Qiu, A., Sakaris, A., Goldberg, G. L., Sandoval, C., et al. (2007). The spectrum of mutations in the PCFT gene, coding for an intestinal folate transporter, that are the basis for hereditary folate malabsorption. *Blood*, *110*(4), 1147–1152.
Zhao, R., Min, S. H., Wang, Y., Campanella, E., Low, P. S., & Goldman, I. D. (2009). A role for the proton-coupled folate transporter (PCFT-SLC46A1) in folate receptor-mediated endocytosis. *The Journal of Biological Chemistry*, *284*(7), 4267–4274.
Zhao, R., Qiu, A., Tsai, E., Jansen, M., Akabas, M. H., & Goldman, I. D. (2008). The proton-coupled folate transporter: Impact on pemetrexed transport and on antifolates activities compared with the reduced folate carrier. *Molecular Pharmacology*, *74*(3), 854–862.
Zhao, R., Russell, R. G., Wang, Y., Liu, L., Gao, F., Kneitz, B., et al. (2001). Rescue of embryonic lethality in reduced folate carrier-deficient mice by maternal folic acid supplementation reveals early neonatal failure of hematopoietic organs. *The Journal of Biological Chemistry*, *276*(13), 10224–10228.
Zhao, R., Shin, D. S., Diop-Bove, N., Ovits, C. G., & Goldman, I. D. (2011). Random mutagenesis of the proton-coupled folate transporter (SLC46A1), clustering of mutations, and the bases for associated losses of function. *The Journal of Biological Chemistry*, *286*(27), 24150–24158.
Zhao, R., Shin, D. S., Fiser, A., & Goldman, I. D. (2012). Identification of a functionally critical GXXG motif and its relationship to the folate binding site of the proton-coupled folate transporter (PCFT-SLC46A1). *American Journal of Physiology Cell Physiology*, *303*(6), C673–C681.
Zhao, R., Unal, E. S., Shin, D. S., & Goldman, I. D. (2010). Membrane topological analysis of the proton-coupled folate transporter (PCFT-SLC46A1) by the substituted cysteine accessibility method. *Biochemistry (Moscow)*, *49*(13), 2925–2931.
Zhao, R., Visentin, M., Suadicani, S. O., & Goldman, I. D. (2013). Inhibition of the proton-coupled folate transporter (PCFT-SLC46A1) by bicarbonate and other anions. *Molecular Pharmacology*, *84*(1), 95–103.

CHAPTER FIVE

Organic Anion-Transporting Polypeptides

Bruno Stieger[*,1], **Bruno Hagenbuch**[†]
[*]Department of Clinical Pharmacology and Toxicology, University Hospital, Zürich, Switzerland
[†]Department of Pharmacology, Toxicology and Therapeutics, The University of Kansas Medical Center, Kansas City, Kansas, USA
[1]Corresponding author: e-mail address: bstieger@kpt.uzh.ch

Contents

Abstract

Organic anion-transporting polypeptides or OATPs are central transporters in the disposition of drugs and other xenobiotics. In addition, they mediate transport of a wide variety of endogenous substrates. The critical role of OATPs in drug disposition has spurred research both in academia and in the pharmaceutical industry. Translational aspects with clinical questions are the focus in academia, while the pharmaceutical industry tries to define and understand the role these transporters play in pharmacotherapy. The present overview summarizes our knowledge on the interaction of food constituents with OATPs and on the OATP transport mechanisms. Further, it gives an update on the available information on the structure–function relationship of the OATPs and, finally, covers the transcriptional and posttranscriptional regulation of OATPs.

Current Topics in Membranes, Volume 73
ISSN 1063-5823
http://dx.doi.org/10.1016/B978-0-12-800223-0.00005-0

1. INTRODUCTION

The liver and the kidney are the two most important organs in the metabolism and excretion of xenobiotics such as drugs or toxins. In addition, the liver constitutes—after the gut wall—a second barrier for the entry of xenobiotics into the systemic circulation. It has been known for a long time that certain toxins like phalloidin (from the death cap *Amanita phalloides*) are organ specific. The liver-specific toxicity of phalloidin has been attributed to a transport system mediating the uptake of this toxin into hepatocytes (Frimmer, 1982). It was found in the perfused rat liver that adding silymarin shortly after phallodin exposure leads to an attenuation of the toxic effect of phalloidin as assessed by potassium efflux from the perfused liver (Weil & Frimmer, 1970). Uptake experiments into isolated rat hepatocytes lead to the postulation of a transport system accepting the bile acid cholate, the cholecystographic agent iodipamide and antamanide, which is another toxin from *A. phalloides* (Petzinger, Joppen, & Frimmer, 1983). This postulated transport system was thought to differ from the transport system responsible for the hepatocellular uptake of bromosulfophthalein (BSP) based on inhibition experiments (Petzinger et al., 1983). Subsequently, this transport system was postulated to be a "multispecific organic anion transporter" (Frimmer & Ziegler, 1988; Ziegler, Frimmer, & Fasold, 1984). Of note, earlier studies on perfused rat liver by a different group provided evidence for a common uptake mechanism for bilirubin, BSP, and indocyanine green that could not be inhibited by glycocholic acid (Scharschmidt, Waggoner, & Berk, 1975). Studies in kidneys demonstrated even earlier that the handling of organic acids and organic bases differs (Lotspeich, 1958), again pointing to the presence of more than one transport system. Hence, by the early to mid-1980s, the stage was set for concepts that functionally discriminated transporters for organic anions from transporters for organic cations (reviewed in van Montfoort et al., 2003).

The method of expression cloning was introduced for transporters by following the uptake activity for the substrate of interest. This lead to the molecular identification of the intestinal sodium-dependent uptake system for D-glucose (SGLT1) (Hediger, Coady, Ikeda, & Wright, 1987). Using BSP as a lead-substrate in conjunction with measurements of chloride-dependent transport activity (Min, Johansen, Campbell, & Wolkoff, 1991), investigators cloned the first sodium-independent organic anion transporter (rOATP1A1) from a rat liver cDNA library (Jacquemin,

Hagenbuch, Stieger, Wolkoff, & Meier, 1994). In addition to the organic anion BSP, this transporter could also mediate the sodium-independent transport of taurocholate and cholate, but independently from chloride. It very soon became clear that rOATP1A1 had to have closely related homologues (Hagenbuch, Scharschmidt, & Meier, 1996), and this lead to the identification of rOATP1A4 by homology screening of a rat brain cDNA library (Noe, Hagenbuch, Stieger, & Meier, 1997). Functional characterization of rOATP1A1 (Bossuyt, Muller, Hagenbuch, & Meier, 1996) and rOATP1A4 (Noe et al., 1997), as well as of their human homolog hOATP1A2 (Bossuyt, Muller, & Meier, 1996) demonstrated their broad substrate specificity and revealed the ability of OATPs to transport drugs.

Today, more than 300 members of the OATP (*SLCO*) transporter superfamily with 11 OATPs expressed in humans are known (Hagenbuch & Stieger, 2013). In a human genome, wide meta-analysis of genes correlating with total serum bilirubin levels, an additional human OATP (hOATP1B7, formerly *SLC21A21* or LST-3TM12) was among other proteins identified to be associated with higher bilirubin levels. In contrast, in another genome-wide association study, hOATP1B7 was not associated with elevated bilirubin levels (Buch et al., 2010). Importantly, this hOATP1B7 has not been reported to be a functionally active OATP so far. Nevertheless, in an analysis of tissue-specific expression of various solute transporters in rats, rOATP1B7 was reported to be highly expressed at the mRNA level exclusively in rat liver (Sreedharan, Stephansson, Schioth, & Fredriksson, 2011). Given the lack of a reported function of hOATP1B7, *SLCO1B7* may be a pseudogene.

Substrates of OATPs constitute endogenous molecules, food constituents, drugs, and toxins (Table 5.1) as summarized in multiple recent reviews (Hagenbuch & Gui, 2008; Hagenbuch & Stieger, 2013; Kalliokoski & Niemi, 2009; Konig, 2011; Kusuhara & Sugiyama, 2009; Roth, Obaidat, & Hagenbuch, 2012; Shitara et al., 2013). The importance of OATPs in drug disposition is now widely accepted and interactions with OATPs are tested during drug development by the pharmaceutical industry as requested by regulatory authorities (Fenner et al., 2012; Giacomini et al., 2010; Huang et al., 2008; Tweedie et al., 2013; Zhang, Huang, & Lesko, 2011; Zhang, Strong, Qiu, Lesko, & Huang, 2006; Zhang, Zhang, Strong, Reynolds, & Huang, 2008). OATPs were identified on their ability to transport BSP, a substrate which was thought to be transported (in part) by a postulated bilirubin transport system (Clarenburg & Kao, 1973; Scharschmidt et al., 1975). Many years after their identification, it was

Table 5.1 Substrate classes of human OATPs

OATP	Endogenous substrates	Xenobiotic substrates	Model substrates
OATP1A2	Bile salts, bilirubin, steroid hormone metabolites, thyroid hormones, and metabolites	Liver function markers, β-blockers, statins, antiviral drugs, antibiotics, mushroom toxins, neuropeptides, anticancer drugs	Bromosulfophthalein Estrone-3-sulfate Fexofenadine
OATP1B1	Bile salts, bilirubin, steroid hormone metabolites, thyroid hormones and metabolites, inflammatory mediators	Liver function markers, mushroom toxins, statins, sartanes, antibiotics, antiviral drugs, anticancer drugs	Bromosulfophthalein Estradiol-17β-glucuronide Estrone-3-sulfate Pitavastatin Atorvastatin Pravastatin Rosuvastatin Valsartan
OATP1B3	Bile salts, bilirubin, steroid hormone metabolites, thyroid hormones, inflammatory mediators	Liver function markers, mushroom toxins, statins, sartans, antibiotics, antiviral drugs, anticancer drugs, peptides	Bromosulfophthalein Cholecystokinin octapeptide Estradiol-17β-glucuronide Valsartan
OATP1C1	Thyroid hormones and metabolites, steroid hormone metabolites	Liver function markers	Thyroxine
OATP2A1	Inflammatory mediators		Prostaglandins (PGE_1, PGE_2, $PGE_{2\alpha}$)
OATP2B1	Steroid hormone metabolites, inflammatory mediators, thyroid hormones	Liver function markers, statins	Bromosulfophthalein Estrone-3-sulfate
OATP3A1	Steroid hormone metabolites, inflammatory mediators, thyroid hormones		PGE_1 PGE_2
OATP4A1	Bile salts, steroid hormone metabolites, thyroid hormones		Triiodothyronine Taurocholate
OATP4C1	Bile salts, steroid hormone metabolites, thyroid hormones		Estrone-3-sulfate Digoxin

For a detailed recommendation of OATP probe substrates and inhibitors, please refer to Brouwer et al. (2013).

indeed demonstrated that hOATP1B1 and hOATP1B3 could mediate bilirubin transport (Briz, Serrano, MacIas, Gonzalez-Gallego, & Marin, 2003; Cui, Konig, Leier, Buchholz, & Keppler, 2001). Furthermore, it was also realized that selected OATPs play a central role in the disposition of drugs in humans. More importantly, identifying OATPs has also significantly contributed to an understanding of the molecular mechanisms of pharmacokinetic drug–drug interactions (Konig, Muller, & Fromm, 2013; Shitara et al., 2013).

2. TRANSPORT OF NATURAL PRODUCTS BY OATPs

Besides the well-established transporter-mediated drug–drug interactions that occur in cases where a transporter either mediates the uptake of two interacting drugs, or where the uptake of one drug is inhibited by the interacting drug, transporter-mediated food–drug interactions have recently become a focus in the field of drug disposition. In particular, polyphenols like flavonoids, which are found in numerous dietary components, fruit juices, and green tea have been studied as potential perpetrators of OATP-mediated drug uptake (Bailey, Dresser, Leake, & Kim, 2007; Dolton, Roufogalis, & McLachlan, 2012; Fuchikami et al., 2006; Wang, Wolkoff, & Morris, 2005). Most of these studies investigated to what extent such dietary components would affect OATP-mediated substrate transport. In 2002, it was demonstrated that grapefruit, orange, and apple juice could inhibit OATP-mediated fexofenadine uptake (Dresser et al., 2002). In follow-up studies, the same group showed that grapefruit juice reduced the oral bioavailability of fexofenadine in healthy human volunteers, possibly by inhibiting hOATP1A2-mediated fexofenadine uptake in the small intestine (Glaeser et al., 2007). Furthermore, they showed that naringin—a component of grapefruit juice—was able to inhibit completely fexofenadine transport mediated by hOATP1A2 expressed in HeLa cells (Bailey et al., 2007). Besides inhibiting fexofenadine transport, naringin also inhibited talinolol transport in hOATP1A2-expressing *Xenopus laevis* oocytes (Shirasaka et al., 2010) and hOATP1A2-mediated uptake of pravastatin and pitavastatin (Shirasaka, Suzuki, Shichiri, Nakanishi, & Tamai, 2011). Naringin also inhibited hOATP2B1-mediated uptake of pitavastatin, but did not affect pravastatin transport by hOATP2B1 (Shirasaka et al., 2011). The inhibition of OATP-mediated BSP transport by the flavonoids apigenin, kaempferol, and quercetin was reported to be of competitive nature for intestinal hOATP1A2 and hOATP2B1 (Mandery et al., 2010), as well as for the liver-expressed hOATP1B1 and hOATP1B3 (Mandery et al., 2012). A standardized extract of milk thistle seeds (silymarin) contains several flavonolignans including silibinin and has been used to treat liver disease since the sixteenth century. Legalon SIL is a drug that contains the two diastereomers silibinin A and silibinin B (also known as silybin A and silybin B) and is used as an antidote for acute amatoxin poisoning (Mengs, Pohl, & Mitchell, 2012). In addition, high-dose intravenous silibinin is currently being investigated in clinical trials for the treatment of hepatitis C infections (Ferenci et al., 2008). It was shown in 2006 that silibinin inhibits hOATP1B3-mediated uptake of amanitin into hOATP1B3-expressing MDCKII cells (Letschert, Faulstich, Keller, & Keppler, 2006). Two recent studies further characterized the interactions of silymarin and some of its major constituents with respect to inhibition of OATP-mediated uptake (Kock, Xie, Hawke, Oberlies, & Brouwer, 2013; Wlcek, Koller, Ferenci, & Stieger, 2013). In these studies, silibinin inhibited OATP-mediated substrate uptake. Kock et al. (2013) determined IC_{50} values for silymarin and three of its major constituents, namely, silybin A, silybin B, and silychristin, on hOATP1B1- and hOATP1B3-mediated estradiol-17β-glucuronide and hOATP2B1-mediated estrone-3-sulfate uptake. Silybin A and silybin B inhibited OATP-mediated substrate uptake with IC_{50} values in the low-micromolar range (Kock et al., 2013). For hOATP1B3- and hOATP2B1-mediated uptake of estrone-3-sulfate, silibinin was a competitive inhibitor with K_i values of 5 μM for hOATP1B3 and 3.6 μM for hOATP2B1 (Wlcek et al., 2013). However, no direct uptake of any of these flavonoids by any OATP has yet been demonstrated.

Another family of flavonoids that have been studied extensively because of their potential beneficial effects on the cardiovascular system (Moore, Jackson, & Minihane, 2009), aging (Khurana, Venkataraman, Hollingsworth, Piche, & Tai, 2013), and cancer (Singh, Shankar, & Srivastava, 2011) are the green tea catechins, such as epicatechin gallate (ECG), and epigallocatechin gallate (EGCG). Both these catechins are modulators of OATP-mediated transport (Roth, Timmermann, & Hagenbuch, 2011). ECG and EGCG inhibited the uptake of estrone-3-sulfate mediated by hOATP1A2, hOATP1B1, and hOATP2B1, while EGCG stimulated estrone-3-sulfate uptake of hOATP1B3. IC_{50} values for the two compounds ranged from about 10 to 100 μM. Direct transport experiments performed with OATP-expressing CHO cells (for hOATP1B1, hOATP1B3, and hOATP2B1) or HEK293 cells (for hOATP1A2) revealed that both ECG, as well as EGCG, are transported substrates of hOATP1A2 and hOATP1B3 (Roth et al., 2011). The signals for hOATP1B1 were not significantly higher than the control cells. Kinetic experiments demonstrated saturable uptake of ECG and EGCG by hOATP1A2 and hOATP1B3 with K_m values between 10 and 39 μM (Roth et al., 2011). These transport data were later confirmed using cytotoxicity experiments where incubation of control CHO cells or CHO cells expressing hOATP1B1 and hOATP1B3 with increasing concentrations of EGCG for 48 h killed the hOATP1B1- and hOATP1B3-expressing cells with IC_{50} values of 7.7 and 3.2 μM, respectively, as compared to 271 μM for wild-type CHO cells (Zhang et al., 2013). The same study also demonstrated that quercetin-3-gallate and three additional synthetic quercetin derivatives preferentially killed CHO cells that expressed hOATP1B1 or hOATP1B3, data consistent with these quercetin derivatives acting as substrates of hOATP1B1 and hOATP1B3 (Zhang et al., 2013). However, additional studies using cell lines expressing individual OATPs are required to determine which of the interacting flavonoids indeed are substrates of the respective OATPs.

3. TRANSPORT MECHANISMS OF OATPs

3.1. OATPs as anion exchangers

OATPs tend to transport amphipathic molecules with a molecular weight of more than 350 Da (Roth et al., 2012). Furthermore, they are capable of mediating the transport of a tightly albumin-bound substance like bilirubin. Hence, the transport mechanism of OATPs should involve the binding of generally amphipathic or even rather hydrophobic substrates followed by a translocation process across the plasma membrane. In the case of bilirubin, this process most likely involves binding and transport of a molecule with two negative charges because the pK_a of both carboxylic groups is most likely below 5.0 (Vega-Hissi, Estrada, Lavecchia, & Pis Diez, 2013), even though others hypothesized that it may be above 8.0 (Mukerjee & Ostrow, 2010). Regardless of this discussion on the protonation state of bilirubin at physiologic pH, BSP is a dianionic molecule. Furthermore, many OATP substrates are anions (Roth et al., 2012). Consequently, transporting anionic OATP substrates across the membrane involves the movement of one or two negative charges against a negative membrane potential of about −35 mV in hepatocytes (Boyer, Graf, & Meier, 1992). This process is energetically unfavorable and not very likely to occur. To date, the transport mechanism(s) of OATPs is not known in detail. However, a considerable body of evidence supporting an anion-exchange transport mechanism has accumulated. Bicarbonate was the first anion to be identified and can act as a counter ion of rOATP1A1. In HeLa cells expressing rOATP1A1, uptake of taurocholate was demonstrated to occur in exchange for bicarbonate (Satlin, Amin, & Wolkoff, 1997). Subsequently, trans-stimulation of uptake of taurocholate and leukotriene by glutathione was demonstrated for rOATP1A1 (Li, Lee, Meier, & Ballatori, 1998). rOATP1A1- and rOATP1A4- mediated uptake of taurocholate was trans-stimulated by preloading *X. laevis* oocytes with taurocholate (Li, Meier, & Ballatori, 2000). In the same study, intracellular glutathione stimulated uptake of the uncharged digoxin via rOATP1A4, suggesting a rather complex transport mechanism for this OATP. Interestingly, *S*-(2,4-dinitrophenyl)-glutathione (DNP-SG) only stimulated taurocholate uptake mediated by rOATP1A4 but not taurocholate uptake mediated by rOATP1A1 (Li et al., 2000). In addition, only rOATP1A4 mediated efflux of this glutathione conjugate. In this context, it should be kept in mind that digoxin is a substrate for rOATP1A4 but not for rOATP1A1 (Noe et al., 1997). Using isolated membrane vesicles with defined sidedness from HeLa cells expressing rOATP1A1, investigators demonstrated asymmetric glutathione transport (Mittur, Wolkoff, & Kaplowitz, 2002). Uptake into inside-out membrane vesicles (representing glutathione efflux from intact cells) was two to threefold higher than in right-side out vesicles (Mittur et al., 2002). Furthermore, this transport activity was insensitive to changes in the membrane potential. A report suggesting that bile acids are cotransported with glutathione by hOATP1B3 (Briz et al., 2006) could not be confirmed by others (Mahagita, Grassl, Piyachaturawat, & Ballatori, 2007). This is in agreement with the fact that substrate transport by hOATP2B1 could not be stimulated by glutathione (Nozawa, Imai, Nezu, Tsuji, & Tamai, 2004). More recently, a study testing various OATPs in stably transfected cell lines found substrate-mediated bicarbonate efflux for all five transporters investigated (Leuthold et al., 2009). Hence, current evidence supports the concept that OATPs, in general, work as organic anion exchangers with bicarbonate likely being a physiologic counterion. This exchange process has not been characterized extensively at a molecular level like reconstituted transporters in liposomes. To the best of our knowledge, there is only one publication so far that describes the electrophysiologic characterization of OATPs. In this study, the authors found that in *X. laevis* oocytes expressing either hOATP1B1 or hOATP1B3, a current across the membrane could be recorded in the presence of different OATP substrates, indicating a net movement of a negative charge into the oocytes (Martinez-Becerra et al., 2011).

3.2. Binding sites of OATPs

While no information of the architecture of the substrate-binding site(s) of any OATP is currently available, there exists considerable evidence for OATPs having multiple binding sites. Kinetic analysis of the transport of estrone-3-sulfate by hOATP1B1 revealed both a high- and a low-affinity transport activity, while the same transporter revealed simple Michaelis–Menten kinetics for estradiol-17β-glucuronide (Tamai et al., 2001). Similarly, hOATP2B1 transports estrone-3-sulfate using two binding sites when expressed in oocytes (Shirasaka, Mori, Shichiri, Nakanishi, & Tamai, 2012), but only one binding site was reported when expressed in HEK-293 cells (Tamai et al., 2001). This discrepancy was explained by a difference in expression levels, although the potential impact of different expression systems (amphibian vs. mammalian cells) was not ruled out. Two binding sites for estrone-3-sulfate on hOATP1B1 were later independently confirmed (Gui & Hagenbuch, 2009; Noe, Portmann, Brun, & Funk, 2007). Recently, different binding sites of hOATP4C1 for estrone-3-sulfate and digoxin were reported (Yamaguchi et al., 2010). Bovine OATP1A2 was also reported, in contrast to hOATP1A2, to have two binding sites for estrone-3-sulfate (Liu et al., 2013).

Additional evidence for the presence of more than one binding site on OATPs was obtained from inhibition experiments and reported first for rOATP1A4, which showed a stimulation of taurocholate transport by estradiol-17β-glucuronide. However, taurocholate did not stimulate the transport of the substrate digoxin (Sugiyama, Kusuhara, Shitara, Abe, & Sugiyama, 2002). Among the human OATPs, hOATP2B1-mediated dehydroepiandrosterone sulfate transport could be increased by coaddition of prostaglandin A_1 or prostaglandin A_2 (Pizzagalli et al., 2003). Low progesterone also stimulated hOATP2B1-mediated estrone-3-sulfate transport, but interestingly, at higher progesterone concentrations, this stimulation was attenuated, demonstrating a biphasic action of progesterone (Grube et al., 2006). The same study found classic inhibition of hOATP2B1-mediated estrone-3-sulfate transport by other steroids, namely, testosterone and mifepristone. However, testosterone has recently been reported to stimulate hOATP2B1 (Karlgren et al., 2012). Some dietary components like the flavonoid rutin stimulated transport of dehydroepiandrosterone sulfate by hOATP1B1, while other flavonoids like biochanin A or luteolin acted as inhibitors (Wang, Wolkoff, et al., 2005). hOATP1B3 transport activity can also be stimulated by green tea extracts or epigallocatechin as measured by estrone-3-sulfate uptake (Roth et al., 2011). Silibinin has recently been tested with several human organic anion transporters and found to be a classic inhibitor of the high-affinity estrone-3-sulfate binding site of hOATP1B1. The low-affinity binding site was stimulated at low silibinin concentrations and inhibited at high silibinin concentrations (Wlcek et al., 2013). Finally, the classic drug and the pregnane X receptor (PXR) agonist clotrimazole stimulated hOATP1B3-, but not hOATP1B1-mediated estradiol-17β-glucuronide transport (Gui et al., 2008). Most interestingly, in the same study, clotrimazole did not affect the hOATP1B3-mediated transport of estrone-3-sulfate and inhibited fluo-3 uptake. As an additional modulating drug, fendiline stimulated hOATPB2B1-mediated estrone-3-sulfate transport (Karlgren et al., 2012).

3.3. Modulation of transport activity

According to *in vitro* experiments, the transport activity of hOATP2B1, which is expressed in the small intestine and many other organs, is stimulated by an extracellular acidic milieu (Kobayashi et al., 2003). Based on an investigation of the pH dependence of 13 rat and human OATPs, all but hOATP1C1 were stimulated by an acidic extracellular pH (Leuthold et al., 2009). Of note, this stimulation was substrate dependent. For example, while hOATP2B1-mediated uptake of prostaglandin E_2 was not stimulated by an extracellular low pH, transport of thyroxine increased at low pH (Leuthold et al., 2009). Similarly, hOATP2B1-mediated pemetrexed transport was strongly stimulated by an acidic pH, while this was not the case for BSP (Visentin, Chang, Romero, Zhao, & Goldman, 2012). Along these lines, hOATP1B1-meditiated transport was also stimulated in a substrate-dependent manner at an acidic pH and may be due to the presence of two different binding sites on this transporter (Leuthold et al., 2009). However, other reasons cannot be excluded. Of note, different statins when tested as hOATP2B1 substrates also show a variable response to extracellular pH (Varma et al., 2011). Kinetic characterization of five OATPs revealed a significantly increased affinity or a trend toward increased affinity at low pH (Leuthold et al., 2009). Mutating a highly conserved histidine in the third transmembrane (TM) domain of rOATP1A1 to a glutamine found at the same position in hOATP1C1 abolished pH sensitivity of transport, while the reverse replacement in hOATP1C1 converted this transporter into a pH-sensitive transporter (Leuthold et al., 2009). Comparing the effect of different pH gradients across the cell membrane on the transport activity of rOATP1A1 revealed that transport stimulation was not proportional to the magnitude of the pH gradient, and the authors suggested a pH-dependent modification of the protonation state of the intracellular portion of the transporter (Marin et al., 2003).

Overall, a detailed mechanistic understanding of the transport mechanism(s) and driving forces for OATPs is currently lacking. There is, however, evidence that OATPs have a rather complex transport mechanism and many (if not most of them) have more than one binding site. In addition, the extracellular pH dependence of OATPs, as seen in the acidic microclimate in the small intestine (Daniel, Fett, & Kratz, 1989), may mean that they are modulated *in vivo* by nearby acid extruders, such as Na^+/H^+-exchangers. Consequently, *in vitro* experiments aimed at determining substrate specificities of OATPs expressed in an acidic environment should be performed at acidic and neutral pH.

4. STRUCTURAL INFORMATION ON OATPs

4.1. General structure

Given that all mammalian OATPs (650–700 amino acids) have an amino acid sequence identity of more than 30% to their more distant relatives (Hagenbuch & Meier, 2003), it is highly likely that they all share the same overall TM domain structure. Initial hydrophobicity analysis of rOATP1A1 and hOATP1A2 predicted the presence of 10–12 TM domains with both N- and C-terminal ends on the cytoplasmic side (Jacquemin et al., 1994; Kullak-Ublick et al., 1995). The intracellular localization of the C-terminal end was experimentally confirmed using an antibody against the C-terminal end of hOATP1B3. Immunostaining was only observed in detergent permeabilized cells (Abe et al., 2001), confirming a topology with both termini on the cytoplasmic side of the membrane. In 2008, the 12-TM domain structure of rOATP1A1 was confirmed using site-directed mutagenesis of putative N-glycosylation sites (Wang, Hata, Xiao, Murray, & Wolkoff, 2008). It was demonstrated that rOATP1A1 was N-glycosylated in the second and fifth extracellular loops, and that the unglycosylated protein was retained intracellularly and thus caused reduced transport function (Wang et al., 2008). Similar results were recently published for hOATP1B1 (Yao et al., 2012). The authors demonstrated that hOATP1B1 is N-glycosylated in the second and fifth extracellular loops, and that the unglycosylated protein is retained in the endoplasmic reticulum. Thus, it is likely that all mammalian OATPs are N-glycosylated in these two extracellular loops. Disulfide bonds can also have an effect on the proper folding and function of a protein. Based on results from site-directed mutagenesis of the 10 conserved cysteine residues in the large extracellular loop 5 of hOATP2B1, all 10 cysteine residues are normally disulfide bonded and these disulfide bonds are important for the proper targeting of hOATP2B1 to the plasma membrane (Hanggi, Grundschober, Leuthold, Meier, & St-Pierre, 2006).

4.2. Role of conserved amino acids

In the absence of a crystal structure for any of the OATPs, homology modeling has been performed by several groups and the respective models were used to test structural hypotheses. The first of such models was described in 2005 and predicted the presence of a central pore in hOATP1B3 and hOATP2B1 that contained several positively charged arginine, lysine, and histidine residues (Meier-Abt, Mokrab, & Mizuguchi, 2005). The role of some of these conserved positively charged amino acids was then tested in site-directed mutagenesis studies using hOATP1B1 (Weaver & Hagenbuch, 2010) and hOATP1B3 (Glaeser, Mandery, Sticht, Fromm, & Konig, 2010; Mandery et al., 2011) as a template. In hOATP1B1, mutations at R57, K361, and R580 all decreased the affinity for estradiol-17β-glucuronide and increased the affinity for BSP, suggesting that these amino acid residues are part of the substrate-binding sites. Mutations of the intracellular K90, H92, and R93 mainly affected the maximal transport activity. Mutants R181K and R580A also reduced surface expression of hOATP1B1 (Weaver & Hagenbuch, 2010). K41, K361, and R580 are important for substrate transport by hOATP1B3 (Glaeser et al., 2010; Mandery et al., 2011), while K399 is necessary for normal surface expression (Mandery et al., 2011). Because hOATP1B1 and hOATP1B3 have overlapping substrate specificities, it is interesting to note that both K361 and R580 are important for the transport function of both hOATP1B1 and hOATP1B3.

4.3. Individual transmembrane domains

According to several studies, certain amino acids in TM domains 2, 6, 8, 9, and 10 are important for the function of hOATP1B1 (Gui & Hagenbuch, 2009; Huang et al., 2013; Li et al., 2012; Miyagawa, Maeda, Aoyama, & Sugiyama, 2009) and hOATP1B3 (Gui & Hagenbuch, 2008). In TM2, charged amino acids D70 and E74 seem critical for estrone-3-sulfate transport (Li et al., 2012). While D70 seems to decrease both high- and low-affinity estron-3-sulfate uptake, E74 seems only important for high-affinity transport. In addition, mutant G76A showed reduced protein expression when compared to wild-type hOATP1B1 and the other mutants (Li et al., 2012). The effect of two tryptophan residues within the OATP family signature at the extracellular half of TM6 of hOATP1B1 was tested by mutating the tryptophan to alanine residues and examining the effect on estrone-3-sulfate transport. While wild-type hOATP1B1 has a high- and a low-affinity binding site for estrone-3-sulfate resulting in biphasic kinetics (Gui & Hagenbuch, 2009; Noe et al., 2007; Tamai et al., 2001), the W258A mutant showed only a single binding component with a K_m value in between the high- and low-affinity values of wild-type hOATP1B1 (Huang et al., 2013). In contrast, the W259A mutant retained biphasic kinetics for estrone-3-sulfate with a higher K_m value for the high-affinity but a similar K_m value for the low-affinity component as compared to wild-type hOATP1B1 (Huang et al., 2013). Several groups utilized a chimera approach and constructed chimeras using hOATP1B1 and hOATP1B3 (Degorter, Ho, Leake, Tirona, & Kim, 2012; Gui & Hagenbuch, 2008, 2009; Miyagawa et al., 2009). Replacing TM8 in hOATP1B1 with that of hOATP1B3 resulted in a protein with an 18-fold higher K_m value for estrone-3-sulfate than that of wild-type hOATP1B1 and abolished estradiol-17β-glucuronide transport (Miyagawa et al., 2009). Replacing TM9 in hOATP1B1 with that of hOATP1B3 did not change the transport of estradiol-17β-glucuronide, but increased the K_m value for estrone-3-sulfate 7.4-fold (Miyagawa et al., 2009). TM10 was identified as being important for substrate transport in both hOATP1B1 (Gui & Hagenbuch, 2009) and hOATP1B3 (Gui & Hagenbuch, 2008). Follow-up site-directed mutagenesis studies identified several amino acids (L545, F546, L550, and S554) within TM10 that when replaced with the hOATP1B3 counterparts converted hOATP1B1 from a biphasic to a monophasic estron-3-sulfate transporter (Gui & Hagenbuch, 2009). Similar experiments with hOATP1B3 resulted in the identification of Y537, S545, and T550 in TM10 as being important for the hOATP1B3-selective substrate CCK-8 (Gui & Hagenbuch, 2008). Finally, in a recent report, hOATP1B1 was converted into a hOATP1B3-like transporter that was able to transport CCK-8 when the three amino acid residues A45 in TM1, L545 in TM10, and T615 in extracellular loop 6 were replaced by the corresponding hOATP1B3 residues (Degorter et al., 2012). Although this triple mutant showed uptake of CCK-8, its intrinsic clearance was about eightfold smaller than the intrinsic clearance by hOATP1B3.

Taken together, available data are consistent with OATPs being 12TM domain proteins with both termini on the cytoplasmic side. Furthermore, OATPs seem to have an aqueous pore that can accept several substrates with overlapping and partially distinct binding sites, which consist in part of conserved positively charged amino acids. Future experiments using higher resolution homology modeling will help to better delineate the location of functionally important amino acid residues until the three-dimensional structure has been elucidated either using crystallography and/or NMR spectroscopy. In addition, the current hypothetical models will have to be improved by investigating which amino acid residues within the postulated TM domains are accessible from the outside or the inside in the absence or presence of substrates.

5. REGULATION OF OATPs

5.1. Transcriptional regulation of OATPs

Regulation of gene expression at the transcriptional level for adaptive purposes is governed by transcription factors, which act as sensors for endogenous ligands and xenobiotics (Germain, Staels, Dacquet, Spedding, & Laudet, 2006). Hence, at the level of the intact organism, transcriptional regulation of gene expression can be studied by applying xenobiotics to animal models and then analyzing the expression of genes of interest. Such studies were performed with knockout mice lacking the genes for nuclear receptors and demonstrated that expression of *SLCOs* can be upregulated by ligands for the arylhydrocarbon receptor, the constitutive androgen receptor (CAR), the PXR, and the farnesoid X receptor (FXR) (Klaassen & Aleksunes, 2010; Meyer zu Schwabedissen & Kim, 2009). However, such studies do not demonstrate that the transcription factors investigated bind directly to the gene of interest. Liver-enriched transcription factors (Schrem, Klempnauer, & Borlak, 2002) belong to different classes of transcriptional regulators. They are also called hepatocyte nuclear factors, and are known to regulate the transcription of *SLCO* genes positively or negatively (reviewed in Geier, Wagner, Dietrich, & Trauner, 2007; Hagenbuch & Stieger, 2013; Klaassen & Aleksunes, 2010; Stieger & Geier, 2011; Svoboda, Riha, Wlcek, Jaeger, & Thalhammer, 2011).

The bile acid sensor FXR can directly and indirectly modulate the expression of *SLCO* genes, leading either to upregulation or to downregulation of gene transcription of the respective genes (Eloranta & Kullak-Ublick, 2008; Godoy et al., 2013). Such studies may also lead to conflicting results. Expression of *SLCO1B1* was reported to be reduced via FXR-dependent repression of HNF-1 (Jung, Elferink, Stellaard, & Groothuis, 2007) and increased via direct upregulation by FXR (Meyer zu Schwabedissen et al., 2010). Other nuclear receptors like PXR, CAR, and LXR are also known to regulate the expression of *SLCO* genes. (Godoy et al., 2013; Hagenbuch & Stieger, 2013; Klaassen & Aleksunes, 2010). Lately, the vitamin D receptor VDR has been added to this list, and it has been shown to activate the expression of *SLCO1A2* (Eloranta, Hiller, Juttner, & Kullak-Ublick, 2012). As ligands for nuclear receptors are also substrates for OATPs (Gui et al., 2008), the transporters may, to some extent, control the access of nuclear receptor ligands to their own genes. Hypoxia leads to the expression of hypoxia-inducible factors, which in turn regulate transcription (Schodel, Mole, & Ratcliffe, 2013). Interestingly, a cancer-specific variant of hOATP1B3 was found to be induced by hypoxia and a promoter construct of *SLCO1B3* was demonstrated to bind HIF-1α (Han, Kim, Thakkar, Kim, & Lee, 2013). Another research group identified two functional HIF-response elements in the promoter of *SLCO1B3* (Ramachandran et al., 2013).

Tissue-specific regulation of protein expression is frequently associated with epigenetic regulation. In mice, a genome-wide DNA-methylation profiling revealed CpG dinucleotides in the area of the transcriptional start site of hOATP1B2 (Imai et al., 2009). An extension of this analysis revealed CpG islands around the translation start sites of mOATP1A1, mOATP1A6, mOATP1C1, and mOATP1A4; comparable findings were obtained for hOATP1B1 and hOATP1B3 (Imai, Kikuchi, Kusuhara, & Sugiyama, 2013). The different methylation patterns are suggested to be involved in the tissue-specific expression of the investigated OATPs. In Caco-2 cells, heterochromatin protein 1β upregulation leads to an upregulation of *SLCO1B3* among other targets (Tell, Wang, Blunier, & Benya, 2011). Expression of a cancer-specific variant of hOATP1B3 in various cancer cell lines is critically dependent on the methylation pattern around the translational start site (Imai, Kikuchi, Tsuruya, et al., 2013), as well as the expression of *SLCO2A1* in human head and neck squamous cell carcinoma (Zolk et al., 2013).

5.2. Posttranslational regulation of OATPs

At the posttranscriptional level, ATP-induced functional downregulation of rOATP1A1 is mediated by serine phosphorylation (Glavy, Wu, Wang,

Orr, & Wolkoff, 2000). This phosphorylation is not associated with subsequent internalization of rOATP1A1. In a subsequent study, S634 and S635 of rOATP1A1 were found to be phosphorylated (Xiao, Nieves, Angeletti, Orr, & Wolkoff, 2006). This study was extended to rOATP1A4 in *X. laevis* oocytes, where PKC-dependent inhibition of transport activity of rOATP1A1 and rOATP1A4 was observed (Guo & Klaassen, 2001). Plasma membrane expression of hOATP1B1 in COS7 and HEK293 cells, on the other hand, was stimulated after treatment of the cells with 8-Br-cAMP, and this stimulation could be blocked by protein kinase A inhibitors (Sun, Ponamgi, Boyer, & Suchy, 2008). Using protein kinase A inhibitors alone, plasma membrane expression of hOATP1B1 was reduced. In contrast, hOATP2B1 stably transfected in MDCK cells could be phosphorylated in a protein kinase C-dependent manner, thereby leading to a rapid internalization of hOATP2B1 (Kock et al., 2010). hOATP1A2 expressed in COS-7 cells was also internalized from the plasma membrane in a protein kinase C-dependent manner by a process requiring ongoing clathrin-dependent endocytosis (Zhou, Lee, Krafczyk, Zhu, & Murray, 2011). According to a recent paper, short-term treatment of either Caco-2 cells or rats with amiodarone leads to an increased surface expression of hOATP2B1 in Caco-2 cells and of rOATP2B1 at the brush border membrane of rat jejunum and ileum as assessed by Western blotting (Segawa et al., 2013).

5.3. Scaffolding of OATPs

Protein–protein interactions leading to scaffolds at the plasma membrane can also be considered a posttranslational regulatory mechanism. A yeast two-hybrid screen using the C-terminus of various drug transporters indeed revealed an interaction of hOATP1A2, hOATP1C1, and hOATP3A1 with PDZ proteins (Kato, Yoshida, Watanabe, Sai, & Tsuji, 2004). PDZ adaptors are well known to be involved in the regulation of transport proteins by regulating their plasma membrane expression level (Sugiura, Shimizu, Kijima, Minakata, & Kato, 2011). rOATP1A1 was shown in a cell-culture system in conjunction with coimmunoprecipitation to interact with PDZK1 (Wang, Wang, et al., 2005). Most interestingly, mice with a disrupted gene for PDZK1 showed a strongly reduced expression of mOATP1A1 at the hepatocyte basolateral membrane, with an increased cytoplasmic localization of mOATP1A1 (Wang, Wang, et al., 2005). These mice display a slightly reduced fractional uptake rate for the prototypic OATP substrate BSP and about a 25% increase in plasma half-life (Wang, Wang, et al., 2005). PDZK1 mediates the trafficking to the plasma membrane by a selective recruitment of microtubule-based motor proteins (Wang, Murray, & Wolkoff, 2014). Others found a reduced estrone-3-sulfate uptake into the portal vein of mice lacking PDZK1 without a change in systemic circulation (Sugiura et al., 2010). Using an Ussing-chamber setup with small intestine from the same mice, uptake of estrone-3-sulfate from the apical lumen into intestinal tissue was significantly reduced. This study confirmed and found interactions of the C-terminus of mouse OATPs with different PDZ proteins (Sugiura et al., 2010). Interactions of human and mouse OATPs with the different adaptor proteins are summarized in Table 5.2 (Sugiura et al., 2010).

In summary, OATPs are extensively regulated at the transcriptional and posttranscriptional levels. Given that many OATPs are xenobiotic transporters and hence involved in the disposition and elimination of potentially toxic xenobiotics, this extensive network of regulatory processes acting on

Table 5.2 Demonstrated interaction of OATPs with adaptor proteins

OATP	PDZK1	PDZK2	NHERF1	NHERF2	IKEPP	References
rOATP1A1	Yes					Wang, Wang, et al. (2005)
mOATP1A1	Yes	Yes	Yes	Yes		Sugiura et al. (2010)
hOATP1A2	Yes		Yes	Yes	Yes	Kato et al. (2004)
mOATP1A4	Yes	Yes	Yes	Yes		Sugiura et al. (2010)
mOATP1A5	Yes	Yes	Yes	Yes		Sugiura et al. (2010)
mOATP1A6	Yes	Yes	Yes	Yes		Sugiura et al. (2010)
mOATP1B2		Yes	Yes	Yes		Sugiura et al. (2010)
hOATP1C1					Yes	Kato et al. (2004)
mOATP1C1	Yes	Yes	Yes	Yes		Sugiura et al. (2010)
mOATP2A1		Yes				Sugiura et al. (2010)
mOATP2B1		Yes		Yes		Sugiura et al. (2010)
hOATP3A1			Yes	Yes	Yes	Kato et al. (2004)
mOATP3A1	Yes	Yes		Yes		Sugiura et al. (2010)

NHERF, Na^+/H^+ exchanger regulatory factor; IKEPP, intestinal and kidney-enriched PDZ protein; 'yes' means that an interaction was demonstrated; an empty cell means that the respective OATP did not interact or an interaction was not demonstrated.

OATPs may contribute to protect organisms from toxic actions of xenobiotics. Regulation of gene expression by xenosensors like the nuclear receptors CAR or PXR is key in the detoxification of xenobiotics by metabolizing as well as exporting them for excretion. Hence, common transcriptional regulatory mechanisms for OATPs as cellular xenobiotic uptake systems with downstream metabolism and metabolite export is an effective functional network preventing cells from being overloaded with potentially harmful compounds.

Although single nucleotide polymorphisms are not in a strict sense a regulatory mechanism because they cannot be modified by cells, there is now ample evidence that polymorphic variants of OATPs are abundant in humans and may contribute significantly to changes in drug disposition. They also constitute risk factors for adverse drug actions. This topic has lately been extensively reviewed (Clarke & Cherrington, 2012; Gong & Kim, 2013; Hagenbuch & Stieger, 2013; Nakanishi & Tamai, 2012; Niemi, 2010; Niemi, Pasanen, & Neuvonen, 2011; Shitara et al., 2013; Stieger & Meier, 2011; Zair, Eloranta, Stieger, & Kullak-Ublick, 2008).

6. OUTLOOK

The role of OATPs in drug and xenobiotic disposition is now well established and accepted (Fenner et al., 2012). Due to their wide expression in most tissues, they govern the access (or exit) of many endogenous and exogenous compounds in a vast variety of cell types and associated tissues and organs. Lately, the importance of transporters, including OATPs, for monitoring the function of organs (e.g., by liver function tests) and for imaging organs has gained marked interest (Hoekstra et al., 2013; Stieger, Heger, de Graaf, Paumgartner, & van Gulik, 2012; Van Beers, Pastor, & Hussain, 2012). Imaging is widely used to diagnose and manage a variety of diseases, including neurologic disorders and cancer. Because OATPs are found to be expressed in many different cancers (Cutler & Choo, 2011; Obaidat, Roth, & Hagenbuch, 2012; Sissung, Reece, Spencer, & Figg, 2012), they are an important diagnostic marker. They may also be used as entry sites for drugs used to fight cancer (Liu & Li, 2014), an important aspect that certainly has a large potential for further anticancer drug development. As OATPs are very important in drug disposition, direct determination of their transport activity *in vivo* is highly desirable. With the rapid progress in the development of better imaging tools such as positron emission tomography or single-photon emission computed tomography (Mairinger, Erker, Muller, & Langer, 2011;

Sugiyama & Yamashita, 2011), such a goal should be achievable. In order to successfully develop novel OATP substrates to be used for *in vivo* transporter function determination, a tight coordination of studies performed *in vitro* (e.g., by using heterologous expression systems) and *in vivo* experiments will be necessary, whereby special consideration has to be given to the overlapping substrate specificity of OATPs. As OATPs are also the entry sites for toxins, (e.g., the ones present in the mushroom *death cap*), their potential as drug targets for the treatment of patients after ingestion of toxic substances could certainly be further developed. Because drug effects and adverse drug actions critically depend on the intracellular drug concentrations (Chu et al., 2013), a major effort has to be made in elucidating the exact transport mechanism as well as the potential driving forces of OATPs. This knowledge will be critical for better understanding and potentially even predicting cellular effects of drugs. Lastly, there is currently no information about the high-resolution structure of OATPs available. Such information is critical for understanding the molecular transport mechanisms of OATPs. The emergence of new tools like cryo-electronmicroscopy (Milne et al., 2013) and its integration with structure determination by NMR or X-ray crystallography will advance this challenging area of OATP research.

ACKNOWLEDGMENTS

B. S. is supported by Grant #310030_144195/1 from the Swiss National Science Foundation. B. H. is supported by National Institutes of Health Grants RR021940 and GM077336.

REFERENCES

Abe, T., Unno, M., Onogawa, T., Tokui, T., Kondo, T. N., Nakagomi, R., et al. (2001). LST-2, a human liver-specific organic anion transporter, determines methotrexate sensitivity in gastrointestinal cancers. *Gastroenterology*, *120*(7), 1689–1699.

Bailey, D. G., Dresser, G. K., Leake, B. F., & Kim, R. B. (2007). Naringin is a major and selective clinical inhibitor of organic anion-transporting polypeptide 1A2 (OATP1A2) in grapefruit juice. *Clinical Pharmacology and Therapeutics*, *81*(4), 495–502.

Bossuyt, X., Muller, M., Hagenbuch, B., & Meier, P. J. (1996). Polyspecific drug and steroid clearance by an organic anion transporter of mammalian liver. *The Journal of Pharmacology and Experimental Therapeutics*, *276*(3), 891–896.

Bossuyt, X., Muller, M., & Meier, P. J. (1996). Multispecific amphipathic substrate transport by an organic anion transporter of human liver. *Journal of Hepatology*, *25*(5), 733–738.

Boyer, J. L., Graf, J., & Meier, P. J. (1992). Hepatic transport systems regulating pHi, cell volume, and bile secretion. *Annual Review of Physiology*, *54*, 415–438.

Briz, O., Romero, M. R., Martinez-Becerra, P., Macias, R. I., Perez, M. J., Jimenez, F., et al. (2006). OATP8/1B3-mediated cotransport of bile acids and glutathione: An export pathway for organic anions from hepatocytes? *The Journal of Biological Chemistry*, *281*(41), 30326–30335.

Briz, O., Serrano, M. A., MacIas, R. I., Gonzalez-Gallego, J., & Marin, J. J. (2003). Role of organic anion-transporting polypeptides, OATP-A, OATP-C and OATP-8, in the human placenta-maternal liver tandem excretory pathway for foetal bilirubin. *The Biochemical Journal*, *371*(Pt. 3), 897–905.

Brouwer, K. L., Keppler, D., Hoffmaster, K. A., Bow, D. A., Cheng, Y., Lai, Y., et al. (2013). In vitro methods to support transporter evaluation in drug discovery and development. *Clinical Pharmacology and Therapeutics*, *94*(1), 95–112.

Buch, S., Schafmayer, C., Volzke, H., Seeger, M., Miquel, J. F., Sookoian, S. C., et al. (2010). Loci from a genome-wide analysis of bilirubin levels are associated with gallstone risk and composition. *Gastroenterology*, *139*(6), 1942–1951.

Chu, X., Korzekwa, K., Elsby, R., Fenner, K., Galetin, A., Lai, Y., et al. (2013). Intracellular drug concentrations and transporters: Measurement, modeling, and implications for the liver. *Clinical Pharmacology and Therapeutics*, *94*(1), 126–141.

Clarenburg, R., & Kao, C. C. (1973). Shared and separate pathways for biliary excretion of bilirubin and BSP in rats. *The American Journal of Physiology*, *225*(1), 192–200.

Clarke, J. D., & Cherrington, N. J. (2012). Genetics or environment in drug transport: The case of organic anion transporting polypeptides and adverse drug reactions. *Expert Opinion on Drug Metabolism & Toxicology*, *8*(3), 349–360.

Cui, Y., Konig, J., Leier, I., Buchholz, U., & Keppler, D. (2001). Hepatic uptake of bilirubin and its conjugates by the human organic anion transporter SLC21A6. *The Journal of Biological Chemistry*, *276*(13), 9626–9630.

Cutler, M. J., & Choo, E. F. (2011). Overview of SLC22A and SLCO families of drug uptake transporters in the context of cancer treatments. *Current Drug Metabolism*, *12*(8), 793–807.

Daniel, H., Fett, C., & Kratz, A. (1989). Demonstration and modification of intervillous pH profiles in rat small intestine in vitro. *The American Journal of Physiology*, *257*(4 Pt. 1), G489–G495.

Degorter, M. K., Ho, R. H., Leake, B. F., Tirona, R. G., & Kim, R. B. (2012). Interaction of three regiospecific amino acid residues is required for OATP1B1 gain of OATP1B3 substrate specificity. *Molecular Pharmaceutics*, *9*(4), 986–995.

Dolton, M. J., Roufogalis, B. D., & McLachlan, A. J. (2012). Fruit juices as perpetrators of drug interactions: The role of organic anion-transporting polypeptides. *Clinical Pharmacology and Therapeutics*, *92*(5), 622–630.

Dresser, G. K., Bailey, D. G., Leake, B. F., Schwarz, U. I., Dawson, P. A., Freeman, D. J., et al. (2002). Fruit juices inhibit organic anion transporting polypeptide-mediated drug uptake to decrease the oral availability of fexofenadine. *Clinical Pharmacology and Therapeutics*, *71*(1), 11–20.

Eloranta, J. J., Hiller, C., Juttner, M., & Kullak-Ublick, G. A. (2012). The SLCO1A2 gene, encoding human organic anion-transporting polypeptide 1A2, is transactivated by the vitamin D receptor. *Molecular Pharmacology*, *82*(1), 37–46.

Eloranta, J. J., & Kullak-Ublick, G. A. (2008). The role of FXR in disorders of bile acid homeostasis. *Physiology (Bethesda)*, *23*, 286–295.

Fenner, K. S., Jones, H. M., Ullah, M., Kempshall, S., Dickins, M., Lai, Y., et al. (2012). The evolution of the OATP hepatic uptake transport protein family in DMPK sciences: From obscure liver transporters to key determinants of hepatobiliary clearance. *Xenobiotica*, *42*(1), 28–45.

Ferenci, P., Scherzer, T. M., Kerschner, H., Rutter, K., Beinhardt, S., Hofer, H., et al. (2008). Silibinin is a potent antiviral agent in patients with chronic hepatitis C not responding to pegylated interferon/ribavirin therapy. *Gastroenterology*, *135*(5), 1561–1567.

Frimmer, M. (1982). Organotropism by carrier-mediated transport. *Trends in Pharmacological Sciences*, *3*, 395–397.

Frimmer, M., & Ziegler, K. (1988). The transport of bile acids in liver cells. *Biochimica et Biophysica Acta, 947*(1), 75–99.

Fuchikami, H., Satoh, H., Tsujimoto, M., Ohdo, S., Ohtani, H., & Sawada, Y. (2006). Effects of herbal extracts on the function of human organic anion-transporting polypeptide OATP-B. *Drug Metabolism and Disposition, 34*(4), 577–582.

Geier, A., Wagner, M., Dietrich, C. G., & Trauner, M. (2007). Principles of hepatic organic anion transporter regulation during cholestasis, inflammation and liver regeneration. *Biochimica et Biophysica Acta, 1773*(3), 283–308.

Germain, P., Staels, B., Dacquet, C., Spedding, M., & Laudet, V. (2006). Overview of nomenclature of nuclear receptors. *Pharmacological Reviews, 58*(4), 685–704.

Giacomini, K. M., Huang, S. M., Tweedie, D. J., Benet, L. Z., Brouwer, K. L., Chu, X., et al. (2010). Membrane transporters in drug development. *Nature Reviews Drug Discovery, 9*(3), 215–236.

Glaeser, H., Bailey, D. G., Dresser, G. K., Gregor, J. C., Schwarz, U. I., McGrath, J. S., et al. (2007). Intestinal drug transporter expression and the impact of grapefruit juice in humans. *Clinical Pharmacology and Therapeutics, 81*(3), 362–370.

Glaeser, H., Mandery, K., Sticht, H., Fromm, M. F., & Konig, J. (2010). Relevance of conserved lysine and arginine residues in transmembrane helices for the transport activity of organic anion transporting polypeptide 1B3. *British Journal of Pharmacology, 159*(3), 698–708.

Glavy, J. S., Wu, S. M., Wang, P. J., Orr, G. A., & Wolkoff, A. W. (2000). Down-regulation by extracellular ATP of rat hepatocyte organic anion transport is mediated by serine phosphorylation of oatp1. *The Journal of Biological Chemistry, 275*(2), 1479–1484.

Godoy, P., Hewitt, N. J., Albrecht, U., Andersen, M. E., Ansari, N., Bhattacharya, S., et al. (2013). Recent advances in 2D and 3D in vitro systems using primary hepatocytes, alternative hepatocyte sources and non-parenchymal liver cells and their use in investigating mechanisms of hepatotoxicity, cell signaling and ADME. *Archives of Toxicology, 87*(8), 1315–1530.

Gong, I. Y., & Kim, R. B. (2013). Impact of genetic variation in OATP transporters to drug disposition and response. *Drug Metabolism and Pharmacokinetics, 28*(1), 4–18.

Grube, M., Kock, K., Karner, S., Reuther, S., Ritter, C. A., Jedlitschky, G., et al. (2006). Modification of OATP2B1-mediated transport by steroid hormones. *Molecular Pharmacology, 70*(5), 1735–1741.

Gui, C., & Hagenbuch, B. (2008). Amino acid residues in transmembrane domain 10 of organic anion transporting polypeptide 1B3 are critical for cholecystokinin octapeptide transport. *Biochemistry, 47*(35), 9090–9097.

Gui, C., & Hagenbuch, B. (2009). Role of transmembrane domain 10 for the function of organic anion transporting polypeptide 1B1. *Protein Science, 18*(11), 2298–2306.

Gui, C., Miao, Y., Thompson, L., Wahlgren, B., Mock, M., Stieger, B., et al. (2008). Effect of pregnane X receptor ligands on transport mediated by human OATP1B1 and OATP1B3. *European Journal of Pharmacology, 584*(1), 57–65.

Guo, G. L., & Klaassen, C. D. (2001). Protein kinase C suppresses rat organic anion transporting polypeptide 1- and 2-mediated uptake. *The Journal of Pharmacology and Experimental Therapeutics, 299*(2), 551–557.

Hagenbuch, B., & Gui, C. (2008). Xenobiotic transporters of the human organic anion transporting polypeptides (OATP) family. *Xenobiotica, 38*(7–8), 778–801.

Hagenbuch, B., & Meier, P. J. (2003). The superfamily of organic anion transporting polypeptides. *Biochimica et Biophysica Acta, 1609*(1), 1–18.

Hagenbuch, B., Scharschmidt, B. F., & Meier, P. J. (1996). Effect of antisense oligonucleotides on the expression of hepatocellular bile acid and organic anion uptake systems in Xenopus laevis oocytes. *The Biochemical Journal, 316*(Pt. 3), 901–904.

Hagenbuch, B., & Stieger, B. (2013). The SLCO (former SLC21) superfamily of transporters. *Molecular Aspects of Medicine, 34*(2–3), 396–412.

Han, S., Kim, K., Thakkar, N., Kim, D., & Lee, W. (2013). Role of hypoxia inducible factor-1alpha in the regulation of the cancer-specific variant of organic anion transporting polypeptide 1B3 (OATP1B3), in colon and pancreatic cancer. *Biochemical Pharmacology*, *86*(6), 816–823.

Hanggi, E., Grundschober, A. F., Leuthold, S., Meier, P. J., & St-Pierre, M. V. (2006). Functional analysis of the extracellular cysteine residues in the human organic anion transporting polypeptide, OATP2B1. *Molecular Pharmacology*, *70*(3), 806–817.

Hediger, M. A., Coady, M. J., Ikeda, T. S., & Wright, E. M. (1987). Expression cloning and cDNA sequencing of the Na+/glucose co-transporter. *Nature*, *330*(6146), 379–381.

Hoekstra, L. T., De Graaf, W., Nibourg, G. A. A., Heger, M., Bennink, R. J., Stieger, B., et al. (2013). Physiological and biochemical basis of clinical liver function tests: A review. *Annals of Surgery*, *257*, 27–36.

Huang, J., Li, N., Hong, W., Zhan, K., Yu, X., Huang, H., et al. (2013). Conserved tryptophan residues within putative transmembrane domain 6 affect transport function of organic anion transporting polypeptide 1B1. *Molecular Pharmacology*, *84*(4), 521–527.

Huang, S. M., Strong, J. M., Zhang, L., Reynolds, K. S., Nallani, S., Temple, R., et al. (2008). New era in drug interaction evaluation: US Food and Drug Administration update on CYP enzymes, transporters, and the guidance process. *Journal of Clinical Pharmacology*, *48*(6), 662–670.

Imai, S., Kikuchi, R., Kusuhara, H., & Sugiyama, Y. (2013). DNA methylation and histone modification profiles of mouse organic anion transporting polypeptides. *Drug Metabolism and Disposition*, *41*(1), 72–78.

Imai, S., Kikuchi, R., Kusuhara, H., Yagi, S., Shiota, K., & Sugiyama, Y. (2009). Analysis of DNA methylation and histone modification profiles of liver-specific transporters. *Molecular Pharmacology*, *75*(3), 568–576.

Imai, S., Kikuchi, R., Tsuruya, Y., Naoi, S., Nishida, S., Kusuhara, H., et al. (2013). Epigenetic regulation of organic anion transporting polypeptide 1B3 in cancer cell lines. *Pharmaceutical Research*, *30*(11), 2880–2890.

Jacquemin, E., Hagenbuch, B., Stieger, B., Wolkoff, A. W., & Meier, P. J. (1994). Expression cloning of a rat liver Na(+)-independent organic anion transporter. *Proceedings of the National Academy of Sciences of the United States of America*, *91*(1), 133–137.

Jung, D., Elferink, M. G., Stellaard, F., & Groothuis, G. M. (2007). Analysis of bile acid-induced regulation of FXR target genes in human liver slices. *Liver International*, *27*(1), 137–144.

Kalliokoski, A., & Niemi, M. (2009). Impact of OATP transporters on pharmacokinetics. *British Journal of Pharmacology*, *158*(3), 693–705.

Karlgren, M., Vildhede, A., Norinder, U., Wisniewski, J. R., Kimoto, E., Lai, Y., et al. (2012). Classification of inhibitors of hepatic organic anion transporting polypeptides (OATPs): Influence of protein expression on drug-drug interactions. *Journal of Medicinal Chemistry*, *55*(10), 4740–4763.

Kato, Y., Yoshida, K., Watanabe, C., Sai, Y., & Tsuji, A. (2004). Screening of the interaction between xenobiotic transporters and PDZ proteins. *Pharmaceutical Research*, *21*(10), 1886–1894.

Khurana, S., Venkataraman, K., Hollingsworth, A., Piche, M., & Tai, T. C. (2013). Polyphenols: Benefits to the cardiovascular system in health and in aging. *Nutrients*, *5*(10), 3779–3827.

Klaassen, C. D., & Aleksunes, L. M. (2010). Xenobiotic, bile acid, and cholesterol transporters: Function and regulation. *Pharmacological Reviews*, *62*(1), 1–96.

Kobayashi, D., Nozawa, T., Imai, K., Nezu, J., Tsuji, A., & Tamai, I. (2003). Involvement of human organic anion transporting polypeptide OATP-B (SLC21A9) in pH-dependent

transport across intestinal apical membrane. *The Journal of Pharmacology and Experimental Therapeutics*, *306*, 703–708.
Kock, K., Koenen, A., Giese, B., Fraunholz, M., May, K., Siegmund, W., et al. (2010). Rapid modulation of the organic anion transporting polypeptide 2B1 (OATP2B1, SLCO2B1) function by protein kinase C-mediated internalization. *The Journal of Biological Chemistry*, *285*(15), 11336–11347.
Kock, K., Xie, Y., Hawke, R. L., Oberlies, N. H., & Brouwer, K. L. (2013). Interaction of silymarin flavonolignans with organic anion-transporting polypeptides. *Drug Metabolism and Disposition*, *41*(5), 958–965.
Konig, J. (2011). Uptake transporters of the human OATP family. Molecular characteristics, substrates, their role in drug-drug interactions, and functional consequences of polymorphisms. *Handbook of Experimental Pharmacology*, *201*, 1–28.
Konig, J., Muller, F., & Fromm, M. F. (2013). Transporters and drug-drug interactions: Important determinants of drug disposition and effects. *Pharmacological Reviews*, *65*(3), 944–966.
Kullak-Ublick, G. A., Hagenbuch, B., Stieger, B., Schteingart, C. D., Hofmann, A. F., Wolkoff, A. W., et al. (1995). Molecular and functional characterization of an organic anion transporting polypeptide cloned from human liver. *Gastroenterology*, *109*, 1274–1282.
Kusuhara, H., & Sugiyama, Y. (2009). In vitro-in vivo extrapolation of transporter-mediated clearance in the liver and kidney. *Drug Metabolism and Pharmacokinetics*, *24*(1), 37–52.
Letschert, K., Faulstich, H., Keller, D., & Keppler, D. (2006). Molecular characterization and inhibition of amanitin uptake into human hepatocytes. *Toxicological Sciences*, *91*(1), 140–149.
Leuthold, S., Hagenbuch, B., Mohebbi, N., Wagner, C. A., Meier, P. J., & Stieger, B. (2009). Mechanisms of pH-gradient driven transport mediated by organic anion polypeptide transporters. *American Journal of Physiology. Cell Physiology*, *296*(3), C570–C582.
Li, N., Hong, W., Huang, H., Lu, H., Lin, G., & Hong, M. (2012). Identification of amino acids essential for estrone-3-sulfate transport within transmembrane domain 2 of organic anion transporting polypeptide 1B1. *PLoS One*, 7(5), e36647.
Li, L., Lee, T. K., Meier, P. J., & Ballatori, N. (1998). Identification of glutathione as a driving force and leukotriene C4 as a substrate for oatp1, the hepatic sinusoidal organic solute transporter. *The Journal of Biological Chemistry*, *273*(26), 16184–16191.
Li, L., Meier, P. J., & Ballatori, N. (2000). Oatp2 mediates bidirectional organic solute transport: A role for intracellular glutathione. *Molecular Pharmacology*, *58*(2), 335–340.
Liu, X., Huang, J., Sun, Y., Zhan, K., Zhang, Z., & Hong, M. (2013). Identification of multiple binding sites for substrate transport in bovine organic anion transporting polypeptide 1a2. *Drug Metabolism and Disposition*, *41*(3), 602–607.
Liu, T., & Li, Q. (2014). Organic anion-transporting polypeptides: A novel approach for cancer therapy. *Journal of Drug Targeting*, *22*(1), 14–22.
Lotspeich, W. D. (1958). Kidney, water and electrolyte metabolism. *Annual Review of Physiology*, *20*, 339–376.
Mahagita, C., Grassl, S. M., Piyachaturawat, P., & Ballatori, N. (2007). Human organic anion transporter 1B1 and 1B3 function as bidirectional carriers and do not mediate GSH-bile acid cotransport. *American Journal of Physiology. Gastrointestinal and Liver Physiology*, *293*(1), G271–G278.
Mairinger, S., Erker, T., Muller, M., & Langer, O. (2011). PET and SPECT radiotracers to assess function and expression of ABC transporters in vivo. *Current Drug Metabolism*, *12*(8), 774–792.
Mandery, K., Balk, B., Bujok, K., Schmidt, I., Fromm, M. F., & Glaeser, H. (2012). Inhibition of hepatic uptake transporters by flavonoids. *European Journal of Pharmaceutical Sciences*, *46*(1–2), 79–85.

Mandery, K., Bujok, K., Schmidt, I., Keiser, M., Siegmund, W., Balk, B., et al. (2010). Influence of the flavonoids apigenin, kaempferol, and quercetin on the function of organic anion transporting polypeptides 1A2 and 2B1. *Biochemical Pharmacology, 80*(11), 1746–1753.

Mandery, K., Sticht, H., Bujok, K., Schmidt, I., Fahrmayr, C., Balk, B., et al. (2011). Functional and structural relevance of conserved positively charged lysine residues in organic anion transporting polypeptide 1B3. *Molecular Pharmacology, 80*(3), 400–406.

Marin, J. J., Mangas, D., Martinez-Diez, M. C., El-Mir, M. Y., Briz, O., & Serrano, M. A. (2003). Sensitivity of bile acid transport by organic anion-transporting polypeptides to intracellular pH. *Biochimica et Biophysica Acta, 1611*(1–2), 249–257.

Martinez-Becerra, P., Briz, O., Romero, M. R., Macias, R. I., Perez, M. J., Sancho-Mateo, C., et al. (2011). Further characterization of the electrogenicity and pH sensitivity of the human organic anion-transporting polypeptides OATP1B1 and OATP1B3. *Molecular Pharmacology, 79*(3), 596–607.

Meier-Abt, F., Mokrab, Y., & Mizuguchi, K. (2005). Organic anion transporting polypeptides of the OATP/SLCO superfamily: Identification of new members in nonmammalian species, comparative modeling and a potential transport mode. *The Journal of Membrane Biology, 208*(3), 213–227.

Mengs, U., Pohl, R. T., & Mitchell, T. (2012). Legalon(R) SIL: The antidote of choice in patients with acute hepatotoxicity from amatoxin poisoning. *Current Pharmaceutical Biotechnology, 13*(10), 1964–1970.

Meyer zu Schwabedissen, H. E., Bottcher, K., Chaudhry, A., Kroemer, H. K., Schuetz, E. G., & Kim, R. B. (2010). Liver X receptor alpha and farnesoid X receptor are major transcriptional regulators of OATP1B1. *Hepatology, 52*(5), 1797–1807.

Meyer zu Schwabedissen, H. E., & Kim, R. B. (2009). Hepatic OATP1B transporters and nuclear receptors PXR and CAR: Interplay, regulation of drug disposition genes, and single nucleotide polymorphisms. *Molecular Pharmaceutics, 6*(6), 1644–1661.

Milne, J. L., Borgnia, M. J., Bartesaghi, A., Tran, E. E., Earl, L. A., Schauder, D. M., et al. (2013). Cryo-electron microscopy—A primer for the non-microscopist. *The FEBS Journal, 280*(1), 28–45.

Min, A. D., Johansen, K. L., Campbell, C. G., & Wolkoff, A. W. (1991). Role of chloride and intracellular pH on the activity of the rat hepatocyte organic anion transporter. *The Journal of Clinical Investigation, 87*(5), 1496–1502.

Mittur, A., Wolkoff, A. W., & Kaplowitz, N. (2002). The thiol sensitivity of glutathione transport in sidedness-sorted basolateral liver plasma membrane and in Oatp1-expressing HeLa cell membrane. *Molecular Pharmacology, 61*(2), 425–435.

Miyagawa, M., Maeda, K., Aoyama, A., & Sugiyama, Y. (2009). The eighth and ninth transmembrane domains in organic anion transporting polypeptide 1B1 affect the transport kinetics of estrone-3-sulfate and estradiol-17beta-D-glucuronide. *The Journal of Pharmacology and Experimental Therapeutics, 329*(2), 551–557.

Moore, R. J., Jackson, K. G., & Minihane, A. M. (2009). Green tea (Camellia sinensis) catechins and vascular function. *The British Journal of Nutrition, 102*(12), 1790–1802.

Mukerjee, P., & Ostrow, J. D. (2010). Review: Bilirubin pKa studies: New models and theories indicate high pKa values in water, dimethylformamide and DMSO. *BMC Biochemistry, 11*, 16.

Nakanishi, T., & Tamai, I. (2012). Genetic polymorphisms of OATP transporters and their impact on intestinal absorption and hepatic disposition of drugs. *Drug Metabolism and Pharmacokinetics, 27*(1), 106–121.

Niemi, M. (2010). Transporter pharmacogenetics and statin toxicity. *Clinical Pharmacology and Therapeutics, 87*(1), 130–133.

Niemi, M., Pasanen, M. K., & Neuvonen, P. J. (2011). Organic anion transporting polypeptide 1B1: A genetically polymorphic transporter of major importance for hepatic drug uptake. *Pharmacological Reviews, 63*, 157–181.

Noe, B., Hagenbuch, B., Stieger, B., & Meier, P. J. (1997). Isolation of a multispecific organic anion and cardiac glycoside transporter from rat brain. *Proceedings of the National Academy of Sciences of the United States of America, 94*(19), 10346–10350.

Noe, J., Portmann, R., Brun, M. E., & Funk, C. (2007). Substrate-dependent drug-drug interactions between gemfibrozil, fluvastatin and other organic anion-transporting peptide (OATP) substrates on OATP1B1, OATP2B1, and OATP1B3. *Drug Metabolism and Disposition, 35*(8), 1308–1314.

Nozawa, T., Imai, K., Nezu, J., Tsuji, A., & Tamai, I. (2004). Functional characterization of pH-sensitive organic anion transporting polypeptide OATP-B in human. *The Journal of Pharmacology and Experimental Therapeutics, 308*, 438–445.

Obaidat, A., Roth, M., & Hagenbuch, B. (2012). The expression and function of organic anion transporting polypeptides in normal tissues and in cancer. *Annual Review of Pharmacology and Toxicology, 52*, 135–151.

Petzinger, E., Joppen, C., & Frimmer, M. (1983). Common properties of hepatocellular uptake of cholate, iodipamide and antamanide, as distinct from the uptake of bromosulfophthalein. *Naunyn-Schmiedeberg's Archives of Pharmacology, 322*(2), 174–179.

Pizzagalli, F., Varga, Z., Huber, R. D., Folkers, G., Meier, P. J., & St-Pierre, M. V. (2003). Identification of steroid sulfate transport processes in the human mammary gland. *The Journal of Clinical Endocrinology and Metabolism, 88*(8), 3902–3912.

Ramachandran, A., Betts, G., Bhana, S., Helme, G., Blick, C., Moller-Levet, C., et al. (2013). An in vivo hypoxia metagene identifies the novel hypoxia inducible factor target gene SLCO1B3. *European Journal of Cancer, 49*(7), 1741–1751.

Roth, M., Obaidat, A., & Hagenbuch, B. (2012). OATPs, OATs and OCTs: The organic anion and cation transporters of the SLCO and SLC22A gene superfamilies. *British Journal of Pharmacology, 165*, 1260–1287.

Roth, M., Timmermann, B. N., & Hagenbuch, B. (2011). Interactions of green tea catechins with organic anion-transporting polypeptides. *Drug Metabolism and Disposition, 39*(5), 920–926.

Satlin, L. M., Amin, V., & Wolkoff, A. W. (1997). Organic anion transporting polypeptide mediates organic anion/HCO_3^- exchange. *The Journal of Biological Chemistry, 272*(42), 26340–26345.

Scharschmidt, B. F., Waggoner, J. G., & Berk, P. D. (1975). Hepatic organic anion uptake in the rat. *The Journal of Clinical Investigation, 56*(5), 1280–1292.

Schodel, J., Mole, D. R., & Ratcliffe, P. J. (2013). Pan-genomic binding of hypoxia-inducible transcription factors. *Biological Chemistry, 394*(4), 507–517.

Schrem, H., Klempnauer, J., & Borlak, J. (2002). Liver-enriched transcription factors in liver function and development. Part I: The hepatocyte nuclear factor network and liver-specific gene expression. *Pharmacological Reviews, 54*(1), 129–158.

Segawa, M., Ogura, J., Seki, S., Itagaki, S., Takahashi, N., Kobayashi, M., et al. (2013). Rapid stimulating effect of the antiarrhythmic agent amiodarone on absorption of organic anion compounds. *Drug Metabolism and Pharmacokinetics, 28*(3), 178–186.

Shirasaka, Y., Kuraoka, E., Spahn-Langguth, H., Nakanishi, T., Langguth, P., & Tamai, I. (2010). Species difference in the effect of grapefruit juice on intestinal absorption of talinolol between human and rat. *The Journal of Pharmacology and Experimental Therapeutics, 332*(1), 181–189.

Shirasaka, Y., Mori, T., Shichiri, M., Nakanishi, T., & Tamai, I. (2012). Functional pleiotropy of organic anion transporting polypeptide OATP2B1 due to multiple binding sites. *Drug Metabolism and Pharmacokinetics, 27*(3), 360–364.

Shirasaka, Y., Suzuki, K., Shichiri, M., Nakanishi, T., & Tamai, I. (2011). Intestinal absorption of HMG-CoA reductase inhibitor pitavastatin mediated by organic anion transporting polypeptide and P-glycoprotein/multidrug resistance 1. *Drug Metabolism and Pharmacokinetics*, *26*(2), 171–179.

Shitara, Y., Maeda, K., Ikejiri, K., Yoshida, K., Horie, T., & Sugiyama, Y. (2013). Clinical significance of organic anion transporting polypeptides (OATPs) in drug disposition: Their roles in hepatic clearance and intestinal absorption. *Biopharmaceutics & Drug Disposition*, *34*(1), 45–78.

Singh, B. N., Shankar, S., & Srivastava, R. K. (2011). Green tea catechin, epigallocatechin-3-gallate (EGCG): Mechanisms, perspectives and clinical applications. *Biochemical Pharmacology*, *82*(12), 1807–1821.

Sissung, T. M., Reece, K. M., Spencer, S., & Figg, W. D. (2012). Contribution of the OATP1B subfamily to cancer biology and treatment. *Clinical Pharmacology and Therapeutics*, *92*(5), 658–660.

Sreedharan, S., Stephansson, O., Schioth, H. B., & Fredriksson, R. (2011). Long evolutionary conservation and considerable tissue specificity of several atypical solute carrier transporters. *Gene*, *478*(1–2), 11–18.

Stieger, B., & Geier, A. (2011). Genetic variations of bile salt transporters as predisposing factors for drug-induced cholestasis, intrahepatic cholestasis of pregnancy and therapeutic response of viral hepatitis. *Expert Opinion on Drug Metabolism & Toxicology*, *7*(4), 411–425.

Stieger, B., Heger, M., de Graaf, W., Paumgartner, G., & van Gulik, T. (2012). The emerging role of transport systems in liver function tests. *European Journal of Pharmacology*, *675*(1–3), 1–5.

Stieger, B., & Meier, P. J. (2011). Pharmacogenetics of drug transporters in the enterohepatic circulation. *Pharmacogenomics*, *12*(5), 611–631.

Sugiura, T., Otake, T., Shimizu, T., Wakayama, T., Silver, D. L., Utsumi, R., et al. (2010). PDZK1 regulates organic anion transporting polypeptide Oatp1a in mouse small intestine. *Drug Metabolism and Pharmacokinetics*, *25*(6), 588–598.

Sugiura, T., Shimizu, T., Kijima, A., Minakata, S., & Kato, Y. (2011). PDZ adaptors: Their regulation of epithelial transporters and involvement in human diseases. *Journal of Pharmaceutical Sciences*, *100*(9), 3620–3635.

Sugiyama, D., Kusuhara, H., Shitara, Y., Abe, T., & Sugiyama, Y. (2002). Effect of 17 beta-estradiol-D-17 beta-glucuronide on the rat organic anion transporting polypeptide 2-mediated transport differs depending on substrates. *Drug Metabolism and Disposition*, *30*(2), 220–223.

Sugiyama, Y., & Yamashita, S. (2011). Impact of microdosing clinical study—Why necessary and how useful? *Advanced Drug Delivery Reviews*, *63*(7), 494–502.

Sun, A. Q., Ponamgi, V. M., Boyer, J. L., & Suchy, F. J. (2008). Membrane trafficking of the human organic anion-transporting polypeptide C (hOATPC). *Pharmaceutical Research*, *25*(2), 463–474.

Svoboda, M., Riha, J., Wlcek, K., Jaeger, W., & Thalhammer, T. (2011). Organic anion transporting polypeptides (OATPs): Regulation of expression and function. *Current Drug Metabolism*, *12*(2), 139–153.

Tamai, I., Nozawa, T., Koshida, M., Nezu, J., Sai, Y., & Tsuji, A. (2001). Functional characterization of human organic anion transporting polypeptide B (OATP-B) in comparison with liver-specific OATP-C. *Pharmaceutical Research*, *18*, 1262–1269.

Tell, R., Wang, Q. T., Blunier, A., & Benya, R. V. (2011). Identification of ChIP-seq mapped targets of HP1beta due to bombesin/GRP receptor activation. *Clinical Epigenetics*, *2*(2), 331–338.

Tweedie, D., Polli, J. W., Berglund, E. G., Huang, S. M., Zhang, L., Poirier, A., et al. (2013). Transporter studies in drug development: Experience to date and follow-up on decision

trees from the International Transporter Consortium. *Clinical Pharmacology and Therapeutics, 94*(1), 113–125.
Van Beers, B. E., Pastor, C. M., & Hussain, H. K. (2012). Primovist, eovist: What to expect? *Journal of Hepatology, 57*(2), 421–429.
van Montfoort, J. E., Hagenbuch, B., Groothuis, G. M., Koepsell, H., Meier, P. J., & Meijer, D. K. (2003). Drug uptake systems in liver and kidney. *Current Drug Metabolism, 4*(3), 185–211.
Varma, M. V., Rotter, C. J., Chupka, J., Whalen, K. M., Duignan, D. B., Feng, B., et al. (2011). pH-sensitive interaction of HMG-CoA reductase inhibitors (statins) with organic anion transporting polypeptide 2B1. *Molecular Pharmaceutics, 8*(4), 1303–1313.
Vega-Hissi, E. G., Estrada, M. R., Lavecchia, M. J., & Pis Diez, R. (2013). Computational chemical analysis of unconjugated bilirubin anions and insights into pKa values clarification. *The Journal of Chemical Physics, 138*(3), 035101.
Visentin, M., Chang, M. H., Romero, M. F., Zhao, R., & Goldman, I. D. (2012). Substrate- and pH-specific antifolate transport mediated by organic anion-transporting polypeptide 2B1 (OATP2B1-SLCO2B1). *Molecular Pharmacology, 81*(2), 134–142.
Wang, P., Hata, S., Xiao, Y., Murray, J. W., & Wolkoff, A. W. (2008). Topological assessment of oatp1a1: A 12-transmembrane domain integral membrane protein with three N-linked carbohydrate chains. *American Journal of Physiology. Gastrointestinal and Liver Physiology, 294*(4), G1052–G1059.
Wang, W. J., Murray, J. W., & Wolkoff, A. W. (2014). Oatp1a1 requires PDZK1 to traffic to the plasma membrane by selective recruitment of microtubule-based motor proteins. *Drug Metabolism and Disposition, 42*(1), 62–69.
Wang, P., Wang, J. J., Xiao, Y., Murray, J. W., Novikoff, P. M., Angeletti, R. H., et al. (2005). Interaction with PDZK1 is required for expression of organic anion transporting protein 1A1 on the hepatocyte surface. *The Journal of Biological Chemistry, 280*(34), 30143–30149.
Wang, X., Wolkoff, A. W., & Morris, M. E. (2005). Flavonoids as a novel class of human organic anion-transporting polypeptide OATP1B1 (OATP-C) modulators. *Drug Metabolism and Disposition, 33*(11), 1666–1672.
Weaver, Y. M., & Hagenbuch, B. (2010). Several conserved positively charged amino acids in OATP1B1 are involved in binding or translocation of different substrates. *The Journal of Membrane Biology, 236*(3), 279–290.
Weil, G., & Frimmer, M. (1970). Effect of silymarin on the isolated perfused rat liver poisoned by phalloidin. *Arzneimittel-Forschung, 20*(6), 862–863.
Wlcek, K., Koller, F., Ferenci, P., & Stieger, B. (2013). Hepatocellular organic anion-transporting polypeptides (OATPs) and multidrug resistance-associated protein 2 (MRP2) are inhibited by silibinin. *Drug Metabolism and Disposition, 41*(8), 1522–1528.
Xiao, Y., Nieves, E., Angeletti, R. H., Orr, G. A., & Wolkoff, A. W. (2006). Rat organic anion transporting protein 1A1 (Oatp1a1): Purification and phosphopeptide assignment. *Biochemistry, 45*(10), 3357–3369.
Yamaguchi, H., Sugie, M., Okada, M., Mikkaichi, T., Toyohara, T., Abe, T., et al. (2010). Transport of estrone 3-sulfate mediated by organic anion transporter OATP4C1: Estrone 3-sulfate binds to the different recognition site for digoxin in OATP4C1. *Drug Metabolism and Pharmacokinetics, 25*(3), 314–317.
Yao, J., Hong, W., Huang, J., Zhan, K., Huang, H., & Hong, M. (2012). N-Glycosylation dictates proper processing of organic anion transporting polypeptide 1B1. *PLoS One, 7*(12), e52563.
Zair, Z. M., Eloranta, J. J., Stieger, B., & Kullak-Ublick, G. A. (2008). Pharmacogenetics of OATP (SLC21/SLCO), OAT and OCT (SLC22) and PEPT (SLC15) transporters in the intestine, liver and kidney. *Pharmacogenomics, 9*(5), 597–624.

Zhang, Y., Hays, A., Noblett, A., Thapa, M., Hua, D. H., & Hagenbuch, B. (2013). Transport by OATP1B1 and OATP1B3 enhances the cytotoxicity of epigallocatechin 3-O-gallate and several quercetin derivatives. *Journal of Natural Products*, *76*(3), 368–373.

Zhang, L., Huang, S. M., & Lesko, L. J. (2011). Transporter-mediated drug-drug interactions. *Clinical Pharmacology and Therapeutics*, *89*(4), 481–484.

Zhang, L., Strong, J. M., Qiu, W., Lesko, L. J., & Huang, S. M. (2006). Scientific perspectives on drug transporters and their role in drug interactions. *Molecular Pharmaceutics*, *3*(1), 62–69.

Zhang, L., Zhang, Y. D., Strong, J. M., Reynolds, K. S., & Huang, S. M. (2008). A regulatory viewpoint on transporter-based drug interactions. *Xenobiotica*, *38*(7–8), 709–724.

Zhou, F., Lee, A. C., Krafczyk, K., Zhu, L., & Murray, M. (2011). Protein kinase C regulates the internalization and function of the human organic anion transporting polypeptide 1A2. *British Journal of Pharmacology*, *162*(6), 1380–1388.

Ziegler, K., Frimmer, M., & Fasold, H. (1984). Further characterization of membrane proteins involved in the transport of organic anions in hepatocytes. Comparison of two different affinity labels: 4,4'-Diisothiocyano-1,2-diphenylethane-2,2'-disulfonic acid and brominated taurodehydrocholic acid. *Biochimica et Biophysica Acta*, *769*(1), 117–129.

Zolk, O., Schnepf, R., Muschler, M., Fromm, M. F., Wendler, O., Traxdorf, M., et al. (2013). Transporter gene expression in human head and neck squamous cell carcinoma and associated epigenetic regulatory mechanisms. *The American Journal of Pathology*, *182*(1), 234–243.

CHAPTER SIX

SLC22, SLC44, and SLC47 Transporters—Organic Anion and Cation Transporters: Molecular and Cellular Properties

Ryan M. Pelis*, Stephen H. Wright†,[1]
*Department of Pharmacology, Dalhousie University, Halifax, Nova Scotia, Canada
†Department of Physiology, University of Arizona, Tucson, Arizona, USA
[1]Corresponding author: e-mail address: shwright@u.arizona.edu

Contents

Abstract

Transporters within the SLC22, SLC44, and SLC47 families of solute carriers mediate transport of a structurally diverse array of organic electrolytes, that is, molecules that are generally charged (cationic, anionic, or zwitterionic) at physiological pH. Transporters in the SLC22 family—all of which are members of the major facilitator superfamily (MFS) of transporters—represent a mechanistically diverse set of processes, including the organic anion transporters (OATs and URAT1) that physiologically operate as organic anion (OA) exchangers, the organic cation transporters (OCTs) that operate as electrogenic uniporters of organic cations (OCs), and the so-called "novel" organic cation transporters (OCTNs) that support Na-cotransport of selected zwitterions. Whereas the OCTNs display a high degree of substrate selectivity, the physiological hallmark of the OATs and OCTs is their multiselectivity—consistent with a principal role in renal and hepatic clearance of a wide array of both endogenous and xenobiotic compounds.

Current Topics in Membranes, Volume 73
ISSN 1063-5823
http://dx.doi.org/10.1016/B978-0-12-800223-0.00006-2

SLC47 consists of members of the multidrug and toxin extruder (MATE) family, which are carriers that are obligatory exchangers and that physiologically support electroneutral H^+ exchange. The MATEs also display a characteristic multiselectivity and are frequently paired with OCTs to mediate transepithelial OC secretion, with the OCTs typically supporting basolateral OC entry and the MATEs supporting apical OC efflux. The SLC44 family contains the choline transporter-like (CTL) transporters. Largely restricted to choline and a limited set of structural congeners, the CTLs appear to support the Na-independent, electrogenic uniport of choline, thereby providing choline for membrane biogenesis. The solution of X-ray crystal structures of representative prokaryotic MFS and MATE transporters has led to the development of homology models of mammalian OAT, OCT, and MATE transporters that, in turn, have supplemented studies of the molecular basis of the complex interactions of ligands with these multiselective proteins.

1. INTRODUCTION

Transporters in the SLC22, SLC44, and SLC47 families of solute carriers (SLCs) predominately support flux of so-called organic electrolytes—a structurally diverse array of compounds that display a negative charge (OAs), a positive charge (OCs), or both negative and positive charges (zwitterions) at physiological pH. Included within this class of chemicals are pharmacologically and toxicologically relevant xenobiotics, as well as endogenous chemicals of physiological importance. Due to their charge at physiological pH, efficient movement of organic electrolytes across plasma membranes requires facilitated transport. Membrane transporters that translocate OAs and OCs can be found in both the ATP-binding cassette (ABC) and SLC families of membrane transporters. But whereas transporters in the ABC family operate via primary active (ATP-dependent) transport, SLC family members function by either facilitated diffusion or secondary active transport, and net flux (influx or efflux) is dictated by the electrochemical gradients of the substrates involved. This review focuses primarily on OA and OC exchangers within the SLC22, 44, and 47 families, and emphasizes the mechanisms of transport, the energetic basis of directionality (uptake vs. efflux), the molecular determinants of substrate binding and translocation, and regulation at the molecular level.

Within the last several years, regulatory agencies including the U.S. Food and Drug Administration and the European Medicines Agency have focused substantial attention on OA and OC transporters with the understanding that their activity profoundly influences the absorption, distribution, metabolism, and excretion of pharmaceutical drugs (e.g., U.S. Food and Drug Administration, 2012a, 2012b). In particular, excretion of OAs and OCs by renal tubule cells and hepatocytes—processes involving SLC exchangers—impacts the body's exposure to drugs (Endres, Hsiao, Chung, & Unadkat, 2006). Because SLC exchangers are multiselective, they are a potential site of drug–drug interactions, and polymorphisms in the genes encoding these transporters can contribute to interindividual variability in drug exposure and response (Sissung et al., 2012). A better understanding of the molecular basis of OA and OC exchanger activity may assist in predicting the role of these important transporters in drug exposure and response. With that in mind, we limit, when possible, the discussion of specific kinetic or selectivity properties to those of the human orthologs of these SLC family members.

2. SLC22 TRANSPORTERS

The SLC22 family of SLCs, which are phylogenetically placed in the major facilitator superfamily (MFS) of transporters (Reddy, Shlykov, Castillo, Sun, & Saier, 2012), includes the following three "subfamilies": the organic cation transporters (OCTs), the "novel" organic cation transporters (OCTNs), and the organic anion transporters (OATs and URAT1). The SLC22 family is comparatively unusual because of the number of distinct transport mechanisms supported by its several subfamilies (Koepsell & Endou, 2004). Whereas the OATs and URATs principally support mediated anion exchange, the primary modes of action of the OCTs and OCTNs are electrogenic facilitated diffusion and Na-dependent cotransport, respectively. Consequently, given the emphasis of the current volume on exchange processes, we limit the discussion of OCTs and OCTNs to brief overviews of their general mechanistic characteristics, followed by a more extended discussion of the exchange activity of the OATs.

2.1. OCTs

There are three OCT homologs: OCTs1, 2, and 3 (SLC22A1, 2, and 3) that were cloned and functionally characterized in the early–mid-1990s (Gründemann, Gorboulev, Gambaryan, Veyhl, & Koepsell, 1994; Kekuda et al., 1998; Okuda, Saito, Urakami, Takano, & Inui, 1996). The human genes for OCT1 and OCT2, both of which contain 11 exons (Eraly, Monte, & Nigam, 2004), occur as a gene pair localized to chromosome 6 (6q26), approximately 90 kb upstream from OCT3 (Eraly,

Hamilton, & Nigam, 2003). There are significant species differences in tissue-specific expression of OCT1 (reviewed in Pelis & Wright, 2011). In humans, OCT1 is most heavily expressed in the sinusoidal (basolateral) membrane of hepatocytes (Bleasby et al., 2006; Schaeffeler et al., 2011), and at more modest levels in other epithelial tissues (including apical expression in both enterocytes (Han et al., 2013), and renal proximal and distal tubule cells (Tzvetkov et al., 2009)). In humans, OCT2 is most heavily expressed in the peritubular (basolateral) membrane of renal proximal tubule cells (Bleasby et al., 2006; Motohashi et al., 2002). Expression of OCT3 is more broadly distributed (e.g., in heart, skeletal muscle, small intestine, as well as neurons, glial cells, and choroid plexus), albeit at modest levels compared to the tissue-specific expression profiles of OCT1 and OCT2 (Bleasby et al., 2006).

2.1.1 Structure

The OCTs (along with all other members of the SLC22 family of transporters) share several structural characteristics with other MFS proteins, including 12 transmembrane-spanning helices (TMHs), cytoplasmic N- and C-termini, and a relatively long cytoplasmic loop between TMHs 6 and 7. Several additional features are unique to members of SLC22, including a long (~110 amino acid residues) extracellular loop between TMHs 1 and 2, as well as a distinguishing sequence motif ([S/T]-IVTE-[F/W]-[D/N]-LVC) before TMH2 (see Schomig et al., 1998). The human orthologs of OCT1, OCT2, and OCT3 contain 554, 555, and 556 amino acid residues, respectively. Consensus sites for phosphorylation mediated by protein kinase A (PKA), protein kinase C (PKC), protein kinase G (PKG), casein kinase 2 (CK2), and/or calcium/calmodulin-dependent protein kinase II (CaMII) are located within or near the long cytoplasmic loop between TMHs 6 and 7, or in the cytoplasmic C-terminal sequence (Burckhardt & Wolff, 2000). In OCT1 and OCT2, the long extracellular loop between TMHs 1 and 2 contains three N-linked glycosylation sites and six conserved cysteine residues. These cysteines influence correct folding, trafficking, plasma membrane insertion, and formation of homo-OCT oligomers (Brast et al., 2011; Keller et al., 2011).

All the SLC22A family members, including the OCTs, presumably share the conserved protein fold of other MFS proteins: 12 TMHs organized as N- and C-terminal bundles of six helices each that display a pseudo twofold symmetry perpendicular to the plane of the membrane. Each bundle of helices contains sequential three helix inverted-topology repeats, a structural motif found in a number of otherwise phylogenetically unrelated families of transport proteins (Pornillos & Chang, 2006) that is consistent with the "alternating access" mechanism of transport (Radestock & Forrest, 2011). The N- and C-terminal halves of MFS proteins define a water-filled cleft lined by amino acid residues from N-terminal TMHs 1, 2, 4, and 5, and C-terminal TMHs 7, 8, 10, and 11, and ligand binding and substrate translocation are speculated to involve interaction with cleft residues. Homology models of OCT1 (Popp et al., 2005) and OCT2 (Zhang, Shirahatti, Mahadevan, & Wright, 2005), based upon the structure of the MFS transporters LacY and GlpT, respectively, were the basis of site-directed studies that identified several "cleft" amino acids in OCTs that influence ligand binding and transport activity. These include a highly conserved (in all OCTs) aspartate residue in TMH 11 at position 475 (in rOCT1) that exerts a profound effect on ligand binding (Gorboulev, Volk, Arndt, Akhoundova, & Koepsell, 1999). These models also provided insight into the structural basis of differential selectivity of OCT1 and OCT2 for selected ligands. For example, rOCT2 has a 40-fold higher affinity for corticosterone than does rOCT1 (Gorboulev, Shatskaya, Volk, & Koepsell, 2005), and rabbit OCT2 has a 40-fold higher affinity for cimetidine than does rbOCT1 (Zhang et al., 2005). In both cases, these relative affinities could be swapped by converting the glutamate found in all OCT2 orthologs in TMH 10 at position 448/447 (in rat/rabbit, respectively) for the glutamine found in all OCT1 orthologs, implicating this site as a defining element for ligand interaction in the OCT translocation pathway.

2.1.2 Mechanism

Whereas all three OCTs support mediated exchange of OCs (e.g., Budiman, Bamberg, Koepsell, & Nagel, 2000; Harper & Wright, 2012), each is also capable of supporting electrogenic uniport of cationic substrates (Budiman et al., 2000; Busch et al., 1996; Wu et al., 2000). Therefore, the OCTs are not obligatory exchangers; the "unloaded" carrier, as well as the loaded carrier, is capable of undergoing the conformational changes associated with substrate translocation (Koepsell, Lips, & Volk, 2007). Based upon functional data, those conformational changes are adequately described by the "alternating access" model (Koepsell et al., 2007). Electroneutral OC:OC exchange is favored when large (i.e., near saturating) substrate concentrations are found on both sides of the membrane; in the absence of such a condition, which probably exemplifies the normal physiological condition, electrogenic uniport is favored (Koepsell et al., 2007). The inside negative membrane potential of cells (including proximal tubule cells and hepatocytes) is the driving force that supports OCT-mediated net accumulation of cationic substrates that in turn supports the role of OCTs as the initial (the "entry") step in renal and hepatic secretion of OCs.

OCT substrates include a structurally diverse array of xenobiotic compounds (including some heavy metals; Schmitt & Koepsell, 2005), as well as a range of endogenous compounds (Nies, Koepsell, Damme, & Schwab, 2011). Whereas the presence of cationic charge facilitates ligand interaction with OCTs (Barendt & Wright, 2002), at least some transported substrates carry little or no charge at physiological pH (e.g., lamivudine (Minuesa et al., 2009)), highlighting the complexity of ligand interactions with these transporters. Indeed, the kinetics of ligand binding (of both substrates and inhibitors) include competitive, noncompetitive, and "mixed-type" interactions (Harper & Wright, 2012). Those observations, combined with the results of site-directed mutagenesis and homology modeling studies, suggest the presence of a large binding surface with multiple (probably overlapping) binding sites within the OCT protein.

2.2. OCTNs

Based on sequence analysis, three members of the human SLC22A family lie within the group of the so-called "Novel" organic cation transporters: OCTN1 (SLC22A4), OCTN2 (SLC22A5), and OCT6 (SLC22A16). Cloned and functionally characterized in the late 1990s (Tamai et al., 1998, 1997), OCTN1 and OCTN2 contain 551–577 amino acids and have ~29–33% sequence identity to hOCT1. As with other members of the SLC22 family, the OCTNs have 12 putative TMHs and a number of consensus sites for phosphorylation mediated by PKA, PKC, and CKII (Pelis & Wright, 2011). OCTN1 is expressed at moderate to high levels in the kidney, lung, and neutrophils, with lower levels of expression in many other tissues (including the retina, brain, prostate, and skeletal muscle) (Bleasby et al., 2006). OCTN2 is heavily expressed in the kidney, pancreas, prostate, skeletal muscle, and pituitary gland, and comparatively well expressed in many other tissues (Bleasby et al., 2006). OCT6 is primarily expressed in the testis and hematopoietic tissues (Enomoto, Wempe, et al., 2002; Gong et al., 2002).

All OCTNs primarily support Na-dependent cotransport of selected zwitterions but also support a Na-independent exchange mode of activity. However, based on comparatively low J_{max}/K_m ratios, the Na-independent modes of OCTN activity probably have minimal impact on mediated solute flux (Gründemann et al., 2005). In human kidney, OCTN1 and OCTN2 are expressed in the apical membrane of proximal tubule cells (Gisler et al., 2003; Tamai et al., 2001) where they support the reabsorption of selected zwitterions via Na-coupled cotransport. In particular, OCTN1 supports high affinity (K_m of 16 μM), Na-dependent transport of the antioxidant, ergothionine (Bacher et al., 2009; Gründemann et al., 2005), and OCTN2 and OCT6 support high affinity (K_m values of 15–20 μM; Bacher, Giersiefer, Bach, Fork, Schomig & Grundemann, 2009; Enomoto et al., 2002), electrogenic (Tamai et al., 2001) transport of L-carnitine. Additional modes of transport activity have been reported for the OCTNs. OCTN1 supports Na-independent transport of tetraethylammonium (TEA), although this may reflect its modest capacity to support TEA/H^+ exchange (Tamai et al., 2004), and OCTN2 supports carnitine/TEA exchange (but only in the presence of Na^+ (Ohashi et al., 2001)). However, as noted earlier, the comparatively low efficiency of these latter modes of OCTN1 and OCTN2 activity calls into question their physiological relevance (Bacher et al., 2009).

2.3. OATs

OAT1 is predominately expressed in the kidney (Lopez-Nieto et al., 1997) where it localizes to basolateral membranes of RPT cells (Hosoyamada, Sekine, Kanai, & Endou, 1999). OAT2 is expressed in sinusoidal membranes of liver hepatocytes (Simonson, Vincent, Roberg, Huang, & Iwanij, 1994). OAT2 is also expressed in the kidney, but subcellular localization exhibits significant species differences. OAT2 localizes to basolateral membranes of human RPTs (Enomoto, Takeda, et al., 2002), but to luminal membranes of rat RPTs (Brzica et al., 2009). In humans, OAT3 is highly expressed in the kidney but is also present in the brain, skeletal muscle (Cha et al., 2001), and adrenal tissue (Asif et al., 2005). In the kidney, OAT3 is expressed in the basolateral membrane of RPT cells (Motohashi et al., 2002). OAT4 is expressed in the kidney, placenta (Cha et al., 2000), and adrenal tissue (Asif et al., 2005). In the kidney, OAT4 localizes to luminal membranes of RPT cells where it is coexpressed in the same cells with OAT1 and OAT3 (Ekaratanawong et al., 2004).

2.3.1 Structure

There are seven OAT homologs in humans: Organic Anion Transporters 1, 2, 3, 4, 5, 7, and 10 (SLC22A6, 7, 8, 11, 10, 9, and 13, respectively). Discussion will focus on OAT1, OAT2, OAT3, and OAT4 as our current understanding of OAT exchange activity comes primarily from work with these transport proteins. The human genes for OAT1 and OAT3 contain 10 exons (Eraly et al., 2004) and occur as a gene pair localized to chromosome 11 (11q12.3), approximately 280 kb upstream from the OAT5–UST3 (SLC22A9) pair, and a further 1.15 Mb from the OAT4–URAT1 pair (Eraly et al., 2003). The human orthologs of OATs share 28–57% sequence homology, with OAT5 and OAT7 being most similar. The first OAT cloned was OAT2 in 1994 (Simonson et al., 1994), followed by OAT1 in 1997 (Lopez-Nieto et al., 1997; Sekine, Watanabe, Hosoyamada, Kanaim, & Endou, 1997) and OAT3 in 1999 (Kusuhara et al., 1999). The OATs contain 541–550 amino acids and share a secondary structure common to other MFS members, including 12 TMHs, intracellular C- and N-termini, the long extracellular loop between TMHs 1 and 2, and the large intracellular loop linking TMHs 6 and 7. The long intracellular loop of OAT1 contains consensus sites for PKA, PKC, CK2, and tyrosine kinase, and many of these consensus sequences are present in other OATs as well (Srimaroeng, Perry, & Pritchard, 2008). Four cysteine residues conserved at homologous positions reside in the long extracellular loop of OATs and, at least for OAT1, are important for membrane targeting and/or stability (Tanaka, Zhou, Kuze, & You, 2004). In addition to conserved cysteine residues, multiple consensus sites for N-linked glycosylation are present in the long extracellular loop of OATs (Srimaroeng et al., 2008). Removal of N-glycosylation in OAT1 and OAT4 impairs membrane expression, and glycosylation at N39 in human OAT1 influences substrate binding and/or translocation (Tanaka, Xu, Zhou, & You, 2004; Zhou et al., 2005). OAT1 forms homo-oligomers in cultured cells and rat kidney (Hong et al., 2005), and coexpression of OAT1 in COS-7 cells with a mini-peptide corresponding to a region of TMH6 disrupts oligomerization, indicating that TMH6 contains a sequence important for oligomerization (Duan, Li, & You, 2011). Furthermore, disruption of OAT1 oligomerization by the mini-peptide is associated with a reduction in membrane expression, suggesting that oligomerization promotes trafficking and/or membrane stability of OAT1 (Duan et al., 2011). OAT4 contains a PDZ interaction motif in its C-terminal end and interacts with the multivalent PDZ proteins PDZK1 and NHERF1 in *in vitro* binding studies (Miyazaki et al., 2005).

All three proteins localize to apical membranes of RPT cells, and *in vitro* studies implicate NHERF-1 and PDZK1 in modulating OAT4 activity by regulating its membrane expression (Miyazaki et al., 2005).

Using the crystal structure of the MFS member *Escherichia coli* glycerol-3-phosphate transporter (GlpT) as a template, Perry and colleagues generated a three-dimensional (3D) homology model of human OAT1 (Perry, Dembla-Rajpal, Hall, & Pritchard, 2006). TMHs 1, 2, 4, and 5 in the N-terminal half, and TMHs 7, 8, 10, and 11 in the C-terminal half, form the putative ligand binding/translocation pathway. The postulated volume of the ligand binding/translocation pathway (830 $Å^3$) can accommodate an individual substrate/inhibitor or, potentially, more than one simultaneously. For example, the OAT1 substrates cidofovir and probenecid have volumes of 158 and 233 $Å^3$, respectively. Amino acids implicated in ligand binding and translocation, including Y230, K431, F438, and R466 (Perry et al., 2006; Rizwan, Krick, & Burckhardt, 2007), line the solvent phase of the ligand binding/translocation pathway in the 3D model. The accessibility of C440 in TMH11 to membrane impermeable thiol-reactive reagents provides additional validation of the OAT1 homology model (Astorga, Wunz, Morales, Wright, & Pelis, 2011). Mutation of Y230 in OAT1 eliminates *para*-aminohippurate transport, but not cidofovir transport, suggesting that OAT1 contains multiple ligand-binding regions, consistent with its multiselectivity (Perry et al., 2006).

2.3.2 Mechanism

OAT1 (Sekine et al., 1997), OAT2 (Kobayashi et al., 2005), OAT3 (Sweet et al., 2003), and OAT4 (Cha et al., 2000) are Na-independent anion exchangers. However, the cellular uptake of OAs by OAT1 and OAT3 is indirectly coupled to the Na^+ gradient via a tertiary active transport process involving Na, K-ATPase, and Na-dependent dicarboxylate cotransport activity (Pritchard, 1988; Shimada, Moewes, & Burckhardt, 1987; Sweet et al., 2003). In this mechanism, Na, K-ATPase establishes and maintains the Na^+ gradient that supports the Na-dependent uptake of the Kreb's cycle intermediate α-ketoglutarate (α-KG), a preferred physiological counter ion of OAT1 and OAT3. The plasma concentration of α-KG in humans is about 8 μM (Rocchiccioli, Leroux, & Cartier, 1984), but the intracellular concentration of α-KG in proximal tubule cells is much higher, in part due to Na-dependent uptake (Pritchard, 1995). OAT1- and OAT3-mediated OA uptake occurs via an exchange mechanism that is stimulated by the steep outwardly directed α-KG gradient. At least for OAT1, the exchange activity can be electroneutral or electrogenic depending on the molecule transported. The exchange of the monocarboxylate *para*-aminohippurate for the dicarboxylate glutarate is 1:1, resulting in a net transfer of positive charge into the cell (Aslamkhan, Han, Walden, Sweet, & Pritchard, 2003). The lack of a voltage change in current-clamped *Xenopus* oocytes when glutarate is applied to the bathing medium suggests that dicarboxylate/dicarboxylate exchange activity by OAT1 is electroneutral (Burckhardt, Wolff, & Burckhardt, 2000). In contrast to OAT1, the stoichiometry and electrogenic operation of the other cloned OATs remain unknown. Extracellular chloride stimulates OAT1 transport activity by increasing its translocation rate (Rizwan et al., 2007). According to site-directed mutagenesis data, R466, which is conserved in all human OATs (except OAT7 in which the homologous amino acid is an asparagine), interacts with chloride to stimulate OAT1 turnover (Rizwan et al., 2007). According to results from preload experiments in *Xenopus* oocytes expressing OAT2, succinate and fumarate, but not glutarate, can serve as counter ions for OAT2-mediated estrone-3-sulfate transport (Kobayashi et al., 2005). Glutarate, OH^-, and Cl^- can serve as counter ions for OAT4-mediated OA transport (Hagos, Stein, Ugele, Burckhardt, & Bahn, 2007).

2.3.3 Selectivity

Ligands of OATs, especially mono- and dicarboxylates, are predominately OAs at physiological pH, although the transporters can also interact with neutral and basic organics (Vanwert, Gionfriddo, & Sweet, 2010). Although the ligand selectivity of OATs is overlapping for some molecules, it is also distinct for others. For example, cidofovir is a relatively specific substrate of OAT1 (Uwai, Ida, Tsuji, Katsura, & Inui, 2007), whereas bumetanide is relatively specific for OAT3 (Hasannejad et al., 2004). Duan, Li, Ai, et al. (2012) screened 727 compounds for inhibition of OAT1 and OAT3 expressed in COS-7 cells. Using pharmacophore modeling, they identified several molecular descriptors that distinguish inhibitors from noninhibitors. For OAT1, the number of hydrogen bond donors, number of rotatable bonds, and topological polar surface area (TPSA) are important, whereas for OAT3, molecular weight, number of hydrogen bond donors and acceptors, TPSA, partition coefficient, and polarizability are discriminating descriptors. These differences suggest that the ligand-binding regions of OAT1 and OAT3 are distinct. Consistent with the apparent structural differences in the OAT-binding regions, C440 in TMH10 of OAT1 is accessible to membrane impermeant thiol-reactive reagents, whereas the homologous cysteine in OAT3 (C428) is not (Astorga et al., 2011). In addition to homolog-specific selectivities in OATs, there is evidence that ligand binding with the intracellular versus extracellular aspects of an individual transporter can differ markedly. For example, glutarate and *para*-aminohippurate are not transported into HEK293 cells expressing OAT4, but when preloaded into the cells, both molecules *trans*-stimulate 6-carboxyfluorescein and estrone-3-sulfate uptake (Hagos et al., 2007). For mouse Oat1, there is little correlation between substrate binding (K_m value) and maximal transport rate (J_{max} value), suggesting that both events (binding and translocation) require different molecular interactions (Kaler et al., 2007). From these data, Kaler et al. (Kaler et al., 2007) proposed a "switching" mechanism for OAT-mediated transport in which the transporter switches from an outward (extracellular) to an inward (intracellular) conformation during the process of substrate translocation. Initial attempts at molecular dynamic simulations with human OAT1 embedded in a phospholipid membrane suggest that the two hemi-domains (the N- and C-terminal halves) that comprise the ligand binding/translocation pathway may in fact move in a manner consistent with the switching hypothesis (Tsigelny et al., 2011). It is possible that during conformational changes associated with the transition between the outward and inward facing conformations a structural rearrangement of amino acids in the ligand-binding region occurs, resulting in asymmetrical binding, or alternatively, that ligand binding to the inwardly versus outwardly directed faces of the transporter involves topologically different regions.

2.3.4 Regulation

OAT activity is under both short-term and long-term regulation. The following discussion is focused on acute regulation of OAT activity by PKA and PKC. The reader is directed to a recent review by Vanwert et al. (2010) for a discussion of transcriptional and epigenetic regulation of OATs. In heterologous cell expression systems, activation of PKC by hormones or nonspecific pharmacological means reduces activity of OAT1 (Li, Duan, & You, 2009; Wolff et al., 2003; You, Kuze, Kohanski, Amsler, & Henderson, 2000; Zhang et al., 2008), OAT3 (Duan, Li, & You, 2010; Takeda, Sekine, & Endou, 2000; Zhang, Suh, Pan, & You, 2012), and OAT4 (Zhou, Hong, & You, 2007) by reducing their expression level at the plasma membrane. Consistent with the reduction in membrane expression, PKC activation reduces their respective maximal transport rates without affecting their Michaelis constants. The inhibition of OAT1 and OAT3 activity elicited by PKC activation is blocked by Gö-6976 (PKCα-selective inhibitor), suggesting that inhibition of OAT activity caused by transporter internalization occurs through activation of the PKCα isoform (Duan et al., 2010; Li et al., 2009). The effect of PKCα activation does not appear to be through direct phosphorylation of the transporter because mutation of the PKC consensus sites in OAT1 does not prevent the PKC-dependent internalization (Wolff et al., 2003). Rather, PKCα activation causes ubiquitination of OAT1 at K297, K303, and K315 that promotes its internalization rate (Li, Zhang, & You, 2013). The dynamin- and clathrin-dependent endocytic pathways contribute to OAT1 internalization (Zhang et al., 2008). The effect of PKC activation on OAT activity appears to be PKC isoform dependent.

Bradykinin causes activation and association of PKCζ, PKCΔ, and PKCε with OAT3 in COS-7 cells, thereby leading to an increase in OAT3 cell surface expression and transport activity via an increased maximal transport rate (Li, Duan, & You, 2010). Both insulin and epidermal growth factor increase surface expression and activity of OAT1 and OAT3 via a complex signaling cascade involving PKA activation (Pelis & Wright, 2011). Using renal cortical slices from mice, Barros et al. (2009) showed that PKCζ activation is downstream of PKA and is likely the end mediator in the signaling cascade involved in insulin and epidermal growth factor stimulation of OAT activity. Parathyroid hormone-related peptide acting through PKA was shown to enhance membrane expression and activity of OAT4 in COS-7 cells, although the study did not investigate involvement of PKCζ (Duan, Li & You, 2012).

Comparatively less is known about the distribution, function, and regulation of OAT5, OAT7, and OAT10. The tissue distribution of OAT5 displays marked species differences. At the mRNA level, OAT5 is expressed exclusively in the human liver (Sun, Wu, van Poelje, & Erion, 2001), whereas it is predominate in rodent kidney (Youngblood & Sweet, 2004). Although human OAT5 has not been functionally characterized, the rat and mouse orthologs support the transport of OAs and likely operate via an exchange mechanism (Anzai et al., 2005; Youngblood & Sweet, 2004). In humans, OAT7 is almost exclusively expressed in the liver where it localizes to sinusoidal membranes of hepatocytes (Shin et al., 2007). OAT7 expressed in *Xenopus laevis* oocytes supports the Na^+-independent transport of estrone sulfate and dehydroepiandrosterone sulfate, and the activity is *trans*-stimulated by butyrate, consistent with an exchange mechanism

(Shin et al., 2007). OAT10 transports a number of OAs, including nictotinate and urate, via an exchange mechanism (Bahn et al., 2008). OAT10 transcript expression is highest in the human kidney and real-time PCR analysis of isolated nephron segments shows localization in RPT and cortical collecting duct cells; Western blots reveal protein expression in brush border membranes of renal tissue (Bahn et al., 2008).

2.4. URAT1

In humans, URAT1 mRNA is exclusively expressed in the kidney (Enomoto, Kimura, et al., 2002). URAT1 has been localized by immunohistochemistry to apical membranes of both human and mouse proximal tubules, but it is absent in distal tubules (Enomoto, Kimura, et al., 2002; Hosoyamada, Ichida, Enomoto, Hosoya, & Endou, 2004). In rat, it also localizes to apical membranes of RPT cells, with weak expression in distal nephron segments as well (Sato et al., 2011). URAT1 colocalizes with PDZK1 in apical membranes of human RPTs (Anzai et al., 2004). *In vitro*, PDZK1 binds the C-terminal PDZ motif of URAT1, and this association increases membrane expression and transport activity of URAT1 (Anzai et al., 2004).

2.4.1 Structure

Cloned in 2002, URAT1 (SLC22A12) is predicted to have the secondary structure common to MFS family members, as described earlier (Enomoto, Kimura, et al., 2002). URAT1 contains 553 amino acids and shares highest sequence identity (52%) with OAT4. The putative long extracellular loop of URAT1 contains three N-linked glycosylation consensus sites and the four cysteines that are conserved at homologous positions in all OAT homologs. A shift in the apparent size on SDS-PAGE gels of mouse Urat1 from crude membranes of mouse kidney following PNGase treatment suggests that one or more of the N-glycosylation sites are used (Hosoyamada et al., 2004), although it is not known if N-glycosylation is important for URAT1 function. URAT1 contains two PKA phosphorylation consensus sites (Enomoto, Kimura, et al., 2002) and a PDZ interaction motif in its C-terminal tail (Anzai et al., 2004).

2.4.2 Mechanism

Numerous studies have examined the mechanism of urate transport using brush-border membrane vesicles (BBMV) isolated from renal cortex. However, because multiple processes within the brush border can support urate transport (e.g., OAT4 in addition to URAT1), here we focus on the URAT1 transport mechanism as determined from (the comparatively few) studies that used the cloned transporter expressed heterologously in cells. Transport of urate by human URAT1 expressed in *Xenopus* oocytes is insensitive to Na^+ removal and membrane depolarization (i.e., high extracellular K^+), but is *trans*-stimulated by Cl^-, consistent with Na^+-independent electroneutral exchange of Cl^- for urate (Cl^-/urate exchange) (Enomoto, Kimura, et al., 2002). In these particular studies, imposition of a pH gradient did not influence urate transport activity, suggesting that URAT1 does not function via urate/OH^- exchange or urate-H^+ cotransport. URAT1-mediated urate uptake is stimulated when *Xenopus* oocytes are preloaded with a variety of monocarboxylates, including lactate, nicotinate, pyrazinoate, and nitrate (Enomoto, Kimura, et al., 2002; Hosoyamada et al., 2004; Sato et al., 2011). Interestingly, URAT1 expressed in oocytes does not mediate the uptake of nicotinate (Lu, Nakanishi, & Tamai, 2013). The ability of nicotinate to *trans*-stimulate URAT1 activity, without itself being a substrate, suggests that, like OAT4, ligand binding to URAT1 is asymmetric. The uptake of urate by URAT1 from filtrate into proximal tubule cells may involve a tertiary active mechanism involving Na, K-ATPase, and Na-dependent monocarboxylate cotransport. Consistent with this energetic mechanism, coexpression in *Xenopus* oocytes of URAT1 with the Na-dependent monocarboxylate cotransporter 1 (SMCT1), which is also expressed in apical membranes of proximal tubule cells, stimulates URAT1-mediated urate transport, but only when nicotinate and lactate are present in the external buffer (Lu et al., 2013).

2.4.3 Selectivity

Urate is a well-known physiological substrate of URAT1, and URAT1 is at least partly involved in regulating plasma urate levels (Enomoto, Kimura, et al., 2002). Accordingly, reduced-function mutations in URAT1 are associated with hypouricemia (Enomoto, Kimura, et al., 2002; Tasic et al., 2011). The Michaelis constants (K_m values) associated with urate transport by the human, mouse and rat orthologs of URAT1 are 371 μM (Enomoto, Kimura, et al., 2002), 1213 μM (Hosoyamada et al., 2004), and 1773 μM (Sato et al., 2011), respectively. Other substrates of URAT1 include the precursor of pyrimidine synthesis orotate (Miura et al., 2011) and the industrial chemical perfluorooctanoate (Yang, Glover, & Han, 2010). A number of OAs *cis*-inhibit URAT1, including the uricosuric drugs probenecid, benzbromarone, and sulfinpyrazone (Enomoto, Kimura, et al., 2002).

2.4.4 Regulation

Several studies have examined the regulation of URAT1 in the kidney. In mice, the expression of Urat1 mRNA (Cheng & Klaassen, 2009) and protein (Hosoyamada et al., 2004) is greater in male than female mice. Both mRNA and protein levels of Urat1 in the kidneys of male mice are reduced following gonadectomy, and this effect is reversed by testosterone replacement, suggesting androgen involvement in regulating Urat1 expression (Hosoyamada, Takiue, Shibasaki, & Saito, 2010). In reporter gene assays, hepatocyte nuclear factor 1α (HNF1α) and 1β (HNF1β) interact with the HNF1 motif in the promoter region of human URAT1 and stimulate its transcription (Kikuchi et al., 2007). Urat1 mRNA level is reduced in the kidneys of Hnf1α null mice (Kikuchi et al., 2007), which further supports the conclusion that Hnf1α is involved in regulating Urat1 expression. The promoter region of mouse Urat1 is hypomethylated in the renal cortex, whereas it is hypermethylated in the medulla, findings that may explain why URAT1 is predominately expressed in proximal tubules (Kikuchi et al., 2007).

3. SLC44A TRANSPORTERS

3.1. CTLs

There are five members of the SLC44A family of transporters (SLC44A1, 2, 3, 4, and 5). Referred to as choline transporter-like (CTL1, 2, 3, 4, and 5) proteins (to contrast them from the Na-dependent "choline transporter" or CHT; SLC5A7), CTL1 was cloned from the highly cholinergic electric lobes of the eel, *Torpedo*, and subsequently in humans and mice (O'Regan et al., 2000). Humans express two splice variants, CTL1a (658 aa) and CTL1b (655 aa) (Yuan, Tie, Tarnopolsky, & Bakovic, 2006). Both transcripts are well expressed in the brain, spinal cord, and lung, and to a lesser extent in many other tissues (including kidney) (Traiffort, O'Regan, & Ruat, 2013). Although typically expressed in the plasma membrane, CTL1 is also found in mitochondria. According to hydropathy analysis, CTL1 contains 9–12 TMHs, with the current consensus being 10 (Michel, Yuan, Ramsubir, & Bakovic, 2006). The sequence contains three potential N-linked glycosylation sites and seven consensus PKC phosphorylation sites. Whereas CTL1 and CTL2 both support relatively high affinity choline transport (K_m values of ~5 to ~50 μM; Traiffort et al., 2013), function of the other CTL homologs remains unknown. Selectivity of CTL1 and CTL2 is quite narrow for choline and its structural congeners, and both transporters are blocked efficiently by hemicholinium-3 (IC_{50}<1 μM; (Traiffort et al., 2013)). The current understanding of the mechanism of CTL-mediated transport is largely restricted to studies of transport in cells (typically tumor cells) that (i) display Na-independent choline transport (thereby eliminating the influence of CHT1) and (ii) express CTL1 mRNA. Together with their wide expression pattern throughout tissues and cell types, SLC44 family members have functional parameters consistent with the hypothesis that these proteins may provide choline for cell membrane phospholipid synthesis (Traiffort et al., 2013).

4. SLC47A TRANSPORTERS

4.1. MATEs

Renal and hepatic secretion of OCs involves the sequential activity of distinct basolateral and apical transporters that mediate OC entry and exit, respectively. Physiological characterization of these processes revealed the entry step to involve electrogenic facilitated uniport and the exit step to involve secondary active OC/H^+ exchange. The former process has long been accepted to involve the OCTs and, in humans, is dominated by activity of OCT2 (Motohashi et al., 2002). The molecular identity of the latter process, OC/H^+ exchange, was unknown until the cloning in 2005 (Otsuka et al., 2005) of the first mammalian members of the multidrug and toxin extruder (MATE) family of transporters, that is, MATE1 and MATE2.

In humans, mRNA for MATE1 is expressed in the kidney and liver (and, although less so, in the heart), and MATE1 protein is found in the apical and canalicular membranes of RPTs and hepatocytes, respectively (Otsuka et al., 2005). The transcript for MATE2 appears to be more widely expressed than MATE1, but expression of its splice variant, MATE2-K, is largely restricted to the kidney. MATE1 is coexpressed with MATE2/2-K in the apical (luminal) membrane of RPT cells (Motohashi et al., 2013).

4.1.1 Structure

Human MATE1 is a 570-aa protein with no N-linked glycosylation sites (in extracellular loops), and three consensus PKG phosphorylation sites in the cytoplasm (and no PKC, PKA, CKII, or CaMII sites). MATE2 is a 602-aa protein, also with no N-linked glycosylation sites, three consensus phosphorylation sites for PKG, and one each for PKC and MAPK. A splice variant of MATE2 that is missing 36 amino acids in the putative intracellular loop between TMHs 4 and 5 is highly expressed in human kidney, resulting in the designation MATE2-K (Masuda et al., 2006). Both MATE2 and MATE2-K have similar functional characteristics that closely parallel their homolog, MATE1 (Komatsu et al., 2011). Based on hydropathy analysis of the 1000+ members of the MATE family of the transport proteins, and the X-ray crystal structure of the prototypic MATE transporter, the OC/Na^+ exchanger of *Vibrio cholerae*, NorM (He, Szewczyk, Karyakin, Evin, & Chang, 2010), the protein fold of MATE transporters is composed of 12 TMHs. Interestingly, use of cysteine scanning and epitope tagging showed that the mammalian MATEs have a terminal 13th TMH that places the C-terminus at the extracellular face of the plasma membrane (Zhang & Wright, 2009). However, elimination of the 13th TMH, either by experimental truncation (Zhang, He, et al., 2012) or as exists in the naturally occurring mouse MATE1 variant that lacks the 13th TMH due to the presence of an early stop codon (Kobara et al., 2008), results in functional proteins with comparatively normal kinetic and selectivity characteristics (Zhang, He, et al., 2012). These observations suggest that MATE transport function resides in the first 12 TMHs. Thus, the role played by the 13th TMH commonly found in mammalian MATE transporters is still unknown.

Supported by the conclusion that the "functional core" of MATE transporters is composed of the first 12 TMHs, the NorM structure of *V. cholerae* was used as a template to produce a homology model of hMATE1 that provides insight into the structural characteristics of SLC47A transporters (Zhang, He, et al., 2012). The protein spans ~50 Å in the plane of the lipid bilayer, arranged as two bundles of six TMHs (TMHs 1–6 and TMHs 7–12), which form a large internal cavity that, in the current structure, is open to the extracellular space. The two halves are related by a pseudo intramolecular twofold axis that is likely a result of gene duplication. These halves are connected by a cytoplasmic loop between TMHs 6 and 7, while the initial helix of each half (TMH1 and TMH7) are preceded by a helical extension from the inner membrane leaflet side. Interestingly, MFS proteins are also characterized by a pseudo twofold symmetry, but the folds of the MFS and NorM (and probably those of the MATE family in general) are structurally distinct and not closely related phylogenetically. Indeed, whereas many secondary active transporters are characterized by a topology that includes inverted repeats (Forrest, Kramer, & Ziegler, 2011), these appear absent in the structure of MATE transporters; the membrane spanning topology of NorM is unique among all known transporters (He et al., 2010).

4.1.2 Mechanism

OC transport mediated by MATE1 and MATE2/2-K heterologously expressed in cultured cells is pH sensitive and *trans*-stimulated by oppositely oriented H^+ gradients, but not influenced by depolarization of the plasma membrane (Dangprapai & Wright, 2011; Otsuka et al., 2005; Tsuda et al., 2007). Membrane vesicles isolated from HEK293 cells that express rMATE1 (Tsuda et al., 2007), hMATE1, or hMATE2-K (Tsuda, Terada, Ueba, et al., 2009) support potential-independent concentrative TEA uptake in the presence of outwardly directed H^+ gradients, providing direct support for the conclusion that mammalian MATEs are electroneutral exchangers of monovalent OCs for H^+.

A substantial body of evidence on the mechanism of renal OC/H^+ exchange is available from past studies that used brush-border membrane preparations isolated from renal cortex. In the light of evidence supporting the conclusion that MATE transporters quantitatively represent the molecular identity of the "OC/H^+ exchange activity" of native renal preparations (e.g., Tsuda, Terada, Mizuno, et al., 2009), it is appropriate to use data obtained with these vesicle preparations to probe the mechanism of MATE-mediated OC/H^+ exchange. With this in mind, "static-head" experiments using rabbit renal BBMV showed that a 10-fold H^+ gradient is required to sustain a 10-fold TEA gradient (oriented in the same direction), an observation that both confirmed the 1:1 coupling of TEA/H^+ exchange and showed that the transporter (presumably reflecting some combination of MATE1 and MATE2/2-K activity) is an obligatory exchanger; only substrate-bound protein can undergo the conformational changes associated with OC translocation (Wright & Wunz, 1987). The kinetic basis of the *cis*- and *trans*-interaction of H^+ with the OC/H^+ exchanger was also examined using renal BBMV (Wright & Wunz, 1988). The inhibition of OC transport into renal BBMV associated with increasing extravesicular (*cis*) [H^+] is a "pure K_m effect," which is consistent with competition between OC and H^+ for a mutually exclusive binding site(s) at the extracellular aspect of the transporter. In contrast, the stimulation of transport associated with increasing intravesicular (*trans*) [H^+] is a "pure J_{max} effect," which is consistent with the requirement of the exchanger to be bound with substrate (e.g., H^+) before it can reorient itself to face the *cis*-aspect of the membrane (i.e., the extracellular face) and thereby access substrate. The influence of *cis* and *trans* H^+ on MATE1-mediated (1-methyl-4-phenylpyridinium) MPP transport has also been examined with results that were consistent with those observed with isolated BBMV (Dangprapai & Wright, 2011).

Several lines of evidence show that the transport of prototypic monovalent OCs, such as TEA and MPP, is insensitive to changes in membrane potential. The rate of TEA uptake into HEK 293 cells expressing MATE1 or MATE2/2K is insensitive to the depolarizing effect of increasing extracellular K^+ (Ohta, Inoue, Hayashi, & Yuasa, 2006; Otsuka et al., 2005). Furthermore, an inside-negative K-diffusion potential has no effect of the rate of TEA uptake into membrane vesicles from HEK 293 cells that express MATE1 or MATE2-K (Tsuda, Terada, Ueba, et al., 2009), or into proteoliposomes containing purified and reconstituted MATE2 or MATE2-K (Komatsu et al., 2011). All these observations are consistent with a 1:1 exchange of H^+ for OC^+. However, MATE1 (but not MATE2-K) supports transport of the diquaternary herbicide, paraquat (Pq^{2+}) (Chen, Zhang, Sorani, & Giacomini, 2007), and a 1:1 exchange of H^+ for Pq^{2+} should be electrogenic. If that is the case, then the inside-negative membrane potential should be a potent inhibitory force that reduces the impact of MATE1 on Pq^{2+} secretion. The influence of membrane potential on MATE1-mediated Pq^{2+} transport has not been addressed, but in isolated renal BBMV, H^+-gradient-driven Pq^{2+} uptake, which presumably reflected MATE activity, was not influenced by K^+-valinomycin diffusion potentials (Wright & Wunz, 1995). Of course, if the effect of membrane potential on MATE1-mediated Pq^{2+} transport is sufficiently small, then the failure to observe it could simply be limited by the resolution of available methods. But a recent experimental observation suggests this is not the case. Pq^{2+} nephrotoxicity in mice was substantially increased when MATE1 was genetically eliminated, indicating that MATE1 plays a quantitatively significant role in renal elimination of Pq^{2+} (Li, Peng, Yang, Wang, & Shu, 2011). MATE1 also transports dibasic agmatine (but not putrescine; Winter, Elmquist, & Fairbanks, 2011). However, with effective pK_a values of 10 and 8, agmatine exists in both divalent and monovalent forms at physiological pH, so it is not clear which species (if not both) serves as substrate. Mechanistically, these observations suggest that the MATE-mediated $OC:H^+$ exchange stoichiometry may be flexible, being 1:1 for monovalent OCs, but 1:2 for divalent OCs.

4.1.3 Selectivity

In a recent study, Wittwer et al. (2013) surveyed >900 prescription drugs as inhibitors of MATE1-mediated transport of the fluorescent cationic dye,

4-(4-(dimethylamino)styryl)-*N*-methylpyridinium (ASP). The authors used these data to identify several physicochemical properties that distinguished effective inhibitors from those of molecules that interacted weakly or not at all with MATE1; good inhibitors of MATE1 tended to be larger, with a higher-molecular weight and a larger number of bonds and ring structures, than poor inhibitors. A recent effort to develop a structural model of inhibitory ligand interaction with MATE1 also found that inhibitory efficacy was correlated with multiple structure features, including increasing lipophilicity, cationic charge, and the number and placement of hydrogen donor/acceptor moieties (Astorga, Ekins, Morales, & Wright, 2012). However, they also found that the structural characteristics of pharmacophores that identified spatial correlations between these features and inhibitory efficacy differed markedly depending on which substrate (i.e., MPP vs. ASP) provided the marker for MATE1 transport activity.

The idea that ligand inhibition of a "multidrug" transporter may be influenced by the structure of the substrate is not new. The kinetics of ligand inhibition of both Pgp- and OATP-mediated transport shows a clear substrate dependency (e.g., Ekins et al., 2002; Garrigues et al., 2002; Roth, Timmermann, & Hagenbuch, 2011; Westholm, Salo, Viken, Rumbley, & Anderson, 2009). Moreover, X-ray crystal structures of P-gp (Aller et al., 2009) and the multidrug-binding transcription regulator, QacR (Schumacher & Brennan, 2003) revealed the presence of multisite binding pockets. Substrate-dependent ligand interaction is, however, a comparatively new idea with respect to inhibition of MATE-mediated transport; its potential impact on understanding, and predicting, drug–drug interaction warrants further study.

5. SYNOPSIS

Exchangers within the SLC22, SLC44, and SLC47 families comprise a mechanistically diverse array of organic solute transporters. Their activity in epithelia regulates tissue and plasma concentrations of important physiological, toxicological, and pharmacological anionic and cationic substrates. In this chapter, we focus on the OAT and MATE transporters, paying particular attention to their transport mechanisms (energetic and kinetic), molecular determinants of ligand binding and translocation, and regulation at the molecular level.

REFERENCES

Aller, S. G., Yu, J., Ward, A., Weng, Y., Chittaboina, S., Zhuo, R., et al. (2009). Structure of P-glycoprotein reveals a molecular basis for poly-specific drug binding. *Science, 323*, 1718–1722.

Anzai, N., Jutabha, P., Enomoto, A., Yokoyama, H., Nonoguchi, H., Hirata, T., et al. (2005). Functional characterization of rat organic anion transporter 5 (Slc22a19) at the apical membrane of renal proximal tubules. *The Journal of Pharmacology and Experimental Therapeutics, 315*, 534–544.

Anzai, N., Miyazaki, H., Noshiro, R., Khamdang, S., Chairoungdua, A., Shin, H. J., et al. (2004). The multivalent PDZ domain-containing protein PDZK1 regulates transport activity of renal urate-anion exchanger URAT1 via its C terminus. *The Journal of Biological Chemistry, 279*, 45942–45950.

Asif, A. R., Steffgen, J., Metten, M., Grunewald, R. W., Muller, G. A., Bahn, A., et al. (2005). Presence of organic anion transporters 3 (OAT3) and 4 (OAT4) in human adrenocortical cells. *Pflügers Archiv, 450*, 88–95.

Aslamkhan, A., Han, Y. H., Walden, R., Sweet, D. H., & Pritchard, J. B. (2003). Stoichiometry of organic anion/dicarboxylate exchange in membrane vesicles from rat renal cortex and hOAT1-expressing cells. *American Journal of Physiology. Renal Physiology, 285*, F775–F783.

Astorga, B., Ekins, S., Morales, M., & Wright, S. H. (2012). Molecular determinants of ligand selectivity for the human multidrug and toxin extrusion proteins, MATE1 and MATE-2K. *Journal of Pharmacology and Experimental Therapeutics, 341*, 743–755.

Astorga, B., Wunz, T. M., Morales, M., Wright, S. H., & Pelis, R. M. (2011). Differences in the substrate binding regions of renal organic anion transporters 1 (OAT1) and 3 (OAT3). *American Journal of Physiology. Renal Physiology, 301*, F378–F386.

Bacher, P., Giersiefer, S., Bach, M., Fork, C., Schomig, E., & Grundemann, D. (2009). Substrate discrimination by ergothioneine transporter SLC22A4 and carnitine transporter SLC22A5: Gain-of-function by interchange of selected amino acids. *Biochimica et Biophysica Acta, 1788*, 2594–2602.

Bahn, A., Hagos, Y., Reuter, S., Balen, D., Brzica, H., Krick, W., et al. (2008). Identification of a new urate and high affinity nicotinate transporter, hOAT10 (SLC22A13). *Journal of Biological Chemistry, 283*, 16332–16341.

Barendt, W. M., & Wright, S. H. (2002). The human organic cation transporter (hOCT2) recognizes the degree of substrate ionization. *Journal of Biological Chemistry, 277*, 22491–22496.

Barros, S. A., Srimaroeng, C., Perry, J. L., Walden, R., Dembla-Rajpal, N., Sweet, D. H., et al. (2009). Activation of protein kinase Czeta increases OAT1 (SLC22A6)- and OAT3 (SLC22A8)-mediated transport. *Journal of Biological Chemistry, 284*, 2672–2679.

Bleasby, K., Castle, J. C., Roberts, C. J., Cheng, C., Bailey, W. J., Sina, J. F., et al. (2006). Expression profiles of 50 xenobiotic transporter genes in humans and pre-clinical species: A resource for investigations into drug disposition. *Xenobiotica, 36*, 963–988.

Brast, S., Grabner, A., Sucic, S., Sitte, H. H., Hermann, E., Pavenstadt, H., et al. (2011). The cysteines of the extracellular loop are crucial for trafficking of human organic cation transporter 2 to the plasma membrane and are involved in oligomerization. *FASEB Journal, 26*, 976–986.

Brzica, H., Breljak, D., Ljubojevic, M., Balen, D., Micek, V., Anzai, N., et al. (2009). Optimal methods of antigen retrieval for organic anion transporters in cryosections of the rat kidney. *Arhiv za Higijenu Rada i Toksikologiju, 60*, 7–17.

Budiman, T., Bamberg, E., Koepsell, H., & Nagel, G. (2000). Mechanism of electrogenic cation transport by the cloned organic cation transporter 2 from rat. *Journal of Biological Chemistry, 275*, 29413–29420.

Burckhardt, G., & Wolff, N. A. (2000). Structure of renal organic anion and cation transporters. *American Journal of Physiology. Renal Physiology, 278*, F853–F866.
Burckhardt, B. C., Wolff, N. A., & Burckhardt, G. (2000). Electrophysiologic characterization of an organic anion transporter cloned from winter flounder kidney (fROAT). *Journal of the American Society of Nephrology, 11*, 9–17.
Busch, A. E., Quester, S., Ulzheimer, J. C., Waldegger, S., Gorboulev, V., Arndt, P., et al. (1996). Electrogenic properties and substrate specificity of the polyspecific rat cation transporter rOCT1. *Journal of Biological Chemistry, 271*, 32599–32604.
Cha, S. H., Sekine, T., Fukushima, J. I., Kanai, Y., Kobayashi, Y., Goya, T., et al. (2001). Identification and characterization of human organic anion transporter 3 expressing predominantly in the kidney. *Molecular Pharmacology, 59*, 1277–1286.
Cha, S. H., Sekine, T., Kusuhara, H., Yu, E., Kim, J. Y., Kim, D. K., et al. (2000). Molecular cloning and characterization of multispecific organic anion transporter 4 expressed in the placenta. *Journal of Biological Chemistry, 275*, 4507–4512.
Chen, Y., Zhang, S., Sorani, M., & Giacomini, K. M. (2007). Transport of paraquat by human organic cation transporters and multidrug and toxic compound extrusion family. *The Journal of Pharmacology and Experimental Therapeutics, 322*, 695–700.
Cheng, X., & Klaassen, C. D. (2009). Tissue distribution, ontogeny, and hormonal regulation of xenobiotic transporters in mouse kidneys. *Drug Metabolism and Disposition, 37*, 2178–2185.
Dangprapai, Y., & Wright, S. H. (2011). Interaction of H+ with the extracellular and intracellular aspects of hMATE1. *American Journal of Physiology. Renal Physiology, 301*, F520–F528.
Duan, P., Li, S., Ai, N., Hu, L., Welsh, W. J., & You, G. (2012). Potent inhibitors of human organic anion transporters 1 and 3 from clinical drug libraries: Discovery and molecular characterization. *Molecular Pharmacology, 9*, 3340–3346.
Duan, P., Li, S., & You, G. (2010). Angiotensin II inhibits activity of human organic anion transporter 3 through activation of protein kinase Calpha: Accelerating endocytosis of the transporter. *European Journal of Pharmacology, 627*, 49–55.
Duan, P., Li, S., & You, G. (2011). Transmembrane peptide as potent inhibitor of oligomerization and function of human organic anion transporter 1. *Molecular Pharmacology, 79*, 569–574.
Duan, P., Li, S., & You, G. (2012). Regulation of human organic anion transporter 4 by parathyroid hormone-related protein and protein kinase A. *International Journal of Biochemistry and Molecular Biology, 3*, 322–327.
Ekaratanawong, S., Anzai, N., Jutabha, P., Miyazaki, H., Noshiro, R., Takeda, M., et al. (2004). Human organic anion transporter 4 is a renal apical organic anion/dicarboxylate exchanger in the proximal tubules. *Journal of Pharmacological Sciences, 94*, 297–304.
Ekins, S., Kim, R. B., Leake, B. F., Dantzig, A. H., Schuetz, E. G., Lan, L. B., et al. (2002). Application of three-dimensional quantitative structure-activity relationships of P-glycoprotein inhibitors and substrates. *Molecular Pharmacology, 61*, 974–981.
Endres, C. J., Hsiao, P., Chung, F. S., & Unadkat, J. D. (2006). The role of transporters in drug interactions. *European Journal of Pharmaceutical Sciences, 27*, 501–517.
Enomoto, A., Kimura, H., Chairoungdua, A., Shigeta, Y., Jutabha, P., Cha, S. H., et al. (2002). Molecular identification of a renal urate anion exchanger that regulates blood urate levels. *Nature, 417*, 447–452.
Enomoto, A., Takeda, M., Shimoda, M., Narikawa, S., Kobayashi, Y., Kobayashi, Y., et al. (2002). Interaction of human organic anion transporters 2 and 4 with organic anion transport inhibitors. *Journal of Pharmacology and Experimental Therapeutics, 301*, 797–802.
Enomoto, A., Wempe, M. F., Tsuchida, H., Shin, H. J., Cha, S. H., Anzai, N., et al. (2002). Molecular identification of a novel carnitine transporter specific to human testis. Insights

into the mechanism of carnitine recognition. *The Journal of Biological Chemistry, 277,* 36262–36271.

Eraly, S. A., Hamilton, B. A., & Nigam, S. K. (2003). Organic anion and cation transporters occur in pairs of similar and similarly expressed genes. *Biochemical and Biophysical Research Communications, 300,* 333–342.

Eraly, S. A., Monte, J. C., & Nigam, S. K. (2004). Novel slc22 transporter homologs in fly, worm, and human clarify the phylogeny of organic anion and cation transporters. *Physiological Genomics, 18,* 12–24.

Forrest, L. R., Kramer, R., & Ziegler, C. (2011). The structural basis of secondary active transport mechanisms. *Biochimica et Biophysica Acta, 1807,* 167–188.

Garrigues, A., Loiseau, N., Delaforge, M., Ferte, J., Garrigos, M., Andre, F., et al. (2002). Characterization of two pharmacophores on the multidrug transporter P-glycoprotein. *Molecular Pharmacology, 62,* 1288–1298.

Gisler, S. M., Pribanic, S., Bacic, D., Forrer, P., Gantenbein, A., Sabourin, L. A., et al. (2003). PDZK1: I. A major scaffolder in brush borders of proximal tubular cells. *Kidney International, 64,* 1733–1745.

Gong, S., Lu, X., Xu, Y., Swiderski, C. F., Jordan, C. T., & Moscow, J. A. (2002). Identification of OCT6 as a novel organic cation transporter preferentially expressed in hematopoietic cells and leukemias. *Experimental Hematology, 30,* 1162–1169.

Gorboulev, V., Shatskaya, N., Volk, C., & Koepsell, H. (2005). Subtype-specific affinity for corticosterone of rat organic cation transporters rOCT1 and rOCT2 depends on three amino acids within the substrate binding region. *Molecular Pharmacology, 67,* 1612–1619.

Gorboulev, V., Volk, C., Arndt, P., Akhoundova, A., & Koepsell, H. (1999). Selectivity of the polyspecific cation transporter rOCT1 is changed by mutation of aspartate 475 to glutamate. *Molecular Pharmacology, 56,* 1254–1261.

Gründemann, D., Gorboulev, V., Gambaryan, S., Veyhl, M., & Koepsell, H. (1994). Drug excretion mediated by a new prototype of polyspecific transporter. *Nature, 372,* 549–552.

Gründemann, D., Harlfinger, S., Golz, S., Geerts, A., Lazar, A., Berkels, R., et al. (2005). Discovery of the ergothioneine transporter. *Proceedings of the National Academy of Sciences of the United States of America, 102,* 5256–5261.

Hagos, Y., Stein, D., Ugele, B., Burckhardt, G., & Bahn, A. (2007). Human renal organic anion transporter 4 operates as an asymmetric urate transporter. *Journal of the American Society of Nephrology, 18,* 430–439.

Han, T. K., Everett, R. S., Proctor, W. R., Ng, C. M., Costales, C. L., Brouwer, K. L., et al. (2013). Organic cation transporter 1 (OCT1/mOct1) is localized in the apical membrane of Caco-2 cell monolayers and enterocytes. *Molecular Pharmacology, 84,* 182–189.

Harper, J. N., & Wright, S. H. (2012). Multiple mechanisms of ligand interaction with the human organic cation transporter, OCT2. *American Journal of Physiology. Renal Physiology, 304,* F56–F67.

Hasannejad, H., Takeda, M., Taki, K., Jung, S. H., Babu, E., Jutabha, P., et al. (2004). Interactions of human organic anion transporters with diuretics. *The Journal of Pharmacology and Experimental Therapeutics, 308,* 1021–1029.

He, X., Szewczyk, P., Karyakin, A., Evin, M., & Chang, G. (2010). Structure of a cation-bound multidrug and toxin compound extrusion (MATE) transporter. *Nature, 467,* 991–994.

Hong, M., Xu, W., Yoshida, T., Tanaka, K., Wolff, D. J., Zhou, F., et al. (2005). Human organic anion transporter hOAT1 forms homooligomers. *Journal of Biological Chemistry, 280,* 32285–32290.

Hosoyamada, M., Ichida, K., Enomoto, A., Hosoya, T., & Endou, H. (2004). Function and localization of urate transporter 1 in mouse kidney. *Journal of the American Society of Nephrology, 15,* 261–268.

Hosoyamada, M., Sekine, T., Kanai, Y., & Endou, H. (1999). Molecular cloning and functional expression of a multispecific organic anion transporter from human kidney. *American Journal of Physiology*, *276*, F122–F128.

Hosoyamada, M., Takiue, Y., Shibasaki, T., & Saito, H. (2010). The effect of testosterone upon the urate reabsorptive transport system in mouse kidney. *Nucleosides, Nucleotides & Nucleic Acids*, *29*, 574–579.

Kaler, G., Truong, D. M., Khandelwal, A., Nagle, M., Eraly, S. A., Swaan, P. W., et al. (2007). Structural variation governs substrate specificity for organic anion transporter (OAT) homologs. Potential remote sensing by OAT family members. *The Journal of Biological Chemistry*, *282*, 23841–23853.

Kekuda, R., Prasad, P. D., Wu, X., Wang, H., Fei, Y. J., Leibach, F. H., et al. (1998). Cloning and functional characterization of a potential-sensitive, polyspecific organic cation transporter (OCT3) most abundantly expressed in placenta. *Journal of Biological Chemistry*, *273*, 15971–15979.

Keller, T., Egenberger, B., Gorboulev, V., Bernhard, F., Uzelac, Z., Gorbunov, D., et al. (2011). The large extracellular loop of organic cation transporter 1 influences substrate affinity and is pivotal for oligomerization. *Journal of Biological Chemistry*, *286*, 37874–37886.

Kikuchi, R., Kusuhara, H., Hattori, N., Kim, I., Shiota, K., Gonzalez, F. J., et al. (2007). Regulation of tissue-specific expression of the human and mouse urate transporter 1 gene by hepatocyte nuclear factor 1 alpha/beta and DNA methylation. *Molecular Pharmacology*, *72*, 1619–1625.

Kobara, A., Hiasa, M., Matsumoto, T., Otsuka, M., Omote, H., & Moriyama, Y. (2008). A novel variant of mouse MATE-1 H^+/organic cation antiporter with a long hydrophobic tail. *Archives of Biochemistry and Biophysics*, *469*, 195–199.

Kobayashi, Y., Ohshiro, N., Sakai, R., Ohbayashi, M., Kohyama, N., & Yamamoto, T. (2005). Transport mechanism and substrate specificity of human organic anion transporter 2 (hOat2 [SLC22A7]). *The Journal of Pharmacy and Pharmacology*, *57*, 573–578.

Koepsell, H., & Endou, H. (2004). The SLC22 drug transporter family. *Pflügers Archiv*, *447*, 666–676.

Koepsell, H., Lips, K., & Volk, C. (2007). Polyspecific organic cation transporters: Structure, function, physiological roles, and biopharmaceutical implications. *Pharmaceutical Research*, *24*, 1227–1251.

Komatsu, T., Hiasa, M., Miyaji, T., Kanamoto, T., Matsumoto, T., Otsuka, M., et al. (2011). Characterization of the human MATE2 proton-coupled polyspecific organic cation exporter. *The International Journal of Biochemistry & Cell Biology*, *43*, 913–918.

Kusuhara, H., Sekine, T., Utsunomiya-Tate, N., Tsuda, M., Kojima, R., Cha, S. H., et al. (1999). Molecular cloning and characterization of a new multispecific organic anion transporter from rat brain. *The Journal of Biological Chemistry*, *274*, 13675–13680.

Li, S., Duan, P., & You, G. (2009). Regulation of human organic anion transporter 1 by ANG II: Involvement of protein kinase Calpha. *American Journal of Physiology. Endocrinology and Metabolism*, *296*, E378–E383.

Li, S., Duan, P., & You, G. (2010). Regulation of human organic anion transporter 3 by peptide hormone bradykinin. *Journal of Pharmacology and Experimental Therapeutics*, *333*, 970–975.

Li, Q., Peng, X., Yang, H., Wang, H., & Shu, Y. (2011). Deficiency of multidrug and toxin extrusion 1 enhances renal accumulation of paraquat and deteriorates kidney injury in mice. *Molecular Pharmaceutics*, *8*, 2476–2483.

Li, S., Zhang, Q., & You, G. (2013). Three ubiquitination sites of organic anion transporter-1 synergistically mediate protein kinase C-dependent endocytosis of the transporter. *Molecular Pharmacology*, *84*, 139–146.

Lopez-Nieto, C. E., You, G., Bush, K. T., Barros, E. J., Beier, D. R., & Nigam, S. K. (1997). Molecular cloning and characterization of NKT, a gene product related to the organic cation transporter family that is almost exclusively expressed in the kidney. *Journal of Biological Chemistry, 272*, 6471–6478.

Lu, Y., Nakanishi, T., & Tamai, I. (2013). Functional cooperation of SMCTs and URAT1 for renal reabsorption transport of urate. *Drug Metabolism and Pharmacokinetics, 28*, 153–158.

Masuda, S., Terada, T., Yonezawa, A., Tanihara, Y., Kishimoto, K., Katsura, T., et al. (2006). Identification and functional characterization of a new human kidney-specific H^+/organic cation antiporter, kidney-specific multidrug and toxin extrusion 2. *Journal of the American Society of Nephrology, 17*, 2127–2135.

Michel, V., Yuan, Z., Ramsubir, S., & Bakovic, M. (2006). Choline transport for phospholipid synthesis. *Experimental Biology and Medicine (Maywood, N.J.), 231*, 490–504.

Minuesa, G., Volk, C., Molina-Arcas, M., Gorboulev, V., Erkizia, I., Arndt, P., et al. (2009). Transport of lamivudine [(-)-beta-L-2′,3′-dideoxy-3′-thiacytidine] and high-affinity interaction of nucleoside reverse transcriptase inhibitors with human organic cation transporters 1, 2, and 3. *Journal of Pharmacology and Experimental Therapeutics, 329*, 252–261.

Miura, D., Anzai, N., Jutabha, P., Chanluang, S., He, X., Fukutomi, T., et al. (2011). Human urate transporter 1 (hURAT1) mediates the transport of orotate. *The Journal of Physiological Sciences, 61*, 253–257.

Miyazaki, H., Anzai, N., Ekaratanawong, S., Sakata, T., Shin, H. J., Jutabha, P., et al. (2005). Modulation of renal apical organic anion transporter 4 function by two PDZ domain-containing proteins. *Journal of the American Society of Nephrology, 16*, 3498–3506.

Motohashi, H., Nakao, Y., Masuda, S., Katsura, T., Kamba, T., Ogawa, O., et al. (2013). Precise comparison of protein localization among OCT, OAT, and MATE in human kidney. *Journal of Pharmaceutical Sciences, 102*, 3302–3308.

Motohashi, H., Sakurai, Y., Saito, H., Masuda, S., Urakami, Y., Goto, M., et al. (2002). Gene expression levels and immunolocalization of organic ion transporters in the human kidney. *Journal of the American Society of Nephrology, 13*, 866–874.

Nies, A. T., Koepsell, H., Damme, K., & Schwab, M. (2011). Organic cation transporters (OCTs, MATEs), in vitro and in vivo evidence for the importance in drug therapy. *Handbook of Experimental Pharmacology, 201*, 105–167.

Ohashi, R., Tamai, I., Nezu, J. J., Nikaido, H., Hashimoto, N., Oku, A., et al. (2001). Molecular and physiological evidence for multifunctionality of carnitine/organic cation transporter OCTN2. *Molecular Pharmacology, 59*, 358–366.

Ohta, K. Y., Inoue, K., Hayashi, Y., & Yuasa, H. (2006). Molecular identification and functional characterization of rat MATE1 as an organic cation/H^+ antiporter in the kidney. *Drug Metabolism and Disposition, 34*, 1868–1874.

Okuda, M., Saito, H., Urakami, Y., Takano, M., & Inui, K.-I. (1996). cDNA cloning and functional expression of a novel rat kidney organic cation transporter, OCT2. *Biochemical and Biophysical Research Communications, 224*, 500–507.

O'Regan, S., Traiffort, E., Ruat, M., Cha, N., Compaore, D., & Meunier, F. M. (2000). An electric lobe suppressor for a yeast choline transport mutation belongs to a new family of transporter-like proteins. *Proceedings of the National Academy of Sciences of the United States of America, 97*, 1835–1840.

Otsuka, M., Matsumoto, T., Morimoto, R., Arioka, S., Omote, H., & Moriyama, Y. (2005). A human transporter protein that mediates the final excretion step for toxic organic cations. *Proceedings of the National Academy of Sciences of the United States of America, 102*, 17923–17928.

Pelis, R. M., & Wright, S. H. (2011). Renal transport of organic anions and cations. *Comprehensive Physiology, 1*, 1795–1835.

Perry, J. L., Dembla-Rajpal, N., Hall, L. A., & Pritchard, J. B. (2006). A three-dimensional model of human organic anion transporter 1: Aromatic amino acids required for substrate transport. *The Journal of Biological Chemistry, 281*, 38071–38079.

Popp, C., Gorboulev, V., Muller, T. D., Gorbunov, D., Shatskaya, N., & Koepsell, H. (2005). Amino acids critical for substrate affinity of rat organic cation transporter 1 line the substrate binding region in a model derived from the tertiary structure of lactose permease. *Molecular Pharmacology, 67*, 1600–1611.

Pornillos, O., & Chang, G. (2006). Inverted repeat domains in membrane proteins. *FEBS Letters, 580*, 358–362.

Pritchard, J. B. (1988). Coupled transport of *p*-aminohippurate by rat kidney basolateral membrane vesicles. *American Journal of Physiology, 255*, F597–F604.

Pritchard, J. B. (1995). Intracellular α-ketoglutarate controls the efficacy of renal organic anion transport. *The Journal of Pharmacology and Experimental Therapeutics, 274*, 1278–1284.

Radestock, S., & Forrest, L. R. (2011). The alternating-access mechanism of MFS transporters arises from inverted-topology repeats. *Journal of Molecular Biology, 407*, 698–715.

Reddy, V. S., Shlykov, M. A., Castillo, R., Sun, E. I., & Saier, M. H., Jr. (2012). The major facilitator superfamily (MFS) revisited. *The FEBS Journal, 279*, 2022–2035.

Rizwan, A. N., Krick, W., & Burckhardt, G. (2007). The chloride dependence of the human organic anion transporter 1 (hOAT1) is blunted by mutation of a single amino acid. *Journal of Biological Chemistry, 282*, 13402–13409.

Rocchiccioli, F., Leroux, J. P., & Cartier, P. H. (1984). Microdetermination of 2-ketoglutaric acid in plasma and cerebrospinal fluid by capillary gas chromatography mass spectroscopy: Application in pediatrics. *Biomedical Mass Spectrometry, 24*, 28.

Roth, M., Timmermann, B. N., & Hagenbuch, B. (2011). Interactions of green tea catechins with organic anion-transporting polypeptides. *Drug Metabolism and Disposition, 39*, 920–926.

Sato, M., Wakayama, T., Mamada, H., Shirasaka, Y., Nakanishi, T., & Tamai, I. (2011). Identification and functional characterization of uric acid transporter Urat1 (Slc22a12) in rats. *Biochimica et Biophysica Acta, 1808*, 1441–1447.

Schaeffeler, E., Hellerbrand, C., Nies, A. T., Winter, S., Kruck, S., Hofmann, U., et al. (2011). DNA methylation is associated with downregulation of the organic cation transporter OCT1 (SLC22A1) in human hepatocellular carcinoma. *Genome Medicine, 3*, 82.

Schmitt, B. M., & Koepsell, H. (2005). Alkali cation binding and permeation in the rat organic cation transporter rOCT2. *The Journal of Biological Chemistry, 280*, 24481–24490.

Schomig, E., Spitzenberger, F., Engelhardt, M., Martel, F., Ording, N., & Gründemann, D. (1998). Molecular cloning and characterization of two novel transport proteins from rat kidney. *FEBS Letters, 425*, 79–86.

Schumacher, M. A., & Brennan, R. G. (2003). Deciphering the molecular basis of multidrug recognition: Crystal structures of the Staphylococcus aureus multidrug binding transcription regulator QacR. *Research in Microbiology, 154*, 69–77.

Sekine, T., Watanabe, N., Hosoyamada, M., Kanaim, Y., & Endou, H. (1997). Expression cloning and characterization of a novel multispecific organic anion transporter. *The Journal of Biological Chemistry, 272*, 18526–18529.

Shimada, H., Moewes, B., & Burckhardt, G. (1987). Indirect coupling to Na^+ of *p*-aminohippuric acid uptake into rat renal basolateral membrane vesicles. *American Journal of Physiology, 253*, F795–F801.

Shin, H. J., Anzai, N., Enomoto, A., He, X., Kim, D. K., Endou, H., et al. (2007). Novel liver-specific organic anion transporter OAT7 that operates the exchange of sulfate conjugates for short chain fatty acid butyrate. *Hepatology, 45*, 1046–1055.

Simonson, G. D., Vincent, A. C., Roberg, K. J., Huang, Y., & Iwanij, V. (1994). Molecular cloning and characterization of a novel liver-specific transport protein. *Journal of Cell Science, 107*, 3–72.

Sissung, T. M., Troutman, S. M., Campbell, T. J., Pressler, H. M., Sung, H., Bates, S. E., et al. (2012). Transporter pharmacogenetics: Transporter polymorphisms affect normal physiology, diseases, and pharmacotherapy. *Discovery Medicine, 13*, 19–34.

Srimaroeng, C., Perry, J. L., & Pritchard, J. B. (2008). Physiology, structure, and regulation of the cloned organic anion transporters. *Xenobiotica, 38*, 889–935.

Sun, W., Wu, R. R., van Poelje, P. D., & Erion, M. D. (2001). Isolation of a family of organic anion transporters from human liver and kidney. *Biochemical and Biophysical Research Communications, 283*, 417–422.

Sweet, D. H., Chan, L. M., Walden, R., Yang, X. P., Miller, D. S., & Pritchard, J. B. (2003). Organic anion transporter 3 [Slc22a8] is a dicarboxylate exchanger indirectly coupled to the Na^+ gradient. *American Journal of Physiology. Renal Physiology, 284*, F763–F769.

Takeda, M., Sekine, T., & Endou, H. (2000). Regulation by protein kinase C of organic anion transport driven by rat organic anion transporter 3 (rOAT3). *Life Sciences, 67*, 1087–1093.

Tamai, I., China, K., Sai, Y., Kobayashi, D., Nezu, J., Kawahara, E., et al. (2001). Na^+-coupled transport of L-carnitine via high-affinity carnitine transporter OCTN2 and its subcellular localization in kidney. *Biochimica et Biophysica Acta, 1512*, 273–284.

Tamai, I., Nakanishi, T., Kobayashi, D., China, K., Kosugi, Y., Nezu, J., et al. (2004). Involvement of OCTN1 (SLC22A4) in pH-dependent transport of organic cations. *Molecular Pharmacology, 1*, 57–66.

Tamai, I., Ohashi, R., Nezu, J., Yabuuchi, H., Oku, A., Shimane, M., et al. (1998). Molecular and functional identification of sodium ion-dependent, high affinity human carnitine transporter OCTN2. *Journal of Biological Chemistry, 273*, 20378–20382.

Tamai, I., Yabuuchi, H., Nezu, J.-I., Sai, Y., Oku, A., Shimane, M., et al. (1997). Cloning and characterization of a novel human pH-dependent organic cation transporter, OCTN1. *FEBS Letters, 419*, 107–111.

Tanaka, K., Xu, W., Zhou, F., & You, G. (2004). Role of glycosylation in the organic anion transporter OAT1. *The Journal of Biological Chemistry, 279*, 14961–14966.

Tanaka, K., Zhou, F., Kuze, K., & You, G. (2004). Cysteine residues in the organic anion transporter mOAT1. *Biochemical Journal, 380*, 283–287.

Tasic, V., Hynes, A. M., Kitamura, K., Cheong, H. I., Lozanovski, V. J., Gucev, Z., et al. (2011). Clinical and functional characterization of URAT1 variants. *PLoS One, 6*, e28641.

Traiffort, E., O'Regan, S., & Ruat, M. (2013). The choline transporter-like family SLC44: Properties and roles in human diseases. *Molecular Aspects of Medicine, 34*, 646–654.

Tsigelny, I. F., Kovalskyy, D., Kouznetsova, V. L., Balinskyi, O., Sharikov, Y., Bhatnagar, V., et al. (2011). Conformational changes of the multispecific transporter organic anion transporter 1 (OAT1/SLC22A6) suggests a molecular mechanism for initial stages of drug and metabolite transport. *Cell Biochemistry and Biophysics, 61*, 251–259.

Tsuda, M., Terada, T., Asaka, J., Ueba, M., Katsura, T., & Inui, K. (2007). Oppositely directed H^+ gradient functions as a driving force of rat H^+/organic cation antiporter MATE1. *American Journal of Physiology. Renal Physiology, 292*, F593–F598.

Tsuda, M., Terada, T., Mizuno, T., Katsura, T., Shimakura, J., & Inui, K. I. (2009). Targeted disruption of the multidrug and toxin extrusion 1 (Mate1) gene in mice reduces renal secretion of metformin. *Molecular Pharmacology, 75*, 1280–1286.

Tsuda, M., Terada, T., Ueba, M., Sato, T., Masuda, S., Katsura, T., et al. (2009). Involvement of human multidrug and toxin extrusion 1 in the drug interaction between cimetidine and metformin in renal epithelial cells. *Journal of Pharmacology and Experimental Therapeutics, 329*, 185–191.

Tzvetkov, M. V., Vormfelde, S. V., Balen, D., Meineke, I., Schmidt, T., Sehrt, D., et al. (2009). The effects of genetic polymorphisms in the organic cation transporters

OCT1, OCT2, and OCT3 on the renal clearance of metformin. *Clinical Pharmacology and Therapeutics*, *86*, 299–306.

U.S. Food and Drug Administration (2012a). *Guidance for Industry: Drug Interaction Studies—Study Design, Data Analysis, Implications for Dosing, and Labeling Recommendations*. Huang, S.-M. and Zhang, L. Draft Guidance, 79 pp.

U.S. Food and Drug Administration (2012b). *Preventable Adverse Drug Reactions: A Focus on Drug Interactions*. U.S. FDA.

Uwai, Y., Ida, H., Tsuji, Y., Katsura, T., & Inui, K. (2007). Renal transport of adefovir, cidofovir, and tenofovir by SLC22A family members (hOAT1, hOAT3, and hOCT2). *Pharmaceutical Research*, *24*, 811–815.

Vanwert, A. L., Gionfriddo, M. R., & Sweet, D. H. (2010). Organic anion transporters: Discovery, pharmacology, regulation and roles in pathophysiology. *Biopharmaceutics & Drug Disposition*, *31*, 1–71.

Westholm, D. E., Salo, D. R., Viken, K. J., Rumbley, J. N., & Anderson, G. W. (2009). The blood-brain barrier thyroxine transporter organic anion-transporting polypeptide 1c1 displays atypical transport kinetics. *Endocrinology*, *150*, 5153–5162.

Winter, T. N., Elmquist, W. F., & Fairbanks, C. A. (2011). OCT2 and MATE1 provide bidirectional agmatine transport. *Molecular Pharmaceutics*, *8*, 133–142.

Wittwer, M. B., Zur, A. A., Khuri, N., Kido, Y., Kosaka, A., Zhang, X., et al. (2013). Discovery of potent, selective MATE1 inhibitors through prescription drug profiling and computational modeling. *Journal of Medicinal Chemistry*, *56*, 781–795.

Wolff, N. A., Thies, K., Kuhnke, N., Reid, G., Friedrich, B., Lang, F., et al. (2003). Protein kinase C activation downregulates human organic anion transporter 1-mediated transport through carrier internalization. *Journal of the American Society of Nephrology*, *14*, 1959–1968.

Wright, S. H., & Wunz, T. M. (1987). Transport of tetraethylammonium by rabbit renal brush-border and basolateral membrane vesicles. *American Journal of Physiology*, *253*, F1040–F1050.

Wright, S. H., & Wunz, T. M. (1988). Mechanism of *cis*- and *trans*-substrate interactions at the tetraethylammonium/H^+ exchanger of rabbit renal brush-border membrane vesicles. *Journal of Biological Chemistry*, *263*, 19494–19497.

Wright, S. H., & Wunz, T. M. (1995). Paraquat^{2+}/H^+ exchange in isolated renal brush-border membrane vesicles. *Biochimica et Biophysica Acta*, *1240*, 18–24.

Wu, X., Huang, W., Ganapathy, M. E., Wang, H., Kekuda, R., Conway, S. J., et al. (2000). Structure, function, and regional distribution of the organic cation transporter OCT3 in the kidney. *American Journal of Physiology. Renal Physiology*, *279*, F449–F458.

Yang, C. H., Glover, K. P., & Han, X. (2010). Characterization of cellular uptake of perfluorooctanoate via organic anion-transporting polypeptide 1A2, organic anion transporter 4, and urate transporter 1 for their potential roles in mediating human renal reabsorption of perfluorocarboxylates. *Toxicological Sciences*, *117*, 294–302.

You, G., Kuze, K., Kohanski, R. A., Amsler, K., & Henderson, S. (2000). Regulation of mOAT-mediated organic anion transport by okadaic acid and protein kinase C in LLC-PK(1) cells. *Journal of Biological Chemistry*, *275*, 10278–10284.

Youngblood, G. L., & Sweet, D. H. (2004). Identification and functional assessment of the novel murine organic anion transporter Oat5 (Slc22a19) expressed in kidney. *American Journal of Physiology. Renal Physiology*, *287*, F236–F244.

Yuan, Z., Tie, A., Tarnopolsky, M., & Bakovic, M. (2006). Genomic organization, promoter activity, and expression of the human choline transporter-like protein 1. *Physiological Genomics*, *26*, 76–90.

Zhang, X., He, X., Baker, J., Tama, F., Chang, G., & Wright, S. H. (2012). Twelve transmembrane helices form the functional core of mammalian MATE1 (multidrug and toxin extruder 1) protein. *The Journal of Biological Chemistry*, *287*, 27971–27982.

Zhang, Q., Hong, M., Duan, P., Pan, Z., Ma, J., & You, G. (2008). Organic anion transporter OAT1 undergoes constitutive and protein kinase C-regulated trafficking through a dynamin- and clathrin-dependent pathway. *The Journal of Biological Chemistry*, *283*, 32570–32579.

Zhang, X., Shirahatti, N. V., Mahadevan, D., & Wright, S. H. (2005). A conserved glutamate residue in transmembrane helix 10 influences substrate specificity of rabbit OCT2 (SLC22A2). *The Journal of Biological Chemistry*, *280*, 34813–34822.

Zhang, Q., Suh, W., Pan, Z., & You, G. (2012). Short-term and long-term effects of protein kinase C on the trafficking and stability of human organic anion transporter 3. *International Journal of Biochemistry and Molecular Biology*, *3*, 242–249.

Zhang, X., & Wright, S. H. (2009). MATE1 has an external COOH terminus, consistent with a 13-helix topology. *American Journal of Physiology. Renal Physiology*, *297*, F263–F271.

Zhou, F., Hong, M., & You, G. (2007). Regulation of human organic anion transporter 4 by progesterone and protein kinase C in human placental BeWo cells. *American Journal of Physiology. Endocrinology and Metabolism*, *293*, E57–E61.

Zhou, F., Xu, W., Hong, M., Pan, Z., Sinko, P. J., Ma, J., et al. (2005). The role of N-linked glycosylation in protein folding, membrane targeting, and substrate binding of human organic anion transporter hOAT4. *Molecular Pharmacology*, *67*, 868–876.

CHAPTER SEVEN

The *SLC24* Family of K^+-Dependent Na^+–Ca^{2+} Exchangers: Structure–Function Relationships

Paul P.M. Schnetkamp[1], Ali H. Jalloul, Guohong Liu, Robert T. Szerencsei

Department of Physiology and Pharmacology, Hotchkiss Brain Institute, University of Calgary, Calgary, Alberta, Canada

[1]Corresponding author: e-mail address: pschnetk@ucalgary.ca

Contents

Abstract

The human *SLC24* gene family contains five members encoding the NCKX1–5 proteins that function as K^+-dependent Na^+–Ca^{2+} exchangers. NCKX proteins have been shown to play critical roles in retinal rod and cone photoreceptors, olfactory neurons, epidermal melanocytes, and the retinal pigment epithelium. NCKX transcripts are also found in many other tissues, in particular throughout the brain, but their specific physiological roles yet need to be elucidated in most cases. Here, we focus on our current knowledge of NCKX transport function as has been described in detail only for *in situ* NCKX1 in the outer segments of retinal rod photoreceptors and on structure–function relationships elucidated for the NCKX2 isoform after expression of its (mutated) cDNA in cell lines.

Current Topics in Membranes, Volume 73
ISSN 1063-5823
http://dx.doi.org/10.1016/B978-0-12-800223-0.00007-4

LIST OF ABBREVIATIONS

CNG cGMP-gated channel
EGTA ethylene glycol tetraacetic acid
HEK293 human embryonic kidney 293 cell line
K_m Michaelis–Menten constant
NCKX Na^+–Ca^{2+} + K^+ exchanger
NCX Na^+–Ca^{2+} exchanger
ROS outer segments of retinal rod photoreceptors
SLC24 Family solute carrier gene family number 24
SLC8 solute carrier gene family number 8
TMS transmembrane segments

1. INTRODUCTION

This chapter concerns the K^+-dependent Na^+–Ca^{2+} exchangers or Na^+–Ca^{2+}+K^+ exchanger (NCKX) proteins, encoded by members of the solute carrier gene family number 24 (*SLC24*). The process of Na^+–Ca^{2+} exchange was first described in the 1960s through the measurement of Na^+-dependent ^{45}Ca fluxes, and the history of Na^+–Ca^{2+} exchange has been reviewed by Blaustein and Lederer (1999). In the late 1960s, measurements of ^{45}Ca fluxes were also used to establish the presence of a rather active Na^+–Ca^{2+} exchange process in the isolated outer segments of retinal rod photoreceptors (ROS) (Schnetkamp, 1980). In 1989, two papers appeared showing that Na^+–Ca^{2+} exchange in ROS differed from that studied in most other tissues, particularly the widely studied Na^+–Ca^{2+} exchange process found in cardiomyocytes. Na^+–Ca^{2+} exchange in ROS required the simultaneous presence of both Ca^{2+} and K^+, and the NCKX protein was shown to operate at a stoichiometry of $4Na^+$ against $1Ca^{2+}$ plus $1K^+$ (Cervetto, Lagnado, Perry, Robinson, & McNaughton, 1989; Schnetkamp, Basu, & Szerencsei, 1989). After the first NCX and NCKX cDNAs were cloned, it became clear that they were the first representatives of two distinct genes families, now known as the solute carrier gene family number 8 (*SLC8*) and *SCL24* families, respectively. We recently published a short review on the history, tissue distribution, and physiology of the five human *SLC24* genes (Schnetkamp, 2013), while more detailed reviews on the physiology of NCX and NCKX proteins have been published recently as well (Altimimi & Schnetkamp, 2007a; Lytton, 2007). Briefly, in addition to the well-defined role of NCKX1 and NCKX2 in retinal rod and cone photoreceptors (Fain, Matthews, Cornwall, & Koutalos, 2001; Paillart,

Winkfein, Schnetkamp, & Korenbrot, 2007; Schnetkamp, 2004), respectively, NCKX2 is also widely expressed in the brain and important for motor learning and memory (Li et al., 2006); NCKX4 has been shown to play a critical role in olfactory neurons (Stephan et al., 2012), and NCKX5 is a pigmentation gene and a genetic determinant for light skin in people of European descent (Lamason et al., 2005; Stokowski et al., 2007). Here, we focus on reviewing the functional and structural studies on NCKX proteins from two time periods. Early studies reported a highly detailed characterization of the *in situ* NCKX in the ROS (now designated NCKX1). These studies primarily used preparations of ROS isolated from bovine or amphibian retinas. Later studies mostly used the human NCKX2 cDNA expressed in cell lines to address structure–function relationships.

2. *IN SITU* FUNCTIONAL PROPERTIES OF NCKX IN RETINAL ROS

The plasma membrane of isolated ROS has been shown to contain only two types of cation transport proteins—the cGMP-gated channels (CNG) and the NCKX1 NCKX (Lagnado, Cervetto, & McNaughton, 1988; Schnetkamp, Szerencsei, & Basu, 1991). When dealing with isolated ROS, the CNG channels are closed due to light-activated hydrolysis of cGMP, which is the excitatory messenger of visual transduction. Furthermore, the density of NCKX proteins in the ROS plasma membrane is unusually high. For instance, in the case of the bovine-isolated ROS, the NCKX1 protein is clearly visible on a protein-stained gel. This leaves us with the unique circumstance of having an easily accessible preparation containing only a single functional cation transporter. Moreover, ROS do not contain any mitochondria or endoplasmic reticulum that act as internal Ca^{2+} stores, but instead contain a large stack of the so-called disk membrane vesicles that sequester some Ca^{2+} by a yet to be identified mechanism operating at a capacity much lower than that of NCKX1 (Schnetkamp, 1995b; Schnetkamp & Szerencsei, 1993).

The first studies that directly addressed Ca^{2+} fluxes in ROS utilized ^{45}Ca transport to show that ROS purified on sucrose gradients and maintained in a Na^+-free and Ca^{2+}-containing media, in a matter of a few seconds, accumulated ^{45}Ca in what proved to be a $^{45}Ca^{2+}$–$^{40}Ca^{2+}$ exchange mechanism (Schnetkamp, 1979). By using the so-called counter transport test, investigators could then quickly release ^{45}Ca from ROS by adding either "cold" ^{40}Ca (representing $^{40}Ca^{2+}$–$^{45}Ca^{2+}$ exchange) or Na^+ (representing Na^+–Ca^{2+} exchange) but not any of the other alkali cations (Schnetkamp, 1980). It was concluded that the ROS plasma membrane contains highly active Na^+–Ca^{2+} exchange that could alter total internal [Ca^{2+}] by ~200 μM/s and displayed a surprisingly low Michaelis–Menten constant (K_m) of ~1 μM for external Ca^{2+}. Moreover, the observation of both Na^+–Ca^{2+} exchange and Ca^{2+}–Ca^{2+} exchange operating at comparable rates presents strong evidence for an alternating access mechanism of transport—now thought to be the predominant mechanism of operation for most SLC transporters. We now know that Na^+–Ca^{2+} exchange in ROS requires and transports K^+, and a conclusive way to demonstrate this was by examining the external Ca^{2+} dependence of simultaneous $^{40}Ca^{2+}$–$^{45}Ca^{2+}$ and $^{86}Rb^+$–K^+ exchange (Schnetkamp, Szerencsei, et al., 1991). Figure 7.1 illustrates such direct measurements of Ca^{2+}, K^+ (using the K^+ analog Rb^+) and Na^+ fluxes mediated by the ROS NCKX. Again, it should be emphasized that these are unique experiments made possible by the fact that no other cation transporters/channels are operating in the plasma membrane of isolated ROS. Such experiments could not be carried out to study NCKX fluxes in other cell types or to study NCKX fluxes in cells transfected with NCKX cDNAs due to the presence of other ion channels/transporters. The presence of self-exchange fluxes, in addition to NCKX, strongly supports the alternating access mechanism for NCKX.

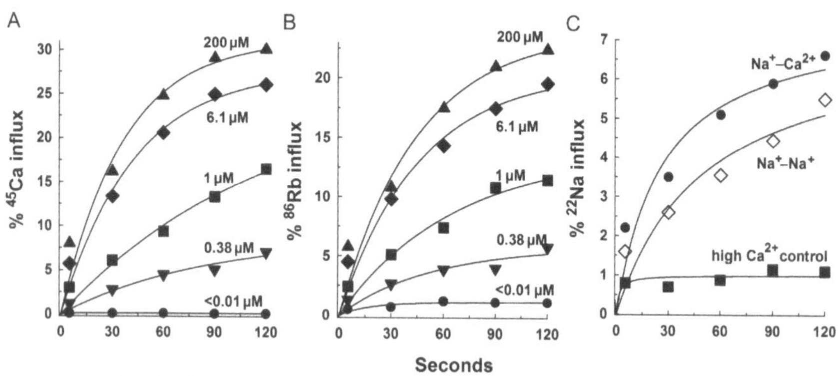

Figure 7.1 $^{45}Ca^{2+}$, $^{86}Rb^+$, and $^{22}Na^+$ fluxes mediated by ROS NCKX1. (A) $^{45}Ca^{2+}$ uptake was measured as a function of indicated external [Ca^{2+}] in isolated bovine ROS enriched in Ca^{2+}. This uptake represented Ca^{2+}–Ca^{2+} exchange. (B) $^{45}Ca^{2+}$ uptake was accompanied by the concurrent influx of K^+ (as measured by uptake of the K^+ analog $^{86}Rb^+$). (C) $^{22}Na^+$ fluxes associated with Na^+–Ca^{2+} exchange and Na^+–Na^+ exchange in bovine ROS are illustrated (high external [Ca^{2+}] is thought to inhibit Na^+ uptake via NCKX and is used as a control). Radioisotope fluxes were measured using a rapid filtration technique. *This figure is taken in modified form from Schnetkamp, Szerencsei, et al. (1991); for further experimental details, consult this reference.*

Although easy to quantitate, radioisotope fluxes have limited kinetic resolution. Subsequent studies on the NCKX in ROS utilized a number of absorption dyes as quantitative indicators of NCKX-mediated fluxes (Schnetkamp, 1986, 1991), fluorescent Ca^{2+}-indicating dyes to measure NCKX-mediated changes in cytosolic-free Ca^{2+} (Schnetkamp, Basu, Li, & Szerencsei, 1991; Schnetkamp, Li, Basu, & Szerencsei, 1991), or the so-called suction electrode method to measure NCKX-mediated currents (Yau & Nakatani, 1984). The suction electrode method makes use of the unique circumstance that in rod photoreceptors the inward NCKX current is compensated by an outward K^+ current via channels located in the rod inner segment, giving rise to a circulating current between the outer and the inner segment measured with the suction electrode. Figure 7.2 illustrates that in ROS isolated in a medium containing physiological Na^+ and Ca^{2+} concentrations, both Ca^{2+} influx and Ca^{2+} efflux could be observed, satisfying another key prediction of the alternating access model, that is, the ability of NCKX to operate in both a Ca^{2+}extrusion (forward exchange) or Ca^{2+} influx (reverse exchange) mode. In this experiment, Ca^{2+} influx or Ca^{2+} extrusion was measured with the extracellular Ca^{2+} indicator Arsenazo III using a dual wavelength spectrophotometer (Schnetkamp, 1986) (note that Ca^{2+} influx was markedly stimulated by low concentrations of K^+). Combined, the above approaches led to the discovery that Na^+–Ca^{2+} exchange in ROS is unlike those described in most other tissues; it not only requires but also transports K^+ with a stoichiometry of $4Na^+$ against $1Ca^{2+}$ and $1K^+$ (Cervetto et al., 1989; Schnetkamp, 1989; Schnetkamp et al., 1989). Because NCKX is thought to operate by the alternating access model, it can operate in both the forward and reverse modes of exchange and can execute self-exchange fluxes. This would imply that the exchanger has a single set of cation-binding sites that can accommodate either $4Na^+$ or $1Ca^{2+}$ and $1K^+$. Figure 7.3 demonstrates (again using Arsenazo III as an extracellular Ca^{2+} indicator) that both forward and reverse exchange in ROS require K^+ and that the internal and the external K^+ dependencies of forward and reverse exchange, respectively, show very similar K_m values of ~2 mM.

3. CHARACTERISTICS OF THE CATION-BINDING SITES OF THE NCKX

Measurement of cation fluxes via the ROS NCKX revealed the following characteristics of the exchanger's cation-binding sites (Lagnado et al., 1988; Lagnado & McNaughton, 1990; Schnetkamp, 1991, 1995a;

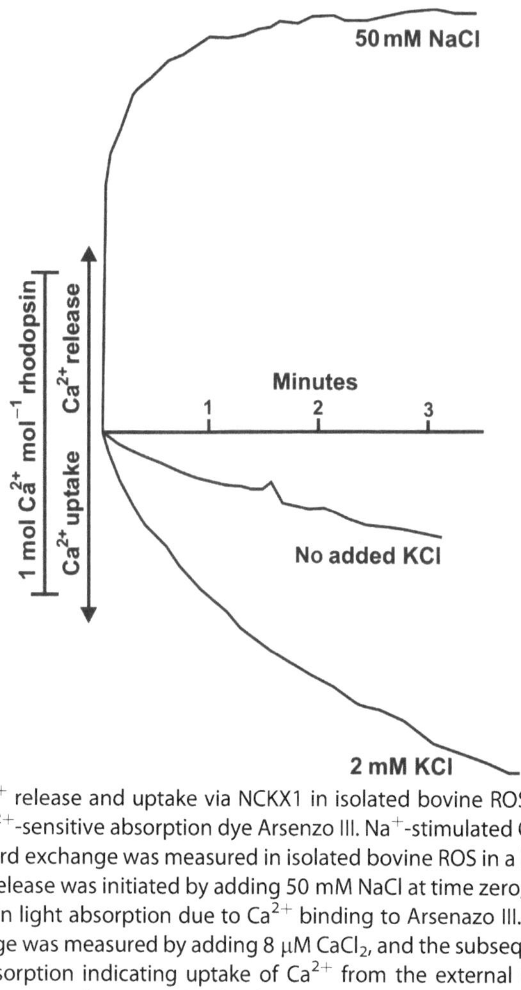

Figure 7.2 Ca^{2+} release and uptake via NCKX1 in isolated bovine ROS measured with the external Ca^{2+}-sensitive absorption dye Arsenzo III. Na^{+}-stimulated Ca^{2+} release representing forward exchange was measured in isolated bovine ROS in a buffered sucrose solution. Ca^{2+} release was initiated by adding 50 mM NaCl at time zero, and is indicated by an increase in light absorption due to Ca^{2+} binding to Arsenazo III. Ca^{2+} uptake via reverse exchange was measured by adding 8 μM $CaCl_2$, and the subsequent decrease in Arsenazo III absorption indicating uptake of Ca^{2+} from the external medium with or without the presence of 2 mM KCl. *This figure is taken from Schnetkamp (1986), and we refer to this reference for further experimental details.*

Schnetkamp, Basu, et al., 1991; Schnetkamp et al., 1989; Schnetkamp, Li, et al., 1991; Schnetkamp & Szerencsei, 1991; Schnetkamp, Szerencsei, et al., 1991; Schnetkamp, Tucker, & Szerencsei, 1995). K_m values reflect those obtained in the absence of competing cations.

1. The K_m values for both internal and external Ca^{2+} were in the 1–4 μM range at pH 7.4 when no competing cations (i.e., Na^{+} or Mg^{2+}) were present. Other divalent cations appear to compete with Ca^{2+} (as indicated by an increase in K_m value measured in their presence), but only Sr^{2+} could replace Ca^{2+} in actual transport. Of the alkali cations, only Na^{+} competes with Ca^{2+}, whereas high concentrations of other alkali

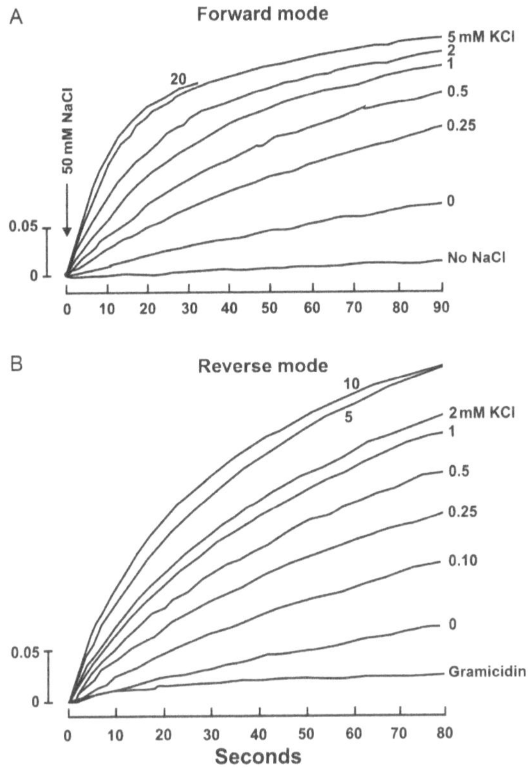

Figure 7.3 K^+ dependence of forward and reverse exchange modes of NCKX1 in isolated bovine ROS. (A) Forward (A) and reverse (B) Na^+–Ca^{2+} exchange were measured by the external Ca^{2+}-indicating dye Arsenazo III as described in Fig. 7.2. In the case of forward exchange, the intracellular K^+ concentration was controlled via the channel ionophore gramicidin. Calibration bar of 0.05 absorbance units represents a Ca^{2+} release of 0.64 mol Ca^{2+}/mol of rhodopsin. *This figure is taken from Schnetkamp et al. (1989), and we refer to this reference for further experimental details.*

cations do not affect the K_m for Ca^{2+} (although K^+ is required for Ca^{2+} transport to occur).

2. Dixon analysis shows that Ca^{2+} and Na^+ compete for a common site, but other alkali cations do not. Protons also compete with Na^+ and Ca^{2+} for this common site but are not transported, that is, no Ca^{2+}–H^+ exchange was observed. The pH dependence of Ca^{2+} transport suggests a binding site similar to the Ca^{2+}-binding site of ethylene glycol tetraacetic acid (EGTA), that is, involving multiple carboxylate moieties such as glutamate or aspartate residues.
3. As mentioned above, Na^+ competes with Ca^{2+} when present on the same side of the membrane and inhibits Ca^{2+} transport. Conversely,

Ca^{2+} inhibits Na^+ transport when present on the same side of the membrane, that is, Ca_o^{2+} inhibits Na_o^+-stimulated Ca^{2+} release. Unlike what is observed for Ca^{2+} transport, other alkali cations (Li^+, K^+) inhibit Na^+ transport as judged by their ability to inhibit Na_o^+-stimulated Ca^{2+} release and Na_i^+-stimulated Ca^{2+} influx. This suggests that Na^+ not only shares a common binding site with Ca^{2+} but also shares common binding site(s) with other alkali cations. However, only Na^+ can engage in Na^+–Ca^{2+} exchange. In the absence of competing cations (i.e., Li^+, K^+, Ca^{2+}) and at pH 7.4, the K_m values for both internal and external Na^+ were in the 30–40 mM range.

4. Only K^+, Rb^+, NH_4^+ and to a much lesser degree Cs^+ (but not Li^+ or organic cations) can activate Ca^{2+} transport and are cotransported with Ca^{2+} (e.g., Fig. 7.1). It should be pointed out that a significant amount of K^+-independent Ca^{2+} influx was associated with reverse exchange measured in media with sucrose or organic cation salts as the major osmotic constituent; this is true for experiments done in ROS (NCKX1) as illustrated in Fig. 7.4A. Because this K^+-independent Ca^{2+} influx was suppressed in the presence of Li^+, LiCl has become the standard major osmotic component in assay media for reverse Na^+–Ca^{2+}+K^+ exchange. In addition, the presence of 50 mM LiCl caused an increase in the K_m for K^+ from 0.4 to 2 mM (Fig. 7.4A). Figure 7.4B and C illustrates a similar phenomenon when measuring reverse exchange in human embryonic kidney 293 cell line (HEK293) cells transfected with the human NCKX2 cDNA; a significant K^+-independent Ca^{2+} influx was observed in choline chloride medium that was not seen in LiCl medium. In four experiments, the K_m value for K^+ was 0.57 mM (SE 0.22, $n=4$) in choline chloride medium, and a 10-fold higher value of 5.2 mM (SE 0.6, $n=4$) was obtained in LiCl medium—a very similar shift as observed in ROS. As noted above for the internal and external K_m values for Na^+ and Ca^{2+}, respectively, the internal and external K_m values for K^+ observed in ROS were similar as well (e.g., Fig. 7.3) and ranged between 1 and 5 mM.

Based on the observations summarized above, we proposed the three-site model for the ROS NCKX illustrated in Fig. 7.5 (Schnetkamp, Szerencsei, et al., 1991). The key feature of this model is that $2Na^+$ ions (but no other alkali cation) share and compete for a common binding site with a single Ca^{2+} ion, and two additional Na^+ ions share common binding sites with other alkali cations but not with organic cations such as choline or tetramethylammonium. Transport will only occur either if $4Na^+$ ions occupy these sites or if these sites are occupied by $1Ca^{2+}$ and $1K^+$. Any other

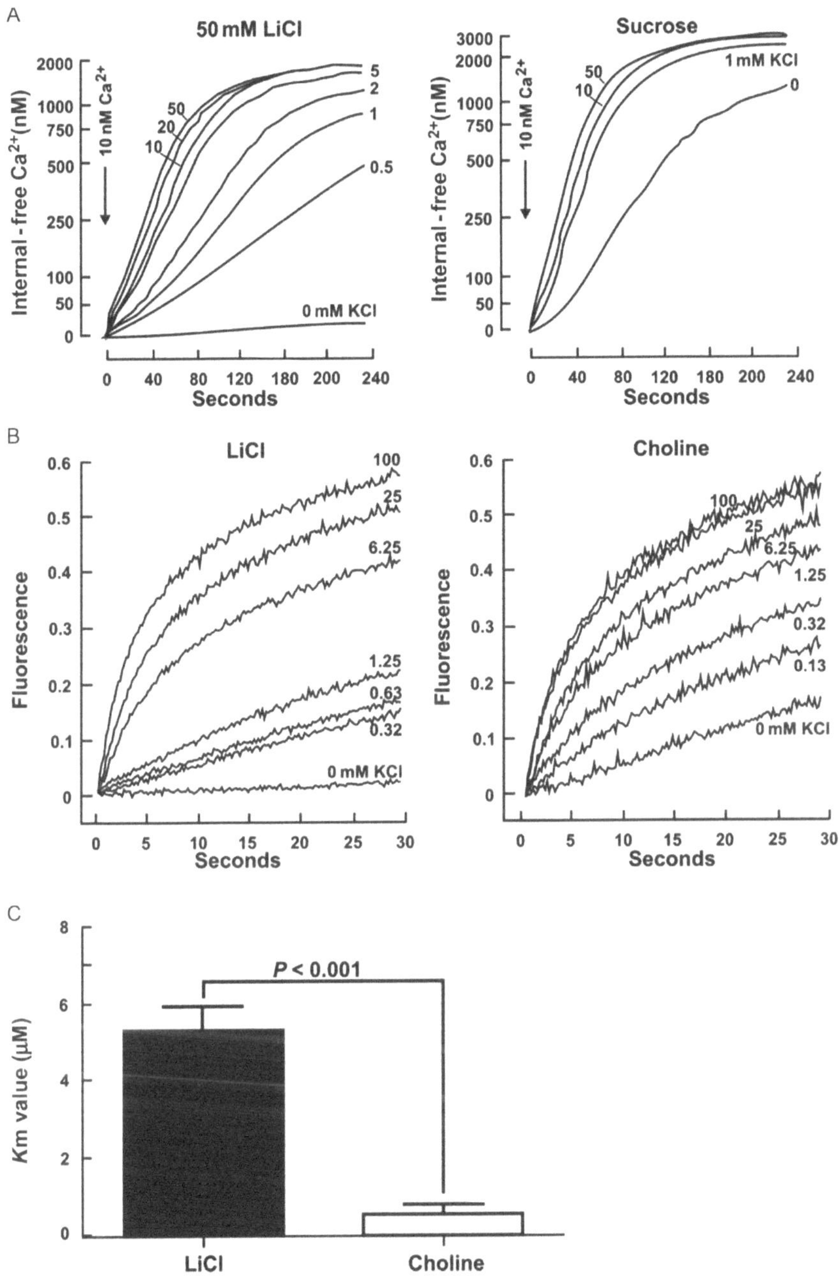

Figure 7.4 The effect of Li^+- on K^+-dependent reverse Na^+–Ca^{2+} exchange. (A) Reverse Na^+–Ca^{2+} exchange in isolated bovine ROS was measured in Ca^{2+}-depleted ROS as a rise in intracellular-free [Ca^{2+}] with the fluorescent intracellular Ca^{2+}-indicating dye fluo-3. Ca^{2+} influx was measured as a function of the extracellular K^+ concentration

(Continued)

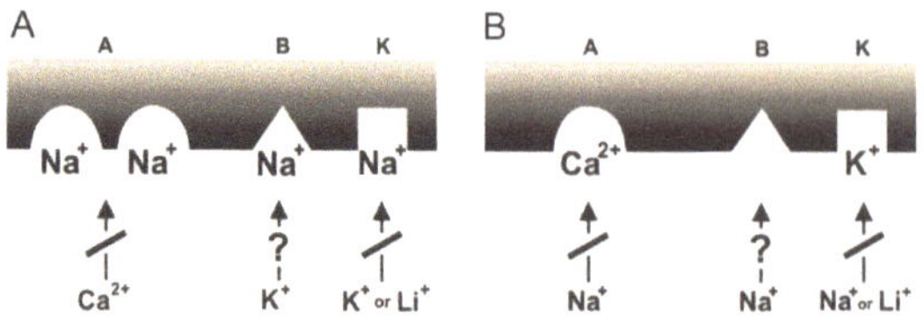

Figure 7.5 A three-site model for NCKX1. Detailed studies on NCKX1 in isolated ROS suggested a three-site model for the outward facing binding sites of NCKX1 as described in the text. (A) In the configuration that the three sites are occupied by $4Na^{+}$ ions resulting in forward exchange, Ca^{2+} can inhibit transport by competing with the $2Na^{+}$ that occupy the A-site, whereas inhibition of transport by other alkali cations reflects competition for the B- and/or K-site. (B) In the configuration that the three sites are occupied by $1Ca^{2+}$ and $1K^{+}$, reverse exchange occurs. Inhibition of reverse exchange by extracellular Na^{+} may reflect competition with Ca^{2+} for the A-site, as well as competition with K^{+} for the K-site (and perhaps B-site), while the effects of extracellular Li^{+} on the K_m for K^{+} reflect competition for the K-site and perhaps B-site. For a full discussion, see Schnetkamp, Szerencsei, et al. (1991).

occupancy pattern does not support transport. A number of other occupancies appear possible (based on the competition experiments, i.e., $1Ca^{2+}$ and $1Na^{+}/Li^{+}$, or $3Na^{+}$ and $1K^{+}/Li^{+}$) but do not result in transport. Some of the more detailed observations that support this model are found here (Schnetkamp, Basu, et al., 1991; Schnetkamp, Li, et al., 1991; Schnetkamp & Szerencsei, 1991; Schnetkamp, Szerencsei, et al., 1991).

Very little has been reported about regulation of transport activity of *in situ* NCKX (e.g., by phosphorylation). The extrusion mode of NCKX1 in ROS is thought to be regulated by an inactivation mechanism that would prevent lowering of cytosolic-free Ca^{2+} to undesirably low values that are predicted by the transport stoichiometry; however, the exact mechanism of this process remains to be elucidated (Sampath, Matthews, Cornwall, & Fain, 1998; Schnetkamp, Basu, et al., 1991; Schnetkamp & Szerencsei, 1993).

Figure 7.4—Cont'd in a buffered sucrose solution with or without 50 mM LiCl present. (B) Reverse Na^{+}–Ca^{2+} exchange was measured in HEK293 cells transfected with human NCKX2 as a rise in intracellular-free [Ca^{2+}] with the fluorescent intracellular Ca^{2+}-indicating dye fluo-4FF; the K^{+} dependence of reverse exchange was measured in either a buffered choline chloride solution or a buffered LiCl solution as indicated. For further experimental details, see Kang, Kinjo, et al. (2005). (C) Bar diagram showing the average (+ standard error of the mean in four experiments) K_m values for the external K^{+} dependence of human NCKX2 expressed in HEK cells. *(A) Adapted from Schnetkamp, Li, et al. (1991).*

4. INTERACTION OF NCKX WITH OTHER PROTEINS

NCKX1 in bovine ROS has been reported to occur as dimers (Schwarzer, Kim, Hagen, Molday, & Bauer, 1997) and binds directly to the cGMP-gated channel (Schwarzer, Schauf, & Bauer, 2000). Similarly, NCKX1 and NCKX2 have been shown to form dimers when expressed in HEK293 cells (Cai, Zhang, & Lytton, 2002; Kang et al., 2003). Both NCKX1 and NCKX2 were found to bind to their respective CNGA subunit of rod and cone cGMP-gated channels when coexpressed in HEK293 cells (Kang et al., 2003). The functional significance of this arrangement is unclear.

5. PURIFICATION OF ROS NCKX1, CLONING OF NCKX cDNAs, AND ANALYSIS OF NCKX SEQUENCES

The NCKX protein was purified from bovine ROS and appeared as a 220-kDa protein when analyzed by SDS-PAGE (Cook & Kaupp, 1988). The purified protein was calculated to have a turnover number of $30Ca^{2+}/s$. However, the *in situ* NCKX protein was calculated to have a turnover number of $\sim 300Ca^{2+}/s$ based on the density of NCKX in the ROS plasma membrane (Cook & Kaupp, 1988) and the maximal NCKX-mediated fluxes observed in intact ROS (Schnetkamp, 1991). Screening of a bovine retinal expression library with an antibody directed against the bovine ROS NCKX protein led to the cloning of the first NCKX cDNA—now designated as bovine NCKX1 (Reiländer et al., 1992). The cDNA encodes a protein of 1199 amino acids with a calculated MW of ~130 kDa and exhibits very limited sequence similarity with NCX1 cloned earlier from dog heart (Nicoll, Longoni, & Philipson, 1990). We now know that the human *SLC24* gene family has five members encoding the NCKX1–5 proteins (Schnetkamp, 2013), while the *SLC8* gene family has three members encoding the NCX1–3 proteins (Quednau, Nicoll, & Philipson, 2004). Most NCKX proteins in the data base range in size between 500 and 700 residues although NCKX1 proteins (in particular mammalian) can be much larger and contain over 1200 residues.

Hydropathy analysis of mammalian (and most other eukaryotic) NCKX sequences (http://www.cbs.dtu.dk/services/TMHMM/) reveals the presence of 12 hydrophobic segments of sufficient length to constitute alpha-helical transmembrane segments (TMS). They are arranged in three clusters. The first cluster is composed of one TMS (close to the N-terminus), while the second and third clusters are composed of five and six putative TMS, respectively, with the last cluster of putative TMS located at the C-terminus. The three clusters are separated by two large hydrophilic loops—the first contains glycosylation sites, and hence, is thought to be located in the extracellular space, while the second large hydrophilic loop separates the two clusters of multiple TMS and is thought to be located in the cytosol. The N-terminal localization of the first TMS and the presence of a cluster of positively charged residues immediately preceding this TMS suggest the presence of a signal anchor or cleavable signal peptide. Analysis by Signal P—a web-based server to predict the presence of signal peptides or signal anchors (http://www.cbs.dtu.dk/services/SignalP-2.0/)—predicts that most NCKX sequences have a signal anchor, while in a few cases (e.g., NCKX4 and NCKX5), a cleavable signal peptide is predicted. Sequence analysis reveals that the putative signal peptide/anchor sequences show no sequence similarity between the five mammalian isoforms (Fig. 7.6A), although for each isoform, the mammalian signal sequences are highly conserved (illustrated here for NCKX4, Fig. 7.6B). Signal peptide cleavage and its importance have been investigated for a few NCKX1 and NCKX2 proteins expressed in cell lines (Kang & Schnetkamp, 2003). For three of the NCKX proteins tested (chicken and human NCKX2 and dolphin NCKX1), the signal peptide was cleaved for about 50% of the total amount of NCKX protein expressed. In contrast, no signal peptide cleavage was observed for the

```
A
                  Signal peptide
hNCKX1     QHLRR--PRGLSSLWAAVSSHQ 20
hNCKX5     TSLPQRLPRATGNSTQCVIS-- 20
hNCKX2     FSISAFSETDTQSTGEASVV-- 20
hNCKX4     -GLFGSLGHKTASASKRVLPD- 20
hNCKX3     EQKELDLMDLIGEDRKWMMA-- 20

B
                  Signal peptide
hNCKX4     ASGLFGSLGHKTASASKRVLPDTW 44
mNCKX4     ASGLFGSLGHKTASAGKHVLLDTW 44
           ***************.*:** ***
```

Figure 7.6 Amino acid sequence alignment of different NCKXs. (A) The alignment of the predicted signal peptide sequences is shown for human NCKX1–5. (signal P v2.0 was used to predict location of the signal peptides). (B) Sequence alignment of the signal peptides of human NCKX4 and mouse NCKX4 is shown. Sequence alignments were made using ClustalW2 (http://www.ebi.ac.uk/Tools/msa/clustalw2/). An asterisk (*) denotes a strictly conserved residue, a colon (:) denotes conserved substitutions, and a period (.) denotes semiconserved substitutions.

chicken NCKX1 protein. This was unexpected because chicken NCKX1 was predicted to have a cleavable signal peptide using Signal P. It was also found that deletion of the signal peptide sequence did not affect protein expression but prevented trafficking to the plasma membrane. This suggests that signal peptide cleavage was a critical step in the correct trafficking to the plasma membrane (Kang & Schnetkamp, 2003). As a result, SDS-PAGE analysis of NCKX2 expressed in cell lines invariable reveals a doublet of which the upper band represents full-length NCKX2 and the lower band represents NCKX2 protein from which the signal peptide was cleaved (Kang & Schnetkamp, 2003). Although no data on signal peptide cleavage for NCKX3–5 have been reported, SDS-PAGE analysis of these proteins expressed in cell lines suggests a similar pattern as observed for NCKX2 (e.g., Ginger et al., 2008). It is interesting to note that NCKX5 is not trafficked to the plasma membrane of epidermal melanocytes, and the same is true for NCKX5 expressed in cell lines (Ginger et al., 2008; Lamason et al., 2005).

Sequence conservation of the large hydrophilic loop—thought to be located in the cytosol—is summarized in Table 7.1. These sequences show

Table 7.1 Sequence comparison of the large intracellular cytoplasmic loop of NCKX1–5

NCKX clone	Length (aa)	Comparison with	Length (aa)	Conservation score (%)
Human NCKX2	210	Human NCKX1	328	31.43
		Human NCKX3	212	10.48
		Human NCKX4	180	16.67
		Human NCKX5	111	11.71
Human NCKX3	212	Human NCKX1	328	12.74
		Human NCKX4	180	38.33
		Human NCKX5	111	10.81
Human NCKX1	328	Mouse NCKX1	392	66.46
Human NCKX2	210	Mouse NCKX2	214	89.05
Human NCKX3	212	Mouse NCKX3	215	98.11
Human NCKX4	180	Mouse NCKX4	199	90.00
Human NCKX5	111	Mouse NCKX5	112	81.08

Conservation scores are shown when the large intracellular loops of various NCKX are compared.

a considerable range in size and are the main reason for the range in the total number of residues noted above for the various NCKX proteins found in the database. Very limited conservation is observed between the loops of the different isoforms with modest conservation observed between NCKX1 and NCKX2 and also between NCKX3 and NCKX4. In contrast, high conservation scores are observed when comparing cytosolic loops of the same isoform for different species.

In contrast to the limited sequence conservation found in the putative signal peptides and the large hydrophilic loops of the five human NCKX isoforms, sequence conservation in the two clusters of putative TMS is highly significant (Fig. 7.7). In particular, the central parts of the two clusters are highly conserved. They are referred to as the alpha 1 and alpha 2 repeats and are thought to have arisen from an ancient gene duplication event. Based on sequence similarity in the two alpha repeats, gene products are identified as putative K^+-dependent Na^+–Ca^{2+} exchanges and are widely distributed in the animal kingdom and in certain salt water algae, but not in plants (Cai & Lytton, 2004a; Emery, Whelan, Hirschi, & Pittman, 2012). More limited sequence similarity (confined to the two alpha repeats) exists with NCX proteins (the human *SLC8* gene family) and putative prokaryotic NCXs. Although many putative NCKX have been identified in the numerous genomic sequence projects, K^+-dependent Na^+–Ca^{2+} exchange activity has only been demonstrated for a handful of vertebrate NCKX gene products (Cai & Lytton, 2004b; Cooper, Winkfein, Szerencsei, & Schnetkamp, 1999; Kraev et al., 2001; Li, Kraev, & Lytton, 2002; Paillart et al., 2007; Poon et al., 2000; Prinsen, Szerencsei, & Schnetkamp, 2000; Tsoi et al., 1998), as well as in one identified and cloned from *Caenorhabditis elegans* (Szerencsei et al., 2000) and two identified and cloned from *Drosophila* (Haug-Collet et al., 1999; Winkfein et al., 2004).

6. A TOPOLOGICAL MODEL FOR THE NCKX2 K^+-DEPENDENT NA^+–CA^{2+} EXCHANGERS

Different algorithms to predict the topology of polytopic membrane proteins like NCKX have been developed and may yield quite different topological models of the arrangement of TMS of NCKX. Even the same algorithm may give different results for the human NCKX1–5 isoforms. An experimental model for the topology of the human NCKX2 protein was obtained based on glycosylation patterns observed when *N*-glycosylation sites were inserted in various positions, and accessibility of inserted cysteine

```
                                            TMS1
hNCKX3    -----ALHEFPNDIFTNEDRRQGAVVLHVLCAIYMFYALAIVCDDFFVPSLEKICERLHL 145
hNCKX4    -----AIHEFPTDLFSNKERQHGAVLLHILGALYMFYALAIVCDDFFVPSLEKICERLHL 118
hNCKX5    -----PSSEFPEGFFTRQERRDGGIIIYFLIIVYMFMAISIVCDEYFLPSLEIISESLGL 104
hNCKX1    LPDLHPKGEYPPDLFSVEERRQGWVVLHVFGMMYVFVALAIVCDEYFVPALGVITDKLQI 490
hNCKX2    ----HAQGDYPKDIFSLEERRKGAIILHVIGMIYMFIALAIVCDEFFVPSLTVITEKLGI 170
               .  ::* .:*: ::*:.* ::::.:  :*:* *::****::*:*:*  * : * :

                  TMS2                         TMS3
hNCKX3    SEDVAGATFMAAGSSAPELFTSVIGVFITKGDVGVGTIVGSAVFNILCIIGVCGLFAGQV 205
hNCKX4    SEDVAGATFMAAGSSTPELFASVIGVFITHGDVGVGTIVGSAVFNILCIIGVCGLFAGQV 178
hNCKX5    SQDVAGTTFMAAGSSAPELVTAFLGVFITKGDIGISTILGSAIYNLLGICAACGLLSNTV 164
hNCKX1    SEDVAGATFMAAGGSAPELFTSLIGVFISHSNVGIGTIVGSAVFNILFVIGTCSLFSREI 550
hNCKX2    SDDVAGATFMAAGGSAPELFTSLIGVFIAHSNVGIGTIVGSAVFNILFVIGMCALFSREI 230
          *:****:******.*:***.::.:****::.::*:.**:***::*:* : . *.*::  :

                    TMS4                         TMS5
hNCKX3    VALSSWCLLRDSIYYTLSVIALIVFIYDEKVSWWESLVLVLMYLIYIVIMKYNACIHQCF 265
hNCKX4    VRLTWWAVCRDSVYYTISVIVLIVFIYDEQIVWWEGLVLIILYVFYILIMKYNVKMQAFF 238
hNCKX5    STLSCWPLFRDCAAYTISAAAVLGIIYDNQVYWYEGALLLLIYGLYVLVLCFDIKINQYI 224
hNCKX1    LNLTWWPLFRDVSFYILDLIMLILFFLDSLIAWWESLLLLLAYAFYVFTMKWNKHIEVWV 610
hNCKX2    LNLTWWPLFRDVSFYIVDLIMLIIFFLDNVIMWWESLLLLTAYFCYVVFMKFNVQVEKWV 290
            *: * : **   * :.   :: :: *. : *:*. :*:  *  *:. : ::  :.  .

                                                  TMS6
hNCKX3    VKWAFTWPLSFVLYFTVPNCNKPRWEKWFMVTFASSTLWIAAFSYMMVWMVTIIGYTLGI 507
hNCKX4    VKWVFTWPLIFLLCVTIPNCSKPRWEKFFMVTFITATLWIAVFSYIMVWLVTIIGYTLGI 448
hNCKX5    IFWVLSLPIITLLFLTTPDCRKKFWKNYFVITFFMSAIWISAFTYILVWMVTITGETLEI 365
hNCKX1    AIYLFLLPIVFPLWLTVPDVRRQESRKFFVFTFLGSIMWIAMFSYLMVWWAHQVGETIGI 968
hNCKX2    VTFLIVFPIVFPLWITLPDVRKPSSRKFFPITFFGSITWIAVFSYLMVWWAHQVGETIGI 530
            : :  *:   * .* *:  :   .::* .**  :  **: *:*::** .   *  *: *

                   TMS7                          TMS8
hNCKX3    PDVIMGITFLAAGTSVPDCMASLIVARQGMGDMAVSNSIGSNVFDILIGLGLPWALQTLA 567
hNCKX4    PDVIMGITFLAAGTSVPDCMASLIVARQGLGDMAVSNTIGSNVFDILVGLGVPWGLQTMV 508
hNCKX5    PDTVMGLTLLAAGTSIPDTIASVLVARKGKGDMAMSNIVGSNVFDMLC-LGIPWFIKTAF 423
hNCKX1    SEEIMGLTILAAGTSIPDLITSVIVARKGLGDMAVSSSVGSNIFDITVGLPVPWLLFSLI 1028
hNCKX2    SEEIMGLTILAAGTSIPDLITSVIVARKGLGDMAVSSSVGSNIFDITVGLPLPWLLYTVI 590
          .: :**:*:******:** ::*::***:* ****:*. :***:**: * :** : :

                        TMS9                      TMS10
hNCKX3    VDYGSYIRLNSRGLIYSVGLLLASVFVTVFGVHLNKWQLDKKLGCGCLLLYGVFLCFSIM 627
hNCKX4    VNYGSTVKINSRGLVYSVVLLLGSVALTVLGIHLNKWRLDRKLGVYVLVLYAIFLCFSIM 568
hNCKX5    INGSAPAEVNSRGLTYITISLNISIIFLFLAVHFNGWKLDRKLGIVCLLSYLGLATLSVL 484
hNCKX1    NG-LQPVPVSSNGLFCAIVLLFLMLLFVISSIASCKWRMNKILGFTMFLLYFVFLIISVM 1087
hNCKX2    HR-FQPVAVSSNGLFCAIVLLFIMLLFVILSIALCKWRMNKILGFIMFGLYFVFLVVSVL 649
                  :.*.**      *   : . . .:    *:::: **   : *  :  .*::
```

Figure 7.7 Amino acid sequence alignment of the two clusters of five TMS of human NCKX1–5. A high degree of sequence conservation is seen in all predicted TMS when compared to the large hydrophilic loops. The highest degree of sequence conservation is observed in the α1 and α2 regions (TMS 2–3 and TMS 7–8, respectively). Sequence alignments were made using ClustalW2 (http://www.ebi.ac.uk/Tools/msa/clustalw2/). An asterisk (*) denotes a strictly conserved residue, a colon (:) denotes conserved substitutions, and a period (.) denotes semiconserved substitutions.

residues to hydrophilic thiol reagents (Kinjo, Szerencsei, Winkfein, Kang, & Schnetkamp, 2003; Szerencsei, Kinjo, & Schnetkamp, 2013). This model (illustrated in Fig. 7.8) places both the N terminus (after signal peptide cleavage) and the C terminus in the extracellular space. The model also contains two clusters of five TMS that are separated by a large hydrophilic loop located in the cytosol. One putative TMS does not span the membrane but is located at the C-terminal end of the cytosolic loop and precedes

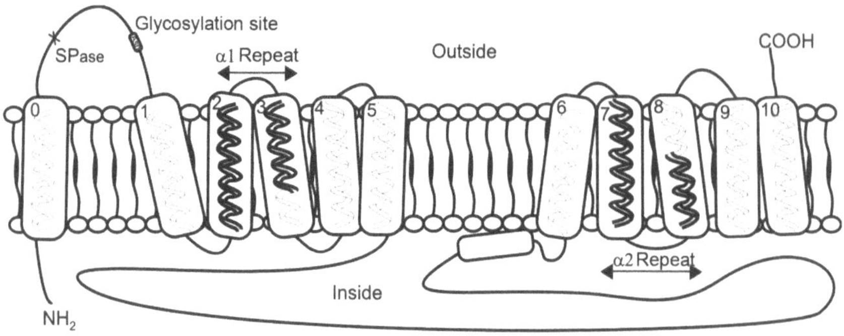

Figure 7.8 Schematic representation of the current topological model of NCKX2. Our current topological model of NCKX proteins consists of two sets of five TMS separated by a large intracellular loop and preceded by an extracellular loop. The extracellular loop contains an *N*-glycosylation site as well as the signal peptide cleavage site (SPase). The TMS contain the highly conserved areas in TMS 2–3 and TMS 7–8 known as the α1 and α2 repeats, respectively (represented by the dark helices). *Based on references Kang et al. (2003), Kang and Schnetkamp (2003), Kinjo et al. (2003), and Szerencsei et al. (2013).*

the second cluster of five TMS. The experimentally derived topological model of human NCKX2 is identical to the topology seen in the X-ray crystal structure recently obtained for the archaebacterial NCX NCX_Mj (Liao et al., 2012).

7. FUNCTIONAL CHARACTERISTICS OF NCKX PROTEINS EXPRESSED IN CELL LINES

A limited number of studies have so far addressed the functional properties of NCKX proteins heterologously expressed in cell lines using ^{45}Ca uptake, whole cell patch clamp to measure NCKX currents, and fluorescent Ca^{2+}-indicating dyes to measure NCKX-mediated changes in free cytosolic $[Ca^{2+}]$. In all cases examined, the putative NCKX proteins indeed displayed K^+-dependent Na^+–Ca^{2+} exchange, and as noted above for NCKX in ROS, a K^+-independent component of reverse exchange was observed when tested in an alkali cation-free medium (i.e., choline Cl vs. LiCl) and LiCl caused a significant increase in the K_m for K^+ (Prinsen et al., 2000; see also Fig. 7.4). Deletion of the two large hydrophilic loops from the bovine NCKX1 sequence did not abolish K^+-dependent Na^+–Ca^{2+} exchange implying that the two sets of TMS (with short connecting loops) are sufficient for transport function (Szerencsei et al., 2000). This is perhaps not surprising given the fact that no sequence similarity in the large hydrophilic loops is observed when comparing the mammalian NCKX1–5 sequences in the database. As reported previously for NCKX1 in ROS, both human and rat NCKX2 expressed in cell lines were shown to operate at a stoichiometry of $4Na^+$ against $1Ca^{2+}$plus $1K^+$, distinct from the $3Na^+$ against $1Ca^{2+}$ (no cotransport of K^+), observed for NCX1 (Dong, Light, French, & Lytton, 2001; Szerencsei, Prinsen, & Schnetkamp, 2001).

K_m values have been determined for various cation dependencies for NCKX1, NCKX2 (Altimimi, Fung, Winkfein, & Schnetkamp, 2010; Altimimi, Szerencsei, & Schnetkamp, 2013; Dong et al., 2001; Prinsen et al., 2000; Sheng, Prinsen, Clark, Giles, & Schnetkamp, 2000; Szerencsei et al., 2001), and to a lesser extent for NCKX3 and NCKX4 (Visser, Valsecchi, Annunziato, & Lytton, 2007). The values obtained compare well for those obtained previously for NCKX1 in ROS (as summarized above), especially when considering the range of methodologies and experimental conditions used. Values for the external Ca^{2+} K_m of NCKX1, NCKX2, NCKX3, and NCKX4 range between 1 and 3 μM, while the internal K_m for Ca^{2+} was 0.5–1 μM. External Na^+ dependencies yielded values of 30–40 mM again, very similar to values obtained for ROS. The internal Na^+ dependence has been reported for NCKX2 to be ~58 mM when measured in a high KCl medium and by using the channel ionophore gramicidin to control intracellular $[Na^+]$; the same value was obtained in bovine ROS adopting the same protocol (our unpublished results). The values reported for the external K_m for K^+ of NCKX2 show a significant range with some studies reporting a value ~40 mM (in LiCl medium) (Dong et al., 2001; Visser et al., 2007), while other studies report lower values between 2 and 10 mM (Kang, Kinjo, Szerencsei, & Schnetkamp, 2005; Prinsen et al., 2000) (see also Fig. 7.4). The reason for the discrepancy is unclear. Interestingly, much lower K_m values of 1–1.5 mM were reported for the K^+ dependence of NCKX3 and NCKX4, and the difference was attributed to a single residue (Thr551 in NCKX2, while the corresponding residue in NCKX3 and NCKX4 is Ala) (Visser et al., 2007).

Few studies have so far examined regulation of NCKX expressed in cell lines. In one study, NCKX2 was shown to display Na^+-dependent inactivation of the Ca^{2+} extrusion mode (Altimimi & Schnetkamp, 2007b), perhaps related to inactivation of the Ca^{2+} extrusion mode described previously for NCKX1 in ROS and discussed above. In another study, Src family kinase-dependent tyrosine phosphorylation of residue Tyr365 NCKX2 was suggested to be involved in endocytosis of NCKX2 reported before in hippocampal neurons (Lee, Ho, & Lee, 2013).

8. SCANNING MUTAGENESIS TO ELUCIDATE RESIDUES IMPORTANT FOR CATION TRANSPORT

Our laboratory has generated a large number of single residue substitutions in human NCKX2 to carry out alanine and cysteine scanning mutagenesis of the two alpha repeats as these are the most conserved elements in the NCKX sequence and likely to be important for cation transport (see above). In addition to residues in the two alpha repeats, all acidic residues found in the two clusters of TMS were substituted as well. Acidic residues are believed to play a role in the binding of the cations, in particular divalent cations such as Ca^{2+}. In general, two types of substitutions were prepared—either conserving the charge (e.g., Asp to Glu) or the size of the substituted residue (Asp to Asn). The first study used $^{45}Ca^{2+}$ uptake associated with reverse exchange in Na^{+}-loaded high five insect cells transfected with the mutant NCKX2 cDNA. Reduced ^{45}Ca uptake reflected reduced transport capacity of the individual mutant NCKX2 proteins because mutant protein expression and plasma membrane trafficking were not affected by any of the residue substitutions studied (Winkfein et al., 2003). For 23 residues located within the two alpha repeat regions, it was found that mutations could lower the maximal transport rate to <20% of WT NCKX2 (Fig. 7.9A), and these fell in the following five groups: (1) five acidic residues (i.e., Glu and Asp), (2) eight polar residues (i.e., Ser, Thr, and Asn), (3) five Gly residues, (4) four small hydrophobic residues, and (5) one proline residue. Many of the acidic and polar residues involved define polar surfaces of TMS 2 and 7 and may contribute oxygen ligands to coordinate Ca^{2+}, while Gly residues are often important for helix–helix contacts or may allow sharp turns and offer flexibility. Twenty of these 23 residues are conserved among all five human NCKX isoforms.

In later studies, the mutant NCKX2 cDNAs were expressed in HEK293 cells and fluorescent Ca^{2+} dyes (e.g., fluo-3, fluo-4, fluo-4FF, fura-2) were used to measure the rate of change in intracellular-free Ca^{2+} associated with reverse exchange as a function of Ca^{2+}, K^{+} (Kang, Kinjo, et al., 2005; Kang, Shibukawa, et al., 2005; Visser et al., 2007), or Na^{+} concentration (Altimimi et al., 2010). These studies focused on identifying residues that are important for the affinity of the NCKX protein for the substrate cations. Of the 26 different mutant NCKX2 proteins examined for shifts in K^{+} and Ca^{2+} affinity, 13 showed a significant decrease in the affinity toward Ca^{2+} and/or K^{+} as summarized in Fig. 7.9B. These mutants include the acidic and polar

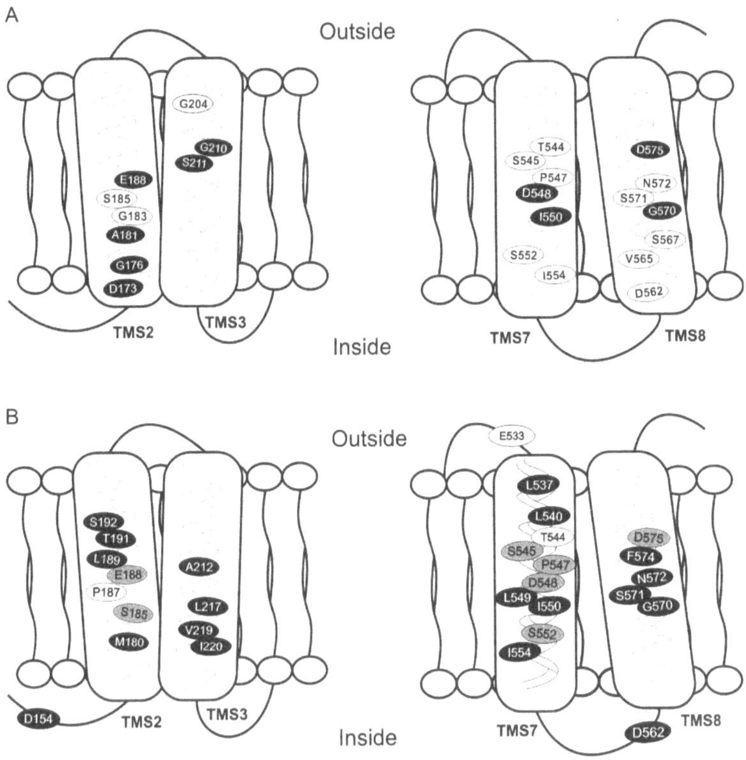

Figure 7.9 Schematic representation of the current topological model of NCKX2 showing residues important for cation transport. (A) Residues for which substitution results in 80–90% (white ovals) or >90% (black ovals) reduction of maximal NCKX activity when the mutated protein is expressed in HEK293 cells (Winkfein et al., 2003). (B) Black ovals with white letters represent residues whose substitution resulted in a shift in Na^+ affinity; white ovals with black letters represent residues whose substitution resulted in shifts in Ca^{2+} and K^+ affinity; gray ovals with black letters represent residues that result in shifts for both Na^+ and Ca^{2+}/K^+ affinities (Altimimi et al., 2010; Kang, Kinjo, et al., 2005; Kang, Shibukawa, Szerencsei, & Schnetkamp, 2005).

residues identified in Fig. 7.9A. Interestingly, in several cases, shifts in Ca^{2+} dependence were accompanied by shifts in K^+ dependence (Kang, Kinjo, et al., 2005). Two additional residues have been identified that might be particularly important for K^+ transport by NCKX2. The Asp575Cys and Asp575Asn substitutions resulted in mutant NCKX2 proteins that showed K^+-independent Na^+–Ca^{2+} exchange, suggesting that Asp575 is particularly important for K^+ binding (Kang, Shibukawa, et al., 2005). In another study, the Thr551Ala substitution led to a large decrease in the K_m for K^+, suggesting that this residue is important in defining the K^+ affinity (Visser et al., 2007). Asp575 is conserved in all NCKX sequences in the database, whereas this position is occupied by an asparagine in all eukaryotic NCX sequences. Based on the above observations and the following points, we concluded that Glu188 and Asp548 are the two main Ca^{2+}-coordinating residues in all eukaryotic NCX and NCKX proteins (Kang, Kinjo, et al., 2005a). (1) These residues are conserved in all eukaryotic NCX and NCKX sequences. (2) They are the only two acidic residues in the TMS of NCKX2 for which eliminating the charge rendered the mutant NCKX2 proteins without any measurable transport function. (3) The charge conservative substitutions of each of these two residues resulted in the largest decrease in Ca^{2+} affinity reported so far. (4) Site-directed disulfide mapping showed that Glu188 and Asp548 are in close proximity (Kinjo, Kang, Szerencsei, Winkfein, & Schnetkamp, 2005). The latter study also demonstrates that many residues important for NCKX2 function as described above and identified in Fig. 7.9 are in close proximity in the native NCKX2 structure (e.g., Ser185, Glu188, Asp548, Ser552, and Asp575).

In a recent study from our laboratory, we scanned >100 single residue substitutions of the human NCKX2 protein for shifts in Na^+ dependencies and identified 31 residues for which substitution resulted in significant shifts in the Na^+ dependency of the mutated protein expressed in HEK293 cells (Altimimi et al., 2010). Twenty-seven of these residues are found in the alpha 1 and 2 repeat regions, which are represented in Fig. 7.9B. Our current model for the cation-binding sites of NCKX proteins incorporates one site where Ca^{2+} and Na^+ compete for binding and other sites where Na^+ and K^+ compete (Fig. 7.5), and this leads to the prediction that many of the same residues that are found important for Ca^{2+} and K^+ transport may also be important for Na^+ transport, and this is indeed observed (Fig. 7.9B). This lends further support for the alternating access model of NCKX-mediated cation transport. For example, the Glu188Asp and Asp548Glu substitutions that resulted in mutant NCKX2 proteins with greatly reduced affinity toward Ca^{2+} also resulted in significant shifts in Na^+ dependency of the mutant NCKX2 proteins. The same is observed for two Asp575 substitutions that led to an apparent abolition of K^+ binding to NCKX as judged by the loss of dependence on K^+. The distribution of residues that defines the apparent Na^+ affinity of NCKX2 would appear to cover much of the width of the membrane, and moreover, residue substitutions could either result in significant increases or decreases in apparent Na^+ affinity (Altimimi et al., 2010).

9. FUTURE PERSPECTIVES

In this review, we have summarized our current knowledge of the transport properties of NCKX proteins, both *in situ* and expressed in cell lines, as well as recent structure–function studies that identified residues important for cation transport and placed these residues within a topological model of the NCKX2 protein. It is gratifying to note that the recently obtained crystal structure of the archaebacterial NCX_Mj has the identical topology of our NCKX2 topological model, and moreover, the two residues equivalent to Glu188 and Asp548 of NCKX2 were shown to be the two primary Ca^{2+}-coordinating residues in NCX_Mj (Liao et al., 2012). This suggests that homology modeling based on the NCX_Mj structure may lead to a good working model for the NCKX2 structure. We can identify two areas in which NCKX research lags far behind that of most other transporter gene families. First, no useful/selective inhibitors of NCKX transport have yet been identified, and this hampers studies on the physiological role of the various NCKX isoforms in many tissues in which NCKX transcripts have been found. Second, the NCKX1–5 proteins have very different large cytosolic loops, which are unlikely to be directly involved in cation transport and presumably contribute to regulation of cation transport.

ACKNOWLEDGMENT

This research was supported through operating Grant MOP-81327 from the Canadian Institutes for Health Research.

REFERENCES

Altimimi, H. F., Fung, E. H., Winkfein, R. J., & Schnetkamp, P. P. (2010). Residues contributing to the Na(+)-binding pocket of the SLC24 Na(+)/Ca(2+)-K(+) exchanger NCKX2. *The Journal of Biological Chemistry*, *285*(20), 15245–15255.

Altimimi, H., & Schnetkamp, P. P. M. (2007a). Na^+/Ca^{2+}-K^+ exchangers (NCKX). Functional properties and physiological roles. *Channels (Austin, Tex)*, *1*(2), 62–69.

Altimimi, H. F., & Schnetkamp, P. P. M. (2007b). Na^+-dependent inactivation of the retinal cone/brain Na^+/Ca^{2+}-K^+ exchanger NCKX2. *The Journal of Biological Chemistry*, *282*, 3720–3729.

Altimimi, H. F., Szerencsei, R. T., & Schnetkamp, P. P. (2013). Functional and structural properties of the NCKX2 Na(+)-Ca (2+)/K (+) exchanger: A comparison with the NCX1 Na (+)/Ca (2+) exchanger. *Advances in Experimental Medicine and Biology*, *961*, 81–94. http://dx.doi.org/10.1007/978-1-4614-4756-6_8.

Blaustein, M. P., & Lederer, W. J. (1999). Sodium/calcium exchange: Its physiological implications. *Physiological Reviews*, *79*(3), 763–854.

Cai, X., & Lytton, J. (2004a). The cation/Ca(2+) exchanger superfamily: Phylogenetic analysis and structural implications. *Molecular Biology and Evolution, 21*(9), 1692–1703.

Cai, X., & Lytton, J. (2004b). Molecular cloning of a sixth member of the K^+-dependent Na^+/Ca^{2+} exchanger gene family, NCKX6. *The Journal of Biological Chemistry, 279*, 5867–5876.

Cai, X., Zhang, K., & Lytton, J. (2002). A novel topology and redox regulation of the rat brain K^+-dependent Na^+/Ca^{2+} exchanger, NCKX2. *The Journal of Biological Chemistry, 277*(50), 48923–48930.

Cervetto, L., Lagnado, L., Perry, R. J., Robinson, D. W., & McNaughton, P. A. (1989). Extrusion of calcium from rod outer segments is driven by both sodium and potassium gradients. *Nature, 337*(6209), 740–743.

Cook, N. J., & Kaupp, U. B. (1988). Solubilization, purification, and reconstitution of the sodium-calcium exchanger from bovine retinal rod outer segments. *The Journal of Biological Chemistry, 263*, 11382–11388.

Cooper, C. B., Winkfein, R. J., Szerencsei, R. T., & Schnetkamp, P. P. M. (1999). cDNA-cloning and functional expression of the dolphin retinal rod Na-Ca+K exchangerNCKX1: Comparison with the functionally silent bovine NCKX1. *Biochemistry, 38*, 6276–6283.

Dong, H., Light, P. E., French, R. J., & Lytton, J. (2001). Electrophysiological characterization and ionic stoichiometry of the rat brain K^+-dependent Na^+/Ca^{2+} exchanger, NCKX2. *The Journal of Biological Chemistry, 276*(28), 25919–25928.

Emery, L., Whelan, S., Hirschi, K. D., & Pittman, J. K. (2012). Protein phylogenetic analysis of Ca(2+)/cation antiporters and insights into their evolution in plants. *Frontiers in Plant Science, 3*, 1. http://dx.doi.org/10.3389/fpls.2012.00001.

Fain, G. L., Matthews, H. R., Cornwall, M. C., & Koutalos, Y. (2001). Adaptation in vertebrate photoreceptors. *Physiological Reviews, 81*(1), 117–151.

Ginger, R. S., Askew, S. E., Ogborne, R. M., Wilson, S., Ferdinando, D., Dadd, T., et al. (2008). SLC24A5 encodes a trans-Golgi network protein with potassium-dependent sodium-calcium exchange activity that regulates human epidermal melanogenesis. *The Journal of Biological Chemistry, 283*(9), 5486–5495.

Haug-Collet, K., Pearson, B., Park, S., Webel, R., Szerencsei, R. T., Winkfein, R. J., et al. (1999). Cloning and characterization of a potassium-dependent sodium/calcium exchanger in *Drosophila*. *The Journal of Cell Biology, 147*(3), 659–669.

Kang, K.-J., Bauer, P. J., Kinjo, T. G., Szerencsei, R. T., Bonigk, W., Winkfein, R. J., et al. (2003). Assembly of retinal rod or cone Na^+/Ca^{2+}-K^+ exchangers oligomers with cGMP-gated channel subunits as probed with heterologously expressed cDNAs. *Biochemistry, 42*(15), 4593–4600.

Kang, K.-J., Kinjo, T. G., Szerencsei, R. T., & Schnetkamp, P. P. M. (2005). Residues contributing to the Ca^{2+} and K^+ binding pocket of the NCKX2 Na^+/Ca^{2+}-K^+ exchanger. *The Journal of Biological Chemistry, 280*, 6823–6833.

Kang, K.-J., & Schnetkamp, P. P. M. (2003). Signal sequence cleavage and plasma membrane targeting of the rod NCKX1 and cone NCKX2 Na^+/Ca^{2+}-K^+ exchangers. *Biochemistry, 42*, 9438–9445.

Kang, K.-J., Shibukawa, Y., Szerencsei, R. T., & Schnetkamp, P. P. M. (2005). Substitution of a single residue, Asp^{575}, renders the NCKX2 K^+-dependent Na^+/Ca^{2+} exchanger independent of K^+. *The Journal of Biological Chemistry, 280*, 6834–6839.

Kinjo, T. G., Kang, K.-J., Szerencsei, R. T., Winkfein, R. J., & Schnetkamp, P. P. M. (2005). Site-directed disulfide mapping of residues contributing to the Ca^{2+} and K^+ binding pocket of the NCKX2 Na^+/Ca^{2+}-K^+ exchanger. *Biochemistry, 44*, 7787–7795.

Kinjo, T. G., Szerencsei, R. T., Winkfein, R. J., Kang, K.-J., & Schnetkamp, P. P. M. (2003). Topology of the retinal cone NCKX2 Na/Ca-K exchanger. *Biochemistry, 42*(8), 2485–2491.

Kraev, A., Quednau, B. D., Leach, S., Li, X. F., Dong, H., Winkfein, R. J., et al. (2001). Molecular cloning of a third member of the potassium-dependent sodium-calcium exchanger gene family, NCKX3. *Journal of Biological Chemistry, 276*(25), 23161–23172.

Lagnado, L., Cervetto, L., & McNaughton, P. A. (1988). Ion transport by the Na:Ca exchange in isolated rod outer segments. *Proceedings of the National Academy of Sciences of the United States of America, 85*, 4548–4552.

Lagnado, L., & McNaughton, P. A. (1990). Electrogenic properties of the Na:Ca exchange. *The Journal of Membrane Biology, 113*, 177–191.

Lamason, R. L., Mohideen, M. A., Mest, J. R., Wong, A. C., Norton, H. L., Aros, M. C., et al. (2005). SLC24A5, a putative cation exchanger, affects pigmentation in zebrafish and humans. *Science, 310*(5755), 1782–1786.

Lee, K. H., Ho, W. K., & Lee, S. H. (2013). Endocytosis of somatodendritic NCKX2 is regulated by Src family kinase-dependent tyrosine phosphorylation. *Frontiers in Cellular Neuroscience*, 7, 14. http://dx.doi.org/10.3389/fncel.2013.00014.

Li, X. F., Kiedrowski, L., Tremblay, F., Fernandez, F. R., Perizzolo, M., Winkfein, R. J., et al. (2006). Importance of K^+-dependent Na^+/Ca^{2+}-exchanger 2, NCKX2, in motor learning and memory. *The Journal of Biological Chemistry, 281*, 6273–8262.

Li, X. F., Kraev, A. S., & Lytton, J. (2002). Molecular cloning of a fourth member of the potassium-dependent sodium-calcium exchanger gene family, NCKX4. *The Journal of Biological Chemistry, 277*(50), 48410–48417.

Liao, J., Li, H., Zeng, W., Sauer, D. B., Belmares, R., & Jiang, Y. (2012). Structural insight into the ion-exchange mechanism of the sodium/calcium exchanger. *Science, 10*(6069), 686–690.

Lytton, J. (2007). Na^+/Ca^{2+} exchangers: Three mammalian gene families control Ca^{2+} transport. *The Biochemical Journal, 406*(3), 365–382.

Nicoll, D. A., Longoni, S., & Philipson, K. D. (1990). Molecular cloning and functional expression of the cardiac sarcolemmal Na^+-Ca^{2+} exchanger. *Science, 250*, 562–565.

Paillart, C., Winkfein, R. J., Schnetkamp, P. P. M., & Korenbrot, J. I. (2007). Functional characterization and molecular cloning of the K^+-dependent Na^+/Ca^+-K^+ exchanger in intact retinal cone photoreceptors. *The Journal of General Physiology, 129*, 1–16.

Poon, S., Leach, S., Li, X. F., Tucker, J. E., Schnetkamp, P. P., & Lytton, J. (2000). Alternatively spliced isoforms of the rat eye sodium/calcium + potassium exchanger NCKX1. *American Journal of Physiology. Cell Physiology, 278*(4), C651–C660.

Prinsen, C. F. M., Szerencsei, R. T., & Schnetkamp, P. P. M. (2000). Molecular cloning and functional expression the potassium-dependent sodium-calcium exchanger from human and chicken retinal cone photoreceptors. *The Journal of Neuroscience, 20*(4), 1424–1434.

Quednau, B. D., Nicoll, D. A., & Philipson, K. D. (2004). The sodium/calcium exchanger family-SLC8. *Pflugers Archiv: European Journal of Physiology, 447*, 543–548.

Reiländer, H., Achilles, A., Friedel, U., Maul, G., Lottspeich, F., & Cook, N. J. (1992). Primary structure and functional expression of the Na/Ca,K-exchanger from bovine rod photoreceptors. *EMBO Journal, 11*, 1689–1695.

Sampath, A. P., Matthews, H. R., Cornwall, M. C., & Fain, G. L. (1998). Bleached pigment produces a maintained decrease in outer segment Ca^{2+} in salamander rods. *The Journal of General Physiology, 111*, 53–64.

Schnetkamp, P. P. M. (1979). Calcium translocation and storage in isolated intact cattle rod outer segments in darkness. *Biochimica et Biophysica Acta, 554*, 441–459.

Schnetkamp, P. P. M. (1980). Ion selectivity of the cation transport system of isolated cattle rod outer segments: Evidence of a direct communication between the rod plasma membrane and the rod disk membranes. *Biochimica et Biophysica Acta, 598*, 66–90.

Schnetkamp, P. P. M. (1986). Sodium-calcium exchange in the outer segments of bovine rod photoreceptors. *The Journal of Physiology, 373*, 25–45.

Schnetkamp, P. P. M. (1989). Na-Ca or Na-Ca-K exchange in the outer segments of vertebrate rod photoreceptors. *Progress in Biophysics and Molecular Biology, 54*, 1–29.
Schnetkamp, P. P. M. (1991). Optical measurements of Na-Ca-K exchange currents in intact outer segments isolated from bovine retinal rods. *The Journal of General Physiology, 98*, 555–573.
Schnetkamp, P. P. M. (1995a). Chelating properties of the Ca^{2+} transport site of the retinal rod Na-Ca+K exchanger: Evidence for a common Ca^{2+} and Na^{+} binding site. *Biochemistry, 34*, 7282–7287.
Schnetkamp, P. P. M. (1995b). How does the retinal rod Na-Ca+K exchanger regulate free cytosolic Ca^{2+}? *The Journal of Biological Chemistry, 270*, 13231–13239.
Schnetkamp, P. P. M. (2004). The SLC24 Na^{+}/Ca^{2+}-K^{+} exchanger family: Vision and beyond. *Pflugers Archiv: European Journal of Physiology, 447*, 683–688.
Schnetkamp, P. P. M. (2013). The SLC24 gene family of Na(+)/Ca(2)(+)-K(+) exchangers: From sight and smell to memory consolidation and skin pigmentation. *Molecular Aspects of Medicine, 34*(2–3), 455–464. http://dx.doi.org/10.1016/j.mam.2012.07.008.
Schnetkamp, P. P. M., Basu, D. K., Li, X.-B., & Szerencsei, R. T. (1991). Regulation of intracellular free Ca^{2+} concentration in the outer segments of bovine retinal rods by Na-Ca-K exchange measured with Fluo-3. II. Thermodynamic competence of transmembrane Na^{+} and K^{+} gradients and inactivation of Na^{+}-dependent Ca^{2+} extrusion. *The Journal of Biological Chemistry, 266*, 22983–22990.
Schnetkamp, P. P. M., Basu, D. K., & Szerencsei, R. T. (1989). Na-Ca exchange in the outer segments of bovine rod photoreceptors requires and transports potassium. *American Journal of Physiology. Cell Physiology, 257*, C153–C157.
Schnetkamp, P. P. M., Li, X.-B., Basu, D. K., & Szerencsei, R. T. (1991). Regulation of free cytosolic Ca^{2+} concentration in the outer segments of bovine retinal rods by Na-Ca-K exchange measured with Fluo-3. I. Efficiency of transport and interactions between cations. *The Journal of Biological Chemistry, 266*, 22975–22982.
Schnetkamp, P. P. M., & Szerencsei, R. T. (1991). Effect of potassium ions and membrane potential on the Na-Ca-K exchanger in isolated intact bovine rod outer segments. *The Journal of Biological Chemistry, 266*, 189–197.
Schnetkamp, P. P. M., & Szerencsei, R. T. (1993). Intracellular Ca^{2+} sequestration and release in intact bovine retinal rod outer segments. *The Journal of Biological Chemistry, 268*, 12449–12457.
Schnetkamp, P. P. M., Szerencsei, R. T., & Basu, D. K. (1991). Unidirectional Na^{+}, Ca^{2+} and K^{+} fluxes through the bovine rod outer segment Na-Ca-K exchanger. *The Journal of Biological Chemistry, 266*, 198–206.
Schnetkamp, P. P. M., Tucker, J. E., & Szerencsei, R. T. (1995). Ca^{2+} influx into bovine retinal rod outer segments mediated by Na-Ca+K exchange. *American Journal of Physiology Cell Physiology, 269*, c1153–c1159.
Schwarzer, A., Kim, T. S. Y., Hagen, V., Molday, R. S., & Bauer, P. J. (1997). The Na/Ca-K exchanger of rod photoreceptor exists as dimer in the plasma membrane. *Biochemistry, 36*(44), 13667–13676.
Schwarzer, A., Schauf, H., & Bauer, P. J. (2000). Binding of the cGMP-gated channel to the Na/Ca-K exchanger in rod photoreceptors. *The Journal of Biological Chemistry, 275*(18), 13448–13454.
Sheng, J.-Z., Prinsen, C. F. M., Clark, R. B., Giles, W. R., & Schnetkamp, P. P. M. (2000). Na^{+}-Ca^{2+}-K^{+} currents measured in insect cells transfected with the retinal cone or rod Na^{+}-Ca^{2+}-K^{+} exchanger cDNA. *Biophysical Journal, 79*, 1945–1953.
Stephan, A. B., Tobochnik, S., Dibattista, M., Wall, C. M., Reisert, J., & Zhao, H. (2012). The Na(+)/Ca(2+) exchanger NCKX4 governs termination and adaptation of the mammalian olfactory response. *Nature Neuroscience, 15*, 131–137.

Stokowski, R. P., Pant, P. V., Dadd, T., Fereday, A., Hinds, D. A., Jarman, C., et al. (2007). A genomewide association study of skin pigmentation in a South Asian population. *American Journal of Human Genetics*, *81*(6), 1119–1132.

Szerencsei, R. T., Kinjo, T. G., & Schnetkamp, P. P. (2013). The topology of the C-terminal sections of the NCX1 Na (+) /Ca (2+) exchanger and the NCKX2 Na (+) /Ca (2+) -K (+) exchanger. *Channels (Austin, Tex.)*, 7(2), 109–114. http://dx.doi.org/10.4161/chan.23898.

Szerencsei, R. T., Prinsen, C. F. M., & Schnetkamp, P. P. M. (2001). The stoichiometry of the retinal cone Na/Ca-K exchanger heterologously expressed in insect cells: Comparison with the bovine heart Na/Ca exchanger. *Biochemistry*, *40*, 6009–6015.

Szerencsei, R. T., Tucker, J. E., Cooper, C. B., Winkfein, R. J., Farrell, P. J., Iatrou, K., et al. (2000). Minimal domain requirement for cation transport by the potassium-dependent Na/Ca-K exchanger: Comparison with an NCKX paralog from *Caenorhabditis elegans*. *The Journal of Biological Chemistry*, *275*(1), 669–676.

Tsoi, M., Rhee, K.-H., Bungard, D., Li, X.-B., Lee, S.-L., Auer, R. N., et al. (1998). Molecular cloning of a novel potassium-dependent sodium-calcium exchanger from rat brain. *The Journal of Biological Chemistry*, *273*(7), 4155–4162.

Visser, F., Valsecchi, V., Annunziato, L., & Lytton, J. (2007). Exchangers NCKX2, NCKX3, and NCKX4: Identification of Thr-551 as a key residue in defining the apparent $K(^+)$ affinity of NCKX2. *The Journal of Biological Chemistry*, *282*(7), 4453–4462.

Winkfein, R. J., Pearson, B., Ward, R., Szerencsei, R. T., Colley, N. J., & Schnetkamp, P. P. M. (2004). Molecular characterization, functional expression and tissue distribution of a second NCKX Na^+/Ca^{2+}-K^+ exchanger from *Drosophila*. *Cell Calcium*, *36*, 147–155.

Winkfein, R. J., Szerencsei, R. T., Kinjo, T. G., Kang, K.-J., Perizzolo, M., Eisner, L., et al. (2003). Scanning mutagenesis of the alpha repeats and of the transmembrane acidic residues of the human retinal cone Na/Ca-K exchanger. *Biochemistry*, *42*(2), 543–552.

Yau, K.-W., & Nakatani, K. (1984). Electrogenic Na-Ca exchange in retinal rod outer segment. *Nature*, *311*, 661–663.

CHAPTER EIGHT

Antiporters of the Mitochondrial Carrier Family

Magnus Monné[*,†], **Ferdinando Palmieri**[*,1]

[*]Department of Biosciences, Biotechnology and Biopharmaceutics, Laboratory of Biochemistry and Molecular Biology, University of Bari, Bari, Italy
[†]Department of Sciences, University of Basilicata, Potenza, Italy
[1]Corresponding author: e-mail address: ferdinando.palmieri@uniba.it

Contents

Abstract

The eukaryotic transport protein family SLC25 consists of mitochondrial carriers (MCs) that are recognized on the sequence level by a threefold repeated and conserved signature motif. The majority of MCs characterized so far catalyzes strict exchanges of substrates across the mitochondrial inner membrane. The substrates are nucleotides, metabolic intermediates, and cofactors that are required in cytoplasmic and matrix metabolism. This review summarizes and discusses the current knowledge of the antiport mechanism(s) of MCs that has been deduced from determining transport characteristics and by analyzing structural, sequence, and mutagenesis data. The mode of transport varies among different MCs with respect to how the substrate translocation depends on the electrical and pH gradients across the mitochondrial inner membrane,

Current Topics in Membranes, Volume 73
ISSN 1063-5823
http://dx.doi.org/10.1016/B978-0-12-800223-0.00008-6

for example, the ADP/ATP carrier is electrogenic (electrophoretic), the GTP/GDP carrier is dependent on the pH gradient, the aspartate/glutamate carrier is dependent on both, and the oxoglutarate/malate carrier is independent of them. The structure of the bovine ADP/ATP carrier consists of a six-transmembrane α-helix bundle with a pseudo-threefold symmetry and a closed matrix gate. By using this structure as a template in homology modeling, residues engaged in substrate binding and the formation of a cytoplasmic gate in MCs have been proposed. The functional importance of the residues of the binding site, the matrix, and the cytoplasmic gates is supported by transport activities of different MCs with single point mutations. Cumulative evidence has been used to postulate a general transport mechanism for MCs.

ABBREVIATIONS

AAC ADP/ATP carrier
AGC aspartate/glutamate carrier
APC ATP-Mg^{2+}/P_i carrier
CAC carnitine/acylcarnitine carrier
CIC citrate carrier
DTC dicarboxylate–tricarboxylate carrier
GC glutamate carrier
Ggc GTP/GDP carrier
MC mitochondrial carrier
MCF mitochondrial carrier family
OGC oxoglutarate carrier
P_i inorganic phosphate
PiC phosphate carrier

1. INTRODUCTION

Mitochondrial carriers (MCs) transport nucleotides, amino acids, carboxylic acids, inorganic ions, and cofactors across the mitochondrial inner membrane, thereby connecting metabolic pathways of the cytoplasm with those of the mitochondrial matrix. For example, MCs provide substrates for oxidative phosphorylation, such as ADP and phosphate (P_i) transported by the ADP/ATP carrier (AAC) and phosphate carrier (PiC), respectively. MCs also transport reducing equivalents of NADH for respiratory complex I by the malate/aspartate shuttle, which employs the following two membrane transporters: the aspartate/glutamate carrier (AGC) and the oxoglutarate carrier (OGC). Moreover, transport steps catalyzed by MCs play roles in gluconeogenesis, thermogenesis, fatty acid and amino acid metabolism, as well as for mitochondrial replication, transcription, and translation. The importance of MCs in intermediary metabolism is illustrated by the fact that severe diseases are caused by mutations in the genes of certain MCs (Lindhurst et al., 2006; Palmieri, 2008, 2013).

MCs are encoded by the nuclear genes of the SLC25 family (Palmieri, 2013), translated in the cytosol, targeted to mitochondria with the help of chaperones (Zara, Ferramosca, Palmisano, Palmieri, & Rassow, 2003; Zara, Ferramosca, Robitaille-Foucher, Palmieri, & Young, 2009; Zara, Palmieri, Mahlke, & Pfanner, 1992), and inserted into the inner membrane with the N- and C-termini in the intermembrane space (Bisaccia, Capobianco, Brandolin, & Palmieri, 1994; Capobianco, Bisaccia, Michel, Sluse, & Palmieri, 1995; Capobianco, Brandolin, & Palmieri, 1991; Palmieri, Bisaccia et al., 1993) by a specific insertion machinery (Endres, Neupert, & Brunner, 1999). Although the majority of the mitochondrial carrier protein family (MCF) members are found in mitochondria, a few of them have been localized to peroxisomes and chloroplasts (Agrimi, Russo, Pierri, & Palmieri, 2012; Agrimi, Russo, Scarcia, & Palmieri, 2012; Palmieri et al., 2009; Palmieri, Rottensteiner et al., 2001).

MCs are characterized by a tripartite structure consisting of three tandemly repeated domains of about 100 amino acids (Saraste & Walker, 1982). Each domain forms two hydrophobic transmembrane segments with connecting loops, and contains a signature sequence motif PX[D/E]XX[K/R]X[K/R] (20–30 residues) [D/E]GXXXX[W/Y/F][K/R]G (PROSITE PS50920, PFAM PF00153, and IPR00193) (Palmieri, 1994, 2004). This motif has been used to identify MCs in eukaryotic genomic sequences; 53 MCF members are found in man, 35 in yeast, and 58 in *Arabidopsis thaliana* (Palmieri, Palmieri, Runswick, & Walker, 1996; Palmieri & Pierri, 2010a; Picault, Hodges, Palmieri, & Palmieri, 2004). About half of these have been functionally characterized, that is, their substrates and possible physiological roles have been identified by direct transport assays (Palmieri, 2004, 2013; Palmieri, Agrimi et al., 2006; Palmieri, Runswick, Fiermonte, Walker, & Palmieri, 2000). Currently, MCs can be divided into four major classes or groups depending on the type of substrates they transport (nucleotides/dinucleotides, carboxylic acids, amino acids, and other substrates) and into subfamilies containing all homologues that transport the same substrate (Palmieri, Pierri, De Grassi, Nunes-Nesi, & Fernie, 2011).

Only the atomic structures of the carboxyatractyloside-inhibited bovine AAC1 and yeast AAC2 and AAC3 have been determined by X-ray crystallography (Pebay-Peyroula et al., 2003: Ruprecht et al., 2014).

These structures have pseudo-threefold symmetry and consist of six-transmembrane α-helices (H1-H6) forming a basket-like structure (Fig. 8.1) closed toward the matrix side by a salt-bridge network formed by the two conserved charged residues that follow the proline in the signature motif. The salt-bridge network is accomplished by the close proximity of the lower part of the odd-numbered α-helices owing to the kink of the α-helices at the level of the prolines of the motifs. On the matrix side, the odd- and even-numbered transmembrane α-helices are connected by small α-helices (h_{12}, h_{34}, h_{56}) that are terminated by the first glycine in the second part of the signature motif. All MCs are thought to have the same basic fold as represented by the structure of bovine AAC1.

Many of the MCF members have been shown experimentally to be antiporters *in vitro* (Table 8.1) (Krämer & Palmieri, 1989; Palmieri, 2013; Palmieri, Lasorsa et al., 2000). When reconstituted into liposomes, these MCs do not transport externally added radioactive substrate unless a counter substrate is loaded inside the proteoliposome. The substrate gradient with a high concentration of cold substrate internally drives the exchange. However, although MCs are considered antiporters based on *in vitro* transport

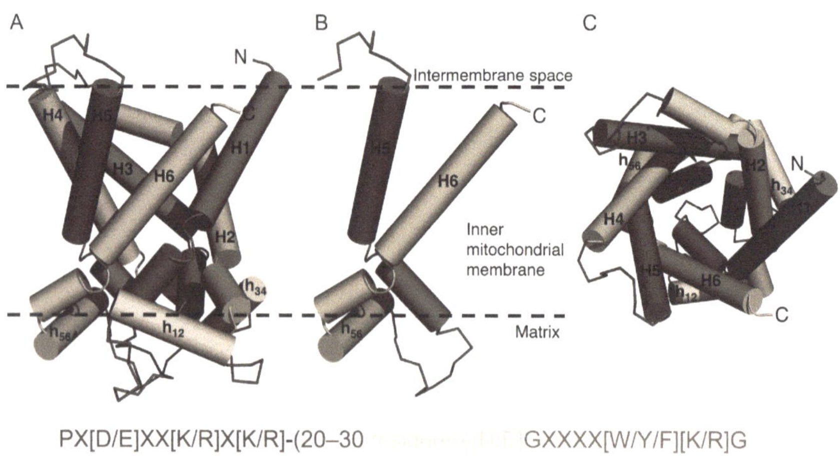

Figure 8.1 Structure of the bovine AAC1. The odd- and even-numbered α-helices are colored in magenta and green, respectively; loops are in red and the matrix α-helices are in cyan. The signature motif of the MCF is shown at the bottom of the figure. (A) Viewed from the lateral membrane side. (B) The last 100-residue repeat. The kinks in the odd-numbered α-helices are induced by the proline in the signature motif. The even-numbered α-helices initiate with the first glycine in the signature motif, and the kinks are induced by the second glycine. (C) View into the central cavity from the intermembrane space side. (See the color plate.)

Table 8.1 MCF members that are typical antiporters

Protein name	Human gene name	*S. saccharomyces* gene name	Main cytosolic substrate	Main matrix substrate	Reference
MCs transporting (deoxy)nucleotides					
AAC1	SLC25A4	ymr056c	ADP	ATP	Klingenberg (2008), Dolce, Scarcia, Iacopetta, and Palmieri (2005)
AAC2	SLC25A5	ybl030c			
AAC3	SLC25A6	ybr085w			
AAC4	SLC25A31				
APC1	SLC25A24	ynl083w	ATP-Mg^{2+}	P_i	Fiermonte et al. (2004)
APC2	SLC25A23				
RIM2	SLC25A33	ybr192w	Py(d)NTP	Py(d)NMP	Floyd et al. (2007), Marobbio, Di Noia, and Palmieri (2006)
PNC1					
GGC1		ydl198c	GTP	GDP	Vozza, Blanco, Palmieri, and Palmieri (2004)
NDT1		yil006w	NAD^+	(d)AMP/(d)GMP	Palmieri et al. (2009), Todisco, Agrimi, Castegna, and Palmieri (2006)
NDT2		yel006w			
CoA carrier	SLC25A42		CoA	(d)NTP, adenosine 3′,5′-diphosphate	Fiermonte, Paradies, Todisco, Marobbio, and Palmieri (2009)
MCs transporting carboxylic acid metabolites					
CIC	SLC25A1	ybr291c	Citrate	Malate	Kaplan, Mayor, Johnston, and Oliveira (1990)

Continued

Table 8.1 MCF members that are typical antiporters—cont'd

Protein name	Human gene name	*S. saccharomyces* gene name	Main cytosolic substrate	Main matrix substrate	Reference
YHM2		ymr241w	α-Ketoglutarate	Citrate	Castegna et al. (2010)
OGC	SLC25A11		Malate (α-ketoglutarate)	α-Ketoglutarate (malate)	Fiermonte, Walker, and Palmieri (1993)
ODC1	SLC25A21	ypl134c	Oxoadipate	α-Ketoglutarate	Fiermonte et al. (2001), Palmieri, Agrimi et al. (2001)
ODC2		yor222w			
DIC	SLC25A10	ylr348c	Malate, P_i	Malate, P_i	Fiermonte, Palmieri et al. (1998), Palmieri et al. (1996, 2008)
SFC1		ygr095w	Succinate	Fumarate	Palmieri, Lasorsa et al. (1997)
OAC1		ykl120w	Oxaloacetate	Sulfate/α-isopropyl-malate	Marobbio, Giannuzzi, Paradies, Pierri, and Palmieri (2008), Palmieri, Vozza, Agrimi et al. (1999)
DTC (in plant)			Oxaloacetate	Citrate, oxaloacetate, α-ketoglutarate	Picault, Palmieri, Pisano, Hodges, and Palmieri (2002)
MCs transporting amino acids or their derivatives					
AGC1	SLC25A12	ypr021c	Glutamate	Aspartate	Cavero et al. (2003), Palmieri, Pardo et al. (2001)
AGC2	SLC25A13				
ORC1	SLC25A15	yor103c	Ornithine	Citrulline	Fiermonte et al. (2003), Hoyos et al. (2003), Palmieri, De Marco et al. (1997)
ORC2	SLC25A2		H^+ (*S. cerevisiae*)	Ornithine (*S. cerevisiae*)	
SAMC	SLC25A26	ynl003c	*S*-Adenosyl-methionine	*S*-Adenosyl-homocysteine	Agrimi et al. (2004), Bouvier et al. (2006), Marobbio, Agrimi, Lasorsa, and Palmieri (2003), Palmieri, Arrigoni, et al. (2006)

characteristics, the transport driving forces *in vivo* are more complex because other factors come into play, such as the proton motive force in different respiratory states and the steady-state concentrations of all substrates, inhibitors, and/or regulators in the matrix and intermembrane space. Another factor that makes the *in vitro* and *in vivo* transport conditions different is that the MCs are in general oriented randomly in proteoliposomes, whereas they are all oriented with the N and C termini in the intermembrane space in mitochondria.

Some MCs are not obligatory antiporters. There is a group of MCs that principally catalyzes uniport transport, such as uncoupling protein 1 (Klingenberg & Winkler, 1985) and the yeast thiamine pyrophosphate carrier (Marobbio et al., 2002), or can catalyze uniport, such as the carnitine/acylcarnitine carrier (CAC) (Iacobazzi, Naglieri, Stanley, Wanders, & Palmieri, 1998; Indiveri et al., 2011; Indiveri, Tonazzi, & Palmieri, 1990, 1991; Indiveri, Tonazzi, Prezioso, & Palmieri, 1991; Palmieri, Lasorsa et al., 1999). Moreover, other MCs are symporters, including the PiC (Dolce, Iacobazzi, Palmieri, & Walker, 1994; Fiermonte, Dolce, & Palmieri, 1998; Kadenbach, Mende, Kolbe, Stipani, & Palmieri, 1982; Runswick, Powell, Nyren, & Walker, 1987) and the glutamate carriers GC1 and GC2 (Fiermonte et al., 2002). In this review, we focus on the MCs that function as antiporters as defined by *in vitro* experiments, and discuss their mode of transport and mechanism.

2. TRANSPORT MODE OF THE MCF ANTIPORTERS

In this section, some MCF members that most likely function as antiporters *in vivo* are described. They exemplify four different transport modes with respect to their dependency on the proton motive force. Transport activities across the mitochondrial inner membrane may be electroneutral or electrophoretic (electrogenic) (i.e., independent or dependent on $\Delta\psi$, respectively), as well as dependent or independent on ΔpH.

2.1. The ADP/ATP carrier

The mitochondrial matrix is the site of oxidative phosphorylation, where ATP is produced from ADP and P_i by ATP synthase in a mechanism involving the respiratory chain complexes and the proton motive force. The matrix-produced ATP is exchanged for cytosolic ADP by the AACs (also called adenine nucleotide translocators) at a ratio of 1:1 (Klingenberg,

2008; Pfaff & Klingenberg, 1968; Pfaff, Klingenberg, & Heldt, 1965). Cytosolic ATP is consumed by a large number of processes and the products ADP and P_i are recycled to the matrix by the AACs and the PiC, respectively. The transport step of the latter carrier costs one proton pumped by the respiratory chain complexes per ATP synthesized (Watt, Montgomery, Runswick, Leslie, & Walker, 2010). The AACs have a narrow substrate specificity because they transport only ADP and ATP. These transporters are inhibited by the specific inhibitors atractyloside and bongkrekic acid (Klingenberg & Buchholz, 1973). The four human AACs have different tissue distributions: AAC1 is found in heart and skeletal muscle, AAC2 is found in proliferating cells, AAC3 is ubiquitous, and AAC4 is found in lung and testis (Dolce et al., 2005; Stepien, Torroni, Chung, Hodge, & Wallace, 1992). They have a K_m for ADP in the micromolar range (De Marcos Lousa, Trézéguet, Dianoux, Brandolin, & Lauquin, 2002), except for AAC4 that has a K_m approximately 10-fold higher.

The $\Delta\psi$ and ΔpH dependencies of the ADP/ATP exchange catalyzed by the mitochondrial AACs have been investigated by using two different ionophores—valinomycin and nigericin (LaNoue, Mizani, & Klingenberg, 1978; Villiers, Michejda, Block, Lauquin, & Vignais, 1979). An artificial $\Delta\psi$ can be created across the membrane of isolated mitochondria or proteoliposomes by adding KCl externally and the K^+-specific ionophore valinomycin (Fig. 8.2A). Valinomycin allows K^+ to pass across the membrane in the direction of the concentration gradient (i.e., inward), thereby creating a charge separation because the Cl^- counter ions remain on the outside. Under these conditions, if transport is dependent on $\Delta\psi$, then the exchange rate of an MC carrying unevenly charged substrates from opposite sides of the membrane should be affected. If a net positive charge is transported inward (or a net negative charge outward) by the exchange, then the transport rate is decreased by the applied $\Delta\psi$. If, on the other hand, a net positive charge is transported outward (or a net negative charge inward), then the transport rate is increased.

An artificial ΔpH gradient can be created across isolated mitochondria or proteoliposomes by the addition of KCl and nigericin on the outside (Fig. 8.2B). Nigericin allows both K^+ and H^+ to pass across the membrane by an exchange mechanism. Because $[K^+]$ is high on the outside and there is a concentration gradient across the membrane, nigericin will in fact exchange external K^+ for internal H^+. In this way, no $\Delta\psi$ is created; rather, a ΔpH with higher $[H^+]$ on the outside is created by the transfer of protons outward.

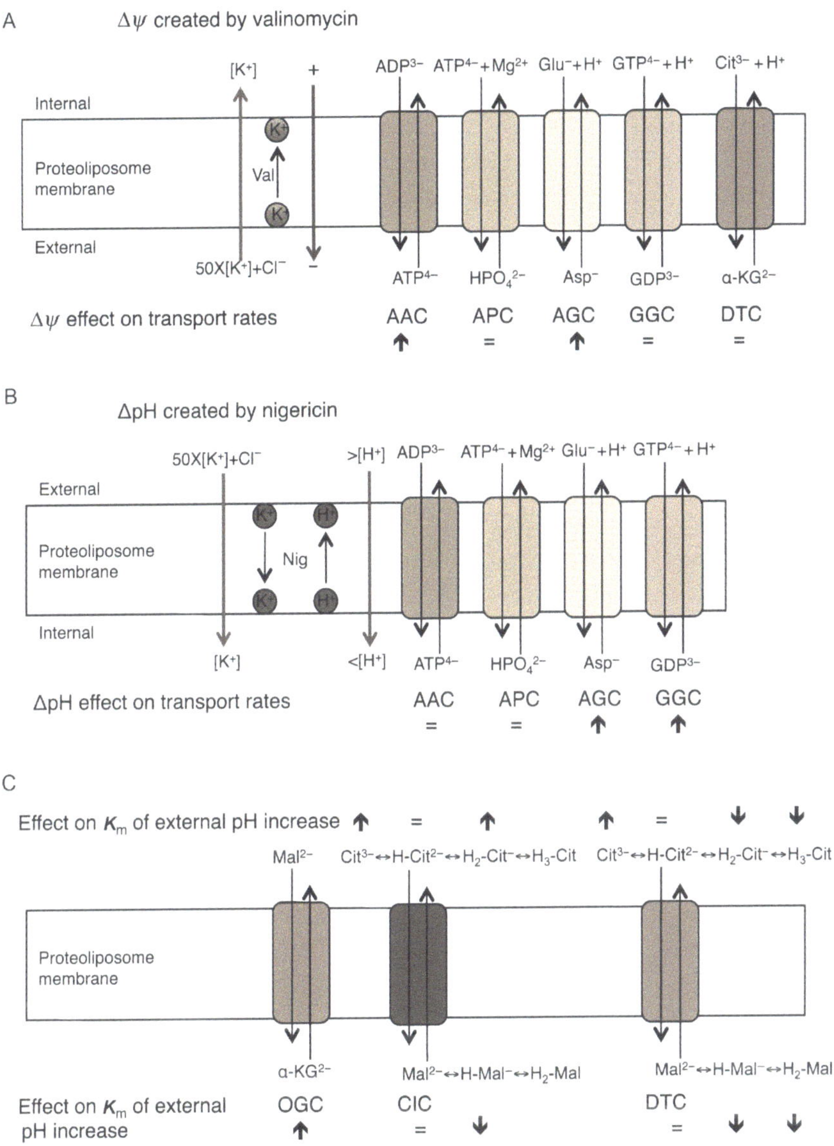

Figure 8.2 Use of valinomycin and nigericin to generate $\Delta\psi$ and ΔpH in proteoliposome transport assays. (A) External addition of KCl and valinomycin (Val, blue filled circles) allows K^+ to pass across the proteoliposome membrane inwardly (blue arrow) while Cl^- remains outside. In this way, the charges are separated across the membrane resulting in a positive net charge internally and a net negative charge externally (green arrow). The effects of $\Delta\psi$ on the transport rates of some antiporters are indicated by equal signs (unaffected) and arrows pointing up (increase). (B) A pH gradient across the proteoliposome membrane can be created by transforming the K^+

(Continued)

Under these experimental conditions, the ADP/ATP exchange catalyzed by the AACs was shown to be dependent on the $\Delta\psi$, but not on the ΔpH *in vitro* (Dolce et al., 2005; LaNoue et al., 1978; Villiers et al., 1979). Therefore, the transport mechanism is electrophoretic, that is, the net negative charge that is transferred from the matrix to the cytosol in the ADP^{3-}/ATP^{4-} exchange moves down the transmembrane charge gradient (Fig. 8.2). These conclusions from the ionophore experiments were confirmed by measuring electrical currents that arise upon ADP/ATP exchange by AAC in reconstituted vesicles (Brustovetsky, Becker, Klingenberg, & Bamberg, 1996; Gropp et al., 1999). These findings suggest that a major driving force for the exchange of cytosolic ADP^{3-} for matrix ATP^{4-} *in vivo* comes from the $\Delta\psi$, which is high in energized animal cells. However, the transport rates *in vivo* also depend on the cytosolic and matrix [ATP] and [ADP]. Active cells are thought to have ATP and ADP concentrations of about 5 and 1 mM, respectively, and the [ATP]/[ADP] ratio fluctuates only slightly despite extreme changes in the energy demand. The AACs work constantly toward equilibrating steady-state concentrations of ADP and ATP between the matrix and the cytosol. The effective ATP concentration could be influenced by Mg^{2+} because of its ability to complex ATP, and the Mg^{2+} concentration might be different inside the matrix with respect to the cytosol.

2.2. The ATP-Mg^{2+}/P_i carrier

The matrix concentration of adenine nucleotides can change upon increased demand for ATP-consuming processes, such as gluconeogenesis from lactate and mitochondrial transcription. Under these conditions, mitochondria require net transport of adenine nucleotides across the inner membrane besides the exchange of mitochondrially produced ATP for cytosolic ADP catalyzed by the AACs. This net transport is catalyzed by the

Figure 8.2—Cont'd gradient (blue arrow) into a H^+ gradient (red arrow) by adding nigericin (Nig, red filled circle). K^+ is preferred by the ionophore on the outside because the relative $[K^+]/[H^+]$ is high, while more H^+ is preferred on the inside because the relative $[K^+]/[H^+]$ is lower. The effects of ΔpH on the transport rates of some antiporters are indicated by equal signs (unaffected) and arrows pointing up (increase). Note that the orientation of the proteoliposome membrane is switched compared to that in panel (A). (C) The effects of changing the external pH on the apparent K_m values of the OGC, CIC, and DTC are indicated by equal signs (unaffected), or arrows pointing up (increase) or down (decrease). (See the color plate.)

ATP-Mg^{2+}/P_i carrier (APC), which exchanges ATP-Mg^{2+} for P_i (Aprille, 1993; Fiermonte et al., 2004), and by the PiC, which recycles back the P_i counter-exchanged with the adenine nucleotides. In man there are four APC isoforms: APC1 (SLC25A24), which is exclusively expressed in testis; APC2 (SLC25A23) and APC3 (SLC25A25), which are almost ubiquitously expressed, although the former is preferentially expressed in kidney, liver, skeletal muscle, and heart, and the latter in lung and brain (Fiermonte et al., 2004); APC4 (SLC25A41), which is expressed in brain, testis, and liver (Traba, Satrústegui, & del Arco, 2009). All human APCs have been localized to mitochondria. APC1-3 have soluble N-terminal EF-hand Ca^{2+}-binding domains, which are found in this subfamily of the MCF and in AGC1 and AGC2; they are thought to regulate the transport activity of the C-terminal domain depending on the cytosolic Ca^{2+} concentration. Therefore, these carriers are also called Ca^{2+}-dependent MCs.

The substrates transported by human APC1 and APC2 were identified by expressing the proteins in *Escherichia coli* and reconstituting them from inclusion bodies into liposomes that were used in transport experiments with radioactive substrates (Fiermonte et al., 2004). APC1 and APC2 are strict antiporters that can exchange the substrates ATP, ADP, AMP, and P_i, and to a much lesser extent dATP, dADP, dAMP, 3′-AMP, 3′,5′-ADP, and pyrophosphate. The transport of ATP is enhanced in the presence of Mg^{2+}. The K_m values of APC1 and APC2 are very similar for the substrates ATP, ADP, AMP, P_i, and ATP-Mg^{2+} (0.3, 0.4, 1, 1.5, and 0.2 mM, respectively). The values indicate that APC1 and APC2 have the highest affinity for ATP-Mg^{2+}.

The transport mode of APC1 and APC2 was demonstrated to be electroneutral for the ATP-Mg^{2+}/P_i exchange (Fiermonte et al., 2004). The initial transport rates of ATP-Mg^{2+} in exchange for ATP-Mg^{2+}, ATP, and P_i in APC-reconstituted liposomes were unaffected when a KCl/valinomycin-generated K^+ gradient was applied across the membrane, indicating that the substrate exchange is not electrophoretic and therefore independent of the $\Delta\psi$. With the pH gradient across the APC proteoliposome membranes generated by KCl and nigericin, the transport rates for the $ATP-Mg^{2+}{}_{ex}/ATP_{in}$ exchange decreased and for the $ATP_{ex}/ATP-Mg^{2+}{}_{in}$ and ATP_{ex}/P_{in} exchanges increased; substrate homoexchanges and the $ATP-Mg^{2+}{}_{ex}/P_{in}$ rates did not change. These results suggest that the inward-directed H^+ gradient, which exists in respiratory active mitochondria, facilitates the electrically unbalanced heteroexchanges of ATP_{ex}/P_{in} and $ATP_{ex}/ATP - Mg^{2+}{}_{in}$ by protons compensating and neutralizing the charge difference. The findings also suggest that the exchange of the major substrates, $ATP^{4-} + Mg^{2+}$ and $HPO_4{}^{2-}$, transported by the APCs (in the direction dependent on the increased or decreased mitochondrial requirement for ATP) is electroneutral (Fig. 8.2), and therefore independent not only of the $\Delta\psi$ but also of the ΔpH.

2.3. The aspartate/glutamate carrier

In glycolysis and other cytoplasmic catabolic pathways, NAD^+ is used as an electron acceptor forming NADH that needs to be reoxidized and recycled as a reductant by complex I in the mitochondrial matrix or by the glycerol-3-phosphate shuttle in the cytosol and intermembrane space in specific tissues. Because there is no MC transporting NADH directly to provide complex I with its substrate, the reducing equivalent of NADH is transported through the malate/aspartate shuttle that involves aspartate/glutamate exchange catalyzed by AGC and α-ketoglutarate/malate exchange catalyzed by OGC. The malate/aspartate shuttle has been reconstituted with purified carriers (Indiveri, Krämer, & Palmieri, 1987). The two human carriers AGC1 (also called aralar1) and AGC2 (also called citrin) have four EF-hand Ca^{2+}-binding motifs in their soluble N-terminal domains (Palmieri, Pardo, et al., 2001). AGC1 is expressed in heart, skeletal muscle, and brain, while AGC2 is expressed in many tissues, particularly the liver (Iijima et al., 2001).

Human AGC1 and AGC2 have been overexpressed in *E. coli*, purified from inclusion bodies and reconstituted into liposomes for functional characterization (Palmieri, Pardo, et al., 2001). Both carriers specifically transport aspartate, glutamate, and cysteinesulfinate by an exchange mechanism, and are regulated by Ca^{2+} binding to the intermembrane space EF-hand domains that activate transport. The K_m values of both AGC1 and AGC2 are about 0.05 and 0.2 mM for aspartate and glutamate uptake, respectively, while the V_{max} is about fourfold higher for AGC2 (about 200 μmol/min/g protein) than AGC1.

The transport mode of AGC was demonstrated to be electrogenic (Palmieri, Pardo, et al., 2001). The transport rates of homoexchanges of glutamate and aspartate in AGC1- and AGC2-reconstituted liposomes with an inward-directed K^+ gradient in the presence of valinomycin were unaffected, while the transport rate of glutamate$_{ex}$/aspartate$_{in}$ exchange was decreased and that of aspartate$_{ex}$/glutamate$_{in}$ was increased. These results indicate that the heteroexchange is electrophoretic and dependent on the $\Delta\psi$. Therefore, *in vivo* AGC1 and AGC2 are likely to catalyze the exchange of cytosolic glutamate + H^+ for matrix aspartate (Fig. 8.2).

2.4. The oxoglutarate carrier

The other membrane component of the malate/aspartate shuttle, apart from AGC, is OGC, which exchanges α-ketoglutarate for malate (Bisaccia, Indiveri, & Palmieri, 1985; Iacobazzi, Palmieri, Runswick, & Walker, 1992; Monné, Miniero, Iacobazzi, Bisaccia, & Fiermonte, 2013; Runswick, Walker, Bisaccia, Iacobazzi, & Palmieri, 1990). The substrate specificity of OGC was determined upon its overexpression in *E. coli* and reconstitution into liposomes followed by transport experiments. OGC also transports malonate, maleate, succinate, oxaloacetate, and 2-oxoadipate (Fiermonte et al., 1993). It has a K_m of about 55 μM for α-ketoglutarate and about 0.2 mM for malate (Indiveri, Palmieri, Bisaccia, & Krämer, 1987; Palmieri, Quagliariello, & Klingenberger, 1972). OGC is involved in the citrate (isocitrate)/oxoglutarate shuttle, the pyruvate/citrate cycling pathway, nitrogen metabolism, and gluconeogenesis from lactate, as well as the malate/aspartate shuttle (Palmieri, 2004). Human OGC is expressed at abundant levels in heart and skeletal muscle, and at moderate levels in brain, placenta, liver, kidney, and pancreas (Huizing et al., 1998).

All substrates of OGC possess two negative charges at physiological pH, and the α-ketoglutarate/malate exchange catalyzed by OGC is electroneutral and independent of pH (Indiveri, Palmieri et al., 1987; Palmieri, Quagliariello, & Klingenberger, 1972). It was shown that OGC-catalyzed α-ketoglutarate homoexchange rates were not pH sensitive. The apparent K_m of OGC for α-ketoglutarate was increased upon raising pH while the V_{max} remained constant, indicating that the substrate binding is affected rather than the transport rate. Thus, OGC transport is independent of ΔpH and $\Delta\psi$ (Fig. 8.2C).

2.5. The citrate carrier

Fatty acid biosynthesis, gluconeogenesis, and the citrate (isocitrate)/oxoglutarate shuttle require the transport of citrate across the mitochondrial inner membrane that is catalyzed by the citrate carrier (CIC) in exchange for malate (Palmieri, Stipani, Quagliariello, & Klingenberg, 1972). Rat CIC has a K_m for malate and citrate of 0.13 and 0.76 mM, respectively (Bisaccia, De Palma, Prezioso, & Palmieri, 1990). The substrate specificity of CIC also includes phosphoenolpyruvate, *cis*-aconitase, and isocitrate, but not *trans*-aconitate or α-ketoglutarate (Bisaccia, De Palma, & Palmieri, 1989; Kaplan, Mayor, & Gremse, 1995; Kaplan et al., 1990). Human CIC is expressed in liver, kidney, pancreas, and in other tissues to a lesser extent (Huizing et al., 1998).

The citrate transport catalyzed by CIC was suggested to be dependent on ΔpH because CIC purified from rat liver mitochondria and reconstituted in liposomes displayed an uptake rate for citrate that was considerably higher at pH 7 than at pH 8 (Bisaccia, De Palma, Dierks, Krämer, & Palmieri, 1993). This pH dependency could be explained by an effect either on the carrier or on the substrate. The K_m for citrate transport was measured at a pH range from 7.0 to 7.8, and the K_m values for the differently charged citrate species were calculated based on the pK_a values for citrate (Bisaccia et al., 1993). The analysis showed that the K_m for citrate^{2-} remained constant in the pH range unlike the other species, suggesting that the effect was linked to protonation of the substrate. The same approach was used for CIC transport of malate and the results showed a constant K_m for malate^{2-}. Based on these results, it was therefore suggested that CIC catalyzes an electroneutral exchange of H^+ + citrate^{3-} for malate^{2-}, an exchange that is therefore dependent on ΔpH (Fig. 8.2C).

2.6. The plant dicarboxylate–tricarboxylate carrier

Fatty acid synthesis, nitrogen assimilation, and the shuttling of reducing equivalents in plants are processes thought to involve the dicarboxylate–tricarboxylate carrier (DTC) that shares a substantial sequence identity with OGC but displays a broader substrate specificity (Picault et al., 2002). DTC from *A. thaliana* and *Nicotiana tabacum* was expressed in *E. coli*, purified, reconstituted into liposomes, and shown to transport both dicarboxylates (such as malate, oxaloacetate, oxoglutarate, and maleate) and tricarboxylates (such as citrate, isocitrate, *cis*-aconitate, and *trans*-aconitate). The K_m of DTC for α-ketoglutarate, malate, and citrate is in the micromolar range. The expression of DTC is found in all plant tissues.

DTC transport of citrate and α-ketoglutarate is pH dependent. The K_m values for the different charged species of citrate and malate were calculated from kinetic transport experiments at different pH values, demonstrating that the K_m was constant for the species with two negative charges (Picault et al., 2002). This finding suggests that H^+ + citrate^{3-} and malate^{2-} are the main substrates for DTC (Fig. 8.2C). Furthermore, the influence of the membrane potential on DTC transport in proteoliposomes was investigated by applying a K^+ diffusion potential in the presence of valinomycin (Fig. 8.2A). The results demonstrated that the DTC-mediated citrate/oxoglutarate exchange was independent of the $\Delta\psi$. Therefore, DTC was suggested to catalyze the electroneutral, ΔpH-dependent 1:1 substrate exchange of H^+ + citrate^{3-}, α-ketoglutarate^{2-}, or malate^{2-}.

2.7. The GTP/GDP carrier (Ggc1p)

Mitochondrial GTP is required for fundamental mitochondrial processes, such as protein and RNA synthesis, as well as iron homeostasis. In many organisms, succinyl-CoA synthetase of the citric acid cycle produces GTP from GDP in mitochondria. An alternative way to generate mitochondrial GTP is by nucleoside diphosphate kinase that catalyzes the transfer of the γ-phosphate of ATP to a GDP molecule. However, in *Saccharomyces cerevisiae*, succinyl-CoA synthetase uses ADP instead of GDP (Przybyla-Zawislak, Dennis, Zakharkin, & McCammon, 1998), and nucleoside diphosphate kinase is localized in the intermembrane space (Amutha & Pain, 2003). In this species, the mitochondrial GTP/GDP carrier Ggc1p (also known as Shm1p and Yhm1p, encoded by ydl198c) catalyzes the exchange of cytosolic GTP for matrix GDP (Vozza et al., 2004).

The biochemical identification and characterization of Ggc1p was performed by expressing the protein in *E. coli* and reconstituting it from inclusion bodies into liposomes that were used in transport experiments (Vozza et al., 2004). Ggc1p is a strict antiporter that exchanges the substrates GTP, GDP, dGTP, dGDP, ITP, and IDP. ITP and IDP are structurally similar to GTP and GDP, the difference being that the former two compounds lack an amino group on the guanosine base. However, according to competitive inhibition experiments, the affinity of Ggc1p for GTP and GDP is about 10-fold higher than for dGTP and dGDP, and 100-fold higher than for IDP and ITP. This finding is also reflected by the K_m values determined for GTP, GDP, and dGTP that are 1.2, 4.5, and 15.9 μM, respectively.

The transport mode of Ggc1p was demonstrated to be electroneutral (Vozza et al., 2004). The transport rates of homo- and heteroexchanges of GDP and GTP in Ggc1p-reconstituted liposomes were unaffected by valinomycin and an inward-directed K^+ gradient, indicating that the GDP/GTP heteroexchange is not electrophoretic and therefore independent of the $\Delta\psi$. When a pH gradient was generated across Ggc1p proteoliposome membranes by adding KCl and nigericin, the transport rate for the GDP_{ex}/GTP_{in} exchange decreased, and that for the GTP_{ex}/GDP_{in} exchange increased; the homoexchange rates were unchanged. These results suggest that the inward-directed H^+ gradient, created by the respiratory chain complexes in mitochondria, facilitates the import of $GTP^{4-}+H^+$ in electroneutral exchange with GDP^{3-}, and the Ggc1p-mediated transport is dependent on ΔpH (Fig. 8.2).

Further support for the conclusion that GTP and GDP are the main substrates for Ggc1p *in vivo* stems from studies of *GGC1Δ* yeast cells where mitochondria have increased levels of GDP and decreased levels of GTP (Vozza et al., 2004). It has been demonstrated that the lack of Ggc1p interferes with cellular iron homeostasis (Lesuisse, Lyver, Knight, & Dancis, 2004)—a defect that can be complemented by the expression of the human mitochondrial nucleoside diphosphate kinase in mitochondria of *GGC1Δ* yeast cells (Gordon, Lyver, Lesuisse, Dancis, & Pain, 2006). That Ggc1p also transports dGDP and dGTP *in vivo* is supported by the fact that it acts as a suppressor of a mutation leading to loss of mitochondrial DNA, and thereby plays a role in maintaining the mitochondrial genome (Kao, Megraw, & Chae, 1996).

3. TRANSPORT MECHANISM OF MCF ANTIPORTERS

3.1. Structure of MCs

For a long time, MCs were thought to be homodimeric. This hypothesis was based on quantifications of how many atractyloside molecules bind to each AAC protein, their apparent size when migrating on gel filtration columns or in blue native gels, and kinetic experiments (Aquila, Eiermann, Babel, & Klingenberg, 1978; Bisaccia et al., 1996; Block & Vignais, 1984; Block, Zaccaï, Lauquin, & Vignais, 1982; Brandolin et al., 1980; Hackenberg & Klingenberg, 1980; Lin, Hackenberg, & Klingenberg, 1980; Palmieri, Indiveri, Bisaccia, & Krämer, 1993; Palmieri, Vozza, Hönlinger et al., 1999; Palmisano et al., 1998; Riccio, Aquila, & Klingenberg, 1975; Schroers, Burkovski, Wohlrab, & Krämer, 1998). However, these results have been re-interpreted, and a large body of evidence is now pointing toward MCs being monomers (Crichton, Harding, Ruprecht, Lee, & Kunji, 2013; Kunji & Crichton, 2010). The X-ray crystallographic structure of bovine AAC1 showed that a single polypeptide chain could form a protein fold by itself, and the crystal packing did not reveal any hint of dimer formation (Pebay-Peyroula et al., 2003). Yeast AAC2 is monomeric in detergent solution and during purification (Bamber, Harding, Butler, &

Kunji, 2006; Bamber, Slotboom, & Kunji, 2007). No inhibitory effect of hypothetical dimer formation could be seen when varying concentrations of mixed active and inactive mutant forms of purified AAC2 were reconstituted into liposomes and transport assays were performed (Bamber, Harding, Monné, Slotboom, & Kunji, 2007). Furthermore, the transport activities of all the mutants used in the complete cysteine-scanning mutagenesis of OGC have revealed that no functionally crucial or conserved dimer interface is found on the bilayer-exposed surface of the OGC structural homology model (Miniero et al., 2011).

3.2. Matrix and cytoplasmic gates in MCs

The inhibitors atractyloside and bongkrekic acid trap the AACs in two different conformations known as the c- and m-state where the carrier is open to the cytoplasmic (intermembrane space) and matrix sides, respectively (Aquila et al., 1978; Brandolin, Dupont, & Vignais, 1985; Buchanan, Eiermann, Riccio, Aquila, & Klingenberg, 1976). The structure of bovine AAC1 revealed the conformation of the carrier inhibited by carboxy–atractyloside where the six-transmembrane α-helix bundle is closed toward the matrix side by a kink in H1, H3, and H5, and a salt-bridge network formed by the charged residues of the threefold repeated signature motif (Fig. 8.3A). These charged residues have been shown to be critical for transport function in AAC (Nelson, Felix, & Swanson, 1998), OGC (Cappello et al., 2007), and CAC (Giangregorio, Tonazzi, Console, Indiveri, & Palmieri, 2010). The AAC1 structure has therefore been seen as representing the carrier with a closed matrix gate as in the c-state.

On the even-numbered α-helices, charged residues have been found near the cytoplasmic side that could form a salt-bridge network closing the cytoplasmic gate (Fig. 8.3C) (Robinson, Overy, & Kunji, 2008). The charged residues of the cytoplasmic and matrix gates are generally conserved in the MCF and are symmetric, that is, amino acids with identical charge occupy the corresponding positions in all three carrier repeats. The residues of the proposed cytosolic gate in OGC have been shown to be functionally important (Miniero et al., 2011).

3.3. The substrate binding site in MCs

The bovine AAC1 structure is in an atractyloside-inhibited form. However, a substrate binding site was proposed based on its approximate location in the 3D structure, and on mutagenesis work that indicated residues crucial for

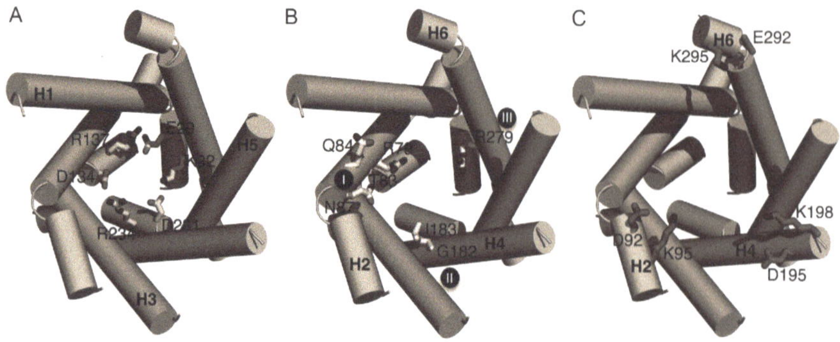

Figure 8.3 Important residues for transport in the central cavity of the bovine AAC1 structure. For simplicity, the loops and matrix α-helices have been removed. (A) The matrix gate is formed by the matrix salt-bridge network that consists of the charged residues of the first half of the signature motif on H1, H3, and H5. (B) The common substrate binding site is formed by three contact points (Roman numbered black filled circles) consisting of residues on H2, H4, and H6 above the matrix gate. (C) The hypothetical cytoplasmic gate is formed in the matrix conformation by a salt-bridge network of the charged residues on top of the three even-numbered transmembrane α-helices. (See the color plate.)

carrier function (Pebay-Peyroula et al., 2003). The definition of the substrate binding site in AAC has later been refined by using molecular dynamics simulations with ADP as a ligand in the carrier cavity (Dehez, Pebay-Peyroula, & Chipot, 2008; Falconi et al., 2006; Wang & Tajkhorshid, 2008). Despite the relatively low sequence identity among many MCs, structural homology models based on the bovine AAC1 structure can be made from reliable alignments due to the presence of the highly conserved signature motifs in all MCF members that define the beginning and the end of the even- and odd-numbered α-helices, respectively. By analyzing cavity residues that display covariation with the substrates of characterized MC subfamilies (Palmieri, Agrimi, et al., 2006 for a review), Robinson and Kunji (2006) proposed a putative common binding site (Fig. 8.3B). This binding site consists of three contact points—one on each even-numbered α-helix between the matrix and the cytoplasmic gates close to the middle of the membrane bilayer. Based on the identity of the residues in contact point II on H4, the following three major classes of substrates can be distinguished: nucleotides, carboxylates, and amino acids. The common binding site corresponds to asymmetric residues, that is, different kinds of amino acids occupy the same corresponding positions within the three tandem repeats (Robinson et al., 2008). The complete cys-scanning mutagenesis studies of OGC, as well as the site-directed mutagenesis data of other carriers, support the crucial role of the contact point residues for function (Briggs, Mincone, & Wohlrab, 1999; Cappello et al., 2006; De Lucas et al., 2008; Giangregorio et al., 2010; Heidkämper, Müller, Nelson, & Klingenberg, 1996; Ma et al., 2007; Stipani et al., 2001; Tonazzi, Console, Giangregorio, Indiveri, & Palmieri, 2012; Wohlrab, Annese, & Haefele, 2002). Experimental evidence that residues in the common binding site directly interact with the substrate has been found in yeast AAC2 and the human ornithine carrier. The mutation R294A of contact point III in AAC2 diminishes the transport rate of ATP without affecting the transport of ADP (Heidkämper et al., 1996), and the substrate specificities of the two human ornithine carriers could be swapped by exchanging the arginine and the glutamine at contact point II of the two isoforms (Monné et al., 2012). Although the common substrate binding site may be mainly responsible for substrate binding, it does not fully explain how different substrates are selected by carriers with identical contact point residues. Other asymmetric residues that are specific for certain MC subfamilies and protrude into the internal carrier cavity in proximity to the three contact points most likely play a role in fine-tuning the substrate specificity of MCs (Palmieri & Pierri, 2010b; Palmieri et al., 2011; Pierri, Palmieri, & De Grassi, 2013). Consequently, potential substrates for an MC with unknown function can be predicted by identifying the small set of residues in its sequence that defines the binding site (Palmieri et al., 2011).

3.4. General transport mechanism for MCs

By combining the data concerning the monomeric state, the evidence for a matrix and cytoplasmic gate, and the location of the substrate binding site, it is possible to suggest a simple model for the antiport transport mechanism of MCs (Palmieri & Pierri, 2010a, 2010b; Palmieri et al., 2011; Robinson et al., 2008) within the framework of the structure of bovine AAC1. This model basically represents the "single binding center-gating pore mechanism" as proposed earlier by Klingenberg (Klingenberg, 1976, 1979) in which the substrate binding site is alternatively accessible from the cytoplasmic or the matrix side. For example, the exchange of cytosolic ADP for matrix ATP catalyzed by AACs includes the following steps (i–vi):

(i) ADP binds to the substrate binding site exposed to the intermembrane space, that is, in the c-state (Fig. 8.3B).

(ii) The ADP–carrier interactions induce conformational changes that involve opening of the matrix gate and closing of the cytoplasmic gate, and result with the substrate bound in the m-state. The conformational change is driven by the binding energy of the substrate to the substrate binding site residues (Fig. 8.3B), and might involve the substrate interfering with the matrix salt-bridge network (Fig. 8.3A) and/or assisting the formation of the cytoplasmic salt-bridge network (Fig. 8.3C). The substrate–carrier interactions could therefore involve both the even-numbered α-helices (binding site and cytoplasmic salt-bridge network) and the odd-numbered α-helices (binding site and matrix salt-bridge network). Moreover, the opening of the matrix gate is likely to involve a switch of the interactions between the relayed charged residues. There may be an occluded state where the substrate is bound and both gates are partially closed, but exactly what the transition- and m-states look like is not known. However, the m-state is probably also pseudo-symmetric because the matrix and cytosolic gate residues are symmetric and, as suggested by the mechanism described here, all three contact points are involved in substrate translocation.

(iii) ADP leaves the substrate binding site of the carrier that is open toward the matrix side. The carrier cannot switch back to the c-state conformation without another bound substrate because it is a strict antiporter.

(iv) ATP binds to the matrix substrate binding site that most likely involves the same residues of the cytoplasmic substrate binding site, but which are positioned in a different conformation.

(v) The ATP-carrier interactions induce the reverse conformational changes leading to closure of the matrix gate and opening of the cytoplasmic gate.

(vi) ATP is released into the intermembrane space and leaves the carrier in the c-state ready for a new cycle of transport.

Mechanistically, MC uniporters and antiporters could work the same except for the requirement in the 1:1 exchange of a counter substrate to induce the retrograde conformational change (step iv). In the case of the uniporters, probably one of the gates (e.g., the cytoplasmic one) is destabilized and only forms transiently during substrate translocation; then it is reopened and the carrier switches directly back to the outside open state without the bound substrate (steps iv–vi). MC uniporters can also perform antiport with increased velocity *in vitro*, because in proteoliposomes, the orientation of the carriers is random and the high counter-substrate concentration drives the opening of the carriers toward the outside.

Thus, the mechanism described here suggests a single substrate binding site in MCs that switches between two or three major conformations, that is, with the binding site open toward the cytosol or the matrix, or occluded from both sides in the transition state. Mutagenesis data from many MCs may be interpreted within the context of the substrate binding site and the transport mechanism as described above (Monné, Palmieri, & Kunji, 2013). Moreover, most of the disease-causing single point mutations are found along the carrier substrate translocation path (Palmieri, 2008; Poduri et al., 2013; Wibom et al., 2009). In the structural homology models of MCs, there are two levels in the carrier cavity where glycines and prolines are symmetric and conserved in transmembrane α-helices (Palmieri & Pierri, 2010b). These residues have been suggested to work either as sites for close helix–helix interactions (Robinson et al., 2008) or as hinges required for opening and closing the cytoplasmic and matrix gates (Palmieri & Pierri, 2010b).

It should be noted that because the transport mechanism described above suggests a single substrate binding site that can bind the substrate in two different conformations, the affinities (K_m) for the same substrate on the cytosolic and matrix sides might be different, as experimentally observed with some carriers in reconstituted liposomes (Palmieri, Indiveri, Bisaccia, & Krämer, 1993). These different substrate affinities will help to maintain different steady-state concentrations of the substrate in the cytosol and in the matrix. In the case where a difference in substrate concentration is not required by the cell, the cytosolic and matrix K_m is probably close to equal.

In the case of carriers that are dependent on ΔpH, it is not completely clear if the proton binds to a residue in the binding site or protonates the substrate. The protonation is probably allowed by the pH gradient and the environment within the substrate binding site causing alteration of local pK_a values. Residues involved in proton coupling have been hypothesized (Kunji & Robinson, 2010), but remain to be established experimentally.

4. KINETIC MECHANISM OF MCs

Several MC antiporters that represent a two-substrate reaction (one substrate binding to the carrier on the outside and one on the inside of the membrane) have been studied kinetically. Most MCs seem to operate according to a sequential (simultaneous) mechanism (Bisaccia et al., 1993; Dierks, Riemer, & Krämer, 1988; Indiveri, Dierks, Krämer, & Palmieri, 1991; Indiveri, Prezioso, Dierks, Krämer, & Palmieri, 1993; Indiveri, Tonazzi, De Palma, & Palmieri, 2001; Sluse, Duyckaerts, Liebecq, & Sluse-Goffart, 1979; Sluse et al., 1991; Stappen & Krämer, 1994), suggesting that both the cytoplasmic and matrix substrates first bind to the carrier (i.e., form a ternary complex) before they are translocated. These results are consistent with the idea that MCs are homodimers with one protomer open toward the cytosol and the other toward the matrix. Upon the binding of the substrate to each protomer (on opposite sides), cooperative conformational changes are induced leading to the exchange of the substrates between the two compartments (Palmieri, Indiveri et al., 1993).

However, the formation of a ternary complex is incompatible with the general transport mechanism of monomeric MCs as described above that corresponds kinetically to a ping-pong mechanism. In this mechanism, one substrate binds to the open carrier and is then translocated across the membrane leaving the carrier open on the opposite side ready for the counter substrate to bind and be translocated. In other words, the substrates are dependent on each other for the conformational change but do not form a ternary complex with the carrier. The only MC found to follow a ping-pong mechanism is CAC (Indiveri, Tonazzi, & Palmieri, 1994)—a finding that might reflect the fact that CAC is able to catalyze uniport transport as well as exchange of substrates. Because of the apparent contradiction between the proposed transport mechanism and the kinetic data, it is possible that either the model for the transport mechanism or the interpretation for the sequential mechanism of MCs is incorrect.

5. SYNOPSIS

Many MCs have been known to be antiporters for a long time, although how they perform this function at the molecular level was unclear. The current model for the transport mechanism of MCs is based on the wealth of information now available about the MCF, including biochemical characterization of a large number of carriers, site-directed mutagenesis studies, identification of disease-causing mutants, genomic sequences, sequence analysis, and structural and functional studies. The picture of the transport mechanism that has emerged applies in part to all MCs and in part to each MC subfamily. A single similarly located binding site and a matrix and cytoplasmic gate are involved in substrate translocation of all MCs by an alternating access mechanism. Specific residues of the binding site, specific substrate–carrier interactions, regulation by N-terminal Ca^{2+}-binding domains, dependency on $\Delta\psi$ and ΔpH, as well as anti-, uni-, and symport modes of transport vary among MC subfamilies. In other words, the driving forces and energies vary, but the structural scaffold and probably also the substrate translocating conformational changes are fundamentally the same for all MCs.

Future investigations are warranted to provide further evidence for the current transport mechanism of MCs and to refine it. The determination of MC structures in the m-state and intermediate conformations within the transport cycle is imperative to assess the dynamics involved in substrate translocation. Many MCF members have not been characterized yet, and several substrates known to be transported across the mitochondrial inner membrane have not been assigned to a transporter. In this respect, the recent finding that uncoupling protein 2 is a transporter for four-carbon metabolites (Vozza et al., 2014) strongly suggests that the function of the previously-named uncoupling proteins 3-5 (encoded by SLC25A9, SLC25A14 and SLC25A27) can also be to transport solutes across the mitochondrial membrane. All the above-mentioned knowledge could contribute to the understanding of the physiological roles of MCs and the molecular basis for the deleterious effects of the constantly increasing number of mutations in MCs involved in human disease.

ACKNOWLEDGMENTS

This work was supported by grants from the Ministero dell'Università e della Ricerca (MIUR), the National Research Council (CNR), the Center of Excellence on Comparative Genomics, Apulia Region, and the Italian Human ProteomeNet no. RBRN07BMCT_009 (MIUR).

REFERENCES

Agrimi, G., Di Noia, M. A., Marobbio, C. M. T., Fiermonte, G., Lasorsa, F. M., & Palmieri, F. (2004). Identification of the human mitochondrial S-adenosylmethionine transporter: Bacterial expression, reconstitution, functional characterization and tissue distribution. *Biochemical Journal, 379*, 183–190.

Agrimi, G., Russo, A., Pierri, C. L., & Palmieri, F. (2012). The peroxisomal NAD(+) carrier of Arabidopsis thaliana transports coenzyme A and its derivatives. *Journal of Bioenergetics and Biomembranes, 44*, 333–340.

Agrimi, G., Russo, A., Scarcia, P., & Palmieri, F. (2012). The human gene SLC25A17 encodes a peroxisomal transporter of coenzyme A, FAD and NAD+. *Biochemical Journal, 443*, 241–247.

Amutha, B., & Pain, D. (2003). Nucleoside diphosphate kinase of Saccharomyces cerevisiae, Ynk1p: Localization to the mitochondrial intermembrane space. *Biochemical Journal, 370*, 805–815.

Aprille, J. R. (1993). Mechanism and regulation of the mitochondrial ATP-Mg/P(i) carrier. *Journal of Bioenergetics and Biomembranes, 25*, 473–481.

Aquila, H., Eiermann, W., Babel, W., & Klingenberg, M. (1978). Isolation of the ADP/ATP translocator from beef heart mitochondria as the bongkrekate-protein complex. *European Journal of Biochemistry, 85*, 549–560.

Bamber, L., Harding, M., Butler, P. J. G., & Kunji, E. R. S. (2006). Yeast mitochondrial ADP/ATP carriers are monomeric in detergents. *Proceedings of the National Academy of Sciences of the United States of America, 103*, 16224–16229.

Bamber, L., Harding, M., Monné, M., Slotboom, D.-J., & Kunji, E. R. S. (2007). The yeast mitochondrial ADP/ATP carrier functions as a monomer in mitochondrial membranes.

Proceedings of the National Academy of Sciences of the United States of America, 104, 10830–10834.

Bamber, L., Slotboom, D.-J., & Kunji, E. R. S. (2007). Yeast mitochondrial ADP/ATP carriers are monomeric in detergents as demonstrated by differential affinity purification. *Journal of Molecular Biology, 371*, 388–395.

Bisaccia, F., Capobianco, L., Brandolin, G., & Palmieri, F. (1994). Transmembrane topography of the mitochondrial oxoglutarate carrier assessed by peptide-specific antibodies and enzymatic cleavage. *Biochemistry, 33*, 3705–3713.

Bisaccia, F., De Palma, A., Dierks, T., Krämer, R., & Palmieri, F. (1993). Reaction mechanism of the reconstituted tricarboxylate carrier from rat liver mitochondria. *Biochimica et Biophysica Acta, 1142*, 139–145.

Bisaccia, F., De Palma, A., & Palmieri, F. (1989). Identification and purification of the tricarboxylate carrier from rat liver mitochondria. *Biochimica et Biophysica Acta, 977*, 171–176.

Bisaccia, F., De Palma, A., Prezioso, G., & Palmieri, F. (1990). Kinetic characterization of the reconstituted tricarboxylate carrier from rat liver mitochondria. *Biochimica et Biophysica Acta, 1019*, 250–256.

Bisaccia, F., Indiveri, C., & Palmieri, F. (1985). Purification of reconstitutively active alpha-oxoglutarate carrier from pig heart mitochondria. *Biochimica et Biophysica Acta, 810*, 362–369.

Bisaccia, F., Zara, V., Capobianco, L., Iacobazzi, V., Mazzeo, M., & Palmieri, F. (1996). The formation of a disulfide cross-link between the two subunits demonstrates the dimeric structure of the mitochondrial oxoglutarate carrier. *Biochimica et Biophysica Acta, 1292*, 281–288.

Block, M. R., & Vignais, P. V. (1984). Substrate-site interactions in the membrane-bound adenine-nucleotide carrier as disclosed by ADP and ATP analogs. *Biochimica et Biophysica Acta, 767*, 369–376.

Block, M. R., Zaccaï, G., Lauquin, G. J., & Vignais, P. V. (1982). Small angle neutron scattering of the mitochondrial ADP/ATP carrier protein in detergent. *Biochemical and Biophysical Research Communications, 109*, 471–477.

Bouvier, F., Linka, N., Isner, J.-C., Mutterer, J., Weber, A. P. M., & Camara, B. (2006). Arabidopsis SAMT1 defines a plastid transporter regulating plastid biogenesis and plant development. *Plant Cell, 18*, 3088–3105.

Brandolin, G., Doussiere, J., Gulik, A., Gulik-Krzywicki, T., Lauquin, G. J., & Vignais, P. V. (1980). Kinetic, binding and ultrastructural properties of the beef heart adenine nucleotide carrier protein after incorporation into phospholipid vesicles. *Biochimica et Biophysica Acta, 592*, 592–614.

Brandolin, G., Dupont, Y., & Vignais, P. V. (1985). Substrate-induced modifications of the intrinsic fluorescence of the isolated adenine nucleotide carrier protein: Demonstration of distinct conformational states. *Biochemistry, 24*, 1991–1997.

Briggs, C., Mincone, L., & Wohlrab, H. (1999). Replacements of basic and hydroxyl amino acids identify structurally and functionally sensitive regions of the mitochondrial phosphate transport protein. *Biochemistry, 38*, 5096–5102.

Brustovetsky, N., Becker, A., Klingenberg, M., & Bamberg, E. (1996). Electrical currents associated with nucleotide transport by the reconstituted mitochondrial ADP/ATP carrier. *Proceedings of the National Academy of Sciences of the United States of America, 93*, 664–668.

Buchanan, B. B., Eiermann, W., Riccio, P., Aquila, H., & Klingenberg, M. (1976). Antibody evidence for different conformational states of ADP, ATP translocator protein isolated from mitochondria. *Proceedings of the National Academy of Sciences of the United States of America, 73*, 2280–2284.

Capobianco, L., Bisaccia, F., Michel, A., Sluse, F. E., & Palmieri, F. (1995). The N- and C-termini of the tricarboxylate carrier are exposed to the cytoplasmic side of the inner mitochondrial membrane. *FEBS Letters, 357*, 297–300.

Capobianco, L., Brandolin, G., & Palmieri, F. (1991). Transmembrane topography of the mitochondrial phosphate carrier explored by peptide-specific antibodies and enzymatic digestion. *Biochemistry, 30*, 4963–4969.

Cappello, A. R., Curcio, R., Valeria Miniero, D., Stipani, I., Robinson, A. J., Kunji, E. R. S., et al. (2006). Functional and structural role of amino acid residues in the even-numbered transmembrane alpha-helices of the bovine mitochondrial oxoglutarate carrier. *Journal of Molecular Biology, 363*, 51–62.

Cappello, A. R., Miniero, D. V., Curcio, R., Ludovico, A., Daddabbo, L., Stipani, I., et al. (2007). Functional and structural role of amino acid residues in the odd-numbered transmembrane alpha-helices of the bovine mitochondrial oxoglutarate carrier. *Journal of Molecular Biology, 369*, 400–412.

Castegna, A., Scarcia, P., Agrimi, G., Palmieri, L., Rottensteiner, H., Spera, I., et al. (2010). Identification and functional characterization of a novel mitochondrial carrier for citrate and oxoglutarate in S. cerevisiae. *Journal of Biological Chemistry, 285*, 17359–17370.

Cavero, S., Vozza, A., Del Arco, A., Palmieri, L., Villa, A., & Blanco, E. (2003). Identification and metabolic role of the mitochondrial aspartate-glutamate transporter in Saccharomyces cerevisiae. *Molecular Microbiology, 50*, 1257–1269.

Crichton, P. G., Harding, M., Ruprecht, J. J., Lee, Y., & Kunji, E. R. S. (2013). Lipid, detergent and Coomassie dye affect the migration of small membrane proteins in blue native gels: Mitochondrial carriers migrate as monomers not dimers. *Journal of Biological Chemistry, 288*, 22163–22173.

Dehez, F., Pebay-Peyroula, E., & Chipot, C. (2008). Binding of ADP in the mitochondrial ADP/ATP carrier is driven by an electrostatic funnel. *Journal of the American Chemical Society, 130*, 12725–12733.

De Lucas, J. R., Indiveri, C., Tonazzi, A., Perez, P., Giangregorio, N., Iacobazzi, V., et al. (2008). Functional characterization of residues within the carnitine/acylcarnitine translocase RX2PANAAXF distinct motif. *Molecular Membrane Biology, 25*, 152–163.

De Marcos Lousa, C., Trézéguet, V., Dianoux, A.-C., Brandolin, G., & Lauquin, G. J.-M. (2002). The human mitochondrial ADP/ATP carriers: Kinetic properties and biogenesis of wild-type and mutant proteins in the yeast S. cerevisiae. *Biochemistry, 41*, 14412–14420.

Dierks, T., Riemer, E., & Krämer, R. (1988). Reaction mechanism of the reconstituted aspartate/glutamate carrier from bovine heart mitochondria. *Biochimica et Biophysica Acta, 43*, 231–244.

Dolce, V., Iacobazzi, V., Palmieri, F., & Walker, J. E. (1994). The sequences of human and bovine genes of the phosphate carrier from mitochondria contain evidence of alternatively spliced forms. *Journal of Biological Chemistry, 269*, 10451–10460.

Dolce, V., Scarcia, P., Iacopetta, D., & Palmieri, F. (2005). A fourth ADP/ATP carrier isoform in man: Identification, bacterial expression, functional characterization and tissue distribution. *FEBS Letters, 579*, 633–637.

Endres, M., Neupert, W., & Brunner, M. (1999). Transport of the ADP/ATP carrier of mitochondria from the TOM complex to the TIM22.54 complex. *EMBO Journal, 18*, 3214–3221.

Falconi, M., Chillemi, G., Di Marino, D., D'Annessa, I., Morozzo della Rocca, B., Palmieri, L., et al. (2006). Structural dynamics of the mitochondrial ADP/ATP carrier revealed by molecular dynamics simulation studies. *Proteins, 65*, 681–691.

Fiermonte, G., De Leonardis, F., Todisco, S., Palmieri, L., Lasorsa, F. M., & Palmieri, F. (2004). Identification of the mitochondrial ATP-Mg/Pi transporter. Bacterial expression, reconstitution, functional characterization, and tissue distribution. *Journal of Biological Chemistry, 279*, 30722–30730.

Fiermonte, G., Dolce, V., David, L., Santorelli, F. M., Dionisi-Vici, C., Palmieri, F., et al. (2003). The mitochondrial ornithine transporter. Bacterial expression, reconstitution, functional characterization, and tissue distribution of two human isoforms. *Journal of Biological Chemistry, 278*, 32778–32783.

Fiermonte, G., Dolce, V., & Palmieri, F. (1998). Expression in Escherichia coli, functional characterization, and tissue distribution of isoforms A and B of the phosphate carrier from bovine mitochondria. *Journal of Biological Chemistry, 273*, 22782–22787.

Fiermonte, G., Dolce, V., Palmieri, L., Ventura, M., Runswick, M. J., Palmieri, F., et al. (2001). Identification of the human mitochondrial oxodicarboxylate carrier. Bacterial expression, reconstitution, functional characterization, tissue distribution, and chromosomal location. *Journal of Biological Chemistry, 276*, 8225–8230.

Fiermonte, G., Palmieri, L., Dolce, V., Lasorsa, F. M., Palmieri, F., Runswick, M. J., et al. (1998). The sequence, bacterial expression, and functional reconstitution of the rat mitochondrial dicarboxylate transporter cloned via distant homologs in yeast and Caenorhabditis elegans. *Journal of Biological Chemistry, 273*, 24754–24759.

Fiermonte, G., Palmieri, L., Todisco, S., Agrimi, G., Palmieri, F., & Walker, J. E. (2002). Identification of the mitochondrial glutamate transporter. Bacterial expression, reconstitution, functional characterization, and tissue distribution of two human isoforms. *Journal of Biological Chemistry, 277*, 19289–19294.

Fiermonte, G., Paradies, E., Todisco, S., Marobbio, C. M. T., & Palmieri, F. (2009). A novel member of solute carrier family 25 (SLC25A42) is a transporter of coenzyme A and adenosine 3',5'-diphosphate in human mitochondria. *Journal of Biological Chemistry, 284*, 18152–18159.

Fiermonte, G., Walker, J. E., & Palmieri, F. (1993). Abundant bacterial expression and reconstitution of an intrinsic membrane-transport protein from bovine mitochondria. *Biochemical Journal, 294*, 293–299.

Floyd, S., Favre, C., Lasorsa, F. M., Leahy, M., Trigiante, G., Stroebel, P., et al. (2007). The insulin-like growth factor-I–mTOR signaling pathway induces the mitochondrial pyrimidine nucleotide carrier to promote cell growth. *Molecular Biology of the Cell, 18*, 3545–3555.

Giangregorio, N., Tonazzi, A., Console, L., Indiveri, C., & Palmieri, F. (2010). Site-directed mutagenesis of charged amino acids of the human mitochondrial carnitine/acylcarnitine carrier: Insight into the molecular mechanism of transport. *Biochimica et Biophysica Acta, 1797*, 839–845.

Gordon, D. M., Lyver, E. R., Lesuisse, E., Dancis, A., & Pain, D. (2006). GTP in the mitochondrial matrix plays a crucial role in organellar iron homoeostasis. *Biochemical Journal, 400*, 163–168.

Gropp, T., Brustovetsky, N., Klingenberg, M., Müller, V., Fendler, K., & Bamberg, E. (1999). Kinetics of electrogenic transport by the ADP/ATP carrier. *Biophysical Journal, 77*, 714–726.

Hackenberg, H., & Klingenberg, M. (1980). Molecular weight and hydrodynamic parameters of the adenosine 5'-diphosphate—Adenosine 5'-triphosphate carrier in Triton X-100. *Biochemistry, 19*, 548–555.

Heidkämper, D., Müller, V., Nelson, D. R., & Klingenberg, M. (1996). Probing the role of positive residues in the ADP/ATP carrier from yeast. The effect of six arginine mutations on transport and the four ATP versus ADP exchange modes. *Biochemistry, 35*, 16144–16152.

Hoyos, M. E., Palmieri, L., Wertin, T., Arrigoni, R., Polacco, J. C., & Palmieri, F. (2003). Identification of a mitochondrial transporter for basic amino acids in Arabidopsis thaliana by functional reconstitution into liposomes and complementation in yeast. *Plant Journal, 33*, 1027–1035.

Huizing, M., Ruitenbeek, W., van den Heuvel, L. P., Dolce, V., Iacobazzi, V., & Smeitink, J. A. (1998). Human mitochondrial transmembrane metabolite carriers: Tissue distribution and its implication for mitochondrial disorders. *Journal of Bioenergetics and Biomembranes, 30*, 277–284.

Iacobazzi, V., Naglieri, M. A., Stanley, C. A., Wanders, R. J., & Palmieri, F. (1998). The structure and organization of the human carnitine/acylcarnitine translocase (CACT1) gene2. *Biochemical and Biophysical Research Communications*, *252*, 770–774.

Iacobazzi, V., Palmieri, F., Runswick, M. J., & Walker, J. E. (1992). Sequences of the human and bovine genes for the mitochondrial 2-oxoglutarate carrier. *DNA Sequence*, *3*, 79–88.

Iijima, M., Jalil, A., Begum, L., Yasuda, T., Yamaguchi, N., Xian Li, M., et al. (2001). Adult-onset type II cintrullinemia. *Advances in Enzyme Regulation*, *41*, 325–342.

Indiveri, C., Dierks, T., Krämer, R., & Palmieri, F. (1991). Reaction mechanism of the reconstituted oxoglutarate carrier from bovine heart mitochondria. *European Journal of Biochemistry*, *198*, 339–347.

Indiveri, C., Iacobazzi, V., Tonazzi, A., Giangregorio, N., Infantino, V., & Convertini, P. (2011). The mitochondrial carnitine/acylcarnitine carrier: Function, structure and physiopathology. *Molecular Aspects of Medicine*, *32*, 223–233.

Indiveri, C., Krämer, R., & Palmieri, F. (1987). Reconstitution of the malate/aspartate shuttle from mitochondria. *Journal of Biological Chemistry*, *262*, 15979–15983.

Indiveri, C., Palmieri, F., Bisaccia, F., & Krämer, R. (1987). Kinetics of the reconstituted 2-oxoglutarate carrier from bovine heart mitochondria. *Biochimica et Biophysica Acta*, *890*, 310–318.

Indiveri, C., Prezioso, G., Dierks, T., Krämer, R., & Palmieri, F. (1993). Kinetic characterization of the reconstituted dicarboxylate carrier from mitochondria: A four-binding-site sequential transport system. *Biochimica et Biophysica Acta*, *1143*, 310–318.

Indiveri, C., Tonazzi, A., De Palma, A., & Palmieri, F. (2001). Kinetic mechanism of antiports catalyzed by reconstituted ornithine/citrulline carrier from rat liver mitochondria. *Biochimica et Biophysica Acta*, *1503*, 303–313.

Indiveri, C., Tonazzi, A., & Palmieri, F. (1990). Identification and purification of the carnitine carrier from rat liver mitochondria. *Biochimica et Biophysica Acta*, *1020*, 81–86.

Indiveri, C., Tonazzi, A., & Palmieri, F. (1991). Characterization of the unidirectional transport of carnitine catalyzed by the reconstituted carnitine carrier from rat liver mitochondria. *Biochimica et Biophysica Acta*, *1069*, 110–116.

Indiveri, C., Tonazzi, A., & Palmieri, F. (1994). The reconstituted carnitine carrier from rat liver mitochondria: Evidence for a transport mechanism different from that of the other mitochondrial translocators. *Biochimica et Biophysica Acta*, *1189*, 65–73.

Indiveri, C., Tonazzi, A., Prezioso, G., & Palmieri, F. (1991). Kinetic characterization of the reconstituted carnitine carrier from rat liver mitochondria. *Biochimica et Biophysica Acta*, *1065*, 231–238.

Kadenbach, B., Mende, P., Kolbe, H. V., Stipani, I., & Palmieri, F. (1982). The mitochondrial phosphate carrier has an essential requirement for cardiolipin. *FEBS Letters*, *139*, 109–112.

Kao, L. R., Megraw, T. L., & Chae, C. B. (1996). SHM1: A multicopy suppressor of a temperature-sensitive null mutation in the HMG1-like abf2 gene. *Yeast*, *12*, 1239–1250.

Kaplan, R., Mayor, J., & Gremse, D. (1995). High level expression and characterization of the mitochondrial citrate transport protein from the yeast Saccharomyces cerevisiae. *Journal of Biological Chemistry*, *270*, 4108–4114.

Kaplan, R. S., Mayor, J. A., Johnston, N., & Oliveira, D. L. (1990). Purification and characterization of the reconstitutively active tricarboxylate transporter from rat liver mitochondria. *Journal of Biological Chemistry*, *265*, 13379–13385.

Klingenberg, M. (1976). The ADP-ATP carrier in mitochondrial membranes. In A. N. Martonosi (Ed.), *The enzymes of biological membranes*, *Vol. 3*, New York: Plenum Press, pp. 383–438.

Klingenberg, M. (1979). The ADP, ATP shuttle of the mitochondrion. *Trends in Biochemical Sciences*, *4*, 249–252.

Klingenberg, M. (2008). The ADP and ATP transport in mitochondria and its carrier. *Biochimica et Biophysica Acta*, *1778*, 1978–2021.

Klingenberg, M., & Buchholz, M. (1973). On the mechanism of bongkrekate effect on the mitochondrial adenine-nucleotide carrier as studied through the binding of ADP. *European Journal of Biochemistry, 38*, 346–358.
Klingenberg, M., & Winkler, E. (1985). The reconstituted isolated uncoupling protein is a membrane potential driven H+ translocator. *EMBO Journal, 4*, 3087–3092.
Krämer, R., & Palmieri, F. (1989). Molecular aspects of isolated and reconstituted carrier proteins from animal mitochondria. *Biochimica et Biophysica Acta, 974*, 1–23.
Kunji, E. R. S., & Crichton, P. G. (2010). Mitochondrial carriers function as monomers. *Biochimica et Biophysica Acta, 1797*, 817–831.
Kunji, E. R. S., & Robinson, A. J. (2010). Coupling of proton and substrate translocation in the transport cycle of mitochondrial carriers. *Current Opinion in Structural Biology, 20*, 440–447.
LaNoue, K., Mizani, S. M., & Klingenberg, M. (1978). Electrical imbalance of adenine nucleotide transport across the mitochondrial membrane. *Journal of Biological Chemistry, 253*, 191–198.
Lesuisse, E., Lyver, E. R., Knight, S. A. B., & Dancis, A. (2004). Role of YHM1, encoding a mitochondrial carrier protein, in iron distribution of yeast. *Biochemical Journal, 378*, 599–607.
Lin, C. S., Hackenberg, H., & Klingenberg, E. M. (1980). The uncoupling protein from brown adipose tissue mitochondria is a dimer. A hydrodynamic study. *FEBS Letters, 113*, 304–306.
Lindhurst, M. J., Fiermonte, G., Song, S., Struys, E., De Leonardis, F., & Schwartzberg, P. L. (2006). Knockout of Slc25a19 causes mitochondrial thiamine pyrophosphate depletion, embryonic lethality, CNS malformations, and anemia. *Proceedings of the National Academy of Sciences of the United States of America, 103*, 15927–15932.
Ma, C., Remani, S., Sun, J., Kotaria, R., Mayor, J. A., Walters, D. E., et al. (2007). Identification of the substrate binding sites within the yeast mitochondrial citrate transport protein. *Journal of Biological Chemistry, 282*, 17210–17220.
Marobbio, C. M. T., Agrimi, G., Lasorsa, F. M., & Palmieri, F. (2003). Identification and functional reconstitution of yeast mitochondrial carrier for S-adenosylmethionine. *EMBO Journal, 22*, 5975–5982.
Marobbio, C. M. T., Di Noia, M. A., & Palmieri, F. (2006). Identification of a mitochondrial transporter for pyrimidine nucleotides in Saccharomyces cerevisiae: Bacterial expression, reconstitution and functional characterization. *Biochemical Journal, 393*, 441–446.
Marobbio, C. M. T., Giannuzzi, G., Paradies, E., Pierri, C. L., & Palmieri, F. (2008). α-Isopropylmalate, a leucine biosynthesis intermediate in yeast, is transported by the mitochondrial oxalacetate carrier. *Journal of Biological Chemistry, 283*, 28445–28553.
Marobbio, C. M. T., Vozza, A., Harding, M., Bisaccia, F., Palmieri, F., & Walker, J. E. (2002). Identification and reconstitution of the yeast mitochondrial transporter for thiamine pyrophosphate. *EMBO Journal, 21*, 5653–5661.
Miniero, D. V., Cappello, A. R., Curcio, R., Ludovico, A., Daddabbo, L., & Stipani, I. (2011). Functional and structural role of amino acid residues in the matrix α-helices, termini and cytosolic loops of the bovine mitochondrial oxoglutarate carrier. *Biochimica et Biophysica Acta, 1807*, 302–310.
Monné, M., Miniero, D. V., Daddabbo, L., Robinson, A. J., Kunji, E. R. S., & Palmieri, F. (2012). Substrate specificity of the two mitochondrial ornithine carriers can be swapped by single mutation in substrate binding site. *Journal of Biological Chemistry, 287*, 7925–7934.
Monné, M., Miniero, D. V., Iacobazzi, V., Bisaccia, F., & Fiermonte, G. (2013). The mitochondrial oxoglutarate carrier: From identification to mechanism. *Journal of Bioenergetics and Biomembranes, 45*, 1–13.
Monné, M., Palmieri, F., & Kunji, E. R. S. (2013). The substrate specificity of mitochondrial carriers: Mutagenesis revisited. *Molecular Membrane Biology, 30*, 149–159.

Nelson, D. R., Felix, C. M., & Swanson, J. M. (1998). Highly conserved charge-pair networks in the mitochondrial carrier family. *Journal of Molecular Biology*, *277*, 285–308.
Palmieri, F. (1994). Mitochondrial carrier proteins. *FEBS Letters*, *346*, 48–54.
Palmieri, F. (2004). The mitochondrial transporter family (SLC25): Physiological and pathological implications. *Pflügers Archives*, *447*, 689–709.
Palmieri, F. (2008). Diseases caused by defects of mitochondrial carriers: A review. *Biochimica et Biophysica Acta*, *1777*, 564–578.
Palmieri, F. (2013). The mitochondrial transporter family SLC25: Identification, properties and physiopathology. *Molecular Aspects of Medicine*, *34*, 465–484.
Palmieri, F., Agrimi, G., Blanco, E., Castegna, A., Di Noia, M. A., & Iacobazzi, V. (2006). Identification of mitochondrial carriers in Saccharomyces cerevisiae by transport assay of reconstituted recombinant proteins. *Biochimica et Biophysica Acta*, *1757*, 1249–1262.
Palmieri, L., Agrimi, G., Runswick, M. J., Fearnley, I. M., Palmieri, F., & Walker, J. E. (2001). Identification in Saccharomyces cerevisiae of two isoforms of a novel mitochondrial transporter for 2-oxoadipate and 2-oxoglutarate. *Journal of Biological Chemistry*, *276*, 1916–1922.
Palmieri, L., Arrigoni, R., Blanco, E., Carrari, F., Zanor, M. I., & Studart-Guimaraes, C. (2006). Molecular identification of an Arabidopsis S-adenosylmethionine transporter. Analysis of organ distribution, bacterial expression, reconstitution into liposomes, and functional characterization. *Plant Physiology*, *142*, 855–865.
Palmieri, F., Bisaccia, F., Capobianco, L., Dolce, V., Fiermonte, G., Iacobazzi, V., et al. (1993). Transmembrane topology, genes, and biogenesis of the mitochondrial phosphate and oxoglutarate carriers. *Journal of Bioenergetics and Biomembranes*, *25*, 493–501.
Palmieri, L., De Marco, V., Iacobazzi, V., Palmieri, F., Runswick, M. J., & Walker, J. E. (1997). Identification of the yeast ARG-11 gene as a mitochondrial ornithine carrier involved in arginine biosynthesis. *FEBS Letters*, *410*, 447–451.
Palmieri, F., Indiveri, C., Bisaccia, F., & Krämer, R. (1993). Functional properties of purified and reconstituted mitochondrial metabolite carriers. *Journal of Bioenergetics and Biomembranes*, *25*, 525–535.
Palmieri, L., Lasorsa, F. M., De Palma, A., Palmieri, F., Runswick, M. J., & Walker, J. E. (1997). Identification of the yeast ACR1 gene product as a succinate-fumarate transporter essential for growth on ethanol or acetate. *FEBS Letters*, *417*, 114–118.
Palmieri, L., Lasorsa, F. M., Iacobazzi, V., Runswick, M. J., Palmieri, F., & Walker, J. E. (1999). Identification of the mitochondrial carnitine carrier in Saccharomyces cerevisiae. *FEBS Letters*, *462*, 472–476.
Palmieri, L., Lasorsa, F. M., Vozza, A., Agrimi, G., Fiermonte, G., & Runswick, M. J. (2000). Identification and functions of new transporters in yeast mitochondria. *Biochimica et Biophysica Acta*, *1459*, 363–369.
Palmieri, L., Palmieri, F., Runswick, M. J., & Walker, J. E. (1996). Identification by bacterial expression and functional reconstitution of the yeast genomic sequence encoding the mitochondrial dicarboxylate carrier protein. *FEBS Letters*, *399*, 299–302.
Palmieri, L., Pardo, B., Lasorsa, F. M., del Arco, A., Kobayashi, K., & Iijima, M. (2001). Citrin and aralar1 are Ca(2+)-stimulated aspartate/glutamate transporters in mitochondria. *EMBO Journal*, *20*, 5060–5069.
Palmieri, L., Picault, N., Arrigoni, R., Besin, E., Palmieri, F., & Hodges, M. (2008). Molecular identification of three Arabidopsis thaliana mitochondrial dicarboxylate carrier isoforms: Organ distribution, bacterial expression, reconstitution into liposomes and functional characterization. *Biochemical Journal*, *410*, 621–629.
Palmieri, F., & Pierri, C. L. (2010a). Mitochondrial metabolite transport. *Essays in Biochemistry*, *47*, 37–52.
Palmieri, F., & Pierri, C. L. (2010b). Structure and function of mitochondrial carriers—Role of the transmembrane helix P and G residues in the gating and transport mechanism. *FEBS Letters*, *584*, 1931–1939.

Palmieri, F., Pierri, C. L., De Grassi, A., Nunes-Nesi, A., & Fernie, A. R. (2011). Evolution, structure and function of mitochondrial carriers: A review with new insights. *Plant Journal, 66*, 161–181.

Palmieri, F., Quagliariello, E., & Klingenberger, M. (1972). Kinetics and specificity of the oxoglutarate carrier in rat-liver mitochondria. *European Journal of Biochemistry, 29*, 408–416.

Palmieri, F., Rieder, B., Ventrella, A., Blanco, E., Do, P. T., & Nunes-Nesi, A. (2009). Molecular identification and functional characterization of Arabidopsis thaliana mitochondrial and chloroplastic NAD+ carrier proteins. *Journal of Biological Chemistry, 284*, 31249–31259.

Palmieri, L., Rottensteiner, H., Girzalsky, W., Scarcia, P., Palmieri, F., & Erdmann, R. (2001). Identification and functional reconstitution of the yeast peroxisomal adenine nucleotide transporter. *EMBO Journal, 20*, 5049–5059.

Palmieri, L., Runswick, M. J., Fiermonte, G., Walker, J. E., & Palmieri, F. (2000). Yeast mitochondrial carriers: Bacterial expression, biochemical identification and metabolic significance. *Journal of Bioenergetics and Biomembranes, 32*, 67–77.

Palmieri, F., Stipani, I., Quagliariello, E., & Klingenberg, M. (1972). Kinetic study of the tricarboxylate carrier in rat liver mitochondria. *European Journal of Biochemistry, 26*, 587–594.

Palmieri, L., Vozza, A., Agrimi, G., De Marco, V., Runswick, M. J., Palmieri, F., et al. (1999). Identification of the yeast mitochondrial transporter for oxaloacetate and sulfate. *Journal of Biological Chemistry, 274*, 22184–22190.

Palmieri, L., Vozza, A., Hönlinger, A., Dietmeier, K., Palmisano, A., Zara, V., et al. (1999). The mitochondrial dicarboxylate carrier is essential for the growth of Saccharomyces cerevisiae on ethanol or acetate as the sole carbon source. *Molecular Microbiology, 31*, 569–577.

Palmisano, A., Zara, V., Hönlinger, A., Vozza, A., Dekker, P. J. T., Pfanner, N., et al. (1998). Targeting and assembly of the oxoglutarate carrier: General principles for biogenesis of carrier proteins of the mitochondrial inner membrane. *Biochemical Journal, 333*, 151–158.

Pebay-Peyroula, E., Dahout-Gonzalez, C., Kahn, R., Trézéguet, V., Lauquin, G. J.-M., & Brandolin, G. (2003). Structure of mitochondrial ADP/ATP carrier in complex with carboxyatractyloside. *Nature, 426*, 39–44.

Pfaff, E., & Klingenberg, M. (1968). Adenine nucleotide translocation of mitochondria. 1. Specificity and control. *European Journal of Biochemistry, 6*, 66–79.

Pfaff, E., Klingenberg, M., & Heldt, H. W. (1965). Unspecific permeation and specific exchange of adenine nucleotides in liver mitochondria. *Biochimica et Biophysica Acta, 104*, 312–315.

Picault, N., Hodges, M., Palmieri, L., & Palmieri, F. (2004). The growing family of mitochondrial carriers in Arabidopsis. *Trends in Plant Science, 9*, 138–146.

Picault, N., Palmieri, L., Pisano, I., Hodges, M., & Palmieri, F. (2002). Identification of a novel transporter for dicarboxylates and tricarboxylates in plant mitochondria. Bacterial expression, reconstitution, functional characterization, and tissue distribution. *Journal of Biological Chemistry, 277*, 24204–24211.

Pierri, C. L., Palmieri, F., & De Grassi, A. (2013). Single-nucleotide evolution quantifies the importance of each site along the structure of mitochondrial carriers. *Cellular and Molecular Life Sciences, 71*, 349–364.

Poduri, A., Heinzen, E., Chitsazzadeh, V., Lasorsa, F., LaCoursiere, C., Martin, E., et al. (2013). SLC25A22 is a novel gene for migrating partial seizures in infancy. *Annals of Neurology, 74*, 873–882.

Przybyla-Zawislak, B., Dennis, R. A., Zakharkin, S. O., & McCammon, M. T. (1998). Genes of succinyl-CoA ligase from Saccharomyces cerevisiae. *European Journal of Biochemistry, 258*, 736–743.

Riccio, P., Aquila, H., & Klingenberg, M. (1975). Purification of the carboxy-atractylate binding protein from mitochondria. *FEBS Letters*, *56*, 133–138.

Robinson, A. J., & Kunji, E. R. S. (2006). Mitochondrial carriers in the cytoplasmic state have a common substrate binding site. *Proceedings of the National Academy of Sciences of the United States of America*, *103*, 2617–2622.

Robinson, A. J., Overy, C., & Kunji, E. R. S. (2008). The mechanism of transport by mitochondrial carriers based on analysis of symmetry. *Proceedings of the National Academy of Sciences of the United States of America*, *105*, 17766–17771.

Runswick, M. J., Powell, S. J., Nyren, P., & Walker, J. E. (1987). Sequence of the bovine mitochondrial phosphate carrier protein: Structural relationship to ADP/ATP translocase and the brown fat mitochondria uncoupling protein. *EMBO Journal*, *6*, 1367–1373.

Runswick, M. J., Walker, J. E., Bisaccia, F., Iacobazzi, V., & Palmieri, F. (1990). Sequence of the bovine 2-oxoglutarate/malate carrier protein: Structural relationship to other mitochondrial transport proteins. *Biochemistry*, *29*, 11033–11040.

Ruprecht, J. J., Hellawell, A. M., Harding, M., Crichton, P. G., McCoy, A. J., & Kunji, E. R. S. (2014). Structures of yeast mitochondrial ADP/ATP carriers support a domain-based alternating-access transport mechanism. *Proceedings of the National Academy of Sciences of the United States of America*, *111*, E426–E434.

Saraste, M., & Walker, J. E. (1982). Internal sequence repeats and the path of polypeptide in mitochondrial ADP/ATP translocase. *FEBS Letters*, *144*, 250–254.

Schroers, A., Burkovski, A., Wohlrab, H., & Krämer, R. (1998). The phosphate carrier from yeast mitochondria. Dimerization is a prerequisite for function. *Journal of Biological Chemistry*, *273*, 14269–14676.

Sluse, F. E., Duyckaerts, C., Liebecq, S., & Sluse-Goffart, M. (1979). Kinetic and binding properties of the oxoglutarate translocator of rat-heart mitochondria. *European Journal of Biochemistry*, *100*, 3–17.

Sluse, F. E., Evens, A., Dierks, T., Duyckaerts, C., Sluse-Goffart, C. M., & Krämer, R. (1991). Kinetic study of the aspartate/glutamate carrier in intact rat heart mitochondria and comparison with a reconstituted system. *Biochimica et Biophysica Acta*, *1058*, 329–338.

Stappen, R., & Krämer, R. (1994). Kinetic mechanism of phosphate/phosphate and phosphate/OH- antiports catalyzed by reconstituted phosphate carrier from beef heart mitochondria. *Journal of Biological Chemistry*, *269*, 11240–11246.

Stepien, G., Torroni, A., Chung, A. B., Hodge, J. A., & Wallace, D. C. (1992). Differential expression of adenine nucleotide translocator isoforms in mammalian tissues and during muscle cell differentiation. *Journal of Biological Chemistry*, *267*, 14592–14597.

Stipani, V., Cappello, A. R., Daddabbo, L., Natuzzi, D., Miniero, D. V., Stipani, I., et al. (2001). The mitochondrial oxoglutarate carrier: Cysteine-scanning mutagenesis of transmembrane domain IV and sensitivity of Cys mutants to sulfhydryl reagents. *Biochemistry*, *40*, 15805–15810.

Todisco, S., Agrimi, G., Castegna, A., & Palmieri, F. (2006). Identification of the mitochondrial NAD+ transporter in Saccharomyces cerevisiae. *Journal of Biological Chemistry*, *281*, 1524–1531.

Tonazzi, A., Console, L., Giangregorio, N., Indiveri, C., & Palmieri, F. (2012). Identification by site-directed mutagenesis of a hydrophobic binding site of the mitochondrial carnitine/acylcarnitine carrier involved in the interaction with acyl groups. *Biochimica et Biophysica Acta*, *1817*, 697–704.

Traba, J., Satrústegui, J., & del Arco, A. (2009). Characterization of SCaMC-3-like/slc25a41, a novel calcium-independent mitochondrial ATP-Mg/Pi carrier. *Biochemical Journal*, *418*, 125–133.

Villiers, C., Michejda, J. W., Block, M., Lauquin, G. J., & Vignais, P. V. (1979). The electrogenic nature of ADP/ATP transport in inside-out submitochondrial particles. *Biochimica et Biophysica Acta*, *546*, 157–170.

Vozza, A., Blanco, E., Palmieri, L., & Palmieri, F. (2004). Identification of the mitochondrial GTP/GDP transporter in Saccharomyces cerevisiae. *Journal of Biological Chemistry, 279*, 20850–20857.

Vozza, A., Parisi, G., De Leonardis, F., Lasorsa, F. M., Castegna, A., Amorese, D., et al. (2014). UCP2 transports C4 metabolites out of mitochondria, regulating glucose and glutamine oxidation. *Proceedings of the National Academy of Sciences of the United States of America, 111*, 960–965.

Wang, Y., & Tajkhorshid, E. (2008). Electrostatic funneling of substrate in mitochondrial inner membrane carriers. *Proceedings of the National Academy of Sciences of the United States of America, 105*, 9598–9603.

Watt, I. N., Montgomery, M. G., Runswick, M. J., Leslie, A. G. W., & Walker, J. E. (2010). Bioenergetic cost of making an adenosine triphosphate molecule in animal mitochondria. *Proceedings of the National Academy of Sciences of the United States of America, 107*, 16823–16827.

Wibom, R., Lasorsa, F. M., Töhönen, V., Barbaro, M., Sterky, F. H., & Kucinski, T. (2009). AGC1 deficiency associated with global cerebral hypomyelination. *New England Journal of Medicine, 361*, 489–495.

Wohlrab, H., Annese, V., & Haefele, A. (2002). Single replacement constructs of all hydroxyl, basic, and acidic amino acids identify new function and structure-sensitive regions of the mitochondrial phosphate transport protein. *Biochemistry, 41*, 3254–3261.

Zara, V., Ferramosca, A., Palmisano, I., Palmieri, F., & Rassow, J. (2003). Biogenesis of rat mitochondrial citrate carrier (CIC): The N-terminal presequence facilitates the solubility of the preprotein but does not act as a targeting signal. *Journal of Molecular Biology, 325*, 399–408.

Zara, V., Ferramosca, A., Robitaille-Foucher, P., Palmieri, F., & Young, J. C. (2009). Mitochondrial carrier protein biogenesis: Role of the chaperones Hsc70 and Hsp90. *Biochemical Journal, 419*, 369–375.

Zara, V., Palmieri, F., Mahlke, K., & Pfanner, N. (1992). The cleavable presequence is not essential for import and assembly of the phosphate carrier of mammalian mitochondria but enhances the specificity and efficiency of import. *Journal of Biological Chemistry, 267*, 12077–12081.

CHAPTER NINE

The Families of Zinc (SLC30 and SLC39) and Copper (SLC31) Transporters

Monika Schweigel-Röntgen[1]

Institute for Muscle Biology & Growth, Leibniz Institute for Farm Animal Biology, Dummerstorf, Germany
[1]Corresponding author: e-mail address: roentgen@fbn-dummerstorf.de

Contents

Current Topics in Membranes, Volume 73
ISSN 1063-5823
http://dx.doi.org/10.1016/B978-0-12-800223-0.00009-8

Abstract

The solute carriers families 30 (SLC30; ZnT), 39 (SLC39; ZIP), and 31 (SLC31; CTR) are involved in the essential maintenance of cellular zinc (Zn^{2+}) and copper (Cu^{2+}) homeostasis, respectively.

ZnTs mediate Zn^{2+} extrusion from cells (SLC30A1) or transport Zn^{2+} into organelles and secretory vesicles/granules (SLC30A2–SLC30A8). SLC39 family members are predominantly localized to the cell membrane where they perform Zn^{2+} uptake and increase the availability of cytosolic Zn^{2+}. SLC39A1 is ubiquitously expressed, whereas other ZIP transporters (e.g., SLC39A2 and SLC39A3) show a more tissue-restricted expression consistent with organ-specific functions of these proteins.

The members A1 (CTR1) and A2 (CTR2) of the SLC31 family of solute carriers belong to a network of proteins that acts to regulate the intracellular Cu^{2+} concentration within a certain range. SLC31A1 is predominantly localized to the plasma membrane, whereas SLC31A2 is mainly found in intracellular membranes of the late endosome and lysosome. The specific function of SLC31A2 is not known. SLC31A1 is ubiquitously expressed and has been characterized as a high-affinity importer of reduced copper (Cu^{+}). Cu^{2+} transport function of CTR proteins is associated with oligomerization; SLC31A1 trimerizes and thereby forms a channel-like structure enabling Cu^{2+} translocation across the cell membrane.

The molecular characteristics and structural details (e.g., membrane topology, conserved Zn^{2+}, and Cu^{2+} binding sites) and mechanisms of translational and posttranslational regulation of expression and/or activity have been described for SLC30 and SLC39 family members, and for SLC31A1.

For SLC31A1, data on tissue-specific functions (e.g., in the intestine, heart, and liver) are also available. A link between SLC31A1, immune function, and disorders such as Alzheimer's disease or cancer makes the protein a candidate therapeutic target.

In secretory tissues (e.g., the mammary gland and pancreas), Zn^{2+} transporters of SLC families 30 and 39 are involved in specific functions such as insulin synthesis and secretion, metallation of digestive proenzymes, and transfer of nutrients into milk. Defective or dysregulated Zn^{2+} metabolism in these organs is associated with disorders such as diabetes and cancer, and impaired Zn^{2+} secretion into milk.

1. INTRODUCTION

Zinc (Zn^{2+}) and copper (Cu^{2+}) are cofactors for a multitude of enzymes (Festa & Thiele, 2011; Leary et al., 2004; Lutsenko, Barnes, Bartee, & Dmitriev, 2007; McCall, Huang, & Fierke, 2000; Vallee &

Auld, 1990). Zn^{2+} is also a structural component of metallothioneins (MTs) and insulin and serves a specific structural role by stabilizing the conformation of various protein domains found in gene regulatory proteins (Brown, Sander, & Argos, 1985; Kadhim et al., 2006; Vallee & Falchuk, 1993). Therefore, Zn^{2+} and Cu^{2+} play an essential role in important biological processes such as growth, development, immune and neuronal functions, reproduction, and metabolism (Beach, Gershwin, Makishima, & Hurley, 1980; Danks, 1988; Festa & Thiele, 2011; Lutsenko et al., 2007; Pena, Lee, & Thiele, 1999; Prasad, Halsted, & Nadimi, 1961; Todd, Elvehjem, & Hart, 1934; Vallee & Falchuk, 1993; Vulpe & Packman, 1995).

Although Zn^{2+} and Cu^{2+} are essential, excess of these trace elements can be toxic to cells (Koh et al., 1996; Lutsenko et al., 2007). For example, excess Cu^{2+} leads to the generation of reactive oxygen species (ROS) and is highly toxic to cells because of its redox-active property (Lutsenko et al., 2007).

These examples clearly illustrate that intracellular Zn^{2+} and Cu^{2+} homeostasis needs to be tightly controlled by influx, intracellular storage/sequestration, and efflux of the ions. Zn^{2+} and Cu^{2+} transport proteins encoded by three solute-linked carrier (SLC) gene families named SLC30 (zinc transporter; ZnT), SLC39 (Zrt- and Irt-like Protein; ZIP), and SLC31 (CTR) are involved in this process.

SLC30 and SLC39 Zn^{2+} transport proteins have opposite roles in the regulation of Zn^{2+} homeostasis. ZnT transporters reduce cytosolic Zn^{2+} availability by mediating either Zn^{2+} efflux from cells or Zn^{2+} uptake into intracellular vesicles. On the other hand, ZIP transporters promote uptake of extracellular Zn^{2+} and release of vesicular Zn^{2+} into the cytosol. Intracellular Zn^{2+} is bound by small cysteine-rich proteins named MTs (Coyle, Philcox, Carey, & Rofe, 2002).

Cellular Cu^{2+} homeostasis is regulated by a complex network of proteins that mediates Cu^{2+} uptake across the plasma membrane, for example, by SLC31 proteins, and its subcellular and systemic distribution, for example, by cytosolic Cu^{2+} chaperons such as ATOX1 and the Cu^{2+}-transporting P-type ATPases, ATP7A and ATP7B (Culotta, Yang, & O'Halloran, 2006; Kim, Nevitt, & Thiele, 2008; Lutsenko et al., 2007; Robinson & Winge, 2010). SLC31A1 (CTR1) is the main transporter for cellular uptake of monovalent copper (Cu^{+}), whereas SLC31A2 (CTR2) appears to be a vacuolar/vesicular transporter—most probably responsible for the recycling of Cu^{2+} from intracellular stores (Bertinato, Swist, Plouffe, Brooks, & L'abbé, 2008; Rees, Lee, & Thiele, 2004; Zhou & Gitschier, 1997).

This chapter is a brief review of the structural and cell biological aspects of SLC30-, SLC39-, and SLC31-mediated Zn^{2+}/Cu^{2+} transport, as well as the function of these proteins in physiological and pathophysiological processes.

2. DISCOVERY OF ZINC TRANSPORTERS

SLC30 and SLC39 together constitute a superfamily of Zn^{2+} transporters. The first mammalian SLC30 genes, *SLC30A1* (*ZnT1*) and *SLC30A2* (*ZnT2*), were identified by transfection of a rat kidney cDNA expression library into a Zn^{2+}-sensitive baby hamster kidney cell line followed by the selection of clones showing resistance to high extracellular Zn^{2+} concentrations (Palmiter, Cole, & Findley, 1996; Palmiter & Findley, 1995). *SLC30A3* (*ZnT3*) was cloned by hybridizing a mouse genomic library to a rat *SLC30A2* cDNA probe (Palmiter, Cole, Quaife, & Findley, 1996). The SLC30A4 (ZnT4) transporter gene was discovered by positional cloning (Huang & Gitschier, 1997). By screening existing sequence and genomic DNA databases, investigators identified *SLC30A5–SLC30A10* members by their sequence homologies to *SLC30A1–SLC30A4* cDNAs (Chimienti, Devergnas, Favier, & Seve, 2004; Huang, Kirschke, & Gitschier, 2002; Kambe et al., 2002; Kirschke & Huang, 2003).

Members of the SLC39 (ZIP) superfamily were first identified in *Arabidopsis* and yeast (Eng, Guerinot, Eide, & Saier, 1998; Rogers, Eide, & Guerinot, 2000; Zhao & Eide, 1996). Computer searches of the human and mouse genome revealed multiple members of the ZIP superfamily (e.g., 15 in human and at least 5 in mice).

3. THE SLC30 (ZnT) FAMILY

At present, 10 members of the SLC30 (ZnT) family have been identified that are responsible for Zn^{2+} transport into organelles or out of cells (Palmiter & Findley, 1995). The transport mechanism is not clear. Ohana et al. (2009) have shown that the proton gradient generated by V-type ATPases drives Zn^{2+} transport into the Golgi apparatus (GA). This finding is in agreement with vH^+-ATPase-dependent Zn^{2+}/H^+ exchange observed in bacteria, yeast, and plants (Guffanti, Wei, Rood, & Krulwich, 2002; Kawachi, Kobae, Mimura, & Maeshima, 2008; MacDiarmid, Milanick, & Eide, 2002).

3.1. Structure, expression, and localization

All SLC30 members show topological characteristics of the cation diffusion facilitator superfamily. They have six conserved transmembrane domains (TMs), a histidine-rich Zn^{2+}-binding region between TMs IV and V, and N and C termini that are both located intracellularly (Ackland & Michalczyk, 2006; Huang & Gitschier, 1997; Palmiter & Findley, 1995). ZnTs typically have long C-terminal tails ranging from 82 amino acids (aa) in ZnT7 to 203 aa in ZnT6. The length and amino acid sequence upstream of the first TM differ largely among the SLC30 members, and this region may contain subcellular targeting signals (Huang & Tepaamorndech, 2013).

Formation of hetero- (SLC30A5/SLC30A6) or homodimers (SLC30A7) has been shown (Suzuki et al., 2005). The Zn^{2+} transporter activity depends on conserved Zn^{2+}-binding sites (e.g., amino acids 455, 595, 599, His451 in SLC30A5) with high identity to the Zn^{2+}-binding site A of the bacterial Zn^{2+} transporter YiiP (Lu, Chai, & Fu, 2009; Ohana et al., 2009). These critical binding sites are not conserved in SLC30A6, thereby making the protein itself unable to transport Zn^{2+} (Fukunaka et al., 2009; Ohana et al., 2009). Fukunaka et al. (2009) proposed that SLC30A6 modulates the Zn^{2+} transport rate.

The expression patterns of ZnTs range from ubiquitous (SLC30A1) to more tissue-specific (SLC30A2, SLC30A3, SLC30A8, and SLC30A10), and different SLC30 proteins exhibit distinct subcellular localization (Table 9.1).

The ubiquitously expressed SLC30A1 functions to mediate Zn^{2+} export from cells and is localized at the plasma membrane (Liuzzi, Bobo, Cui, McMahon, & Cousins, 2003; Qin, Thomas, Fontaine, & Colvin, 2009; Yu, Kirschke, & Huang, 2007). In addition to SLC30A1, specific isoforms of SLC30A2 (human isoform 2), SLC30A5 (human isoform b), and human SLC30A4 have also been found on the plasma membrane (Cragg et al., 2005; Lopez & Kelleher, 2009; Overbeck, Uciechowski, Ackland, Ford, & Rink, 2008; Valentine et al., 2007).

SLC30A2 (Lopez & Kelleher, 2009), SLC30A3 (Cole, Wenzel, Kafer, Schwartzkroin, & Palmiter, 1999), SLC30A4 (Huang & Gitschier, 1997), SLC30A5 (Kambe et al., 2002), SLC30A6 (Huang et al., 2002), SLC30A7 (Kirschke & Huang, 2003), and SLC30A8 (Chimienti et al., 2004), however, are localized within intracellular membranes and remove Zn^{2+} from the cytosol to vesicles or organelles for secretion, storage, or binding to

Table 9.1 Characteristics of the SLC30 family of zinc transporters (See the color plate.)

Gene	Organ/tissue distribution		Subcellular localization							Link to disease
	Widespread	Predominant expression	PM	SG	Vesicles	GA	Mito	C	N	
SLC30										
A1	× (ubiquitous)	Tissues involved in Zn^{2+} acquisition such as small intestine, renal tubule, and placenta	×							Embryonic death, Alzheimer's disease (AD)
A2		Mammary gland, prostate, retina, placenta, testis, seminal vesicles, pancreas, small intestine, kidney	×	×	Endosomal, lysosomal		×			Breast cancer
A3		Brain, testis, pancreas			Synaptic					AD
A4	×	Mammary gland, placenta, prostate, brain, kidney	×	×	Endosomal					Lethal milk syndrome, prostate cancer, AD
A5	×	Heart, placenta, prostate, ovary, testis, small intestine, thymus, bone	×	×	×	×				
A6	×	Brain, lung, small intestine, liver			×	×				AD
A7	×	Intestine, stomach, pancreas, prostate, testis, retina, muscle			×	×				
A8		Pancreas, thyroid, testis, adrenal gland		×						Diabetes (type 1 and 2)
A9	×							×	×	
A10		Brain, retina, liver	Unknown							

PM, plasma membrane; SG, secretory granules; GA, Golgi apparatus; Mito, mitochondria; C, cytosol; N, nucleus.

Zn^{2+}-sensitive proteins (Haase & Beyersmann, 2002; Huang et al., 2002; Palmiter, Cole, & Findley, 1996; Wenzel, Cole, Born, Schwartzkroin, & Palmiter, 1997; Table 9.1).

3.2. Functional roles and regulation of SLC30 zinc transporters in mammary gland and pancreas

In specialized secretory tissues such as mammary gland and pancreas, SLC30 members A2, A4, A5, and A8 reside in the membrane of secretory vesicles/granules and mediate Zn^{2+} uptake for exocytosis (reviewed in Kelleher, McCormick, Velasquez, & Lopez, 2011).

For example, in epithelial cells of the lactating mammary gland, SLC30A2 (human) and SLC30A4 (mice) transport Zn^{2+} into secretory vesicles and are thus responsible for the Zn^{2+} content of milk (Chowanadisai, Lonnerdal, & Kelleher, 2006; Kelleher & Lönnerdal, 2003; Liuzzi et al., 2003; Lopez & Kelleher, 2009). A nonsense mutation at arginine codon 297 of *SLC30A4* resulting in premature protein termination was reported to cause the lethal milk syndrome (Huang & Gitschier, 1997), an inherent disease of Zn^{2+} metabolism in mice (Lee, Shay, & Cousins, 1992). The lethal milk mouse is characterized by reduced Zn^{2+} secretion from the mammary gland that leads to a 34% reduction of dams' milk Zn^{2+} content and lethal Zn^{2+} deficiency in the pups (Ackland & Mercer, 1992; Piletz & Ganschow, 1978). In women with low milk Zn^{2+} content, a mis-sense mutation in SLC30A2 that substitutes a conserved N-terminal histidine at amino acid 54 with arginine (H54R) has been found (Chowanadisai et al., 2006). Expression of the H54R variant reduced Zn^{2+} secretion from HEK293 cells by 30%, most probably as a consequence of perinuclear, aggresomal accumulation of the mutated SLC30A2. Mutations in the N terminus (L23P) or C terminus (R340C) of SLC30A2 lead to mis-targeting of the protein followed by dysregulation of cellular Zn^{2+} homeostasis and elevated oxidative stress (Seo & Kelleher, 2010). Zn^{2+} hyperaccumulation and increased generation of ROS may increase the risk of breast cancer for women who carry these ZnT2 variants (Cui, Vogt, Olson, Glass, & Rohan, 2007; Taylor et al., 2007).

The lactogenic hormone prolactin regulates Zn^{2+} metabolism of the lactating mammary gland through transcriptional upregulation of ZnT2 expression via the JAK2/STAT5 signaling pathway (Qian, Lopez, Seo, & Kelleher, 2009). In addition, prolactin increases Zn^{2+} accumulation in secretory vesicles and their intracellular movement to the cell membrane for Zn^{2+} secretion into the milk (Lopez & Kelleher, 2009). Lactation increases expression of ZnT4 in mouse (Huang & Gitschier, 1997).

At the organism level, Zn^{2+} homeostasis is maintained through a balance between intestinal absorption, tissue turnover, and the bile–pancreatic secretions (Guo et al., 2010; Langmade, Ravindra, Daniels, & Andrews, 2000). Zn^{2+} enters the intestinal tract via zymogen granules that contain enzymes necessary for digestion. SLC30A2 mediates Zn^{2+} transport into zymogen granules of pancreatic acinar cells where it binds to and activates digestive enzymes (Guo et al., 2010). Knockdown of SLC30A2 mRNA in AR42J pancreatic acinar cells decreased $^{65}Zn^{2+}$ content in zymogen granules by 15% (Guo et al., 2010). In addition, a transient Zn^{2+} accumulation in the cytoplasm accompanied by upregulation of Zn^{2+}-binding methallothionein and Zn^{2+} exporter SLC30A1 was observed.

Pancreatic SLC30A1 and SLC30A2 are sensitive to decreased or elevated Zn^{2+} intake; deficient dietary Zn^{2+} reduces and high dietary Zn^{2+} up regulates its expression, respectively (Guo et al., 2010). Zn^{2+}-dependent transcription of *SLC30A1* and *SLC30A2* involves the metal response element (MRE) binding transcription factor 1 (MTF-1) and downstream MREs in their promotor regions (Guo et al., 2010; Langmade et al., 2000). MTF-1 contains six zinc fingers (Cys_2His_2) and functions as an intracellular Zn^{2+} sensor. After Zn^{2+} binding, MTF-1 is activated to bind to MREs, resulting in increased transcription of genes. The same regulatory mechanism has been shown for Zn^{2+}-dependent regulation of SLC30A1 expression in the intestine, brain, and placenta (Chowanadisai, Kelleher, & Lonnerdal, 2005; Helston et al., 2007; Langmade et al., 2000).

Dexamethasone, a glucocorticoid analog, increased Zn^{2+} uptake into zymogen granules, and thus the amount of Zn^{2+} available for export from the cells along with digestive proenzymes via exocytosis (Guo et al., 2010). In both rat AR42J cells and mice pancreas, applying dexamethasone stimulated SLC30A2 mRNA and protein expression (Guo et al., 2010). The signaling pathway for glucocorticoid-dependent upregulation of SLC30A2 required interaction of dimerized glucocorticoid receptors with STAT5.

Zn^{2+} also effects endocrine functions of the pancreas, for example, the biosynthesis and secretion of insulin. Among the SLC30 Zn^{2+} transporters, SLC30A8 is most abundant within pancreatic β-cells and localized in the membrane of the insulin containing secretory vesicles (Lemaire et al., 2009). Expression of *SLC30A8* in pancreatic β-cells is associated with the expression of the β-cell-restricted transcription factor Pdx-1 (Pound et al., 2011) which is reduced in the islets of diabetic mice (Tamaki et al., 2009). SLC30A8 mediates Zn^{2+} transport into insulin granules of the β-cells that is important for insulin crystallization (Chimienti et al., 2004; Kelleher et al., 2011; Lemaire et al., 2009).

SLC30A8$^{-/-}$ mice had a complete normal glucose tolerance when fed a standard diet but became glucose intolerant or diabetic after being fed a high-fat diet (Lemaire et al., 2009). A risk allele (R325W) of *SLC30A8* has been found to be a susceptibility locus of type-2 diabetes (Scott et al., 2007; Sladek et al., 2007). It seems most probable that a combination of genetic predisposition and environmental stressors such as high-fat diet, is responsible for the development of type-2 diabetes. Autoimmunity that targets the C terminus of SLC30A8 in particular, has been identified as a pathogenetic mechanism involved in the development of type-1 diabetes (Wenzlau et al., 2007).

3.3. Neuronal Zn^{2+} transport by SLC30A3

SLC30A3 transports Zn^{2+} into synaptic vesicles in the brain (Palmiter, Cole, Quaife, et al., 1996). Interestingly, *SLC30A3* expression in the brain is independent of the availability of dietary Zn^{2+} (Chowanadisai et al., 2005), and may be important in conservation of brain Zn^{2+} during Zn^{2+} deficiency. However, *SLC30A3* expression is downregulated during aging, and this process proceeds much faster in patients suffering from the neurodegenerative disorder Alzheimer's disease (AD) (Adlard, Parncutt, Finkelstein, & Bush, 2010). An increased expression of other ZnTs (SLC30A1, SLC30A4, and SLC30A6) has also been observed in AD patients (Lovell, Smith, Xiong, & Markesbery, 2005; Lyubartseva, Smith, Markesbery, & Lovell, 2010). Thus, it seems likely that a loss of synaptic accumulation and subsequent release of Zn^{2+} play a role in AD pathogenesis (Adlard et al., 2010).

3.4. SLC30A9: A coactivator of nuclear receptors

SLC30A9 shows a cation efflux motif and, thus, is a putative Zn^{2+} transporter. However, in contrast to all other ZnTs, SLC30A9 has been found in the cytosol and in nuclear fractions of cells (Sim & Chow, 1999). SLC30A9 functions as a coactivator of nuclear receptors that regulate transcriptional activation of target genes in proliferating cells and after hormonal stimulation, for example, by estrogen (Chen, Kim, & Stallcup, 2005).

3.5. Other Zn^{2+} transporters of the SLC30 family

Expression patterns of SLC30 members A5–A7 are given in Table 9.1. These Zn^{2+} transporters are mainly localized on the membranes of the GA and cytosolic vesicles. They allocate cytosolic Zn^{2+} into the secretory pathway for sequestration and/or for supply to proteins (e.g., alkaline phosphatases) that need Zn^{2+} for their structural formation or activities (Kambe et al., 2002; Kirschke & Huang, 2003; Suzuki et al., 2005).

4. THE SLC39 (ZIP) FAMILY

Proteins belonging to the SLC39 family are involved in Zn^{2+} uptake and localize mainly to the plasma membrane (Dufner-Beattie, Langmade, Wang, Eide, & Andrews, 2003; Gaither & Eide, 2000). Consistent with the ability of ZIPs to promote Zn^{2+} uptake, overexpression of human and mice ZIP1, ZIP2, and ZIP3 in various cell systems (e.g., HEK293, K562, and prostate cells) increased $^{65}Zn^{2+}$ accumulation (~2- to 4-fold) (Dufner-Beattie, Langmade, et al., 2003; Gaither & Eide, 2000; Wang et al., 2004). The specific mechanism for this transport is not known, but it is temperature dependent and shows Michaelis–Menten kinetics ($K_m \sim 2$–3 μM; $V_{max} = 9.5$–10.5 pmol zinc/min/mg protein) (Dufner-Beattie, Langmade, et al., 2003; Gaither & Eide, 2000; Taylor, Morgan, Johnson, & Nicholson, 2005), thus pointing to secondary active transport. Human ZIP2-mediated Zn^{2+} uptake was stimulated by HCO_3^-, suggesting cotransport of Zn^{2+} together with HCO_3^- (Gaither & Eide, 2000). In contrast to the human forms, mouse ZIP1 and ZIP2 show a relatively high selectivity for Zn^{2+}. Mouse ZIP3, however, is less specific for Zn^{2+} and also transports Cu^{2+}, Cd^{2+}, Mn^{2+}, Co^{2+}, Ag^{2+}, Mg^{2+}, and Ni^{2+} (but not iron).

4.1. Structure–function relationships of SLC39 proteins

Most ZIP proteins have eight TMs, a long intracellular loop between TMs III and IV and a very short extracellular C terminus (Dufner-Beattie, Langmade, et al., 2003; Eide, 2006; Liuzzi & Cousins, 2004; Taylor & Nicholson, 2003).

The length and sequence of the intracellular loop region are not well conserved but always contain a histidine-rich domain (HRD) (HX, $n = 3$–6)

(Gaither & Eide, 2000; Rogers et al., 2000; Taylor et al., 2005; Taylor & Nicholson, 2003). This sequence was postulated to serve as a metal-ion-binding site in ZIP proteins (Eng et al., 1998; Rogers et al., 2000), but its specific function is unknown. An $(HX)_6$ motif is a characteristic feature of proteins that require protein:protein interactions for their functions. Thus, the HX*n* motif could function to facilitate macromolecular interactions via metal binding (Eng et al., 1998). Deletion of the HRD in the yeast ZIP transporter TjZNT1 increased its specificity for Zn^{2+} (Nishida, Mizuno, & Obata, 2008). The authors propose that the HRD could be involved in ion specificity of TjZNT1. In the *Arabidopsis* metal transporter IRT1, a mutation of aspartate 136 (D136A) located at the beginning of the TM III–IV intracellular loop increased the specificity for Zn^{2+} by eliminating its ability to transport Fe^{2+} and Mn^{2+} (Rogers et al., 2000). In addition, mutations (D100A and E103A) that change the specificity of IRT1 for Zn^{2+} are located in an extracellular loop between TMs II and III (Rogers et al., 2000).

The long extracellular N terminus is also histidine rich (2–34 residues). These histidine residues may be important for Zn^{2+} transport, perhaps by binding Zn^{2+} and increasing the local concentration of Zn^{2+} available for transport. In addition, three potential glycosylation sites (e.g., residues N192, N219, and N272 in ZIP4, and residues N77, N87, and N102 in ZIP14) are found in the N-terminal domain (Taylor et al., 2005; Wang et al., 2004). SLC39 proteins that show defects in N-glycosylation fail to accumulate sufficiently on the cell surface—an effect that results from misfolding and/or mislocalization of these proteins. For example, acrodermatitis eneropathica-associated mutations of conserved residues within TM I (G340D), TM III (L382P and G384R), and TM VIII (G643R) of ZIP4 change nonpolar residues to charged (D, R) or helix-disrupting (P) amino acids that disrupt protein folding (Wang et al., 2004).

The greatest degree of homology over the ZIP family is found in TMs IV–VIII (Gaither & Eide, 2000). In particular, a 12-aa signature sequence (HSVFEGLAVGLQ) typical for SLC39 Zn^{2+} transporters is localized to TM IV (Eng et al., 1998). In ZIP3, the second valine is a leucine instead. Human SLC39 family members A4–A8, A10, and A12–A14 belong to the LIV1 subfamily of ZIP transporters that have a conserved metalloprotease motif (HEXPHEXGD in ZIP8, and EEXPHEXGD in ZIP14) in TM V (Taylor et al., 2005). Interestingly, ZIP14 mutants lacking one TM (TM VIII) or five TMs (TM IV–VIII) show normal location in the plasma membrane indicating that they are not needed for cellular processing

(Taylor et al., 2005). TMs IV and V are predicted to be amphipathic helixes and contain conserved histidine, serine, and glycine residues (Dufner-Beattie, Langmade, et al., 2003; Eng et al., 1998; Gaither & Eide, 2000; Rogers et al., 2000; Taylor & Nicholson, 2003). Therefore, they are supposed to form the transmembrane aqueous channel in the transporter through which substrate metal ions, for example, Zn^{2+}, can pass (Eng et al., 1998; Rogers et al., 2000). Conserved histidine and serine residues within TM IV and V (H197, S198, H224) are suggested to comprise part of a metal-binding site in the center of the membrane (Dufner-Beattie, Langmade, et al., 2003; Eng et al., 1998; Gaither & Eide, 2000; Rogers et al., 2000). Mutations of these histidine residues or of adjacent polar/charged residues (serine, glutamic acid) eliminated transport function of the *Arabidopsis* transporter IRT1 (Rogers et al., 2000). An acrodermatitis eneropathica-associated mutant (G539R in TM V) of ZIP4 has wild-type affinity for Zn^{2+}, but 70% reduced V_{max} values (Wang et al., 2004). The G539R mutation replaces a nonpolar residue at the beginning of TM V with a positively charged residue. As Zn^{2+} is also positively charged, this may interfere with transport due to electrostatic repulsion (Wang et al., 2004).

4.2. Expression of SLC39A1–SLC39A3

Zip1 mRNA has been detected in most adult rat and all embryonic mouse tissues (Lioumi et al., 1999). Expression of *zip1* was detected in all adult mouse tissues (except for pancreas) with highest levels in the intestine and ovary (Dufner-Beattie, Langmade, et al., 2003). *ZIP1* mRNA is ubiquitously expressed in adult human tissues and human cell lines (Gaither & Eide, 2001).

zip2 and *zip3* expression was tissue-restricted with highest levels in skin, liver, ovary for *zip2*, and in the testis for *zip3*—expression patterns consistent with cell-specific functions of these proteins. *zip1–zip3* transcripts were detectable in mouse adult intestine and embryonic visceral yolk sac (Dufner-Beattie, Langmade, et al., 2003). However, expression of these three genes was not regulated by dietary Zn^{2+} deficiency in these organs, although they are responsible for Zn^{2+} absorption from diet and Zn^{2+} supply to the embryo. Alternatively, as for other Zn^{2+} transporters, hormonal regulation may play a role. For example, stimulation of human *ZIP1* expression by prolactin and testosterone has been demonstrated in cultured prostate cells (Costello, Liu, Zou, & Franklin, 1999).

4.3. Functional role and regulation of SLC39A4

In contrast to *zip1-3*, mouse *zip4*—the ortholog of human *ZIP4* that is mutated in the Zn^{2+} metabolism-disorder acrodermatitis enteropathica (Wang et al., 2001)—is regulated by Zn^{2+} availability in mouse intestine and visceral yolk sac (Dufner-Beattie, Wang, et al., 2003). Zn^{2+} deficiency causes a marked increase in intestinal *zip4* expression (Dufner-Beattie, Wang, et al., 2003) that is consistent with its main role in Zn^{2+} absorption (Kury, Devilder, Avet-Loiseau, Dreno, & Moisan, 2001; Wang, Zhou, Kuo, Zemansky, & Gitschier, 2002). Upregulation of *zip4* also explains the finding of increased intestinal Zn^{2+} uptake activity during periods of dietary Zn^{2+} deficiency in mammals (Cousins & McMahon, 2000; Liuzzi, Blanchard, & Cousins, 2001). ZIP4 also undergoes posttranslational regulation in response to changing Zn^{2+} levels; the protein's subcellular localization is regulated in a Zn^{2+}-dependent manner. In Zn^{2+} deficiency, ZIP4 migrates to the plasma membrane, whereas it is endocytosed and associated with intracellular compartments when enough Zn^{2+} is available (Dufner-Beattie, Wang, et al., 2003; Kim et al., 2004). Intact mechanisms of Zn^{2+} sensing, proper folding, and glycosylation are important for correct localization of ZIP4 to the plasma membrane (Kim et al., 2004; Wang et al., 2004).

4.4. A role of SLC39 family proteins in cancer

A changed expression of SLC39 family proteins, and thus dysregulation of cellular Zn^{2+} metabolism, has been shown to be associated with pancreatic, prostate, and breast cancers (Desouki, Geradts, Milton, Franklin, & Costello, 2007; Li et al., 2007; Taylor et al., 2007). This is not surprising because Zn^{2+} is a cofactor of enzymes such as matrix metalloproteases (MMPs) and carbonic anhydrase, that are involved in cell proliferation, angiogenesis, and metastasis of cancer (Jeong & Eide, 2013).

In particular, SLC39A6 (LIV-1) has been implicated in metastatic breast cancer; SLC39A6 was shown to be estrogen regulated and highly expressed in estrogen-receptor-positive breast cancers that spread to the lymph nodes (Taylor et al., 2007). SLC39A6 is a downstream target of the transcription factor STAT3 and essential for the nuclear localization of the transcription factor Snail, which has a role in the development of cancer and downregulates the expression of genes associated with cell adhesion (Yamashita et al., 2004). Like the other members of the LIV1 subfamily of ZIP transporters, SLC39A6 has a metalloprotease motif in its TM V, and MMPs are well known to be involved in cancer invasion (Taylor & Nicholson,

2003; Zhang, Chen, Yao, & Li, 2010). Estrogen-receptor-positive breast cancers with resistance to antihormones (e.g., tamoxifen and fulvestrant) show a more aggressive phenotype. SLC39A7 and SLC39A8 are upregulated in antihormone-resistant cells, suggesting a role in the development of resistance (Taylor & Nicholson, 2003). In accordance with this assumption, depletion of SLC39A7 prevented the activation of signaling pathways such as EGFR, Src, and IGF1-R, all of which are known to be activated by Zn^{2+} and responsible for increased proliferation and invasion of antihormone-resistant tumors (Taylor & Nicholson, 2003).

Increased SLC39A4 expression in pancreatic cells is associated with pancreatic cancer (Li et al., 2007). SLC39A4 was considerably overexpressed (1.5- to 53.6-fold) in 94% of clinical pancreatic adenocarcinoma specimens compared to normal pancreatic tissue, and in several pancreatic cancer cell lines (Panc-1, BxPC-3, Hs766T, ASPC-1, Capan-1, HPAF-II, and PL45) compared to control HPDE cells (Li et al., 2007). Overexpression of SLC39A4 was accompanied by higher Zn^{2+} content of pancreatic tumor cells and tissues, increased cell proliferation, and tumor volume (Li et al., 2007). The mechanism(s) of SLC39A4-mediated pancreatic cancer growth include increased expression of neuropilin (NRP)-1, VEGF, MMP-2, and MMP-9 (Zhang, Chen, et al., 2010), and activation of IL-6/STAT3 pathway via the Zn^{2+}-dependent transcription factor CREB (Zhang, Bharadwaj, et al., 2010).

In contrast, decreased Zn^{2+} accumulation due to decreased expression of SLC39A1 protein or redistribution of SLC39A3 to lysosomal vesicles is associated with progression of prostate cancer (Desouki et al., 2007; Huang, Kirschke, & Zhang, 2006).

5. THE ROLE OF SLC31A1 (CTR1) IN CU^{2+} UPTAKE AND BODY CU^{2+} HOMEOSTASIS

SLC31A1 (CTR1) has been characterized as a high-affinity transporter specific for reduced Cu^{+}. Copper accumulates in cells overexpressing human CTR1 (Lee, Prohaska, Dagenais, Glover, & Thiele, 2000; Zhou & Gitschier, 1997).

5.1. Localization and expression pattern

In agreement with its role as the main mechanism for uptake of extracellular Cu^{2+} (Eisses, Chi, & Kaplan, 2005), the protein is predominantly found at the plasma membrane (reviewed in Kaplan & Lutsenko, 2009; Nevitt, Öhrvik, &

Thiele, 2012) and expressed in all organs and tissues examined (Lee et al., 2000; Zhou & Gitschier, 1997). CTR1 has been localized to cytoplasmic vesicular compartments in some carcinoma cell lines, including A549, H441, and HepG2 (Klomp, Tops, Van Denberg, Berger, & Klomp, 2002).

In mice, the highest levels of CTR1 have been found in the liver, small intestine, heart, and kidneys (Kuo, Gybina, Pyatskowit, Gitschier, & Prohaska, 2006; Lee, Prohaska, & Thiele, 2001). In addition, CTR1 levels are elevated in the mammary gland tissue obtained from pregnant or lactating mice (Kuo et al., 2006).

5.2. Results from mice with organ-specific CTR1 knockout

Generating mice with CTR1 knocked out in the intestine, liver, and heart allows for the investigation of tissue-specific functions of the protein.

Mice with intestine-specific *Slc31a1* knockout developed systemic Cu^{2+} deficiency and ataxia, and died prior to weaning (Nose, Kim, & Thiele, 2006). CTR1-mediated intestinal copper transport is thus critical to meet the Cu^{2+} demands of neonatal development, and to absorb Cu^{2+} from the diet. Confirming these functions, CTR1 is localized at the luminal side of intestinal epithelial cells of adult rat, mouse, and pig (Nose et al., 2010). However, in mammalian enterocytes, the subcellular localization of CTR1 seems to depend on the degree of cell differentiation. CTR1 is localized at the basolateral membrane of duodenal cells from suckling mice and in human models of intestinal cells (e.g., Caco2 and T84), and might play a role in acquiring Cu^{2+} from blood for their intracellular processes (Kuo et al., 2006; Zimnicka, Maryon, & Kaplan, 2007).

Specific *Slc31a1* knockout in the liver and heart of mice results in severe defects in both tissue Cu^{2+} accumulation and Cu^{2+}-dependent biochemical pathways in these organs (Kim, Son, Bailey, & Lee, 2009, Kim et al., 2010).

5.3. Interorgan signaling of Cu^{2+} status

Compared to wild-type mice, hepatic *Slc31a1* knockout mice have higher urinary Cu^{2+} secretion, although copper levels are similar in other organs (Kim et al., 2009). In liver biliary epithelial cells, CTR1 mediates Cu^{2+} uptake across the basolateral membrane, followed by Cu^{2+} delivery into the bile via ATP7B-mediated Cu^{2+} export across the apical membrane. Increased renal Cu^{2+} excretion is indicative of a compensatory mechanism to prevent hyperaccumulation of Cu^{2+} in tissues under conditions of decreased hepatobiliary Cu^{2+} excretion (Kim et al., 2009).

Another example of interorgan signaling comes from a cardiomyocyte-specific *Slc31a1* knockout mouse (Kim et al., 2010). This mouse displays increased expression of the ATP7A Cu^{2+} exporter in the intestine and liver, along with increased levels of serum Cu^{2+}. In addition, serum obtained from these mice greatly increase ATP7A expression when applied to cultured human umbilical vein endothelial cells or CaCo-2 cells. These results suggest that an unknown factor is secreted from CTR1-lacking cardiomyocytes into blood, and promotes ATP7A expression in the intestine and liver (Kim et al., 2010).

6. STRUCTURE AND MODE OF ACTION OF SLC31 FAMILY MEMBERS

The need for copper in important cell functions such as growth, development, and metabolism (Kim et al., 2008) is highly conserved across all phylogenetic levels. As such, the yeast *Saccharomyces cerevisiae* has served as a crucial model for identifying Cu^{2+} transporters. The human *CTR1* cDNA was isolated based on its capacity to rescue Cu^{2+} transport in yeast cells with defective Cu^{2+} influx systems (Zhou & Gitschier, 1997). Human genes *Slc31a1* (with five exons of 4744 bps) and *Slc31a2* (with four exons of 1785 bp) are located on chromosome 9q32. Translation of *Slc31a1* and *Slc31a2* transcripts results in proteins of 190 and 143 aa, respectively. SLC31 family members are unique in their amino acid sequences, and the sequence homology among family members is only 33% (Kim, Wu, & Lee, 2013). However, they share several structural features (Fig. 9.1), such as three predicted TMs, multiple potential metal-binding residues (e.g., methionine, histidine, cysteine), and the ability to form multimeric complexes (De Feo, Aller, & Unger, 2007; Dumay, Debut, Mansour, & Saier, 2006; Kaplan & Lutsenko, 2009; Lee, Pena, Nose, & Thiele, 2002; Petris, 2004).

CTR1 is composed of the following three major domains: an extracellular N-terminal domain, three TMs, and a cytosolic C-terminal tail (Fig. 9.1). The long N-terminal domain contains two methionine-rich clusters at residues 7–12 and 40–45, and two histidine clusters at residues 3–6 and 22–24 (De Feo et al., 2007; Dumay et al., 2006; Kaplan & Lutsenko, 2009; Larson et al., 2010; Petris, 2004). Another methionine-rich metal-binding motif (MX3M) exists within the second TM segment (Fig. 9.1; De Feo, Mootien, Siluvai, Blackburn, & Unger, 2009; Puig, Lee, Lau, & Thiele, 2002).

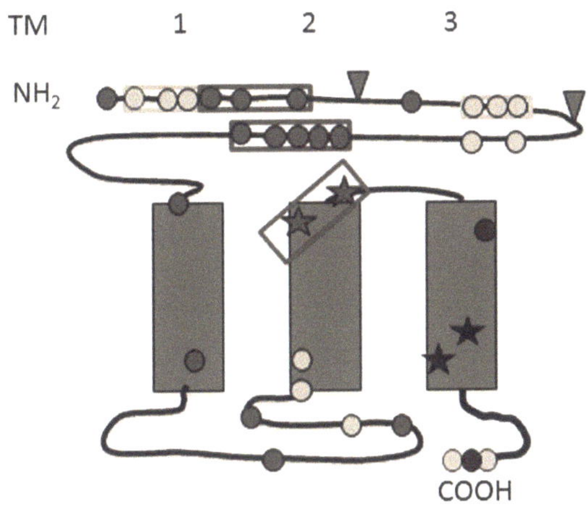

Figure 9.1 Model of SLC31A1 membrane topology. The protein possesses three transmembrane domains (TM) with the N and C termini in an outside-in configuration. Potential methionine (●), histidine (○), and cysteine (●) metal-binding residues are marked. The long N terminus contains two methionine (●) and two histidine (○) clusters that are important for copper acquisition from plasma, and for Cu^{2+} reduction to Cu(I). The methionine (★)-rich MX3M-motif in the second TM is essential for SLC31A1 copper transport activity and posttranslational regulation. N- and O-glycosylation sites at the N terminus are shown (▼). O-glycosylation at Thr27 protects the transporter against cleavage of a ~30-aa fragment from the N terminus. The functional transporter assembles as a homotrimer and two conserved glycine residues (★) in the third TM are critically involved in this process. (See the color plate.)

6.1. Channel-like architecture based on oligomerization

According to biochemical analysis and structural studies, CTR1 assembles into a compact homotrimer with a minimal membrane-spanning pore of 9 Å, allowing for the passage of copper ions across the lipid bilayer (Aller & Unger, 2006; De Feo et al., 2007, 2009; Lee, Pena, et al., 2002). The third TM domain contains a conserved GXXXG (GG4) motif (Fig. 9.1) that is known to be involved in oligomerization of other integral membrane proteins (De Feo et al., 2007). The GG4 motif is critical in forming the functional CTR1 multimer. In particular, it forms the tight interface between TM3 and TM1 (Aller, Eng, De Feo, & Unger, 2004; De Feo, Mootien, & Unger, 2010). According to data on human CTR1, the multimeric complex changes conformation in conjunction with Cu^{2+} transport. A rotational movement around the pore central axes followed by the movement of the cytoplasmic part of the helixes away from the pore center have been predicted (Eisses & Kaplan, 2002; Schushan, Barkan, Haliloglu, & Ben-Tal, 2010; Sinani, Adle, Kim, & Lee, 2007). Excess Cu^{2+} and cisplatin (an anticancer drug) stabilize the multimeric CTR1 complexes (Guo, Smith, & Petris, 2004; Sinani et al., 2007). Stabilization of the CTR1 homotrimer involves binding of cisplatin to the methionine-rich clusters in the extracellular N terminus followed by formation of cross-links between CTR1 polypeptides (Guo, Smith, & Petris, 2004). The appearance of stable CTR1 complexes may be an important mechanism underlying CTR1-mediated Cu^{2+} and cisplatin uptake, control of CTR1 activity, and/or subcellular trafficking. In fission yeast, heteromultimerization of CTR4 and CTR5 is important for surface trafficking of the transporter (Beaudoin, Thiele, Labbé, & Puig, 2011). Interestingly, hormones such as estrogen, progesterone, and insulin also induce CTR1 multimerization (Hardman et al., 2006).

6.2. Mechanism of SLC31-mediated copper transport

SLC31 family members do not possess a domain that indicates ATP as an energy source for Cu^{2+} transport. Consistently, inhibitors of ATP synthesis do not abolish human CTR1-mediated Cu^{2+} transport (Lee, Pena, et al., 2002).

Therefore, the electrochemical gradient of Cu^{2+} or other ions (e.g., K^{+}) across the plasma membrane may be the driving force for CTR1-mediated Cu^{2+} uptake. However, considering that only a small fraction of Cu^{2+} in both cytoplasm and plasma is unbound, the transmembrane ionic gradient of free Cu^{2+} is small. Lee, Pena, et al. (2002) showed that the initial rate of Cu^{2+} uptake in HEK293 cells overexpressing human CRT1 was 10-fold greater when cells were suspended in HEPES-buffered salt solution (HBSS) than when cells were incubated in Dulbecco's minimal essential medium (DMEM) with 10% fetal bovine serum. Supplementing HBSS with albumin or histidine (yielding cupric-histidine) diminished Cu^{2+} uptake to the levels seen in DMEM (Lee, Pena, et al., 2002). These results indicate competition for ionic Cu^{2+} between CTR1 and other Cu^{2+}-binding molecules, for example, other plasma proteins, amino acids, or ions. More than 90% of serum Cu^{2+} is bound as Cu^{2+}-containing ferroxidase ceruloplasmin (Cp) (Hellman & Gitlin, 2002). However, Cu^{2+} uptake in organs and tissues of Cp knockout mice is unchanged (Meyer, Durley, Prohaska, & Harris, 2001). Therefore, it seems more likely that albumin and, to a lesser extent, transcuprein or transcuprein-like proteins and transferrin are major carriers of the exchangeable pool of plasma Cu^{2+}.

CTR1 comprises rather specific, high-affinity Cu^{2+} transport. Other metals such as Zn^{2+}, Fe^{2+}, Mn^{2+}, and Cd^{2+} hardly inhibit accumulation of $^{64}Cu^{2+}$ in wild-type and CTR1 overexpressing HEK293 cells (Lee, Petris, & Thiele, 2002). However, Cu^{2+} competes with silver (a metal ion that possesses similar characteristics to Cu^{+}) but not with divalent metals for CTR1-mediated uptake. On the other hand, Cu^{2+} uptake is stimulated by ascorbic acid, a Cu^{2+} reductant, suggesting that reduced Cu^{+} is the substrate of CTR1 (Lee, Petris, et al., 2002). This assumption is consistent with data on yeast showing that yeast FRE metalloreductases reduce Cu^{2+} and facilitate yCTR1-mediated transport of the ion (Georgatsou, Mavrogiannis, Fragiadakis, & Alexandraki, 1997). Candidates in mammalian cells that possess cupric-reductase activity are the six-transmembrane epithelia antigen of the prostate (Steap) protein family (Ohgami, Campagna, McDonald, & Fleming, 2006) and inducible duodenal cytochrome *b* (Dcytb or Cybrd1) (McKie et al., 2001; Wyman, Simpson, McKie, & Sharp, 2008). The latter is predominantly localized to the duodenal brush-border membrane (McKie et al., 2001) and has been shown to reduce Cu^{2+} complexes and Cu^{2+}-histidine when overexpressed in MDCK cells (Wyman et al., 2008). Steap members 2, 3, and 4 also act as Cu^{2+} reductases and, moreover, have been shown to stimulate uptake of the ion (Ohgami et al., 2006). Steap 2–4 proteins are widely expressed and localized in the plasma membrane and the intracellular endosomal compartment. Altogether, these results point to a physiological role of these metalloreductases in CTR1- and possibly CTR2-mediated Cu^{2+} translocation.

Copper ions or copper bound to extracellular ligands might be captured by CTR1 or other cell-surface molecule(s) such as metalloreductases and adaptors to be specifically processed for CTR1-mediated translocation. Putative mechanisms that protect Cu^{+} for its transport by CTR1 include binding to a molecule that has a high affinity to the CTR1 N terminus, or formation of metalloreductase-CTR1 complexes for directly passing the reduced Cu^{+}.

Several residues and motifs shown in Fig. 9.1 are involved in Cu^{2+} transport by CTR1 (De Feo et al., 2007; Eisses & Kaplan, 2005; Kaplan & Lutsenko, 2009; Puig et al., 2002). These motifs may interact with plasma Cu^{2+} carrier(s) and metalloreductase(s) and/or control gating of the channel formed within the CTR1 multimeric complex. According to experimental data, the last methionine cluster at the extracellular N terminus and two methionines at the second TM of CTR1 are essential for Cu^{2+} binding and translocation (Puig et al., 2002; Schushan et al., 2010). The methionine-rich metal-binding motif (MX3M; Fig. 9.1) within the second TM has been shown to facilitate Cu^{2+} uptake via formation of Cu–S linkages (De Feo et al., 2009; Puig et al., 2002). Haas, Putterman, White, Thiele, and Franz (2011) recently showed that the first N-terminal histidine-rich motif of human CTR1 (Fig. 9.1) plays a role in copper acquisition from the plasma followed by reduction to Cu^{+}.

7. REGULATION OF COPPER ACQUISITION BY SLC31A1

Strict regulation of uptake, distribution, and excretion of Cu^{2+} is necessary for maintaining optimal activities of copper-dependent cellular pathways.

7.1. Regulation of CTR1 expression by Cu^{2+}

ctr1 expression is regulated at the transcriptional and translational levels (Bertinato et al., 2008). High extracellular Cu^{2+} levels reduce hCTR1 mRNA expression in small-cell lung cancer (SCLC) cells by 30% (25 μM Cu^{2+}) and 60% (100 μM Cu^{2+}), respectively (Song et al., 2008). On the other hand, depletion of Cu^{2+} from the SCLCs medium by using the Cu^{2+} chelator BCS leads to a twofold increase of hCTR1 mRNA expression (Song et al., 2008). Regulation of basal and Cu^{2+}-responsive hCTR1 mRNA expression is mediated by the transcription factor specificity protein 1 (SP1), which interacts with three SP1 binding sites (GC boxes) localized in the proximal region of the *hCTR1* promotor (Song et al., 2008). Knockdown of SP1 expression by siRNA or its overexpression by transfection leads to respective changes of hCTR1 mRNA expression—downregulation and upregulation, respectively (Liang, Tsai, Lee, Savaraj, & Kuo, 2012). SP1 contains three zinc-finger domains at its C terminus that bind to GC-rich DNA sequences. The zinc-finger domains in SP1 seem to function as Cu^{2+} sensors, and its deletion renders hCTR1 insensitive to changes of extracellular Cu^{2+} (Liang, Tsai, et al., 2012; Song et al., 2008).

Basal expression of CTR1 in the intestine is regulated by the transcription factor HIF2α (hypoxia-inducible factor 2α). Chemical inhibition of HIF2α in Caco-2 cells, treatment of HuTu 80 cells with HIF2α-siRNA, and intestine-specific HIF2α-knockout in mice all decreased the basal expression of CTR1 mRNA (Pourvali et al., 2012).

Hypoxia, a stress condition found within the microenvironment of injured tissue, has been shown to increase CTR1 mRNA and protein levels in Caco-2 cells (Pourvali et al., 2012). In addition, hypoxia stimulates the expression of CTR1 and Cu^{2+} uptake in RAW264.7 macrophage cells and in primary mouse peritoneal macrophages (White et al., 2009). The associated elevation of cytosolic Cu^{2+} induces trafficking of ATP7A from the perinuclear *trans*-Golgi network to cytoplasmic vesicles (White et al., 2009). Via ATP7A, Cu^{2+} is then delivered to the secretory pathway and targeted to ceruloplasmin (White et al., 2009).

Interferon-γ also was found to increase CTR1 expression, and thus Cu^{2+} uptake in macrophages (Zheng et al., 2010). The increase in intracellular Cu^{2+} levels triggers translocation of ATP7A from the Golgi to cytoplasmic vesicles partially identical with the phagosomes (Zheng et al., 2010). These data support a significant role for cellular Cu^{2+} distribution in innate immunity that was first observed several decades ago (Prohaska & Lukasewycz, 1990). A similar regulation of the Cu^{2+} transporters was observed in microglial cells that are clustered around amyloid plaques in a mouse model of AD and in interferon-γ-stimulated microglial cells (Zheng et al., 2010). Cu^{2+} sequestration by microglia under these conditions may be a mechanism for either acquiring copper for neurons, or protecting them from copper toxicity.

Cu^{2+}-deficient animals exhibit higher CTR1 expression in a tissue-specific manner. For instance, total and apical membrane CTR1 levels are elevated due to increased protein stability in the intestine of mice fed with a copper-deficient diet (Kuo et al., 2006; Nose et al., 2010).

7.2. Posttranslational modification

Posttranslational regulation causes changes in the cellular localization of CTR1 (van den Berghe & Klomp, 2010). At high extracellular Cu^{2+} concentrations, CTR1 is rapidly internalized by endocytosis in order to prevent intracellular Cu^{2+} accumulation (Liang, Stockton, Savaraj, & Kuo, 2009; Molloy & Kaplan, 2009). In some cell types, this internalization is followed by degradation of the protein (Guo, Smith, Lee, Thiele, & Petris, 2004; Petris, Smith, Lee, & Thiele, 2003). According to *in vitro* mutagenesis experiments, intramembranous methionines Met-150 and Met-154 (Fig. 9.1), which are essential for transport activity, are also required for Cu^{2+}-stimulated endocytosis and degradation of CTR1 (Guo, Smith, Lee, et al., 2004). The latter process is mediated by the N-terminal methionine clusters (Guo, Smith, Lee, et al., 2004).

In lactating rats, Cu^{2+} transport into the milk was stimulated by suckling and hyperprolactinemia (Kelleher & Lönnerdal, 2006). Prolactin has no effect on CTR1 mRNA expression or protein abundance, but results in relocalization of CTR1 to the cell membrane (Kelleher & Lönnerdal, 2006). In addition, insulin has been shown to alter the intracellular localization of CTR1 in the placental cell line Jeg-3 (Hardman et al., 2006). Together these data suggest that modulation of CTR1 trafficking is important in both hormonal and Cu^{2+}-dependent regulation of CTR1-mediated Cu^{2+} entry.

In yeast, binding of Cu^{2+} to Cys residues in the C-terminal cytosolic domain of yCTR1 is essential for rapid, concentration-dependent inactivation in response to excess Cu^{2+}, but not for high-affinity Cu^{2+} transport (Wu, Sinani, Kim, & Lee, 2009; Xiao, Loughlin, George, Howlett, & Wedd, 2004). In other words, Cys residues in the CTR1 C terminus are responsible for lowering its functional activity under conditions of excess Cu^{2+}. This regulatory mechanism might be conserved in mammalian CTR Cu^{2+} transporters including CTR1 with similar Cys residues (Fig. 9.1). Rapid inactivation may prevent excessive accumulation of toxic Cu^{2+} levels (Molloy & Kaplan, 2009), for example, by limiting Cu^{2+} entry into the body through the intestine (Nose et al., 2010).

N- and O-glycosylation occurs at the extracellular N terminus of human CTR1 (Eisses & Kaplan, 2002; Maryon, Molloy, & Kaplan, 2007) that is similar to that seen in heavily O-glycosylated yeast CTR1 (Dancis, Haile, Yuan, & Klausner, 1994). Removing N-linked glycosylation at residue 15 did not alter localization of CTR1, but significantly reduced Cu^{2+} transport activity (Maryon, Molloy, Zimnicka, & Kaplan, 2007). O-glycosylation at Thr27 protects CTR1 against N-terminal cleavage of approximately 30 aa (Maryon, Molloy, & Kaplan, 2007). The biological function of this cleaved variant and the protease responsible are currently unclear, but this posttranslational modification may endow novel biochemical properties to the transporter. Alternatively, the proteolytic fragment that is released might serve a specific function in the extracellular milieu.

8. THE SLC31A2 (CTR2) COPPER TRANSPORTER

CTR2 is the second member of the mammalian SLC31 family of Cu^{2+} transporters and shows structural homology to the second transmembrane domain of CTR1 that is essential for Cu^{2+} transport (Zhou &

Gitschier, 1997). Its overall organization mirrors that of CTR1 with three putative TMs and an intramembranous MX3M motif, but without an extended N-terminal domain. According to biochemical studies, CTR2 functions as an oligomeric protein, but structural details are currently lacking. Highest levels of the CTR2 protein are found in the rat heart and placenta (Bertinato et al., 2008). Based on immunocytochemical experiments, CTR2 localizes predominantly to the endosome and lysosome in mammalian cells (van den Berghe et al., 2007) and in multiple cell lines (HEK293T, HeLa, U20S, COS7) in contrast to that seen for CTR1 (Bertinato et al., 2008; van den Berghe et al., 2007). Comparison of hCTR1 and hCTR2 primary amino acid sequences revealed a conserved dileucine motif in the C terminus of hCTR2 but not hCTR1. This motif is frequently involved in endocytosis of integral membrane proteins including ATP7A and could explain the exclusive intracellular localization of hCTR2.

Hence, CTR2 appears responsible for moving Cu^{2+} from vesicular organelles to the cytoplasm (Rees et al., 2004). CTR2-mediated Cu^{2+} transport might serve as a mechanism for Cu^{2+} recycling from intracellular stores following the degradation of cuproenzymes (Bertinato et al., 2008). A small percentage of CTR2 ($\sim$5%) also exists at the plasma membrane and mediates Cu^{2+} transport into the cytoplasm (Bertinato et al., 2008; van den Berghe et al., 2007). Overexpression of CTR2 in Cu^{2+}-depleted COS-7 cells resulted in the hyperaccumulation of Cu^{2+} when the cells were subsequently exposed to Cu^{2+}-enriched media (Bertinato et al., 2008). Posttranslational regulation of CTR2 function has not been reported. Also, it is not known whether CTR1 and CTR2 form heteromultimers, which might provide yet another level of regulation for controlling Cu^{2+} transport.

9. A ROLE OF SLC31 FAMILY MEMBERS IN DISEASE

9.1. Cancer

CTR1 is an important factor in determining toxicity of platinum-based anticancer drugs such as cisplatin (Howell, Safaei, Larson, & Sailor, 2010; Ishida, Lee, Thiele, & Herskowitz, 2002; Kuo, Chen, Song, Savaraj, & Ishikawa, 2007). Cisplatin and related drugs appear to bind with high affinity to Cu^{2+}-binding residues and domains of CTR1 and other molecules involved in Cu^{2+} metabolism (Crider, Holbrook, & Franz, 2010; Guo, Smith, & Petris, 2004). Compared with wild-type cells, mice embryonic fibroblasts with *CTR1*-knockout display only 30–35% of cisplatin transport activity (Holzer, Manorek, & Howell, 2006). The cytotoxic effect of cisplatin is proportional to the total platinum content inside the cancerous cells. Therefore, a high level of CTR1 expression is linked to a good prognosis and chemotherapy response (Chen et al., 2008, 2012; Lee et al., 2011; Liang, Long, et al., 2012). Ishida et al. (2002) investigated CTR1 expression in 91 patients with ovarian cancer and found a positive correlation between high CTR1 expression levels and longer disease-free survival times. In patients with stage III non-SCLC, overexpression of CTR1 was associated with a better chemotherapy response, and an increase in progression-free and overall survival times (Chen et al., 2012). CTR1 expression can be modulated by the cellular pool of bioavailable Cu^{2+}; for example, cisplatin-resistant cell lines and cell lines established from patients who had failed in platinum-based chemotherapy were resensitized by Cu^{2+} lowering agents that increase CTR1 expression (Liang, Long, et al., 2012). Glutathione functions as a Cu^{2+} chelator, and its overexpression leads to upregulation of CTR1, increased cisplatin uptake, and cell death (Chen et al., 2008). In contrast, CTR2 is associated with cisplatin insensitivity, and higher relative levels indicate a poorer cancer patient outcome, for example, shorter progression-free survival (Blair et al., 2011; Lee et al., 2011). Thus, determination of the ratio between CTR1 and CTR2 expression might allow to predict the patient responses to platinum-based drug therapies.

9.2. Immune function

The role of Cu^{2+} in immune function has become a topic of considerable focus (Prohaska & Lukasewycz, 1990). In response to bacterial infection, macrophages upregulate *SLC31A1*, exhibit increased surface CTR1, and accumulate Cu^{2+} within their phagosomes via ATP7A (reviewed in Hodgkinson & Petris, 2012). An upregulation of CTR1, CTR2, and ATP7A has been observed in murine bone-derived macrophages infected with *Salmonella typhimurium* or treated with LPS (Achard et al., 2012).

9.3. Neurodegenerative diseases

Chronic perturbations to copper homeostasis result in neuronal degeneration (reviewed in Nevitt et al., 2012). Microglia has been shown to upregulate SLC31A1 in response to interferon-γ activation. Increased Cu^{2+} uptake and trafficking by microglia may have a neuroprotective role in AD (Zheng et al., 2010). Similarly, upregulation of CTR1 in astrocytes has a protective effect in Cu^{2+} toxicity (reviewed by Nevitt et al., 2012).

10. CONCLUSION

The SLC30/39 families of Zn^{2+} transporters and the SLC31 family of Cu^{2+} transporters are important components of a network of proteins that balance the homeostasis of these essential trace elements within a narrow range between toxicity and deficiency.

10.1. SLC30 and SLC39 Zn^{2+} transporters

For most SLC30 and SLC39 Zn^{2+} transporters, their topology, expression pattern, subcellular localization, and general role in Zn^{2+} efflux, sequestration of intracellular Zn^{2+}, or Zn^{2+} uptake are clear. Members of both families have been shown to be regulated by transcriptional, translational, and posttranslational mechanisms in response to changes in Zn^{2+} availability and by hormones. Currently, the localization and specific function(s) of SLC30A10 are unknown. SLC30A9 shows a cation efflux motif and, thus, is a putative Zn^{2+} transporter. However, in contrast to all other ZnTs, SLC30A9 has been found in the cytosol and in nuclear fractions of cells. Its exact role in activation of transcription has to be clarified. Multimeric assembly has been predicted and experimentally shown for SLC30 members. The specific function of heterodimer formation, for example, to modulate Zn^{2+} transport, is an interesting topic for further research. The mechanisms of Zn^{2+} transport for most SLC30 and SLC39 members are unknown but appear to involve Zn^{2+}/H^{+} exchange and cotransport of Zn^{2+} with bicarbonate, respectively. The presence of a multitude of SLC30 and SLC39 proteins with different tissue expression profiles and subcellular localizations points to functional differences and/or differences in Zn^{2+} specificity of these transporters. Also, some may not be involved in transport directly but operate as chaperons or modulators of transport activity.

Because Zn^{2+} transporters are involved in a multitude of physiological processes, it is not surprising that defective or dysregulated transporters are involved in the development and/or progression of diseases such as diabetes, cancer, or AD. Thus, a better understanding of the precise role of Zn^{2+} transporters in physiological and pathophysiological pathways will be important for the prevention, diagnosis/prognosis, and treatment of such diseases.

10.2. The SLC31 Cu^{2+} transporters

Experimental data from studies on cells overexpressing SLC31A1 or mouse models harboring an *Slc31a1* knockout confirm that the protein plays an essential role in Cu^{2+} translocation across cellular membranes into the cytosol. Specific functions of SLC31A2 are far from clear, but the protein appears to mediate transport of Cu^{2+} from vesicular organelles into the cytoplasm. Structural details of SLC31A2 have yet to be explored.

Multimerization of three SLC31A1 molecules is critical in forming functional Cu^{2+} transporters and might also play roles in CTR1 expression and posttranslational regulation. It would be interesting to know if SLC31A1 and SLC31A2 also form heteromultimers. Another question is how Cu^{2+} is transferred from the oligomeric CTR1 to intracellular metallochaperones, and where in the cell this occurs.

Cu^{2+} homeostasis is not only regulated at the cellular level. There is also evidence for interorgan communication of Cu^{2+} status. For example, under conditions of decreased hepatobiliary Cu^{2+} excretion, a compensatory elevation of renal Cu^{2+} excretion is observed. In this regard, regulated cleavage of a peptide from the CTR1 N terminus may play a role. It will be important to determine whether this peptide exists in serum, and whether it regulates Cu^{2+} homeostasis when applied to the extracellular milieu.

The specific roles of SLC31 members in copper-linked human disorders, including immune dysfunction, cancer, and neurodegenerative diseases, are only poorly characterized but are important future research topics.

REFERENCES

Achard, M. E., Stafford, S. L., Bokil, N. J., Chartres, J., Bernhardt, P. V., Schembri, M. A., et al. (2012). Copper redistribution in murine macrophages in response to Salmonella infection. *Biochemical Journal, 444*, 51–57.

Ackland, M. L., & Mercer, J. E. (1992). The murine mutation, lethal milk, results in production of zinc-deficient milk. *The Journal of Nutrition, 122*, 1214–1218.

Ackland, M. L., & Michalczyk, A. (2006). Zinc deficieny and its inherent disorders—A review. *Genes & Nutrition, 1*, 41–50.

Adlard, P. A., Parncutt, J. M., Finkelstein, D. I., & Bush, A. I. (2010). Cognitive loss in zinc transporter-3 knock-out mice: A phenocopy for the synaptic and memory deficits of Alzheimer's disease? *Journal of Neuroscience, 30*, 1631–1636.

Aller, S. G., Eng, E. T., De Feo, C. J., & Unger, V. M. (2004). Eukaryotic CTR copper uptake transporters require two faces of the third transmembrane domain for helix packing, oligomerization, and function. *Journal of Biological Chemistry, 279*, 53435–53441.

Aller, S. G., & Unger, V. M. (2006). Projection structure of the human copper transporter CTR1 at 6-A resolution reveals a compact trimer with a novel channel like architecture.

Proceedings of the National Academy of Sciences of the United States of America, 103, 3627–3632.
Beach, R. S., Gershwin, M. E., Makishima, R. K., & Hurley, L. S. (1980). Impaired immunologic ontogeny in postnatal zinc deprivation. *The Journal of Nutrition, 110*, 805–815.
Beaudoin, J., Thiele, D. J., Labbé, S., & Puig, S. (2011). Dissection of the relative contribution of the Schizosaccharomyces pombe Ctr4 and Ctr5 proteins to the copper transport and cell surface delivery functions. *Microbiology, 157*, 1021–1031.
Bertinato, J., Swist, E., Plouffe, L. J., Brooks, S. P., & L'abbé, M. R. (2008). Ctr2 is partially localized to the plasma membrane and stimulates copper uptake in COS-7 cells. *Biochemical Journal, 409*, 731–740.
Blair, B. G., Larson, Ch. A., Adams, P. L., Abada, P. B., Pesce, E., Safaei, R., et al. (2011). Coper transporter 2 regulates endocytosis and controls tumor growth and sensitivity to cispltin in vivo. *Molecular Pharmacology, 79*, 157–166.
Brown, R. S., Sander, C., & Argos, P. (1985). The primary structure of transcription factor IIIA has 12 consecutive repeats. *FEBS Letters, 186*, 271–274.
Chen, Y. H., Kim, J. H., & Stallcup, M. R. (2005). GAC63, a GRIP1-dependent nuclear receptor coactivator. *Molecular and Cellular Biology, 25*, 5965–5972.
Chen, H. H., Song, I. S., Hossain, A., Choi, M. K., Yamane, Y., Liang, Z. D., et al. (2008). Elevated glutathione levels confer cellular sensitization to cisplatin toxicity by up-regulation of copper transporter hCtr1. *Molecular Pharmacology, 74*, 697–704.
Chen, H. H., Yan, J. J., Chen, W. C., Kuo, M. T., Lai, Y. H., Lai, W. W., et al. (2012). Predictive and prognostic value of human copper transporter 1 (hCtr1) in patients with stage III non-small-cell lung cancer receiving first-line platinum-based doublet chemotherapy. *Lung Cancer, 75*, 228–234.
Chimienti, F., Devergnas, S., Favier, A., & Seve, M. (2004). Identification and cloning of a beta-cell-specific zinc transporter, ZnT-8, localized into insulin secretory granules. *Diabetes, 53*, 2330–2337.
Chowanadisai, W., Kelleher, S. L., & Lonnerdal, B. (2005). Zinc deficiency is associated with increased brain zinc import and LIV-1 expression and decreased ZnT-1 expression in neonatal rats. *The Journal of Nutrition, 135*, 1002–1007.
Chowanadisai, W., Lonnerdal, B., & Kelleher, S. L. (2006). Identification of a mutation in SLC30A2 (ZnT2) in women with low milk zinc concentration that results in transient neonatal zinc deficiency. *Journal of Biological Chemistry, 281*, 39699–39707.
Cole, T. B., Wenzel, H. J., Kafer, K. E., Schwartzkroin, P. A., & Palmiter, R. D. (1999). Elimination of zinc from synaptic vesicles in the intact mouse brain by disruption of the ZnT3 gene. *Proceedings of the National Academy of Sciences of the United States of America, 96*, 1716–1721.
Costello, L. C., Liu, Y., Zou, J., & Franklin, R. B. (1999). Evidence for a zinc uptake transporter in human prostate cancer cells which is regulated by prolactin and testosterone. *Journal of Biological Chemistry, 274*, 17499–17504.
Cousins, R. J., & McMahon, R. J. (2000). Integrative aspects of zinc transporters. *The Journal of Nutrition, 130*, 1384S–1387S.
Coyle, P., Philcox, J. C., Carey, L. C., & Rofe, A. M. (2002). Metallothionein: The multipurpose protein. *Cellular and Molecular Life Sciences, 59*, 627–647.
Cragg, R. A., Phillips, S. R., Piper, J. M., Varma, J. S., Campbell, F. C., Mathers, J. C., et al. (2005). Homeostatic regulation of zinc transporters in the human small intestine by dietary zinc supplementation. *Gut, 54*, 469–478.
Crider, S. E., Holbrook, R. J., & Franz, K. J. (2010). Coordination of platinum therapeutic agents to met-rich motifs of human copper transport protein1. *Metallomics, 2*, 74–83.
Cui, Y., Vogt, S., Olson, N., Glass, A. G., & Rohan, T. E. (2007). Levels of zinc, selenium, calcium, and iron in benign breast tissue and risk of subsequent breast cancer. *Cancer Epidemiology, Biomarkers & Prevention, 16*, 1682–1685.

Culotta, V. C., Yang, M., & O'Halloran, T. V. (2006). Activation of superoxide dismutases: Putting the metal to the pedal. *Biochimica et Biophysica Acta, 1763*, 747–758.
Dancis, A., Haile, D., Yuan, D. S., & Klausner, R. D. (1994). The *Saccharomyces cerevisiae* copper transport protein (Ctr1p). Biochemical characterization, regulation by copper, and physiologic role in copper uptake. *Journal of Biological Chemistry, 269*, 25660–25667.
Danks, D. M. (1988). Copper deficiency in humans. *Annual Review of Nutrition, 8*, 235–257.
De Feo, C. J., Aller, S. G., & Unger, V. M. (2007). A structural perspective on copper uptake in eukaryotes. *Biometals, 20*, 705–716.
De Feo, C. J., Mootien, S., Siluvai, G. S., Blackburn, N. J., & Unger, V. M. (2009). Three-dimensional structure of the human copper transporter hCTR1. *Proceedings of the National Academy of Sciences of the United States of America, 106*, 4237–4422.
De Feo, C. J., Mootien, S., & Unger, V. M. (2010). Tryptophan scanning analysis of the membrane domain of CTR-copper transporters. *Journal of Membrane Biology, 234*, 113–123.
Desouki, M. M., Geradts, J., Milton, B., Franklin, R. B., & Costello, L. C. (2007). hZip2 and hZip3 zinc transporters are down regulated in human prostate adenocarcinomatous glands. *Molecular Cancer, 6*, 37–43.
Dufner-Beattie, J., Langmade, S. J., Wang, F., Eide, D., & Andrews, G. K. (2003). Structure, function and regulation of a subfamily of mouse zinc transporter genes. *Journal of Biological Chemistry, 278*, 50142–50150.
Dufner-Beattie, J., Wang, F., Kuo, Y. M., Gitschier, J., Eide, D., & Andrews, G. K. (2003). The acrodermatitis enteropathica gene ZIP4 encodes a tissue-specific, zinc regulated zinc transporter in mice. *Journal of Biological Chemistry, 278*, 33474–33481.
Dumay, Q. C., Debut, A. J., Mansour, N. M., & Saier, M. H. (2006). The copper transporter (Ctr) family of Cu^{+} uptake systems. *Journal of Molecular Microbiology and Biotechnology, 11*, 10–19.
Eide, D. J. (2006). Zinc transporters and cellular trafficking of zinc. *Biochimica et Biophysica Acta, 1763*, 711–722.
Eisses, J. F., Chi, Y., & Kaplan, J. H. (2005). Stable plasma membrane levels of hCTR1 mediate cellular copper uptake. *Journal of Biological Chemistry, 280*, 9635–9639.
Eisses, J. F., & Kaplan, J. H. (2002). Molecular characterization of hCTR1, the human copper uptake protein. *Journal of Biological Chemistry, 277*, 29162–29171.
Eisses, J. F., & Kaplan, J. H. (2005). The mechanism of copper uptake mediated by human CTR1: A mutational analysis. *Journal of Biological Chemistry, 280*, 37159–37168.
Eng, B. H., Guerinot, M. L., Eide, D., & Saier, M. H., Jr. (1998). Sequence analyses and phylogenetic characterization of the ZIP family of metal ion transport proteins. *The Journal of Membrane Biology, 166*, 1–7.
Festa, R. A., & Thiele, D. J. (2011). Copper: An essential metal in biology. *Current Biology, 21*, R877–R883.
Fukunaka, A., Suzuki, T., Kurokawa, Y., Yamazaki, T., Fujiwara, N., Ishihara, K., et al. (2009). Demonstration and characterization of the heterodimerization of ZnT5 and ZnT6 in the early secretory pathway. *Journal of Biological Chemistry, 284*, 30798–30806.
Gaither, L. A., & Eide, D. J. (2000). Functional expression of the human hZIP2 zinc transporter. *Journal of Biological Chemistry, 275*, 5560–5564.
Gaither, L. A., & Eide, D. J. (2001). Eukaryotic zinc transporters and their regulation. *Biometals, 14*, 251–270.
Georgatsou, E., Mavrogiannis, L. A., Fragiadakis, G. S., & Alexandraki, D. (1997). The yeast fre1p/Fre2p cupric reductases facilitate copper uptake and are regulated by copper-modulated Mac1p activator. *Journal of Biological Chemistry, 272*, 13786–13792.
Guffanti, A. A., Wei, Y., Rood, S. V., & Krulwich, T. A. (2002). An antiport mechanism for a member of the cation diffusion facilitator family: Divalent cations efflux in exchange for K^{+} and H^{+}. *Molecular Microbiology, 45*, 145–153.

Guo, L., Lichten, L. A., Ryu, M. S., Liuzzi, J. P., Wang, F., & Cousins, R. J. (2010). STAT5-glucocorticoid receptor interaction and MTF-1 regulate the expression of ZnT2 (Slc30a2) in pancreatic acinar cells. *Proceedings of the National Academy of Sciences of the United States of America*, *107*, 2818–2823.

Guo, Y., Smith, K., Lee, J., Thiele, D. J., & Petris, M. J. (2004). Identification of methionine-rich clusters that regulate copper-stimulated endocytosis of the human Ctr1 copper transporter. *Journal of Biological Chemistry*, *279*, 17428–17433.

Guo, Y., Smith, K., & Petris, M. J. (2004). Cisplatin stabilizes a multimeric complex of the human Ctr1 copper transporter: Requirement for the extracellular methionine-rich clusters. *Journal of Biological Chemistry*, *279*, 46393–46399.

Haas, K. L., Putterman, A. B., White, D. R., Thiele, D. J., & Franz, K. J. (2011). Model peptides provide new insights into the role of histidine residues as potential ligands in human cellular copper acquisition via Ctr1. *Journal of the American Chemical Society*, *133*, 4427–4437.

Haase, H., & Beyersmann, D. (2002). Intracellular zinc distribution and transport in C6 rat glioma cells. *Biochemical and Biophysical Research Communications*, *296*, 923–928.

Hardman, B., Manuelpillai, U., Wallace, E. M., Monty, J. F., Kramer, D. R., Kuo, Y. M., et al. (2006). Expression, localisation and hormone regulation of the human copper transporter hCTR1 in placenta and choriocarcinoma Jeg-3 cells. *Placenta*, *27*, 968–977.

Hellman, N. E., & Gitlin, J. D. (2002). Ceruloplasmin metabolism and function. *Annual Review of Nutrition*, *22*, 439–458.

Helston, R. M., Phillips, S. R., McKay, J. A., Jackson, K. A., Mathers, J. C., & Ford, D. (2007). Zinc transporters in the mouse placenta show a coordinated regulatory response to changes in dietary zinc intake. *Placenta*, *28*, 437–444.

Hodgkinson, V., & Petris, M. J. (2012). Copper homeostasis at the host–pathogen interface. *Journal of Biological Chemistry*, *287*, 13549–13555.

Holzer, A. K., Manorek, G. H., & Howell, S. B. (2006). Contribution of the major copper influx transporter CTR1 to the cellular accumulation of cisplatin, carboplatin, and oxaliplatin. *Molecular Pharmacology*, *70*, 1390–1394.

Howell, S. B., Safaei, R., Larson, C. A., & Sailor, M. J. (2010). Copper transporters and the cellular pharmacology of the platinum-containing cancer drugs. *Molecular Pharmacology*, *77*, 887–894.

Huang, L., & Gitschier, J. (1997). A novel gene involved in zinc transport is deficient in the lethal milk mouse. *Nature Genetics*, *17*, 292–297.

Huang, L., Kirschke, C. P., & Gitschier, J. (2002). Functional characterization of a novel mammalian zinc transporter, ZnT6. *Journal of Biological Chemistry*, *277*, 26389–26395.

Huang, L., Kirschke, C. P., & Zhang, Y. (2006). Decreased intracellular zinc in human tumorigenic prostate epithelial cells: A possible role in prostate cancer progression. *Cancer Cell International*, *6*, 10. http://dx.doi.org/10.1186/1475-2867-6-10.

Huang, L., & Tepaamorndech, S. (2013). The SLC30 family of zinc transporters—A review of current understanding of their biological and pathophysiological roles. *Molecular Aspects of Medicine*, *34*, 548–560.

Ishida, S., Lee, J., Thiele, D. J., & Herskowitz, I. (2002). Uptake of the anticancer drug cisplatin mediated by the copper transporter Ctr1 in yeast and mammals. *Proceedings of the National Academy of Sciences of the United States of America*, *99*, 14298–14302.

Jeong, J., & Eide, D. J. (2013). The SLC39 family of zinc transporters. *Molecular Aspects of Medicine*, *34*, 612–619.

Kadhim, H. M., Ismail, S. H., Hussein, K. I., Bakir, I. H., Sahib, A. S., Khalaf, B. H., et al. (2006). Effects of melatonin and zinc on lipid profile and renal function in type 2 diabetic patients poorly controlled with metformin. *Journal of Pineal Research*, *41*, 189–193.

Kambe, T., Narita, H., Yamaguchi-Iwai, Y., Hirose, J., Amano, T., Sugiura, N., et al. (2002). Cloning and characterization of a novel mammalian zinc transporter, zinc transporter 5, abundantly expressed in pancreatic beta cells. *Journal of Biological Chemistry, 7*, 19049–19055.

Kaplan, J. H., & Lutsenko, S. (2009). Copper transport in mammalian cells: Special care for a metal with special needs. *Journal of Biological Chemistry, 284*, 25461–25465.

Kawachi, M., Kobae, Y., Mimura, T., & Maeshima, M. (2008). Deletion of a histidine-rich loop of AtMTP1, a vacuolar Zn^{2+}/H^{+} antiporter of Arabidopsis thaliana, stimulates the transport activity. *Journal of Biological Chemistry, 283*, 8374–8383.

Kelleher, S. L., & Lönnerdal, B. (2003). Zn transporter levels and localization change throughout lactation in rat mammary gland and are regulated by Zn in mammary cells. *The Journal of Nutrition, 133*, 3378–3385.

Kelleher, S. L., & Lönnerdal, B. (2006). Mammary gland copper transport is stimulated by prolactin through alterations in Ctr1 and Atp7A localization. *American Journal of Physiology. Regulatory, Integrative and Comparative Physiology, 291*, R1181–R1191.

Kelleher, S. L., McCormick, N. H., Velasquez, V., & Lopez, V. (2011). Zinc in specialized secretory tissues: Roles in the pancreas, prostate, and mammary gland. *Advances in Nutrition: An International Review Journal, 2*, 101–111.

Kim, B. E., Nevitt, T., & Thiele, D. J. (2008). Mechanisms for copper acquisition, distribution and regulation. *Nature Chemical Biology, 4*, 176–185.

Kim, H., Son, H. Y., Bailey, S. M., & Lee, J. (2009). Deletion of hepatic Ctr1 reveals its function in copper acquisition and compensatory mechanisms for copper homeostasis. *American Journal Physiology. Gastrointestinal and Liver Physiology, 296*, G356–G364.

Kim, B. E., Turski, M. L., Nose, Y., Casad, M., Rockman, H. A., & Thiele, D. J. (2010). Cardiac copper deficiency activates a systemic signaling mechanism that communicates with the copper acquisition and storage organs. *Cell Metabolism, 11*, 353–363.

Kim, B. E., Wang, F., Dufner-Beattie, J., Andrews, G. K., Eide, D. J., & Petris, M. J. (2004). Zn^{2+}-stimulated endocytosis of the mZIP4 zinc transporter regulates its location at the plasma membrane. *Journal of Biological Chemistry, 279*, 4523–4530.

Kim, H., Wu, X., & Lee, J. (2013). SLC31 (CTR) family of copper transporters in health and disease. *Molecular Aspects of Medicine, 34*, 561–570.

Kirschke, C. P., & Huang, L. (2003). ZnT7, a novel mammalian zinc transporter, accumulates zinc in the Golgi apparatus. *Journal of Biological Chemistry, 278*, 4096–4102.

Klomp, A. E., Tops, B. B., Van Denberg, I. E., Berger, R., & Klomp, L. W. (2002). Biochemical characterization and subcellular localization of human copper transporter 1 (hCTR1). *Biochemical Journal, 364*, 497–505.

Koh, J., Suh, S. W., Gwag, B. J., He, Y. Y., Hsu, C. Y., & Choi, D. W. (1996). The role of zinc in selective neuronal death after transient global cerebral ischemia. *Science, 272*, 1013–1016.

Kuo, M. T., Chen, H. H., Song, I. S., Savaraj, N., & Ishikawa, T. (2007). The roles of copper transporters in cisplatin resistance. *Cancer and Metastasis Reviews, 26*, 71–83.

Kuo, Y. M., Gybina, A. A., Pyatskowit, J. W., Gitschier, J., & Prohaska, J. R. (2006). Copper transport protein (Ctr1) levels in mice are tissue specific and dependent on copper status. *British Journal of Nutrition, 136*, 21–26.

Kury, S., Devilder, M. C., Avet-Loiseau, H., Dreno, B., & Moisan, J. P. (2001). Expression pattern, genomic structure and evaluation of the human SLC30A4 gene as a candidate for acrodermatitis enteropathica. *Human Genetics, 109*, 178–185.

Langmade, S. J., Ravindra, R., Daniels, P. J., & Andrews, G. K. (2000). The transcription factor MTF-1 mediates metal regulation of the mouse ZnT1 gene. *Journal of Biological Chemistry, 275*, 34803–34809.

Larson, C. A., Adams, P. L., Jandial, D. D., Blair, B. G., Safaei, R., & Howell, S. B. (2010). The role of the N-terminus of mammalian copper transporter 1 in the cellular accumulation of cisplatin. *Biochemical Pharmacology*, *80*, 448–454.

Leary, S. C., Kaufman, B. A., Pellecchia, G., Guercin, G. H., Mattman, A., Jaksch, M., et al. (2004). Human SCO1 and SCO2 have independent, cooperative functions in copper delivery to cytochrome c oxidase. *Human Molecular Genetics*, *13*, 1839–1848.

Lee, Y. Y., Choi, C. H., Do, I. G., Song, S. Y., Lee, W., Park, H. S., et al. (2011). Prognostic value of the copper transporters, CTR1 and CTR2, in patients with ovarian carcinoma receiving platinum-based chemotherapy. *Gynecologic Oncology*, *122*, 361–365.

Lee, J., Pena, M. M., Nose, Y., & Thiele, D. J. (2002). Biochemical characterization of the human copper transporter Ctr1. *Journal of Biological Chemistry*, *277*, 4380–4387.

Lee, J., Petris, M. J., & Thiele, D. J. (2002). Characterization of mouse embryonic cells deficient in the ctr1 high affinity copper transporter. Identification of a Ctr1-independent copper transport system. *Journal of Biological Chemistry*, *277*, 40253–40259.

Lee, J., Prohaska, J. R., Dagenais, S. L., Glover, T. W., & Thiele, D. J. (2000). Isolation of a murine copper transporter gene, tissue specific expression and functional complementation of a yeast copper transport mutant. *Gene*, *254*, 87–96.

Lee, J., Prohaska, J. R., & Thiele, D. J. (2001). Essential role for mammalian copper transporter Ctr1 in copper homeostasis and embryonic development. *Proceedings of the National Academy of Sciences of the United States of America*, *98*, 6842–6847.

Lee, D. Y., Shay, N. F., & Cousins, R. J. (1992). Altered zinc metabolism occurs in murine lethal milk syndrome. *The Journal of Nutrition*, *122*, 2233–2238.

Lemaire, K., Ravier, M. A., Schraenen, A., Creemers, J. W., Van de Plas, R., Granvik, M., et al. (2009). Insulin crystallization depends on zinc transporter ZnT8 expression, but is not required for normal glucose homeostasis in mice. *Proceedings of the National Academy of Sciences of the United States of America*, *106*, 14872–14877.

Li, M., Zhang, Y., Liu, Z., Bharadwaj, U., Wang, H., Wang, X., et al. (2007). Abberant expression of zinc transporter ZIP4 (SLC39A4) significantly contributes to human pancreatic cancer pathogenesis and progression. *Proceedings of the National Academy of Sciences of the United States of America*, *104*, 18636–18641.

Liang, Z. D., Long, Y., Tsai, W.-B., Fu, S., Kurzrock, R., Gagea-Iurascu, M., et al. (2012). Mechanistic basis for overcoming platinum resistance using copper chelating agents. *Molecular Cancer Therapeutics*, *11*, 2483–2494.

Liang, Z. D., Stockton, D., Savaraj, N., & Kuo, M. T. (2009). Mechanistic comparison of human high-affinity copper transporter 1-mediated transport between copper ion and cisplatin. *Molecular Pharmacology*, *76*, 843–853.

Liang, Z. D., Tsai, W.-B., Lee, M.-Y., Savaraj, N., & Kuo, M. T. (2012). Specificity protein 1 (SP1) oscillation is involved in copper homeostasis maintenance by regulating human high-affinity copper transporter 1 expression. *Molecular Pharmacology*, *81*, 455–464.

Lioumi, M., Ferguson, C. A., Sharpe, P. T., Freeman, T., Marenholz, I., Mischke, D., et al. (1999). Isolation and characterization of human and mouse ZIRTL, a member of the IRT1 family of transporters, mapping within the epidermal differentiation complex. *Genomics*, *62*, 272–280.

Liuzzi, J. P., Blanchard, R. K., & Cousins, R. J. (2001). Differential regulation of zinc transporter 1, 2, and 4 mRNA expression by dietary zinc in rats. *The Journal of Nutrition*, *131*, 46–52.

Liuzzi, J. P., Bobo, J. A., Cui, L., McMahon, R. J., & Cousins, R. J. (2003). Zinc transporters 1, 2 and 4 are differently expressed and localized in rats during pregnancy and lactation. *The Journal of Nutrition*, *133*, 342–351.

Liuzzi, J. P., & Cousins, R. J. (2004). Mammalian zinc transporters. *Annual Review of Nutrition*, *24*, 151–172.

Lopez, V., & Kelleher, S. L. (2009). Zinc transporter-2 (ZnT2) variants are localized to distinct subcellular compartments and functionally transport zinc. *Biochemical Journal, 422*, 43–52.

Lovell, M. A., Smith, J. L., Xiong, S., & Markesbery, W. R. (2005). Alterations in zinc transporter protein-1 (ZnT-1) in the brain of subjects with mild cognitive impairment, early, and late-stage Alzheimer's disease. *Neurotoxicity Research*, 7, 265–271.

Lu, M., Chai, J., & Fu, D. (2009). Structural basis for autoregulation of the zinc transporter YiiP. *Nature Structural & Molecular Biology, 16*, 1063–1067.

Lutsenko, S., Barnes, N. L., Bartee, M. Y., & Dmitriev, O. Y. (2007). Function and regulation of human copper-transporting ATPases. *Physiological Reviews, 87*, 1011–1046.

Lyubartseva, G., Smith, J. L., Markesbery, W. R., & Lovell, M. A. (2010). Alterations of zinc transporter proteins ZnT-1, ZnT-4 and ZnT-6 in preclinical Alzheimer's disease brain. *Brain Pathology, 20*, 343–350.

MacDiarmid, C. W., Milanick, M. A., & Eide, D. J. (2002). Biochemical properties of vacuolar zinc transport systems of Saccharomyces cerevisiae. *Journal of Biological Chemistry, 277*, 39187–39194.

Maryon, E. B., Molloy, S. A., & Kaplan, J. H. (2007). O-linked glycosylation at threonine 27 protects the copper transporter hCTR1 from proteolytic cleavage in mammalian cells. *Journal of Biological Chemistry, 282*, 20376–20387.

Maryon, E. B., Molloy, S. A., Zimnicka, A. M., & Kaplan, J. H. (2007). Copper entry into human cells: Progress and unanswered questions. *Biometals, 20*, 355–364.

McCall, K. A., Huang, C., & Fierke, C. A. (2000). Function and mechanism of zinc metalloenzymes. *The Journal of Nutrition, 130*, 1437S–1446S.

McKie, A. T., Barrow, D., Latunde-Dada, G. O., Rolfs, A., Sager, G., Mudaly, E., et al. (2001). An iron-regulated ferric reductase associated with the absorption of dietary iron. *Science, 291*, 1755–1759.

Meyer, L. A., Durley, A. P., Prohaska, J. R., & Harris, Z. L. (2001). Copper transport and metabolism are normal in a ceruloplasminemic mice. *Journal of Biological Chemistry, 276*, 36857–36861.

Molloy, S. A., & Kaplan, J. H. (2009). Copper-dependent recycling of hCTR1, the human high affinity copper transporter. *Journal of Biological Chemistry, 284*, 29704–29713.

Nevitt, T., Öhrvik, H., & Thiele, D. J. (2012). Charting the travels of copper in eukaryotes from yeast to mammals. *Biochemistry and Biophysics Acta, 1823*, 1580–1593.

Nishida, S., Mizuno, T., & Obata, H. (2008). Involvement of histidine-rich domain of ZIP family transporer TjZNT1 in metal ion specificity. *Plant Physiology and Biochemistry, 46*, 601–606.

Nose, Y., Kim, B. E., & Thiele, D. J. (2006). Ctr1 drives intestinal copper absorption and is essential for growth, iron metabolism, and neonatal cardiac function. *Cell Metabolism, 4*, 235–244.

Nose, Y., Wood, L. K., Kim, B. E., Prohaska, J. R., Fry, R. S., Spears, J. W., et al. (2010). Ctr1 is an apical copper transporter in mammalian intestinal epithelial cells in vivo that is controlled at the level of protein stability. *Journal of Biological Chemistry, 285*, 32385–32392.

Ohana, E., Hoch, E., Keasar, C., Kambe, T., Yifrach, O., Hershfinkel, M., et al. (2009). Identification of the Zn^{2+} binding site and mode of operation of a mammalian Zn^{2+} transporter. *Journal of Biological Chemistry, 284*, 17677–17686.

Ohgami, R. S., Campagna, D. R., McDonald, A., & Fleming, M. D. (2006). The Steap proteins are metalloreductases. *Blood, 108*, 1388–1394.

Overbeck, S., Uciechowski, P., Ackland, M. L., Ford, D., & Rink, L. (2008). Intracellular zinc homeostasis in leukocyte subsets is regulated by different expression of zinc exporters ZnT-1 to ZnT-9. *Journal of Leukocyte Biology, 83*, 368–380.

Palmiter, R. D., Cole, T. B., & Findley, S. D. (1996). ZnT2, a mammalian protein that confers resistance to zinc by facilitating vesicular sequestration. *EMBO Journal, 15*, 1784–1791.

Palmiter, R. D., Cole, T. B., Quaife, C. J., & Findley, S. D. (1996). ZnT-3, a putative transporter of zinc into synaptic vesicles. *Proceedings of the National Academy of Sciences of the United States of America, 93*, 14934–14939.

Palmiter, R. D., & Findley, S. D. (1995). Cloning and functional characterization of a mammalian zinc transporter that confers resistance to zinc. *EMBO Journal, 14*, 639–649.

Pena, M. M., Lee, J., & Thiele, D. J. (1999). A delicate balance homeostatic control of copper uptake and distribution. *British Journal of Nutrition, 129*, 1251–1260.

Petris, M. J. (2004). The SLC31 (Ctr) copper transporter family. *Pflügers Archive—European Journal of Physiology, 447*, 752–755.

Petris, M. J., Smith, K., Lee, J., & Thiele, D. J. (2003). Copper-stimulated endocytosis and degradation of the human copper transporter hCtr1. *Journal of Biological Chemistry, 278*, 9639–9646.

Piletz, J. E., & Ganschow, R. E. (1978). Zinc deficiency in murine milk underlies expression of the lethal milk (lm) mutation. *Science, 199*, 181–183.

Pound, L. D., Hang, Y., Sarkar, S. A., Wang, Y., Milam, L. A., Oeser, J. K., et al. (2011). The pancreatic islet beta cell-enriched transcription factor Pdx-1 regulates *Slc30a8* gene transcription through an intronic enhancer. *Biochemical Journal, 433*, 95–105.

Pourvali, K., Matak, P., Latunde-ada, G. O., Solomou, S., Mastrogiannaki, M., Peyssonnaux, C., et al. (2012). Basal expression of copper transporter 1 in intestinal epithelial cells is regulated by hypoxia-inducible factor 2α. *FEBS Letters, 586*, 2423–2427.

Prasad, A. S., Halsted, J. A., & Nadimi, M. (1961). Syndrome of iron deficiency anaemia, hepatosplenomegaly, hypogonadism, dwarfism and geophagia. *American Journal of Medicine, 31*, 532–546.

Prohaska, J. R., & Lukasewycz, O. A. (1990). Effects of copper deficiency on the immune system. *Advances in Experimental Medicine and Biology, 262*, 123–143.

Puig, S., Lee, J., Lau, M., & Thiele, D. J. (2002). Biochemical and genetic analyses of yeast and human high affinity copper transporters suggest a conserved mechanism for copper uptake. *Journal of Biological Chemistry, 277*, 26021–26030.

Qian, L., Lopez, V., Seo, Y. A., & Kelleher, S. L. (2009). Prolactin regulates ZNT2 expression through the JAK2/STATS signaling pathway in mammary cells. *American Journal of Physiology. Cell Physiology, 297*, C369–C377.

Qin, Y., Thomas, D., Fontaine, C. P., & Colvin, R. A. (2009). Silencing of ZnT1 reduces Zn^{2+} efflux in cultured cortical neurons. *Neuroscience Letters, 450*, 206–210.

Rees, E. M., Lee, J., & Thiele, D. J. (2004). Mobilization of intracellular copper stores by the ctr2 vacuolar copper transporter. *Journal of Biological Chemistry, 279*, 54221–54229.

Robinson, N. J., & Winge, D. R. (2010). Copper metallochaperones. *Annual Review of Biochemistry, 79*, 537–562.

Rogers, E. E., Eide, D. J., & Guerinot, M. L. (2000). Altered selectivity in an Arabidopsis metal transporter. *Proceedings of the National Academy of Sciences of the United States of America, 97*, 12356–12360.

Schushan, M., Barkan, Y., Haliloglu, T., & Ben-Tal, N. (2010). C(alpha)-trace model of the transmembrane domain of human copper transporter 1, motion and functional implications. *Proceedings of the National Academy of Sciences of the United States of America, 107*, 10908–10913.

Scott, L. J., Mohlke, K. L., Bonnycastle, L. L., Willer, C. J., Li, Y., Duren, W. L., et al. (2007). A genome-wide association study of type 2 diabetes in Finns detects multiple susceptibility variants. *Science, 316*, 1341–1345.

Seo, Y. A., & Kelleher, S. L. (2010). Functional analysis of two single nucleotide polymorphisms in SLC30A2 (ZnT2): Implications for mammary gland function and breast disease in women. *Physiological Genomics, 42A*, 219–227.

Sim, D. L., & Chow, V. T. (1999). The novel human HUEL (C4orf1) gene maps to chromosome 4p12-p13 and encodes a nuclear protein containing the nuclear receptor interaction motif. *Genomics, 59*, 224–233.

Sinani, D., Adle, D. J., Kim, H., & Lee, J. (2007). Distinct mechanisms for Ctr1-mediated copper and cisplantin transport. *Journal of Biological Chemistry, 282*, 26775–26785.
Sladek, R., Rocheleau, G., Rung, J., Dina, C., Shen, L., Serre, D., et al. (2007). A genome-wide association study identifies novel risk loci for type 2 diabetes. *Nature, 445*, 881–885.
Song, I. S., Chen, H. H., Aiba, I., Hossain, A., Liang, Z. D., Klomp, L. W., et al. (2008). Transcription factor Sp1 plays an important role in the regulation of copper homeostasis in mammalian cells. *Molecular Pharmacology, 74*, 705–713.
Suzuki, T., Ishihara, K., Migaki, H., Nagao, M., Yamaguchi-Iwai, Y., & Kambe, T. (2005). Two different zinc transport complexes of cation diffusion facilitator proteins localized in the secretory pathway operate to activate alkaline phosphatases in vertebrate cells. *Journal of Biological Chemistry, 280*, 30956–30962.
Tamaki, M., Fujitani, Y., Uchida, T., Hirose, T., Kawamori, R., & Watada, H. (2009). Downregulation of ZnT8 expression in pancreatic beta-cells of diabetic mice. *Islets, 1*, 124–128.
Taylor, K. M., Morgan, H. E., Johnson, A., & Nicholson, R. I. (2005). Structure-functin analysis of a novel member of the LIV-1 subfamily of zinc transporters, ZIP14. *FEBS Letters, 579*, 427–432.
Taylor, K. M., Morgan, H. E., Smart, K., Zahari, N. M., Pumford, S., Ellis, I. O., et al. (2007). The emerging role of the LIV-1 subfamily of zinc transporters in breast cancer. *Molecular Medicine, 13*, 396–406.
Taylor, K. M., & Nicholson, R. I. (2003). The LZT proteins; the LIV-1 subfamily of zinc transporters. *Biochimica et Biophysica Acta, 1611*, 16–30.
Todd, W. R., Elvehjem, C. A., & Hart, E. B. (1934). Zinc in the nutrition of the rat. *American Journal of Physiology, 107*, 146–156.
Valentine, R. A., Jackson, K. A., Christie, G. R., Mathers, J. C., Taylor, P. M., & Ford, D. (2007). ZnT5 variant B is a bidirectional zinc transporter and mediates zinc uptake in human intestinal Caco-2 cells. *Journal of Biological Chemistry, 282*, 14389–14393.
Vallee, B. L., & Auld, D. S. (1990). Zinc coordination, function, and structure of zinc enzymes and other proteins. *Biochemistry, 29*, 5647–5659.
Vallee, B. L., & Falchuk, K. H. (1993). The biochemical basis of zinc physiology. *Physiological Reviews, 73*, 79–118.
van den Berghe, P. V., Folmer, D. E., Malingre, H. E., van Beurden, E., Klomp, A. E., van de Sluis, B., et al. (2007). Human copper transporter 2 is localized in late endosomes and lysosomes and facilitates cellular copper uptake. *Biochemical Journal, 407*, 49–59.
van den Berghe, P. E., & Klomp, L. W. (2010). Posttranslational regulation of copper transporters. *Journal of Biological Inorganic Chemistry, 15*, 37–46.
Vulpe, C. D., & Packman, S. (1995). Cellular copper transport. *Annual Review of Nutrition, 15*, 293–322.
Wang, F., Kim, B.-E., Dufner-Beattie, J., Petris, M. J., Andrews, G., & Eide, D. J. (2004). Acrodermatitis enteropathica mutations affect transport activity, localization and zinc-responsive trafficking of the mouse ZIP4 zinc transporter. *Human Molecular Genetics, 13*, 563–571.
Wang, K., Pugh, E. W., Griffen, S., Doheny, K. F., Mostafa, W. Z., Al-Aboosi, M. M., et al. (2001). Homozygosity mapping places the acrodermatitis enteropathica gene on chromosomal region 8q24.3. *American Journal of Human Genetics, 68*, 1055–1060.
Wang, K., Zhou, B., Kuo, Y. M., Zemansky, J., & Gitschier, J. (2002). A novel member of a zinc transporter family is defective in acrodermatitis enteropathica. *American Journal of Human Genetics, 71*, 66–73.
Wenzel, H. J., Cole, T. B., Born, D. E., Schwartzkroin, P. A., & Palmiter, R. D. (1997). Ultrastructural localization of zinc transporter-3 (ZnT-3) to synaptic vesicle membranes

within mossy fiber boutons in the hippocampus of mouse and monkey. *Proceedings of the National Academy of Sciences of the United States of America, 94*, 12676–12681.

Wenzlau, J. M., Juhl, K., Yu, L., Moua, O., Sarkar, S. A., Gottlieb, P., et al. (2007). The cation efflux transporter ZnT8 (Slc30A8) is a major autoantigen in human type 1 diabetes. *Proceedings of the National Academy of Sciences of the United States of America, 104*, 17040–17045.

White, C., Kambe, T., Fulcher, Y. G., Sachdev, S. W., Bush, A. I., Fritsche, K., et al. (2009). Copper transport into the secretory pathway is regulated by oxygen in macrophages. *Journal of Cell Science, 122*, 1315–1321.

Wu, X., Sinani, D., Kim, H., & Lee, J. (2009). Copper transport activity of yeast Ctr1 is down-regulated via its C terminus in response to excess copper. *Journal of Biological Chemistry, 284*, 4112–4122.

Wyman, S., Simpson, R. J., McKie, A. T., & Sharp, P. A. (2008). Dcytb (Cybrd1) functions as both a ferric and a cupric reductase in vitro. *FEBS Letters, 582*, 1901–1906.

Xiao, Z., Loughlin, F., George, G. N., Howlett, G. J., & Wedd, A. G. (2004). C-terminal domain of the membrane copper transporter Ctr1 from *Saccharomyces cerevisiae* binds four Cu(I) ions as a cuprous-thiolate polynuclear cluster: Sub-femtomolar Cu(I) affinity of three proteins involved in copper trafficking. *Journal of the American Chemical Society, 126*, 3081–3090.

Yamashita, S., Miyagi, C., Fukada, T., Kagara, N., Che, Y. S., & Hirano, T. (2004). Zinc transporter LIV1 controls epithelial-mesenchymal transition in zebrafish gastrula organizer. *Nature, 429*, 298–302.

Yu, Y. Y., Kirschke, C. P., & Huang, L. (2007). Immunohistochemical analysis of ZnT1, 4, 5, 6, and 7 in the mouse gastrointestinal tract. *Journal of Histochemistry and Cytochemistry, 55*, 223–234.

Zhang, Y., Bharadwaj, U., Logsdon, C. D., Chen, C., Yao, Q., & Li, M. (2010). ZIP4 regulates pancreatic cancer cell growth by activating IL-6/STAT3 pathway through zinc finger transcription factor CREB. *Clinical Cancer Research, 16*, 1423–1430.

Zhang, Y., Chen, C., Yao, Q., & Li, M. (2010). ZIP4 upregulates the expression of neuropilin-1, vascular endothelial growth factor, and matrix metalloproteases in pancreatic cancer cell lines and xenografts. *Cancer Biology & Therapy, 9*, 236–242.

Zhao, H., & Eide, D. (1996). The *ZRT2* gene encodes the low affinity zinc transporter in *Saccharomyces cerevisiae*. *The Journal of Biological Chemistry, 271*, 23203–23210.

Zheng, Z., White, C., Lee, J., Peterson, T. S., Bush, A. I., Sun, G. Y., et al. (2010). Altered microglial copper homeostasis in a mouse model of Alzheimer's disease. *Journal of Neurochemistry, 114*, 1630–1638.

Zhou, B., & Gitschier, J. (1997). hCTR1: A human gene for copper uptake identified by complementation in yeast. *Proceedings of the National Academy of Sciences of the United States of America, 94*, 7481–7486.

Zimnicka, A. M., Maryon, E. B., & Kaplan, J. H. (2007). Human copper transporter hCTR1 mediates basolateral uptake of copper into enterocytes: Implications for copper homeostasis. *Journal of Biological Chemistry, 282*, 26471–26480.

CHAPTER TEN

The SLC37 Family of Sugar-Phosphate/Phosphate Exchangers

Janice Y. Chou[*,1], Brian C. Mansfield[*,†]

[*]Section on Cellular Differentiation, Program on Developmental Endocrinology and Genetics, Eunice Kennedy Shriver National Institute of Child Health and Human Development, National Institutes of Health, Bethesda, Maryland, USA
[†]Foundation Fighting Blindness, Columbia, Maryland, USA
[1]Corresponding author: e-mail address: chouja@mail.nih.gov

Contents

Abstract

The SLC37 family members are endoplasmic reticulum (ER)-associated sugar-phosphate/phosphate (P_i) exchangers. Three of the four members, SLC37A1, SLC37A2, and SLC37A4, function as P_i-linked glucose-6-phosphate (G6P) antiporters catalyzing G6P:P_i and P_i:P_i exchanges. The activity of SLC37A3 is unknown. SLC37A4, better known as the G6P transporter (G6PT), has been extensively characterized, functionally and structurally, and is the best characterized family member. G6PT contains 10 transmembrane helices with both N and C termini facing the cytoplasm. The primary *in vivo* function of the G6PT protein is to translocate G6P from the cytoplasm into the ER lumen where it couples with either the liver/kidney/intestine-restricted glucose-6-phosphatase-α (G6Pase-α or G6PC) or the ubiquitously expressed G6Pase-β (or G6PC3) to hydrolyze G6P to glucose and P_i. The G6PT/G6Pase-α complex maintains interprandial glucose homeostasis, and the G6PT/G6Pase-β complex maintains neutrophil energy homeostasis and functionality. G6PT is highly selective for G6P and is

Current Topics in Membranes, Volume 73
ISSN 1063-5823
http://dx.doi.org/10.1016/B978-0-12-800223-0.00010-4

competitively inhibited by cholorogenic acid and its derivatives. Neither SLC37A1 nor SLC37A2 can couple functionally with G6Pase-α or G6Pase-β, and the antiporter activities of SLC37A1 or SLC37A2 are not inhibited by cholorogenic acid. Deficiencies in G6PT cause glycogen storage disease type Ib (GSD-Ib), a metabolic and immune disorder. To date, 91 separate *SLC37A4* mutations, including 39 missense mutations, have been identified in GSD-Ib patients. Characterization of missense mutations has yielded valuable information on functionally important residues in the G6PT protein. The biological roles of the other SLC37 proteins remain to be determined and deficiencies have not yet been correlated to diseases.

1. INTRODUCTION

The solute carrier (SLC) gene series consists of over 50 gene families that are predicted to encode membrane-bound transporters (He, Vasiliou, & Nebert, 2009). The SLC37 family consists of the following four sugar-phosphate exchange (SPX) proteins (Table 10.1): SLC37A1 (SPX1), SLC37A2 (SPX2), SLC37A3 (SPX3), and SLC37AA4 (SPX4) (Bartoloni & Antonarakis, 2004; Chou, Jun, & Mansfield, 2013). These proteins were originally grouped into the SLC37 family based on sequence homology to the bacterial organo-phosphate:phosphate (P_i) exchangers (Pao, Paulsen, & Saier, 1998), which are members of the major facilitator superfamily (MFS) (Reddy, Shlykov, Castillo, Sun, & Saier, 2012). The MFS transporters are single-polypeptide secondary carriers capable of transporting small solutes in response to chemiosmotic ion gradients, as opposed to primary carriers that create the gradients (ProSite PS50850). In the Transport Classification Database (TCDB), the SLC37 transporters are members of family 2.A.1.4 (http://www.tcdb.org/). Within the SLC37 family, the gene localization, structure, and amino acid compositions vary significantly, suggesting that the proteins have evolved independently and do not arise through gene duplication (Bartoloni & Antonarakis, 2004; Chou et al., 2013). The best characterized member is SLC37A4, which is better known as the glucose-6-phosphate (G6P) transporter (G6PT) (Chou et al., 2013; Chou, Matern, Mansfield, & Chen, 2002). In this review, we use G6PT when describing the SLC37A4 protein. Within the family, SLC37A1 and SLC37A2 are the most closely related at the protein level, while the G6PT protein is the most distant member (Fig. 10.1) (Bartoloni & Antonarakis, 2004; Chou et al., 2013; Corpet, 1988).

Table 10.1 The SLC37 family of sugar-phosphate/phosphate exchangers

Human gene name	**SLC37A1**	**SLC37A2**	**SLC37A3**	**SLC37A4**
Protein name(s)	SLC37A1, SPX1	SLC37A2, SPX2	SLC37A3, SPX3	SLC37A4, G6PT, SPX4
Known substrate	Glucose-6-phosphate	Glucose-6-phosphate	Unknown	Glucose-6-phosphate
Link to disease	None reported	None reported	None reported	GSD-Ib/non-GSD-Ia (OMIM 232220/232240)
Human gene locus	21q22.3	11q24.2	7q34	11q23.3
Sequence accession ID	NM_018964	NM_198277	NM_207113	NM_001467
Mouse model status (IMSR)[a]	Targeted mutation and gene trap—phenotype not reported	Targeted mutation—phenotype not reported	Targeted mutation and gene trap—phenotype not reported	Targeted mutation—GSD-Ib phenotype
Major protein isoform	533-amino acid	505-amino acid	494-amino acid	429-amino acid
Number of isoforms[b,c]	1	4	3	2

[a]The International Mouse Search Resource (IMSR): http://www.findmice.org/.
[b]NCBI blast for amino acid identity and isoform prediction: http://blast.ncbi.nlm.nih.gov/Blast.cgi.
[c]Nextprot for isoform prediction: http://www.nextprot.org/.

The G6PT first came to attention as a result of studies of type I glycogen storage diseases (GSD-I), a group of autosomal recessive disorders with an overall incidence of approximately 1 in 100,000 (Chou et al., 2002). Deficiencies in G6PT cause GSD type Ib (GSD-Ib, MIM232220), representing 15% of all GSD-I cases (Chou, Jun, & Mansfield, 2010a, 2010b; Chou et al., 2002). The primary *in vivo* function of G6PT is to translocate G6P from the cytoplasm into the lumen of the endoplasmic reticulum (ER) where it is hydrolyzed by a glucose-6-phosphatase (G6Pase) into glucose and P_i (Chou et al., 2002, 2010a,

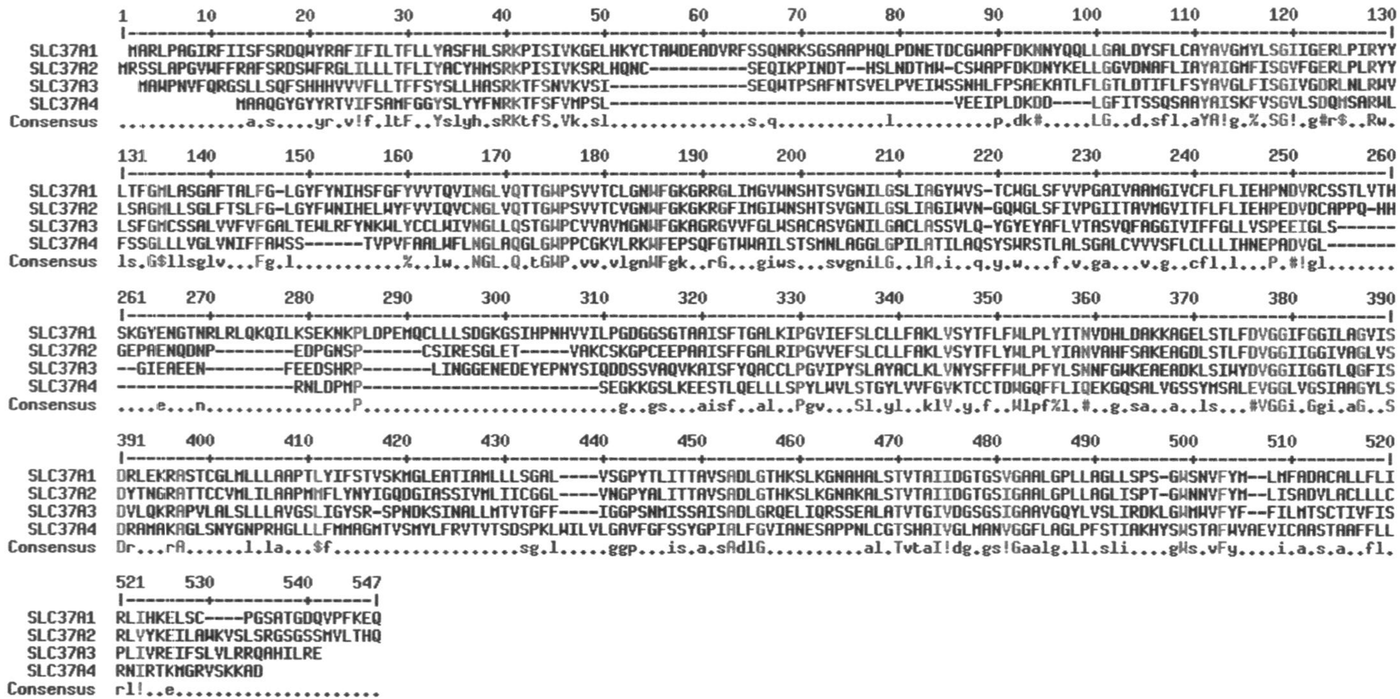

Figure 10.1 Alignment of the amino acid sequences of human SLC37A1, SLC37A2, SLC37A3, and SLC37A4/G6PT by Multalin (http://multalin.toulouse.inra.fr/multalin) (Corpet, 1988). The amino acid sequences are GENBANK accession numbers NP_061837.3 (SLC37A1), NP_938018.1 (SLC37A2), AAH46567.1 (SLC37A3), and CAG33014.1 (SLC37A4/G6PT). In the consensus sequence, ! represents either I or V; $ represents either L or M; % represents either F or Y; and # represents N, D, Q, E, B, or Z. (See the color plate.)

2010b) (Fig. 10.2). This transport activity is dependent on the ability of G6PT to form a functional complex with a G6Pase (Lei et al., 1996). In the absence of a G6Pase, the transport of G6P is minimal. There are two enzymatically active G6Pases. G6Pase-α (or G6PC) expression is restricted to the liver, kidney, and intestine (Lei, Shelly, Pan, Sidbury, & Chou, 1993), while G6Pase-β (or G6PC3) is expressed ubiquitously (Shieh, Pan, Mansfield, & Chou, 2003).

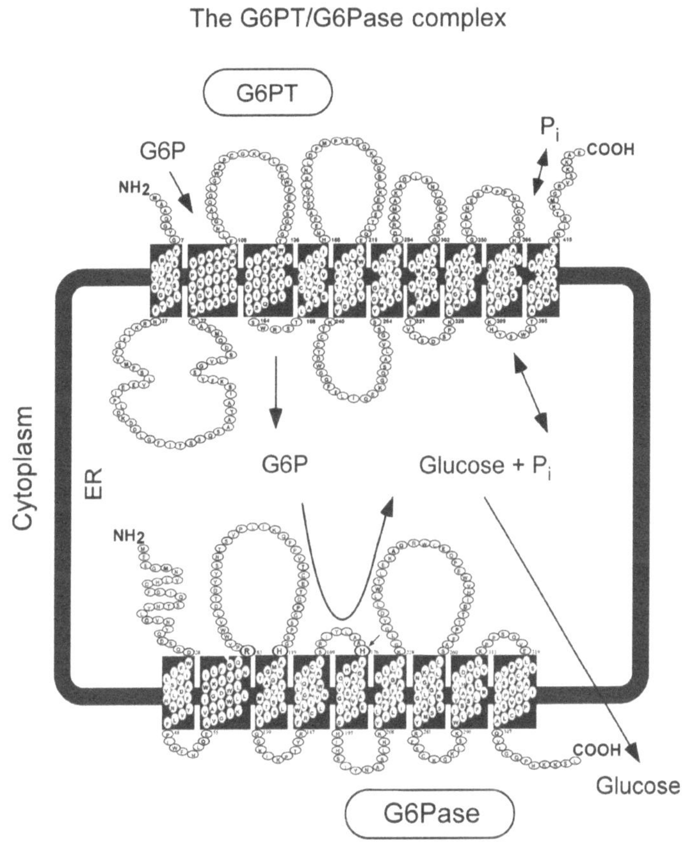

Figure 10.2 The G6PT/G6Pase complex. The diagram shows a cross-section of the ER. The cell cytoplasm lies outside the ER membrane that encircles the ER lumen. In the gluconeogenic tissues of the liver, kidney, and intestine, G6PT couples with G6Pase-α, while in nongluconeogenic tissues (including neutrophils and macrophages), G6PT couples with G6Pase-β. The proteins are shown as spatially separated for clarity, but evidence suggests that they reside in physical contact with each other as a G6PT/ G6Pase complex. The G6PT/G6Pase-α complex expressed in gluconeogenic tissues maintains interprandial blood glucose homeostasis. The analogous G6PT/G6Pase-β complex, which is ubiquitously expressed, maintains neutrophil/macrophage energy homeostasis and functionality.

Pathogenic mutations of G6Pase-α result in GSD-Ia (Chou et al., 2002, 2010b), a metabolic disorder characterized by the hallmark of fasting hypoglycemia. While GSD-Ia and GSD-Ib have similar metabolic characteristics, GSD-Ib differs by the presence of neutropenia. Notably, pathogenic mutations of G6Pase-β do not lead to the systemic metabolic disease associated with G6Pase-α mutations, but instead result in a severe congenital neutropenia syndrome type 4 (SCN4) (Boztug et al., 2009; Chou et al., 2010a, 2010b). Because of the functional coupling between G6PT and G6Pase-α/G6Pase-β, GSD-Ib provides the link between GSD-Ia and SCN4, being a G6P metabolic disease with neutropenia. The G6PT/G6Pase-α complex is critical for the maintenance of interprandial glucose homeostasis (Chou et al., 2002, 2010a, 2010b). The analogous G6PT/G6Pase-β complex plays a critical role in the maintenance of neutrophil (Cheung et al., 2007; Jun et al., 2010) and macrophage (Jun, Cheung, Lee, Mansfield, & Chou, 2012) energy homeostasis and functionality. Three members of the SLC37 family (SLC37A1, SLC37A2, and SLC37A4/G6PT) function as P_i-linked G6P antiporters capable of both homologous (P_i:P_i) and heterologous (G6P:P_i) exchanges, although they show differential inhibition by cholorogenic acid (Chen, Pan, Nandigama et al., 2008; Pan et al., 2011). In contrast, SLC37A3 lacks the antiporter activity (Pan et al., 2011). The molecular genetics and therapies for G6PT-deficiency/GSD-Ib have been reviewed recently (Chou et al., 2010a, 2010b, 2013) and will not be covered here. While structure–function studies have not been reported for the SLC37A1, SLC37A2, and SLC37A3 proteins, the SLC37A4/G6PT protein has been extensively characterized and is the focus of this review.

2. G6PT (SLC37A4 OR SPX4)

G6PT was the first SLC37 member identified and is the best characterized protein in this family (Chou et al., 2002, 2010a, 2010b, 2013). Human *SLC37A4* is a single copy gene consisting of nine exons (Gerin, Veiga-da-Cunha, Noel, & Van Schaftingen, 1999; Hiraiwa, Pan, Lin, Moses, & Chou, 1999) on chromosome 11q23 (Annabi et al., 1998). The G6PT proteins from human (Gerin, Veiga-da-Cunha, Achouri, Collet, & Van Schaftingen, 1997), mouse (Lin, Annabi, Hiraiwa, Pan, & Chou, 1998), and rat (Lin et al., 1998) are predicted to be very similar, with mouse and rat proteins sharing 95% and 93% amino acid sequence identity to the human protein, respectively. Transcription of the human *SLC37A4* gene gives rise to two alternately spliced transcripts, G6PT and variant G6PT (vG6PT) that encode nonglycosylated polypeptides of 429 and 451 amino acids, respectively, and reflect the absence or presence of exon-7 encoded sequences (Gerin et al., 1997; Hiraiwa et al., 1999; Lin, Pan, & Chou, 2000). The *SLC37A4* gene is ubiquitously expressed (Lin et al., 1998), while the variant *SLC37A4* is expressed primarily in the brain, heart, and skeletal muscle (Lin et al., 2000). The significance of these expression patterns is not understood. Both G6PT and vG6PT are equally active in G6P uptake (Lin et al., 2000) and so further discussion will refer to both species as G6PT, although most experiments use G6PT, not the variant.

2.1. The function of the G6PT protein

The primary *in vivo* function of the G6PT protein is to translocate G6P from the cytoplasm into the lumen of the ER (see Fig. 10.2). Inside the lumen, G6P is hydrolyzed to glucose and P_i either by G6Pase-α (Chou et al., 2010a, 2010b; Lei et al., 1993) or G6Pase-β (Chou et al., 2010a, 2010b; Shieh et al., 2003). G6Pase-α (Ghosh, Shieh, Pan, Sun, & Chou, 2002) and G6Pase-β (Ghosh, Shieh, Pan, & Chou, 2004) are ER transmembrane proteins with their active sites situated inside the ER lumen. Both G6Pases depend upon the G6PT for their activity in two distinct ways. First, G6PT provides the enzyme substrate by transporting cytoplasmic G6P into the ER lumen. Second, G6PT provides a physical interaction that increases G6P transport activity, presumably by some allosteric mechanism. This functional coupling of G6PT and G6Pase-α activities has been demonstrated in several ways. The coupling was first revealed by characterizing hepatic microsomal G6P uptake activity in GSD-Ia mice (G6Pase-$\alpha^{-/-}$, G6PT$^{+/+}$) (Lei et al., 1996). While hepatic microsomes from wild-type mice were observed to transport G6P efficiently, hepatic microsomes from GSD-Ia mice transported G6P inefficiently despite the presence of the transporter (Lei et al., 1996) (Fig. 10.3A). However, the G6P transport activity could be increased markedly if hepatic G6Pase-α activity was restored in GSD-Ia mice via gene transfer (Zingone et al., 2000) (Fig. 10.3A).

Cell-based activity assays for recombinant G6PT proteins have also been developed (Chen, Lin, Pan, Hiraiwa, & Chou, 2000; Chen, Pan, Shieh, & Chou, 2002; Hiraiwa et al., 1999). Similarly, these assays showed that G6Pase-α-expressing G6PT-deficient microsomes (G6Pase-$\alpha^{+/+}$/G6PT$^{-/-}$) exhibited little or no G6P uptake activity (Fig. 10.3B), G6PT-expressing G6Pase-α-deficient microsomes (G6Pase-$\alpha^{-/-}$/G6PT$^{+/+}$) exhibited low levels of G6P uptake activity (Fig. 10.3B), and G6Pase-α/G6PT-expressing microsomes (G6Pase-$\alpha^{+/+}$/G6PT$^{+/+}$) had markedly

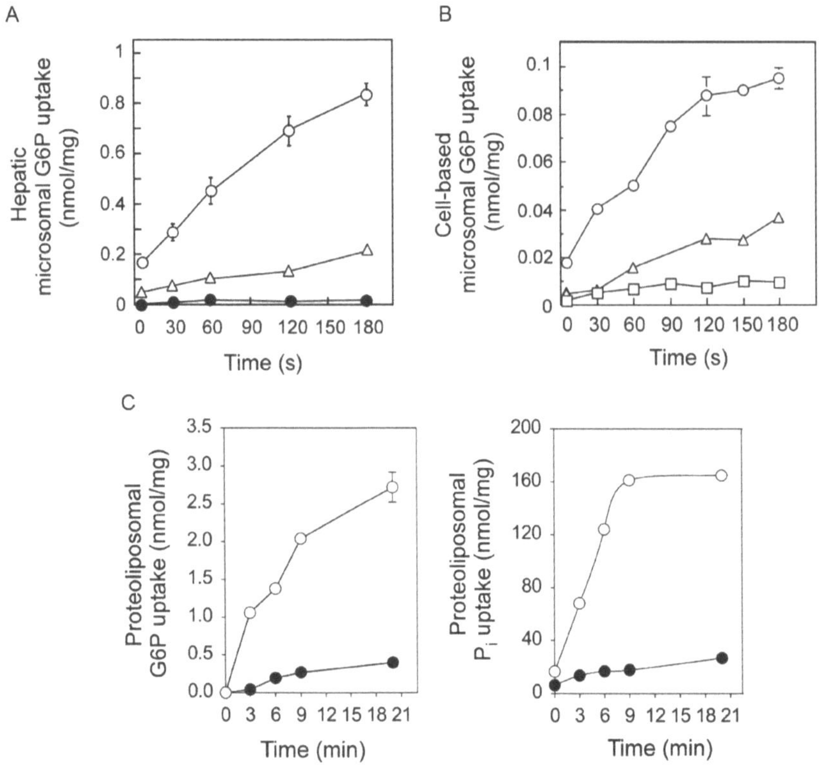

Figure 10.3 The function of G6PT. (A) Uptake of G6P into hepatic microsomes of wild-type mice (○), GSD-Ia mice (•), or GSD-Ia mice at day 14 post Ad-G6Pase-α infusion (△). (B) The cell-based microsomal G6P uptake activity. Uptake of G6P into microsomes prepared from COS-1 cells transfected with cDNA encoding G6PT alone (△), G6Pase-α alone (□), or G6PT and G6Pase-α (○). (C) The antiporter activity of G6PT. G6P uptake (left) and P_i uptake (right) were determined in proteoliposomes expressing G6PT preloaded with 50 mM P_i (○) or with MOPS buffer (•). Data are presented as the mean ± SEM.

increased G6P transport activity (Fig. 10.3B). This suggests that the role of a coexpressed G6Pase-α is to increase the P_i concentration in the ER lumen to generate a driving P_i gradient required for antiporter activity of the G6PT protein. Further evidence came from studying reconstituted proteoliposomes preloaded with P_i. Using this system, G6PT was shown to be a P_i-linked G6P antiporter that catalyzes G6P:P_i and P_i:P_i exchanges (Chen, Pan, Nandigama, et al., 2008) (Fig. 10.3C). Preloading of proteoliposomes with P_i eliminates the need for a coexpressing G6Pase, and the P_i-loaded G6PT-proteoliposomes transport both G6P and P_i efficiently without a coexpressing G6Pase (Fig. 10.3C) (Chen, Pan, Nandigama, et al., 2008).

When the 23 *SLC37A4* mutations identified in GSD-Ib patients were characterized using the cell-based assay and proteoliposome assay (see Fig. 10.8) (Chen, Pan, Lee, Peng, & Chou, 2008) similar results have been seen for 22 of the 23 *SLC37A4* mutations, confirming the validity of each assay system. Interestingly, the outstanding mutation, p.Q133P, identified in a patient originally classified as a nonGSD-Ia (Veiga-da-Cunha et al., 1999), is devoid of G6P transport activity regardless of the assay methods, but retains 5% wild-type P_i transport activity (Chen, Pan, Lee, et al., 2008). The results suggest that further characterization of the G6P and P_i transport activities of G6PT mutations may yield insights into this genetic disorder. It is possible that while the active sites for the G6P and P_i transport activity overlap, they may also depend on residues that are not shared.

In addition to GSD-Ia (G6Pase-α deficiency), and GSD-Ib (G6PT deficiency), early biochemical classifications of GSD-I subtypes, based on limited kinetic analysis of the activity of the G6PT/G6Pase-α complex in patient's hepatic microsomes, implied there was a third form of GSD-I—GSD-Ic (Nordlie, Sukalski, Munoz, & Baldwin, 1983) characterized by a lack of P_i transport activity (Chou et al., 2002). Initially proposed to represent a third gene implicated in GSD-I, more recent genotyping of GSD-Ic or nonGSD-Ia patients has revealed deleterious *SLC37A4* mutations (Galli et al., 1999; Janecke et al., 2000; Veiga-da-Cunha et al., 1998, 1999), consistent with the dual role of G6PT as both a G6P and P_i transporter.

2.2. Competitive inhibitors for G6PT

G6Pase-α is a nonspecific phosphatase and the substrate specificity of the G6PT/G6Pase-α complex is conferred by the high specificity of G6PT for G6P (Arion, Wallin, Carlson, & Lange, 1972). Chlorogenic acid, the ester of caffeic acid and (−)quinic acid (Fig. 10.4), is a reversible, competitive inhibitor for G6PT (Arion et al., 1998; Hemmerle et al., 1997). S3483 (Arion et al., 1998) and S4048 (Herling et al., 1999) (Fig. 10.4) are cholorogenic acid derivatives that are 2–3 orders of magnitude more potent as competitive inhibitors for G6PT. These inhibitors have been used in mechanistic studies on the metabolic defects in GSD-1 (Bandsma et al., 2001; Grefhorst et al., 2010). As expected, G6P transport activity in mouse hepatic microsomes (Lei et al., 1996) and microsomes and proteoliposomes containing recombinant G6PT protein is markedly inhibited by chlorogenic acid (Chen, Pan, Nandigama, et al., 2008; Hiraiwa et al., 1999).

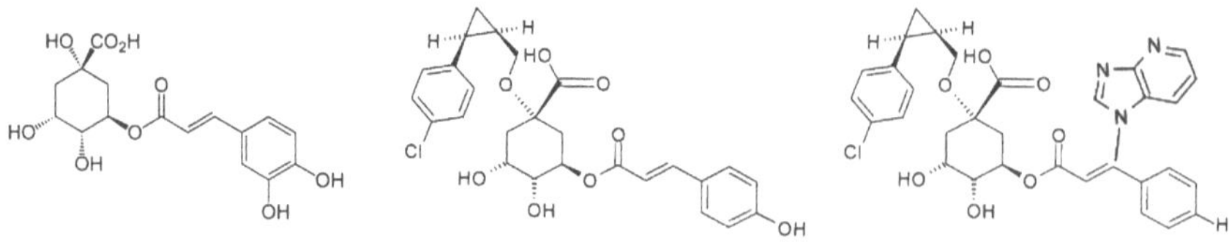

Figure 10.4 The structure of chlorogenic acid, S3483, and S4048.

2.3. Transmembrane topology of the G6PT protein

Hydropathy profile analysis predicts that G6PT contains either 10 (Fig. 10.5) (Hoffman & Stoffel, 1993) or 12 (Gerin et al., 1997) putative transmembrane helices. Along with G6PT, both the glycerol-3-phosphate transporter (GlpT) and the bacterial hexose-6-phosphate transporter (UhpT) are members of the organo-phosphate:P_i antiporter family (He et al., 2009; Pao et al., 1998). The X-ray crystallography structure of *Escherichia coli* GlpT has been solved and shown to contain 12 transmembrane helices (Huang, Lemieux, Song, Auer, & Wang, 2003). Homology modeling (Almqvist, Huang, Hovmöller, & Wang, 2004) lends support to G6PT having a similar 12-helix structure. To address the transmembrane structure of G6PT experimentally, investigators have performed mutational glycosylation and protease protection analyses. Protease protection confirmed that G6PT contains an even number (but not the exact number) of transmembrane helices, with both N and C termini facing the cytoplasm (Pan, Lin, & Chou, 1999). Because G6PT is a nonglycosylated protein, mutational glycosylation was used to differentiate between the 10 and 12 transmembrane domain models. The region of G6PT corresponding to amino acid residues 50–71 is predicted to form a transmembrane segment in the 12-domain model that would be resistant to glycosylation, while it is predicted to lie within a 51-residue lumenal loop available for glycosylation (Landolt-Marticorena & Reithmeier, 1994; Nilsson & von Heijne, 1993) in the 10-domain model (Fig. 10.5). It has been shown that glycosylation of membrane proteins is critical for their membrane targeting and folding (Goder & Spiess, 2001; Parodi, 2000), and incorrect glycosylation can cause misfolding and preferential degradation of the misfolded proteins (Caramelo & Parodi, 2007; Molinari, 2007). It was predicted that G6PT-p.T53N and G6PT-p.S55N glycoprotein mutants would be misfolded and degraded in the ER (Almqvist et al., 2004) if there were 12 helices, but expressed without disruption if the 10 helix model was correct. Contrary to the predictions based on protein homology, transient expression of G6PT, G6PT-p.T53N,

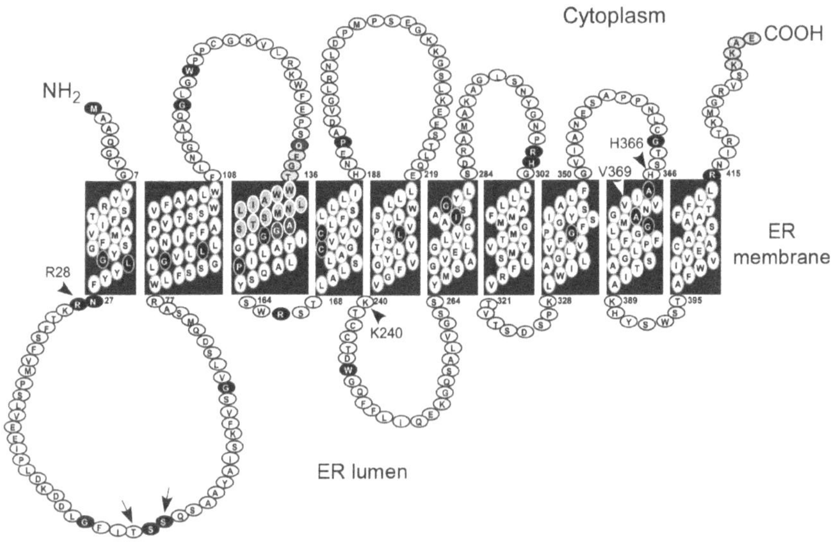

Figure 10.5 The topology of human G6PT. The 429 amino acid residues are denoted by circles. The 34 amino acid residues altered by the 39 missense *SLC37A4* mutations identified in GSD-Ib patients are shown in black. The signature motif (composed of amino acids 133–149) and the ER retention signal KKAE (composed of amino acids 426–429) are denoted by shaded circles. Three residues (p.Q133, p.A148, and p.G149) in the signature motif altered by missense *SLC37A4* mutations identified in GSD-Ib patients are shown in dark shaded circles. The residues mutagenized for generating the glycosylation mutants p.T53N and p.S55N are denoted by arrows. Amino acids corresponding to essential residues in UhpT are indicated by arrowheads.

and G6PT-p.S55N mutants showed that all were expressed at similar levels (Fig. 10.6A) (Pan et al., 1999, 2009), suggesting that the 10-helical model of G6PT is correct.

Protease sensitivity assays were then used to confirm that the expressed proteins were embedded in the ER membrane. Intact microsomes expressing the wild-type or mutant G6PTs were subjected to limited trypsin digestion and analyzed by Western blot (Pan et al., 2009). G6PT wild-type and p.S55N (or p.T53N) mutant migrated as 36-kDa and 40-kDa polypeptides, respectively, prior to trypsin digestion and 36-kDa/16-kDa and 40-kDa/23-kDa polypeptides, respectively, following limited trypsin digestion (Fig. 10.6B). In the presence of the glycosylation inhibitor tunicamycin, the G6PT-p.S55N (or p.T53N) mutant directed the synthesis of a 36-kDa polypeptide (Fig. 10.6B), similar to, the size of wild-type G6PT (Pan et al., 1999). As expected, the G6PT-p.S55N mutant synthesized in the presence of tunicamycin migrated as 36-kDa/16-kDa polypeptides after

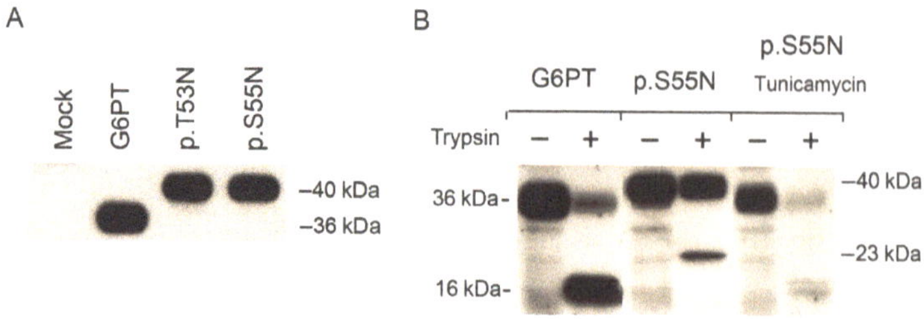

Figure 10.6 Protease sensitivity of G6PT wild-type and glycosylation mutants. Intact microsomes were isolated from COS-1 cells infected with Ad-G6PT, Ad-G6PT-p.T53N, or Ad-G6PT-p.S55N (Pan, Chen, Lee, & Chou, 2009). (A) Western blot analysis of the synthesis of G6PT, G6PT-p.T53N, and G6PT-p.S55N. (B) Western blot analysis of G6PT and G6PT-p.S55N after limited trypsin digestion in the absence and presence of tunicamycin.

limited trypsin digestion similar to, wild-type G6PT (Fig. 10.6B). These data further supported a 10-domain model for the G6PT protein (Pan et al., 1999, 2009).

According to homology modeling (Almqvist et al., 2004), amino acid residues in G6PT that correspond to the amino acid residues essential for the activity of the 12-domain UhpT (Lloyd & Kadner, 1990) would also play a vital role in G6PT if it contained 12 domains. The amino acid residues absolutely required for transport activities in UhpT include p.R46 and p. R275 that are proposed to form the substrate-binding site (Fann et al., 1998), and p.D388 and p.K391 that are proposed to form an intra-helical salt bridge (Hall, Fann, & Maloney, 1999). The corresponding residues in G6PT are p.R28, p.K240, p.H366, and p.V369 (Fig. 10.5). Site-directed mutagenesis and transient expression analyses showed that while amino acid residue p.R28 is an essential residue in G6PT, p.K240, p.H366, and p.V369 are not (Pan et al., 2009), indicating that the structural requirements of G6PT and UhpT differ. In UhpT, the p.R46K and p.R275K mutations destabilize the protein (Fann et al., 1998). In contrast, the p.R28C, p.R28H, p.K240C, and p.K240R mutants support the synthesis of wild-type levels of G6PT proteins (Chen et al., 2000; Pan et al., 2009). Taken together, the experimental data solidly support the conclusion that G6PT is anchored in the ER by 10 transmembrane helices (Fig. 10.5).

2.4. Structure–function analyses of G6PT

G6PT deficiency gives rise to GSD-Ib, an autosomal recessive metabolic and immune disorder with an incidence of 1 in 700,000 (Chou et al., 2002,

2010a, 2010b). The GSD-Ib metabolic phenotype is impaired glucose homeostasis, characterized by fasting hypoglycemia, hepatomegaly, nephromegaly, growth retardation, hyperlipidemia, and lactic acidemia. The GSD-Ib immune phenotype is neutropenia and neutrophil dysfunction. The latter is characterized by impaired respiratory bust, chemotaxis, calcium mobilization, and phagocytic activities. A large number of mutations have been identified in the *SLC37A4* gene of GSD-Ib patients. Characterizing these mutations may yield insights to this genetic disorder including the genotype–phenotype relationship for *SLC37A4* mutations.

2.4.1 Mutations that cause GSD-Ib

To date, 91 separate mutations have been identified in the *SLC37A4* gene of GSD-Ib/nonGSD-Ia patients (Dissanayake, Jayasinghe, Thilakaratne, & Jayasekara, 2011; Qiu, Lu, Wang, Qiu, & Wei, 2011; reviewed in Chou et al., 2013). These include 39 missense, 11 nonsense, 21 insertion/deletion (including 2 codon deletion mutations), and 19 splicing mutations that are distributed across the entire coding and exon–intron junction regions with the exception of the small exon VII (Fig. 10.7). There is also one gross deletion mutation of sequences upstream of exon 7. Characterization of missense and codon deletion mutations that result in single amino acid alterations provides valuable information on functionally important residues in a protein. In G6PT, 31 missense and 2 codon deletion mutations have been characterized functionally via site-directed mutagenesis and transient expression assays and were shown to abolish or reduce greatly microsomal G6P uptake activity (Chen et al., 2000, 2002; Chen, Pan, Lee, et al., 2008; Hiraiwa et al., 1999). These mutations are grouped into the following three categories: helical (H), nonhelical that includes cytoplasmic (C) and luminal (L) loops, and terminal that includes N- (p.MIV) and C- (p.R415X) terminal domains (Fig. 10.8).

Among the 18 helical mutations, 9 missense and 2 codon deletion mutations are devoid of microsomal G6P uptake activity, while 7 missense mutations retain residual activities (Fig. 10.8). Among the 14 nonhelical missense mutations, 9 are devoid of microsomal G6P uptake activity (Fig. 10.8). It is interesting to note that of the 7 missense mutations characterized in luminal loop-1, 6 are null mutations—a finding consistent with this loop being critical for G6P uptake activity. Similarly, the N-terminus appears critical for function. The p.MIV N-terminal mutation that lacks the entire N-terminal domain (amino acids 1–7) and part of helix 1 (amino acids 8–16) is devoid of microsomal G6P transport activity (Chen et al., 2002).

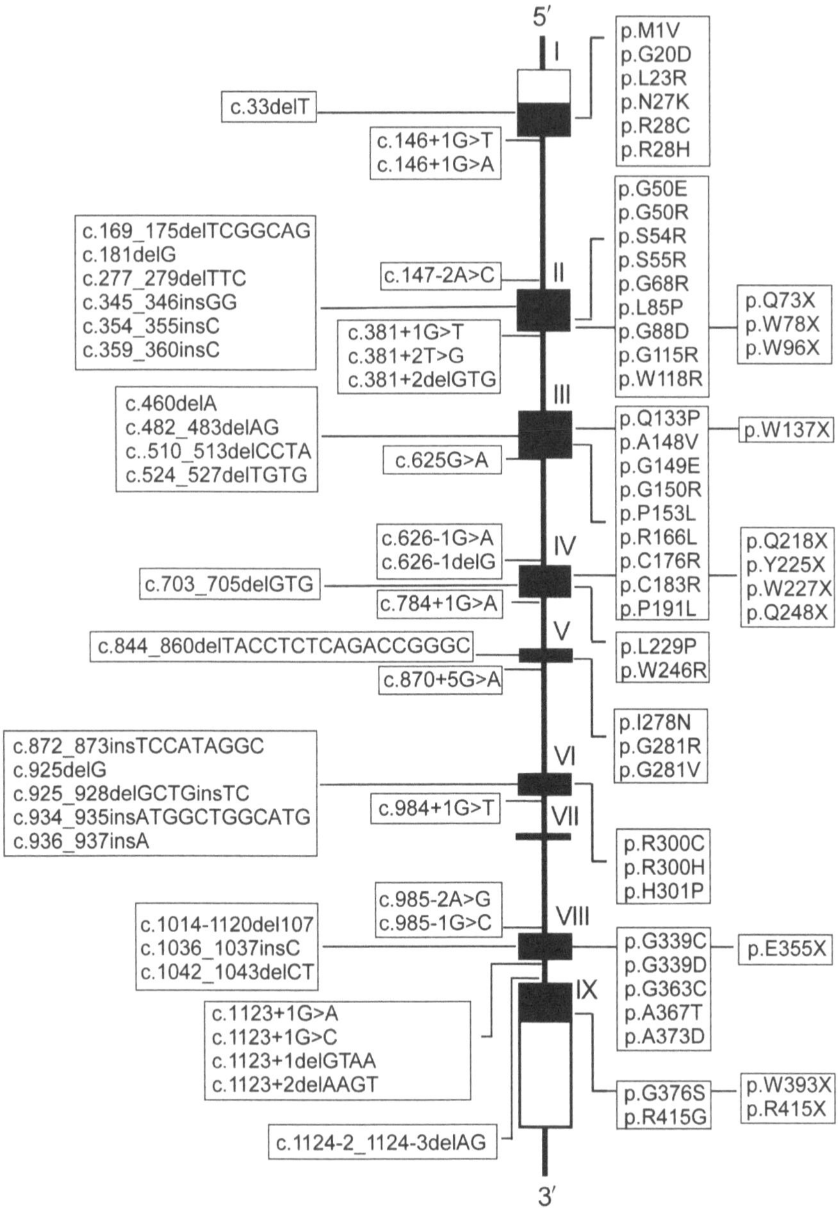

Figure 10.7 Mutations identified in the *SLC37A4* gene of GSD-Ib patients. The *SLC37A4* gene is shown as a line diagram with the nine exons marked as boxes I to IX. Black boxes represent coding regions, and white boxes represent the 5′- and 3′-untranslated regions of the *SLC37A4* transcript. The positions of all known mutations are listed from left to right as insertion/deletion, splicing, missense, and nonsense mutations.

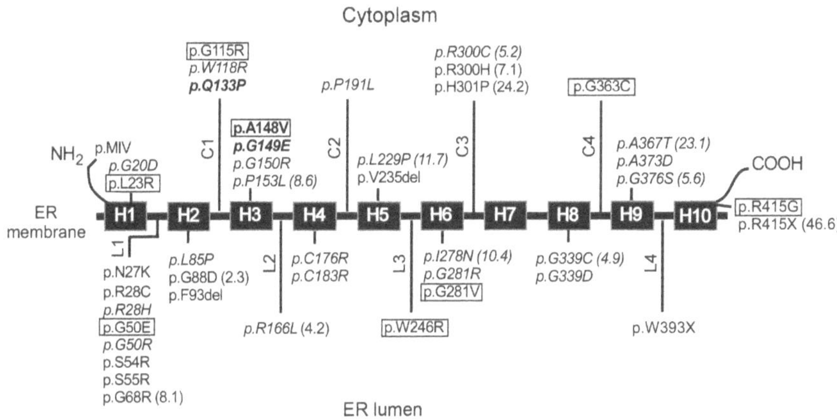

Figure 10.8 A summary of mutations of the *SLC37A4* gene that affect G6P uptake activity. The G6PT protein is shown embedded in the ER membrane. The 10-helical transmembrane domains of G6PT are marked as boxes H1 to H10, the cytoplasmic loops are marked as C1 to C4, and the luminal loops are marked as L1 to L4. Mutations that have been shown to obliterate G6PT activity are listed without parentheses after. Mutants that retain residual activity are listed with the percent of wild-type transport activity retained noted in parentheses. The three naturally occurring signature motif mutations (p.Q133P, p.A148V, and p.G149E) identified to date in GSD-Ib patients are shown in boldface. Mutations that have not been functionally characterized are boxed. Mutations that have been characterized by both cell-based assays and proteoliposomal assays are shown in italic.

However, surprisingly, the p.R415X C-terminal mutation that lacks the entire C-terminal tail containing the ER retention signature motif (KKAE) (Jackson, Nilsson, & Peterson, 1990, 1993) at amino acids 426–429 (Fig. 10.5) retains 47% of wild-type G6PT activity (Fig. 10.8) (Chen et al., 2000).

The following seven helical mutations lying within transmembrane domains destabilize the G6PT protein: p.G20D (H1), p.F93del (H2), p. C176R (H4), p.V235del (H5), p.I278N (H6), p.G339C (H8), and p.G339D (H8). p.G50R is the only nonhelical mutation that destabilizes the G6PT protein (Chen et al., 2002). The results suggest the structural integrity of the transmembrane helices is critical for the stability of the G6PT protein. The p.W393X nonsense mutation was identified in the *SLC37A4* gene of a GSD-Ib patient (Hiraiwa et al., 1999). Transient expression analysis showed that the p.W393X mutant supported the synthesis of little or no G6PT. Two additional nonsense mutations (p.E401X and p.T408X) that delete or disrupt the structure of helix 10 were generated. Again, transient expression assays showed that the p.E401X and p.T408X mutants supported the synthesis of little or no G6PT protein (Chen et al., 2000). While the data show that G6PT mutants lacking an intact helix-10 are misfolded and undergo degradation within cells, additional nonsense mutations need to be characterized to delineate the role of transmembrane helices in the stability of G6PT.

2.4.2 The N- and C-terminal domains of G6PT

The N-terminal cytoplasmic domain of human G6PT consists of seven amino acids (Fig. 10.5). To examine the role of the N domain in G6P transport function, mutants were generated with sequential deletions (Chen et al., 2002). The p.A3M, p.Q4M, p.G5M, p.Y6M, p.G7M, and p.Y8M mutants retained 44.5%, 20.2%, 15.1%, 17.4%, 16.5%, and 8.7% of wild-type G6PT activity, respectively. Because all these N-terminal deletion mutants expressed similar amounts of G6PT proteins as the wild-type transporter, the N domain appears essential for optimal microsomal G6P uptake, but not G6PT stability.

The C-terminal cytoplasmic domain (amino acids 415–429) of human G6PT consists of 15 amino acids (Fig. 10.5) and appears important for protein stability. The naturally occurring R415X mutant lacking the entire cytoplasmic tail is degraded more rapidly than wild-type G6PT (Chen et al., 2000). To determine the minimal length of the cytoplasmic tail required for G6PT folding and/or stability, G6PT mutants were generated that lacked the last 14 (p.N416X), 13 (p.I417X), 12 (R418X), 11 (p.T419X), and 10 (p.K420X) amino acids (Chen et al., 2000). The length of the first three residues at amino acids 415–417 of the cytoplasmic tail appeared to correlate with stability, with p.R415X<p.N416<p.I417X to wild-type G6PT stability, while p.R418X, p.T419X, and p.K420X mutants supported the synthesis of similar amounts of proteins as wild-type G6PT (Chen et al., 2000).

2.4.3 The signature motif in G6PT

The organo-phosphate:P_i antiporter family members (G6PT, GlpT, and uhpT), the G6P receptor (uhpC), and the phosphoglycerate transporter (PgtP) share a signature motif (ProSite PDOC00726), which lies between amino acids 133–149 in human G6PT (Fig. 10.9A). One *SLC37A4* mutation identified in GSD-Ib/nonGSD-Ia patients, p.Q133P (Veiga-da-Cunha et al., 1999), which alters the first amino acid in this motif, was shown to be devoid of G6P transport activity but retains 5% wild-type P_i transport activity (Chen, Pan, Lee, et al., 2008; Chen et al., 2002).

A

```
         133                  149
hG6PT QFGTW WAILS TSMNL AG
mG6PT QFGTW WAVLS TSMNL AG
rG6PT QFGTW WAVLS TSMNL AG
 uhpC ERGGW WALWN TAHNV GG
 uhpT KRGTF LGFWN ISHNL GG
 GlpT ERGGI VSVWN CAHNV GG
 PgtP ERGRV GAFWN ISHNV GG
```

B

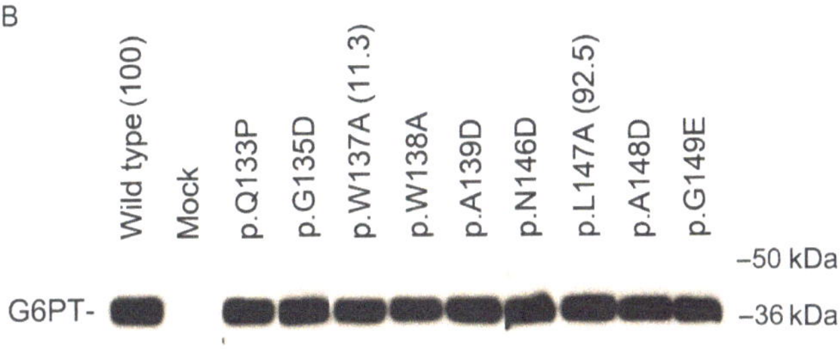

Figure 10.9 The signature motif in G6PT. (A) Alignment of the signature motifs of mammalian G6PT, uhpT, uhpC, GlpT, and PgtP. Amino acid residues mutated in human G6PT are shown in bold. (B) Western blot analysis of synthesis and microsomal G6P uptake activity of the G6PT signature motif mutants. Mutations that have been shown to obliterate G6PT activity are listed without parentheses after. Mutants that retain residual activity are listed with the percent of wild-type transport activity retained noted in parentheses.

The other *SLC37A4* mutation identified in GSD-Ib patients, p.G149E (Hiraiwa et al., 1999), which alters the last amino acid in this motif, was shown to be a null mutation that completely abolished microsomal G6P and P_i uptake (Chen et al., 2002). Although the p.A148V *SLC37A4* mutation identified in a GSD-Ib patient has not been functionally characterized, the p.A148D mutation was shown to be null (Fig. 10.9B). To evaluate the role of the signature motif in G6PT activity, G6PT mutants involving the invariant residues (p.G135D and p.N146D), the near invariant residues (p.L147A and p.A148D), and the uhpC invariant residues (p.W137A, p.W138A, and p.A139D) were generated and functionally characterized (Pan et al., 2003). None of the signature mutants destabilize the G6PT protein (Fig. 10.9B). However, the p.G135D, p.W138A, p.A139D, p.N146D, and p.A148D mutants, along with p.Q133P and p.G149E mutants are devoid of microsomal G6P uptake activity (Fig. 10.9B), suggesting that in G6PT the signature motif is a functional element required for optimal microsomal G6P transport.

3. SLC37A1, SLC37A2, AND SLC37A3

SLC37A1 or SPX1 (Bartoloni & Antonarakis, 2004; Bartoloni et al., 2000; Chou et al., 2013; Iacopetta et al., 2010; Pan et al., 2011), SLC37A2 or SPX2 (Bartoloni & Antonarakis, 2004; Chou et al., 2013; Kim, Tillison, Zhou, Wu, & Smas, 2007; Pan et al., 2011; Takahashi et al., 2000), and SLC37A3 or SPX3 (Bartoloni & Antonarakis, 2004; Chou et al., 2013; Pan et al., 2011) are the less characterized members of the SLC37 family. Human SLC37A1, SLC37A2, and SLC37A3 are similar to G6PT in being ER-associated transmembrane proteins (Pan et al., 2011).

3.1. SLC37A1 (SPX1)

The human *SLC37A1* gene consists of 19 coding exons on chromosome 21q22.3 that encodes a 533 amino acid polypeptide with a calculated molecular weight of 58 kDa (Table 10.1) (Bartoloni et al., 2000). The SLC37A1 protein shares 59% homology to SLC37A2, 35% homology to SLC37A3, and 22% homology to SLC37A4 (Fig. 10.1) (Bartoloni & Antonarakis, 2004; Chou et al., 2013). The SLC37A1 protein also shares 30% sequence identity to GlpT, suggesting that SLC37A1 could be a mammalian glycerol-3-phosphate transporter (Bartoloni et al., 2000). The *SLC37A1* gene is widely expressed, with the highest transcript levels in adult kidney, bone marrow, intestine, spleen, and liver (Bartoloni et al., 2000). In the liver and kidney, the *SLC37A1* transcripts constitute less than 2% of the levels seen for *SLC37A4* (Pan et al., 2011). In the intestine, pancreas, and macrophage, the *SLC37A1* transcripts constitute 60%, 69%, and 43%, respectively, of the levels seen for *SLC37A4* (Chou et al., 2013). In neutrophils, the *SLC37A1* transcripts are significantly higher (2.8-fold) than that of *SLC37A4*. The expression of the *SLC37A1* transcript is upregulated by epidermal growth factor in breast cancer and endometrial cancer cells (Iacopetta et al., 2010). One hypothesis is that SLC37A1 is involved in phospholipid biosynthesis (Iacopetta et al., 2010), which has a role in the proliferation of tumor cells. However, to date, there is no direct evidence for a role of SLC37A1 in breast cancer, or any other disease. Because the levels of the *SLC37A1* transcript in neutrophils is higher than that of *SLC37A4*, which is known to play a vital role in neutrophil homeostasis (Chou et al., 2010a, 2010b, 2013), neutrophils may be an attractive cell type to initiate a search for the biological role of SLC37A1.

3.2. SLC37A2 (SPX2)

The human *SLC37A2* gene consists of 18 coding exons on chromosome 11q24.2 that encodes four alternative spliced transcripts (Table 10.1). The 505-residue SLC37A2 shares 59% homology to SLC37A1, 36% homology to SLC37A3, and 23% homology to SLC37A4 (Fig. 10.1) (Bartoloni & Antonarakis, 2004; Chou et al., 2013). The murine *Slc37a2* was first identified as a cAMP-inducible gene in RAW264 macrophages (Takahashi et al., 2000). The 510 amino acid murine Slc37a2 protein contains three potential N-linked glycosylation sites and migrates as a heterogeneous species of 55–75 kDa (Kim et al., 2007). In this respect, it differs from the non-glycosylated SLC37A1, SLC37A3, and SLC37A4. The *Slc37a2* transcript in mice is primarily expressed in the spleen, thymus, and macrophages (Kim et al., 2007). Compared to *Slc37a4*, the *Slc37a2* transcript is highly expressed in neutrophils and macrophages. Human SLC37A2 is post-translationally modified by N-linked glycosylation, and the human SLC37A2 transcript increases markedly during differentiation of THP-1 human monocytes to macrophages (Kim et al., 2007). The tissue-specific expression of SLC37A2 in macrophages and neutrophils may point to a biological transport role in these cells.

3.3. SLC37A3 (SPX3)

The human *SLC37A3* gene consists of 17 coding exons on chromosome 7q34 that encodes three alternatively spliced transcripts (Table 10.1) (Bartoloni & Antonarakis, 2004). The *SLC37A3* transcripts are expressed in liver, kidney, intestine, pancreas, neutrophils, and macrophages (Pan et al., 2011). Interestingly, the levels of the *SLC37A3* transcript in the pancreas and neutrophils are significantly higher than the levels of other SLC37 members (Chou et al., 2013), suggesting a yet undiscovered role for this protein in the pancreas and neutrophils. Indeed, all SLC37 members seem to be expressed well in neutrophils, suggesting a possible common role in the immune system.

3.4. Functional analysis of the SLC37A1, SLC37A2, and SLC37A3 proteins

Using reconstituted proteoliposomes, investigators demonstrated that SLC37A1 and SLC37A2 are P_i-linked G6P antiporters capable of G6P:P_i and P_i:P_i exchanges, as seen with G6PT (Fig. 10.10A) (Pan et al., 2011). The SLC37A3 isoform of 494 amino acids lacks G6P antiporter activity

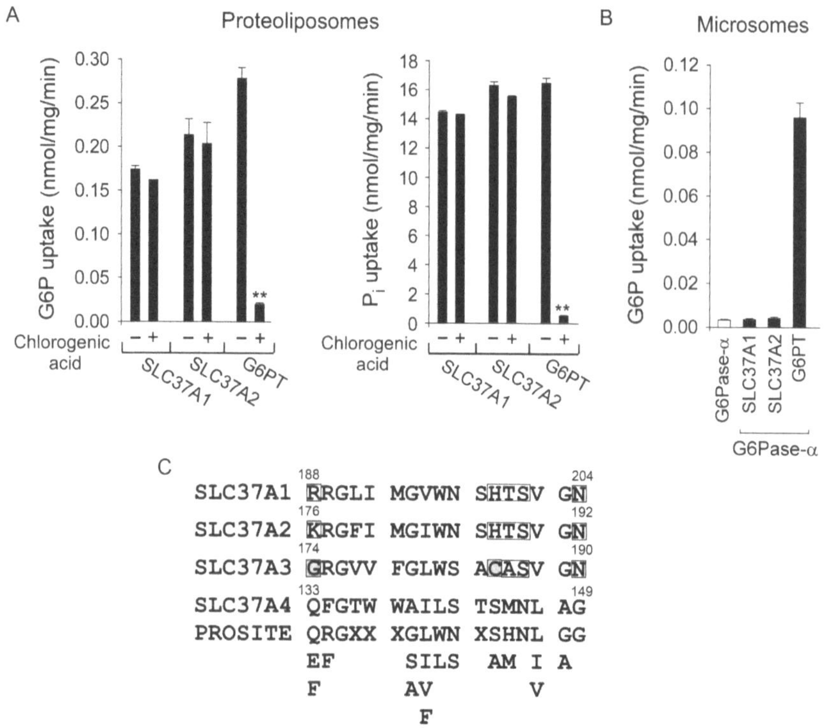

Figure 10.10 The activities of SLC37A1, SLC37A2, and G6PT, and the organo-phosphate:P_i antiporter family consensus signature motif. (A) The antiporter activity. The activity was determined in 50 mM P_i-preloaded proteoliposomes expressing SLC37A1, SLC37A2, or G6PT in the absence or presence of 2 mM chlorogenic acid. Data are presented as the mean ± SEM. (B) Effects of G6Pase-α on microsomal G6P transport activity of SLC37A1, SLC37A2, or G6PT. (C) The organo-phosphate:P_i antiporter family signature motif, ProSite PDOC00726, present in the SLC37 members. The motif lies at residues 188–204 in SLC37A1, residues 176–192 in SLC37A2, residues 174–190 in SLC37A3, and residues 133–149 in SLC37A4/G6PT protein sequences. Only SLC37A4 shows an exact match with the ProSite motif. While only one residue (p.G135) is invariant across the family, 12 of the 17 residues in the ProSite motif are conserved in all members. The five nonconserved residues that cluster at the N and C termini of the motif are boxed. The two residues in SLC37A3 (p.G174 and p.C185) with significantly different characteristics are boxed and shaded. The redundancy at each position of the ProSite motif is shown. X denotes any amino acid.

(Pan et al., 2011), and there are no other reported activities for the protein. The G6PT-mediated microsomal G6P uptake activity, which is co-dependent upon G6Pase-α activity (Lei et al., 1996) and sensitive to cholorgenic acid inhibition (Arion et al., 1998), is the rate-limiting step in endogenous glucose production (Arion, Lange, & Ballas, 1976). Despite

SLC37A1, SLC37A2, and G6PT/SLC47A4 being P_i-linked G6P antiporters, the transport activity of SLC37A1 and SLC37A2 differs from that of G6PT. First, the transport activities of SLC37A1 and SLC37A2 are not sensitive to chlorogenic acid inhibition (Pan et al., 2011) (Fig. 10.10A). Second, neither SLC37A1 nor SLC37A2 can couple functionally with either G6Pase-α (Fig. 10.10B) or G6Pase-β to mediate microsomal G6P uptake (Pan et al., 2011). Studies have shown that in human cell lines, there is an additional microsomal G6P transporter activity that is insensitive to chlorogenic acid inhibition (Leuzzi et al., 2001). In contrast, to G6PT that is highly selective for G6P, the G6PT-unrelated microsomal transport activity is equally responsive to G6P and glucose 1-phosphate (Leuzzi et al., 2001). Presently, it is not known whether these activities are in fact due to the SLC37 family proteins or not. Taken together, multiple sugar-phosphate transporters exist in the ER, although the physiological substrate for SLC37A1 and SLC37A2 remains to be determined.

Interestingly, the 17-residue signature motif (ProSite PDOC00726) shared by G6PT, GlpT, uhpT, uhpC, and PgtP (Fig. 10.9A) is also shared by SLC37A1, SLC37A2, and SLC37A3 (Fig. 10.10C). While SLC37A4/G6PT shows an exact match to the consensus sequence, SLC37A1 and SLC37A2 differ from the consensus in five positions but contain residues with similar characteristics (Fig. 10.10C). Between SLC37A1 and SLC37A2, this motif is almost entirely conserved. The signature motif in SLC37A3 also differs from SLC37A4 at the same five positions but has two residues with significantly different characteristics, including a small noncharged Gly in place of the charged residue at position 1, and a Cys at position 12 (Fig. 10.10C). It is possible that this motif in SLC37A3 forms a different tertiary structure that may contribute to the lack of antiporter activity of this protein. It would be of interest to determine the impact of mutations in these residues on the transport activity of SLC37A3.

4. SYNOPSIS

A deficiency in G6PT causes the genetic disorder GSD-Ib, characterized by impaired glucose homeostasis, neutropenia, and myeloid dysfunction. A large number of *SLC37A4* mutations identified in GSD-Ib patients have been functionally characterized and shown to reduce or abolish G6PT activity. G6PT is anchored in the ER by 10 transmembrane helices with both N and C termini facing the cytoplasm. Despite the identification of cholorogenic acid and its derivatives as potent competitive inhibitors for

G6PT, the G6P-binding motif in the G6PT protein remains to be identified. SLC37A1, SLC37A2, and SLC37A3 are all predicted to contain 10–12 transmembrane domains. However, very little is known about the other SLC37 proteins beyond the P:P and G6P:P exchange activities of SLC37A1 and SLC37A2. The inability of SLC37A1 and SLC37A2 to couple functionally with a G6Pase, compounded with their lack of sensitivity to inhibition by cholorogenic acid, suggests that they are not the physiological ER G6P transporter. Despite expressing similar or higher levels than G6PT in key tissues, including neutrophils and macrophages, SLC37A1, SLC37A2, and SLC37A3 do not appear to have roles in regulating neutrophil/macrophage glucose metabolism or blood glucose homeostasis. Future structure analysis of these proteins may yield clues to their function. The mouse model of GSD-Ib that mimics the phenotype of human GSD-Ib has been generated (Chen et al., 2003) and is now being used to develop new therapies for this disorder. Mouse models of the other SLC37 family members do exist but their phenotypes remain to be characterized. Studying these models may give valuable insights into the biological and functional roles of the other SLC37 family members. In summary, there is still much to learn about the other ER transmembrane proteins of the SLC37 family.

ACKNOWLEDGMENTS

This research was supported by the Intramural Research Program of the Eunice Kennedy Shriver National Institute of Child Health and Human Development, National Institutes of Health.

Conflict of interest: The authors declare no conflict of interest.

REFERENCES

Almqvist, J., Huang, Y., Hovmöller, S., & Wang, D. N. (2004). Homology modeling of the human microsomal glucose 6-phosphate transporter explains the mutations that cause the glycogen storage disease type Ib. *Biochemistry*, *43*(29), 9289–9297.

Annabi, B., Hiraiwa, H., Mansfield, B. C., Lei, K.-J., Ubagai, T., Polymeropoulos, M. H., et al. (1998). The gene for glycogen storage disease type 1b maps to chromosome 11q23. *American Journal of Human Genetics*, *62*(2), 400–405.

Arion, W. J., Canfield, W. K., Ramos, F. C., Su, M. L., Burger, H. J., Hemmerle, H., et al. (1998). Chlorogenic acid analogue S 3483: A potent competitive inhibitor of the hepatic and renal glucose-6-phosphatase systems. *Achieves of Biochemistry and Biophysics*, *351*(2), 279–285.

Arion, W. J., Lange, A. J., & Ballas, L. M. (1976). Quantitative aspects of relationship between glucose 6-phosphate transport and hydrolysis for liver microsomal glucose-6-phosphatase system. Selective thermal inactivation of catalytic component in situ at acid pH. *Journal of Biological Chemistry*, *251*(16), 6784–6790.

Arion, W. J., Wallin, B. K., Carlson, P. W., & Lange, A. J. (1972). The specificity of glucose 6-phosphatase of intact liver microsomes. *Journal of Biological Chemistry, 247*(8), 2558–2565.

Bandsma, R. H., Wiegman, C. H., Herling, A. W., Burger, H. J., ter Harmsel, A., Meijer, A. J., et al. (2001). Acute inhibition of glucose-6-phosphate translocator activity leads to increased de novo lipogenesis and development of hepatic steatosis without affecting VLDL production in rats. *Diabetes, 50*(11), 2591–2597.

Bartoloni, L., & Antonarakis, S. E. (2004). The human sugar-phosphate/phosphate exchanger family SLC37. *Pflügers Archiv, 447*(5), 780–783.

Bartoloni, L., Wattenhofer, M., Kudoh, J., Berry, A., Shibuya, K., Kawasaki, K., et al. (2000). Cloning and characterization of a putative human glycerol 3-phosphate permease gene (SLC37A1 or G3PP) on 21q22.3: Mutation analysis in two candidate phenotypes, DFNB10 and a glycerol kinase deficiency. *Genomics, 70*(2), 190–200.

Boztug, K., Appaswamy, G., Ashikov, A., Schäffer, A. A., Salzer, U., Diestelhorst, J., et al. (2009). A syndrome with congenital neutropenia and mutations in G6PC3. *New England Journal of Medicine, 360*(1), 32–43.

Caramelo, J. J., & Parodi, A. J. (2007). How sugars convey information on protein conformation in the endoplasmic reticulum. *Seminars in Cell and Developmental Biology, 18*(6), 732–742.

Chen, L.-Y., Lin, B., Pan, C.-J., Hiraiwa, H., & Chou, J. Y. (2000). Structural requirements for the stability and microsomal transport activity of the human glucose-6-phosphate transporter. *Journal of Biological Chemistry, 275*(44), 34280–34286.

Chen, S. Y., Pan, C. J., Lee, S., Peng, W., & Chou, J. Y. (2008). Functional analysis of mutations in the glucose-6-phosphate transporter that cause glycogen storage disease type Ib. *Molecular Genetics and Metabolism, 95*(4), 220–223.

Chen, S. Y., Pan, C. J., Nandigama, K., Mansfield, B. C., Ambudkar, S. V., & Chou, J. Y. (2008). The glucose-6-phosphate transporter is a phosphate-linked antiporter deficient in glycogen storage disease type Ib and Ic. *FASEB Journal, 22*(7), 2206–2213.

Chen, L.-Y., Pan, C.-J., Shieh, J.-J., & Chou, J. Y. (2002). Structure-function analysis of the glucose-6-phosphate transporter deficient in glycogen storage disease type Ib. *Human Molecular Genetics, 11*(25), 3199–3207.

Chen, L.-Y., Shieh, J.-. J., Lin, B., Pan, C.-J., Gao, J.-. L., Murphy, P. M., et al. (2003). Impaired glucose homeostasis, neutrophil trafficking and function in mice lacking the glucose-6-phosphate transporter. *Human Molecular Genetics, 12*(19), 2547–2558.

Cheung, Y. Y., Kim, S. Y., Yiu, W. H., Pan, C. J., Jun, H. S., Ruef, R. A., et al. (2007). Impaired neutrophil activity and increased susceptibility to bacterial infection in mice lacking glucose-6-phosphatase-beta. *Journal of Clinical Investigation, 117*(3), 784–793.

Chou, J. Y., Jun, H. S., & Mansfield, B. C. (2010a). Neutropenia in type Ib glycogen storage disease. *Current Opinion in Hematology, 17*(1), 36–42.

Chou, J. Y., Jun, H. S., & Mansfield, B. C. (2010b). Glycogen storage disease type I and G6Pase-β deficiency: Etiology and therapy. *Nature Reviews. Endocrinology, 6*(12), 676–688.

Chou, J. Y., Jun, H. S., & Mansfield, B. C. (2013). The SLC37 family of phosphate-linked sugar phosphate antiporters. *Molecular Aspects of Medicine, 34*(2–3), 601–611.

Chou, J. Y., Matern, D., Mansfield, B. C., & Chen, Y.-T. (2002). Type I glycogen storage diseases: Disorders of the glucose-6-phosphatase complex. *Current Molecular Medicine, 2*(12), 121–143.

Corpet, F. (1988). Multiple sequence alignment with hierarchical clustering. *Nucleic Acids Research, 16*(22), 10881–10890.

Dissanayake, V. H., Jayasinghe, J. D., Thilakaratne, V., & Jayasekara, R. W. (2011). A novel mutation in SLC37A4 gene in a Sri Lankan boy with glycogen storage disease type Ib associated with very early onset neutropenia. *Journal of Molecular and Genetic Medicine: An International Journal of Biomedical Research, 5*, 262–263.

Fann, M. C., Davies, A. H., Varadhachary, A., Kuroda, T., Sevier, C., Tsuchiya, T., et al. (1998). Identification of two essential arginine residues in UhpT, the sugar phosphate antiporter of *Escherichia coli*. *Journal of Membrane Biology*, *164*(2), 187–195.

Galli, L., Orrico, A., Marcolongo, P., Fulceri, R., Burchell, A., Melis, D., et al. (1999). Mutations in the glucose-6-phosphate transporter (G6PT) gene in patients with glycogen storage diseases type 1b and 1c. *FEBS Letters*, *459*(2), 255–258.

Gerin, I., Veiga-da-Cunha, M., Achouri, Y., Collet, J. F., & Van Schaftingen, E. (1997). Sequence of a putative glucose-6-phosphate translocase, mutated in glycogen storage disease type 1b. *FEBS Letters*, *419*(2–3), 235–238.

Gerin, I., Veiga-da-Cunha, M., Noel, G., & Van Schaftingen, E. (1999). Structure of the gene mutated in glycogen storage disease type Ib. *Gene*, *227*(2), 189–195.

Ghosh, A., Shieh, J.-J., Pan, C.-J., & Chou, J. Y. (2004). Histidine-167 is the phosphate acceptor in glucose-6-phosphatase-β forming a phosphohistidine-enzyme intermediate during catalysis. *Journal of Biological Chemistry*, *279*(13), 12479–12483.

Ghosh, A., Shieh, J.-J., Pan, C.-J., Sun, M.-S., & Chou, J. Y. (2002). The catalytic center of glucose-6-phosphatase: His^{176} is the nucleophile forming the phosphohistidine-enzyme intermediate during catalysis. *Journal of Biological Chemistry*, *277*(36), 32837–32842.

Goder, V., & Spiess, M. (2001). Topogenesis of membrane proteins: Determinants and dynamics. *FEBS Letters*, *504*(3), 87–93.

Grefhorst, A., Schreurs, M., Oosterveer, M. H., Cortés, V. A., Havinga, R., Herling, A. W., et al. (2010). Carbohydrate-response-element-binding protein (ChREBP) and not the liver X receptor α (LXRα) mediates elevated hepatic lipogenic gene expression in a mouse model of glycogen storage disease type 1. *Biochemical Journal*, *432*(2), 249–254.

Hall, J. A., Fann, M. C., & Maloney, P. C. (1999). Altered substrate selectivity in a mutant of an intrahelical salt bridge in UhpT, the sugar phosphate carrier of *Escherichia coli*. *Journal of Biological Chemistry*, *274*(10), 6148–6153.

He, L., Vasiliou, K., & Nebert, D. W. (2009). Analysis and update of the human solute carrier (SLC) gene superfamily. *Human Genomics*, *3*(2), 195–206.

Hemmerle, H., Burger, H. J., Below, P., Schubert, G., Rippel, R., Schindler, P. W., et al. (1997). Chlorogenic acid and synthetic chlorogenic acid derivatives: Novel inhibitors of hepatic glucose-6-phosphate translocase. *Journal of Medicinal Chemistry*, *40*(2), 137–145.

Herling, A. W., Burger, H., Schubert, G., Hemmerle, H., Schaefer, H., & Kramer, W. (1999). Alterations of carbohydrate and lipid intermediary metabolism during inhibition of glucose-6-phosphatase in rats. *European Journal of Pharmacology*, *386*(1), 75–82.

Hiraiwa, H., Pan, C.-J., Lin, B., Moses, S. W., & Chou, J. Y. (1999). Inactivation of the glucose-6-phosphate transporter causes glycogen storage disease type 1b. *Journal of Biological Chemistry*, *274*(9), 5532–5536.

Hoffman, K., & Stoffel, W. (1993). TMbase—A database of membrane spanning protein segments. *Biological Chemistry Hoppe-Seyler*, *347*, 166–170.

Huang, Y., Lemieux, M. J., Song, J., Auer, M., & Wang, D. N. (2003). Structure and mechanism of the glycerol-3-phosphate transporter from Escherichia coli. *Science*, *301*(5633), 616–620.

Iacopetta, D., Lappano, R., Cappello, A. R., Madeo, M., De Francesco, E. M., Santoro, A., et al. (2010). SLC37A1 gene expression is up-regulated by epidermal growth factor in breast cancer cells. *Breast Cancer Research and Treatment*, *122*(3), 755–764.

Jackson, M. R., Nilsson, T., & Peterson, P. A. (1990). Identification of a consensus motif for retention of transmembrane proteins in the endoplasmic reticulum. *EMBO Journal*, *9*(10), 3153–3162.

Jackson, M. R., Nilsson, T., & Peterson, P. A. (1993). Retrieval of transmembrane proteins to the endoplasmic reticulum. *Journal of Cell Biology*, *121*(2), 317–333.

Janecke, A. R., Linder, M., Erdel, M., Mayatepek, E., Moslinger, D., Podskarbi, T., et al. (2000). Mutation analysis in glycogen storage disease type 1 non-a. *Human Genetics*, *107*(3), 285–289.

Jun, H. S., Cheung, Y. Y., Lee, Y. M., Mansfield, B. C., & Chou, J. Y. (2012). Glucose-6-phosphatase-β, implicated in a congenital neutropenia syndrome, is essential for macrophage energy homeostasis and functionality. *Blood*, *119*(17), 4047–4055.

Jun, H. S., Lee, Y. M., McDermott, D. H., DeRavin, S. S., Murphy, P. M., Mansfield, B. C., et al. (2010). Lack of glucose recycling between endoplasmic reticulum and cytoplasm underlies cellular dysfunction in glucose-6-phosphatase-beta-deficient neutrophils in a congenital neutropenia syndrome. *Blood*, *116*(15), 2783–2792.

Kim, J. Y., Tillison, K., Zhou, S., Wu, Y., & Smas, C. M. (2007). The major facilitator superfamily member Slc37a2 is a novel macrophage- specific gene selectively expressed in obese white adipose tissue. *American Journal of Physiology. Endocrinology and Metabolism*, *293*(1), E110–E120.

Landolt-Marticorena, C., & Reithmeier, R. A. F. (1994). Asparagine-linked oligosaccharides are localized to single extracytosolic segments in multi-span membrane glycoproteins. *Biochemical Journal*, *302*(Pt 1), 253–260.

Lei, K.-J., Chen, H., Pan, C.-J., Ward, J. M., Mosinger, B., Lee, E. J., et al. (1996). Glucose-6-phosphatase dependent substrate transport in the glycogen storage disease type 1a mouse. *Nature Genetics*, *13*(2), 203–209.

Lei, K.-J., Shelly, L. L., Pan, C.-J., Sidbury, J. B., & Chou, J. Y. (1993). Mutations in the glucose-6-phosphatase gene that cause glycogen storage disease type 1a. *Science*, *262*(5133), 580–583.

Leuzzi, R., Fulceri, R., Marcolongo, P., Bánhegyi, G., Zammarchi, E., Stafford, K., et al. (2001). Glucose 6-phosphate transport in fibroblast microsomes from glycogen storage disease type 1b patients: Evidence for multiple glucose 6-phosphate transport systems. *Biochemical Journal*, *357*(Pt 2), 557–562.

Lin, B., Annabi, B., Hiraiwa, H., Pan, C.-J., & Chou, J. Y. (1998). Cloning and characterization of cDNAs encoding a candidate glycogen storage disease type 1b protein in rodents. *Journal of Biological Chemistry*, *273*(48), 31656–31670.

Lin, B., Pan, C.-J., & Chou, J. Y. (2000). Human variant glucose-6-phosphate transporter is active in microsomal transport. *Human Genetics*, *107*(5), 526–529.

Lloyd, A. D., & Kadner, R. J. (1990). Topology of the *Escherichia coli* UhpT sugar-phosphate transporter analyzed by using TnphoA fusions. *Journal of Bacteriology*, *172*(4), 1688–1693.

Molinari, M. (2007). N-glycan structure dictates extension of protein folding or onset of disposal. *Nature Chemical Biology*, *3*(6), 313–320.

Nilsson, I. M., & von Heijne, G. (1993). Determination of the distance between the oligosaccharyltransferase active site and the endoplasmic reticulum membrane. *Journal of Biological Chemistry*, *268*(8), 5798–5801.

Nordlie, R. C., Sukalski, K. A., Munoz, J. M., & Baldwin, J. J. (1983). Type 1c, a novel glycogenosis. *Journal of Biological Chemistry*, *258*, 9739–9744.

Pan, C. J., Chen, S. Y., Jun, H. S., Lin, S. R., Mansfield, B. C., & Chou, J. Y. (2011). SLC37A1 and SLC37A2 are phosphate-linked, glucose-6-phosphate antiporters. *PLoS One*, *6*(9), e23157.

Pan, C. J., Chen, S. Y., Lee, S., & Chou, J. Y. (2009). Structure-function study of glucose-6-phosphate transporter, an eukaryotic antiporter deficient in glycogen storage disease type Ib. *Molecular Genetics and Metabolism*, *96*(1), 32–37.

Pan, C. J., Chen, L. Y., Mansfield, B. C., Salani, B., Varesio, L., & Chou, J. Y. (2003). The signature motif in human glucose-6-phosphate transporter is essential for microsomal transport of glucose-6-phosphate. *Human Genetics*, *112*(4), 430–433.

Pan, C.-J., Lin, B., & Chou, J. Y. (1999). Transmembrane topology of human glucose-6-phosphate transporter. *Journal of Biological Chemistry*, *274*(20), 13865–13869.

Pao, S. S., Paulsen, I. T., & Saier, M. H., Jr. (1998). Major facilitator superfamily. *Microbiology and Molecular Biology Reviews*, *62*(1), 1–34.

Parodi, A. J. (2000). Protein glucosylation and its role in protein folding. *Annual Review of Biochemistry*, *69*, 69–93.

Qiu, Z. Q., Lu, C. X., Wang, W., Qiu, J. J., & Wei, M. (2011). Mutation in the SLC37A4 gene of glycogen storage disease type Ib in 15 families of the mainland of China. *Zhonghua Er Ke Za Zhi. Chinese Journal of Pediatrics*, *49*(3), 203–208.

Reddy, V. S., Shlykov, M. A., Castillo, R., Sun, E. I., & Saier, M. H., Jr. (2012). The major facilitator superfamily (MFS) revisited. *FEBS Journal*, *279*(11), 2022–2035.

Shieh, J.-J., Pan, C.-J., Mansfield, B. C., & Chou, J. Y. (2003). Glucose-6-phosphate hydrolase, widely expressed outside the liver, can explain age-dependent resolution of hypoglycemia in glycogen storage disease type Ia. *Journal of Biological Chemistry*, *278*(47), 47098–47103.

Takahashi, Y., Miyata, M., Zheng, P., Imazato, T., Horwitz, A., & Smith, J. D. (2000). Identification of cAMP analogue inducible genes in RAW264 macrophages. *Biochimica et Biophysica Acta*, *1492*(2–3), 385–394.

Veiga-da-Cunha, M., Gerin, I., Chen, Y.-T., de Barsy, T., de Lonlay, P., Dionisi-Vici, C., et al. (1998). A gene on chromosome 11q23 coding for a putative glucose-6-phosphate translocase is mutated in glycogen-storage disease types Ib and Ic. *American Journal of Human Genetics*, *63*(4), 976–983.

Veiga-da-Cunha, M., Gerin, I., Chen, Y.-T., Lee, P. J., Leonard, J. V., Maire, I., et al. (1999). The putative glucose 6-phosphate translocase gene is mutated in essentially all cases of glycogen storage disease type I non-a. *European Journal of Human Genetics*, 7(16), 717–723.

Zingone, A., Hiraiwa, H., Pan, C.-J., Lin, B., Chen, H., Ward, J. M., et al. (2000). Correction of glycogen storage disease type 1a in a mouse model by gene therapy. *Journal of Biological Chemistry*, *275*(2), 828–832.

CHAPTER ELEVEN

SLC41 Transporters—Molecular Identification and Functional Role

Monika Schweigel-Röntgen*[,1], Martin Kolisek[†]
*Institute for Muscle Biology & Growth, Leibniz Institute for Farm Animal Biology, Dummerstorf, Germany
[†]Institute of Veterinary Physiology, Free University Berlin, Berlin, Germany
[1]Corresponding author: e-mail address: roentgen@fbn-dummerstorf.de

Contents

Abstract

The solute carrier family 41 (SLC41) encompasses three members A1, A2, and A3. Based on their distant homology to the bacterial Mg^{2+} channel MgtE, all have been linked to Mg^{2+} transport. There is only very limited knowledge on the molecular biology and exact functions of SLC41A2 and SLC41A3. SLC41A1 is ubiquitously expressed and data on its functional and molecular properties, regulation, complex-forming ability, and spectrum of binding partners are available. SLC41A1 was recently identified as being the Na^+/Mg^{2+} exchanger (NME)—a predominant Mg^{2+} efflux system. Mg^{2+}-dependent and hormonal regulation of NME activity is now known to depend on the intracellular N terminus of SLC41A1 that is involved in Mg^{2+} sensing and contains phosphorylation

Current Topics in Membranes, Volume 73
ISSN 1063-5823
http://dx.doi.org/10.1016/B978-0-12-800223-0.00011-6

sites for protein kinase (PK) A and PKC. Data showing a link between SLC41A1 and human disorders such as Parkinson's disease, nephronophthisis (induced by the null mutation c.698G>T in renal *SLC41A1*), and preeclampsia make the protein a candidate therapeutic target.

1. INTRODUCTION

Magnesium (Mg^{2+}), in particular, its free intracellular form, plays an important role as a cofactor for many enzymes, as a modulator of ion channels, and in numerous other processes such as protein synthesis, nucleic acid stability, neuromuscular excitability, secretion of hormones, and in antagonizing the actions of Ca^{2+} (Bara & Guiet-Bara, 2001; Garban, 1997; Hartwig, 2001; Jóna et al., 2001; Mooren et al., 2001; Watanabe & Dreyfus, 1972; White & Hartzell, 1989). Therefore, the regulation of extracellular and intracellular Mg^{2+} levels by transmembrane and transepithelial transport processes is critical for numerous cellular and organ functions. At a functional level, the presence of such active plasma membrane Mg^{2+} transport mechanisms has been indicated for a long time, but the identification of their molecular entities has started only recently (reviewed in Quamme, 2010).

Among Mg^{2+} transporters are the three members A1, A2, and A3 of solute carrier family 41 (SLC41) with which this review deals. All of them are distantly homologous to the unique bacterial Mg^{2+} transport protein MgtE (Smith, Thompson, & Maguire, 1995), suggesting that they also may be involved in transport of the ion across cell membranes (Quamme, 2010; Wabakken, Rian, Kveine, & Aasheim, 2003). Indeed, the group of Quamme (Goytain & Quamme, 2005a, 2005b; Quamme, 2010) demonstrated that all three SLC41 isoforms mediate electrogenic transport of Mg^{2+} when cloned from renal mouse distal tubule cells and heterologously expressed in *Xenopus laevis* oocytes. Furthermore, in a *Salmonella enterica* strain (MM281) displaying disruption of all natural Mg^{2+} transporters, overexpression of hSLC41A1 functionally substitutes these transporters and thereby restores bacterial growth (Kolisek et al., 2008).

Using a Mag-Fura2-based fluorescence spectroscopic assay on HEK293 cells overexpressing human SLC41A1, our group performed the first direct studies showing SLC41A1-linked Mg^{2+} transport in a mammalian cell system (Kolisek et al., 2008; Kolisek, Nestler, Vormann, & Schweigel-Röntgen, 2012). According to our data, SLC41A1 is a functional Mg^{2+}

carrier that normally mediates efflux of the ion (Kolisek et al., 2008) and functions as a Na^+/Mg^{2+} exchanger (NME). The NME is the best characterized Mg^{2+} transporter of nonepithelial cells and tissues (Günther & Vormann, 1995; Hintz, Günzel, & Schlue, 1999; Xu & Willis, 1994; Yoshimura et al., 1995). The NME was first described in chicken and turkey red blood cells by the laboratory of Günther (Günther & Vormann, 1985; Günther, Vormann, & Förster, 1984). This laboratory (Büttner, Günther, Schäfer, & Vormann, 1998; Günther & Vormann, 1990a) and others (Feray & Garay, 1986; Flatman & Smith, 1996; Xu & Willis, 1994) studied the Na^+-dependent Mg^{2+} transporter in numerous vertebrate erythrocytes. Over the years, NME activity has been identified in the majority of examined mammalian cell types (Cefaratti & Ruse, 2007; Fagan & Romani, 2001; Handy, Gow, Ellis, & Flatman, 1996; He, Yao, Savoia, & Touyz, 2005; Schweigel, Park, Etschmann, & Martens, 2006; Tessman & Romani, 1998).

Regulated Mg^{2+} efflux by the NME is essential for the maintenance of free intracellular magnesium concentrations ($[Mg^{2+}]_i$) at levels between 0.4 and 1.2 mM (Schweigel et al., 2006; Standley & Standley, 2002), and for Mg^{2+} transport across intestinal, renal, placental, and mammary epithelia (Leonhard-Marek, Stumpff, Brinkmann, Breves, & Martens, 2005; Quamme, 1989; Shaw, Mughal, Maresh, & Sibley, 1991; Wolf et al., 2009). Additionally, hormones/mediators activate the NME to induce an $[Mg^{2+}]_i$ decrease, which directly or indirectly influences cellular transport mechanisms and physiological functions (Fagan & Romani, 2001; He et al., 2005; Kolisek, Sponder, et al., 2013). NME disruption or dysregulation may alter systemic and intracellular Mg^{2+} homeostasis and participate in the pathogenesis of various diseases and associated events such as primary hypertension (Kisters, Krefting, Barenbrock, Spieker, & Rahn, 1999), ischemic heart disease (Altura & Altura, 1995), diabetes (Saris, Mervaala, Karppanen, Khawaja, & Lewenstam, 2000), cystic fibrosis (Vormann, 1992), neurodegeneration (Barbagallo, Belvedere, Di Bella, & Dominguez, 2011; Lovitt et al., 2010), inflammation (Weglicki & Phillips, 1992), and preeclampsia/eclampsia (Weaver, 1986).

Although members A2 and A3 of the SLC41 family have also been linked to Mg^{2+} transport, their molecular and biochemical properties (including membrane topology, cellular localization, transcriptomics, and proteomics) and physiology compared to SLC41A1 are only poorly explored. Therefore, this review concentrates on the molecular and functional characteristics of SLC41A1, in addition to its regulation and involvement in the pathophysiology of human diseases.

2. SLC41A1—TISSUE DISTRIBUTION AND FUNCTIONAL CHARACTERISTICS

2.1. SLCA1 expression profile

Transcripts of the human and mice gene *SLC41A1* have been identified in various organs and tissues. It is abundantly expressed in heart, testis, skeletal muscle, brain, kidney, liver, the adrenal and thyroid glands, and also in prostate and ovaries; lower levels have been detected in all other tested tissues, for example, colon and small intestine (Goytain & Quamme, 2005a). Transcript has been shown to be expressed in numerous cell lines such as HEK-293, Tom-1, BV173, Reh, DT40, Jurkat, and JVM-13 (Kolisek et al., 2008; M. Kolisek et al., unpublished data; Mandt, Song, Scharenberg, & Sahni, 2011; Wabakken et al., 2003). Therefore, we can assume that *SLC41A1* is ubiquitously expressed.

Functional data support such ubiquitous expression. NME activity is evident in various avian and mammalian cell types, for example, hepatocytes (Cefaratti & Ruse, 2007; Fagan & Romani, 2001; Tessman & Romani, 1998), thymocytes (Günther & Vormann, 1992), cardiac myocytes (Handy et al., 1996; Tashiro, Inoue, & Konishi, 2009), smooth muscle cells (He et al., 2005; Tashiro & Konishi, 1997), neurons and glial cells (Hintz et al., 1999; Stout, Li-Smerin, Johnson, & Reynolds, 1996), rat sublingual acini (Zhang & Melvin, 1995), ruminal epithelial cells (Schweigel et al., 2006; Schweigel, Vormann, & Martens, 2000), mammary epithelial cells (Wolf et al., 2009), trophoblast cells (Standley & Standley, 2002), and human platelets (Okorodudu, Yang, & Elghetany, 2001; Yoshimura et al., 1995).

2.2. Functional characteristics of Na^+-dependent Mg^{2+} efflux

In the presence of physiological transmembrane Na^+ ($[Na^+]_i < [Na^+]_e$) and Mg^{2+} ($[Mg^{2+}]_i \leq [Mg^{2+}]_e$) gradients, the NME mediates efflux of intracellular Mg^{2+} in exchange for influx of extracellular Na^+ (Schweigel et al., 2006). The main driving force for this exchange is the inwardly directed electrochemical gradient of Na^+; the transporter is indirectly dependent on the Na^+/K^+ pump and ATP (Günther & Vormann, 1995; Okorodudu et al., 2001; Yoshimura et al., 1995). The stoichiometry of the exchanger seems to vary among cell types and possibly according to the functional state. In most cases, an electroneutral ratio of 2 Na^+ to 1 Mg^{2+} has been reported (Günther & Vormann, 1990b; Günther et al., 1984). For human erythrocytes, a stoichiometric ratio of either 3 Na^+ to

1 Mg^{2+} (Féray & Garay, 1988) or 1 Na^{+} to 1 Mg^{2+} (Frenkel, Graziani, & Schatzmann, 1989) has been calculated.

Mg^{2+} extrusion via the NME is strictly dependent on extracellular Na^{+} (Fagan & Romani, 2001; Schweigel et al., 2006; Wolf, Di Francesco, Covacci, & Cittadani, 1995; Yoshimura et al., 1995). Substitution of extracellular Na^{+} with K^{+}, Li^{+}, or choline^{+} hardly supports Mg^{2+} efflux (Cefaratti, Romani, & Scarpa, 1998; Kubota et al., 2003) and reduces Mg^{2+} extrusion by 91% (Kolisek et al., 2012) or even 97% (Schweigel et al., 2006). An exception is described by Zhang and Melvin (1995). For the NME in rat sublingual acini, K^{+} can partly substitute for Na^{+} and elicit only a 53% decrease in the Mg^{2+} efflux rate. The NME of adult porcine and rat erythrocytes has the capability to transport Mn^{2+} instead of Na^{+} (Büttner et al., 1998). As expected for a carrier system, Mg^{2+} efflux is saturable and its activation by extracellular Na^{+} follows a Michael–Menten relationship with reported K_m values between 8 and 30 mM (Cefaratti, Romani, & Scarpa, 2000; Cefaratti et al., 1998; Feray & Garay, 1986; Frenkel et al., 1989; Günther & Vormann, 1985, 1989; Schweigel et al., 2006; Tashiro, Tursun, & Konishi, 2005).

In some cells (e.g., rat and ferret erythrocytes, heart muscle cells, and ruminal epithelial cells), the Mg^{2+} efflux via NME can be switched to Mg^{2+} influx by reversing the Na^{+} gradient (Flatman & Smith, 1991; Günther & Vormann, 1995; Kolisek et al., 2008; Schweigel et al., 2006, 2000). In comparison to wild-type controls, SLC41A1-overexpressing HEK293 cells have a high intracellular [Na^{+}] (about 40 mM) but reduced intracellular [Mg^{2+}] (about 0.1 mM). Therefore, increasing the extracellular Mg^{2+} concentration will reduce the driving force for electroneutral Mg^{2+} efflux and possibly promote Mg^{2+} influx (Kolisek et al., 2008). The physiological significance of reverse NME activity is not clear, but it might compensate for defects in Mg^{2+} uptake and associated cell growth (Mandt et al., 2011).

Depending on cell type, free intracellular Mg^{2+} level, and the method used to determine Mg^{2+} efflux (e.g., flux studies with ^{28}Mg, determination of intra- or extracellular [Mg^{2+}] and [Na^{+}] by atomic absorption spectrometry or fluorescence probes), transport capacity of the NME varies considerably. Rodent erythrocytes showed NME transport rates of about 6.6 mM/l cells/15 min under nearly V_{max} conditions (after preloading with 12 mM Mg^{2+}). The Mg^{2+} efflux rate was considerably lower in pig, dog, and cat erythrocytes (about 2 mM/l cells/15 min) and very low in red cells from horse (0.59 mM/l cells/15 min), sheep (0.18 mM/l cells/15 min), and cattle (0.03 mM/l cells/15 min) (Büttner et al., 1998). In human erythrocytes, Na^+-dependent Mg^{2+} efflux is only 0.095 mM/l cells/15 min after high-Mg^{2+} loading. After preloading ruminal epithelial cells with 6 mM Mg^{2+}, we found a NME-mediated Mg^{2+} efflux of 9.4 ± 4.7 mM/l cells/15 min (Schweigel et al., 2006). Similar high transport rates have been found only with Mg^{2+}-loaded rat (Günther & Vormann, 1989) and hamster (Willis, Xu, & Zhao, 1992; Xu & Willis, 1994) red cells, which, by the way, show reduced activity of other Na^+-dependent mechanisms, for example, the Na^+/H^+ exchanger. The higher efflux rate in epithelial cells seems to reflect its capability to maintain a normal $[Mg^{2+}]_i$ of 0.6–1.2 mM in spite of disturbances induced by marked Mg^{2+} influx (37.5–42 μM/min) accompanying Mg^{2+} absorption (Schweigel, Lang, & Martens, 1999; Schweigel & Martens, 2003). Indeed, a rapid $[Mg^{2+}]_i$ increase in these cells occurs in Na^+-free medium and after application of an anti-NME antibody (Schweigel et al., 2006).

The considerable variation of functional properties such as maximum transport capacity, reversibility, capability of Mn^{2+} transport, and stoichiometry of NME-dependent Mg^{2+} transport among cell types points to the existence of various isoforms of SLC41A1.

2.3. Toward identification of SLC41A1 as the NME

On the basis of its structural similarity to the bacterial Mg^{2+} transporter MgtE, Wabakken et al. (2003) suggested that SLC41A1 may be responsible for translocating Mg^{2+} across the membrane of eukaryotic cells. This assumption was first tested by Goytain and Quamme (2005a), who used the two-microelectrode voltage-clamp technique to functionally characterize SLC41A1 cloned from mouse renal distal tubule cells and heterologously expressed in *X. laevis* oocytes. With this method, the authors described an Mg^{2+}-related, inwardly directed current. Although SLC41A1 prefers Mg^{2+} as a substrate, it also mediates electrogenic transport of Fe^{2+}, Zn^{2+}, Cu^{2+}, Co^{2+}, and Cd^{2+}, but not Ca^{2+}. The authors also found that mice fed a low-Mg^{2+} diet upregulated transcript levels of *SLC41A1* in some tissues. The authors concluded "that SLC41A1 is a regulated Mg^{2+} transporter that might be involved in magnesium homeostasis in epithelial cells" (Goytain & Quamme, 2005a).

However, experiments performed by our group (Kolisek et al., 2008) on SLC41A1 expressed in a mammalian tetracycline-inducible HEK293 cell system using the whole-cell patch-clamp technique did not give rise to any measurable Mg^{2+} currents. Instead, overexpression of SLC41A1 caused a secondary activation of an endogenous ATP-sensitive Cl^- current that was completely inhibited by the Cl^--channel blocker DIDS. The latter experimental manipulation did not uncover any additional currents indicating that the SLC41A proteins are not constitutively active ion channels.

However, in the same study (Kolisek et al., 2008), we used a Mag-Fura-2-based ratiometric assay and measured significant changes in the free intracellular Mg^{2+} ($[Mg^{2+}]_i$) concentration due to transmembrane net Mg^{2+} movement. These $[Mg^{2+}]_i$ changes only occurred in SLC41A1-overexpressing cells, and their direction and extent were dependent on both the transmembrane gradient for Mg^{2+} and the expression level of SLC41A1. Our experiments were designed to promote Mg^{2+} uptake; cells were Mg^{2+}-depleted by incubating them in totally Mg^{2+}-free media and then performing experiments under conditions with an inwardly directed Mg^{2+} gradient ($[Mg^{2+}]_i \ll [Mg^{2+}]_e$: 2, 5, or 10 mM). This procedure leads to an $[Mg^{2+}]_i$ increase due to an SLC41A1-independent linear component (most probably mediated by the Mg^{2+} channel TRPM7), and an SLC41A1-mediated temperature-sensitive component. The latter was insensitive to cobalt(III)hexaammine—a known inhibitor of bacterial/mitochondrial (Kolisek et al., 2003; Kucharski, Lubbe, & Maguire, 2000) and mammalian (Schweigel, Kuzinski, Deiner, & Kolisek, 2009; Wolf et al., 2009) Mg^{2+} channels. In contrast to wild-type HEK293 cells, SLC41A1-overexpressing cells were Mg^{2+} deficient and showed a significant loss of Mg^{2+} when incubated in Mg^{2+}-free solution ($[Mg^{2+}]_i > [Mg^{2+}]_e$: 0 mM). Measurements of total $[Mg^{2+}]$ via atomic absorption spectrometry confirmed net Mg^{2+} transport; there was a 12.7% increase or 25.6% loss of total Mg^{2+} after a 20-min incubation in medium containing 5 mM or zero Mg^{2+}, respectively (Kolisek et al., 2008). Together, these data support the conclusion that SLC41A1 is a functional Mg^{2+} carrier rather than a channel, and provide the first evidence that the protein mediates Mg^{2+} extrusion.

In a recent second study using the same tetracycline-inducible HEK293 cell system, we tested whether SLC41A1 functions as one of the postulated Mg^{2+} efflux systems (Kolisek et al., 2012). The postulated Mg^{2+} efflux systems include an anion-dependent pathway (Günther & Vormann, 1989; Ödblom & Handy, 1999) and a putative NME (Günther & Vormann, 1989; Tashiro et al., 2005). Experiments were carried out on HEK293 cells, Mg^{2+} loaded by a 20-min incubation in a solution containing 10 mM Mg^{2+}. The subsequent SLC41A1-mediated Mg^{2+} extrusion was not directly linked to anion (Cl^-, HCO_3^-) transport. However, the extrusion showed typical features of NME such as (1) activation by increased $[Mg^{2+}]_i$ (Kubota et al., 2003), (2) strict dependence on extracellular Na^+ (Schweigel et al., 2006; Wolf et al., 1995), (3) inhibition by imipramine and quinidine (Schweigel et al., 2009), and (4) stimulation via 3′,5′-cyclic monophosphate (cAMP)-dependent PKA (Schweigel et al., 2009). More details on NME inhibitors and regulation via signal cascades are given in Section 4.

It can be concluded that SLC41A1 unquestionably represents the NME—the predominant Mg^{2+}-efflux system in mammalian cells (Fagan & Romani, 2001; Schweigel et al., 2006).

3. MOLECULAR CHARACTERISTICS OF SLC41A1

SLC41A1 consists of 513 amino acids with a molecular mass of ~56 kDa. In both human and mouse cells, SLC41A1 was predicted and experimentally confirmed to be a protein integral to the cytoplasmic membrane (Goytain & Quamme, 2005a; Kolisek et al., 2008, 2012; Mandt et al., 2011).

Initially, a model of SLC41A1 topology comprising 10 transmembrane helices (TM) with the N-inside/C-inside configuration was predicted (Fig. 11.1). Using various membrane protein prediction software programs, a nine TM "N-inside–C-outside" model was also predicted for SLC41A1 topology (Fig. 11.1). Independent studies achieved a consensus concerning the intracellular orientation of the N terminus (Kolisek et al., 2008, 2012; Mandt et al., 2011). However, for the C terminus, the extracellular orientation was assumed from epitope-tagging studies (Mandt et al., 2011). An N-inside/C-outside model of SLC41A1 transmembrane topology possessing 11 TM domains was adopted (Fig. 11.1). With regard to the C-terminus orientation, this model was not confirmed by another independent study performed by Sponder, Rutschmann, and Kolisek (2013). These authors propose a cytosolic orientation of the C terminus based on data acquired with the split-ubiquitin functional assay.

It was shown that SLC41A1 forms hetero-oligomeric protein complexes with high molecular mass *in vivo* (Kolisek et al., 2008; Nestler et al., 2013). Most of the identified binding partners were proteins integral to membranes of the endoplasmic reticulum (ER) and Golgi apparatus (GA) that are involved in proteoneogenesis, proper folding, maturation, secretion, anterograde transport, and in the regulation of apoptosis (Nestler et al., 2013). Potential SLC41A1-binding partners include the following:

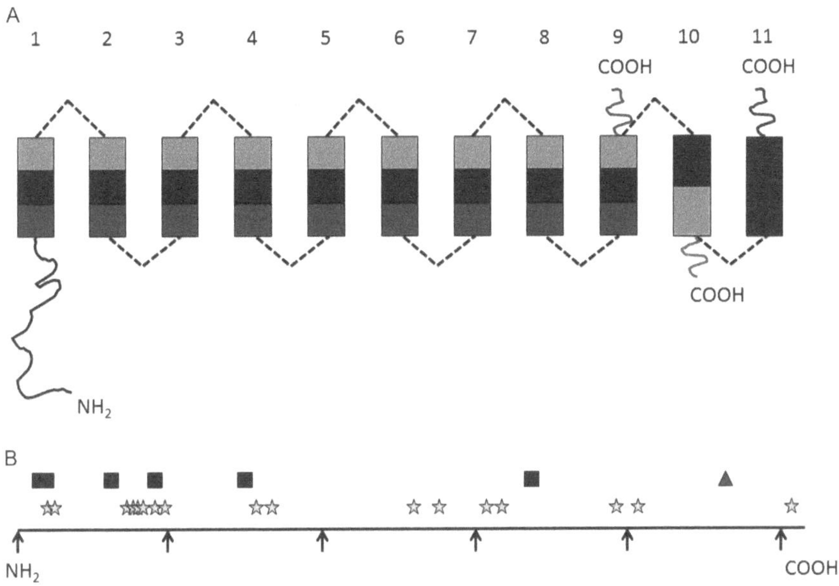

Figure 11.1 (A) Possible models of SLC41A1 membrane topology. The N terminus is intracellular in all three models. The model predicted as being the most likely (gray color) possesses 10 transmembrane domains (TM) with the termini in the "inside-in" configuration. The model possessing 11 TM is depicted in red and the C terminus is exposed extracellularly. The nine TM "inside-out" model is shown in blue. (B) Posttranslational modifications of SLC41A1. Green stars indicate predicted phosphorylation sites, red squares indicate ubiquitination hotspots, and the blue triangle denotes an N-glycosylation site. Arrows indicate sequence increments of 100 amino acids. (See the color plate.)

emopamil-binding protein (EBP), B-cell receptor-associated protein 31 (BCAP31), IER3IP1, PPIB, UPF0480 protein C15orf24, SPINT2, C14orf1/PEBP28, NIFIE14, YIPF6, KCP2, CRACM1, MTCH2, ACCA1, UBB, ATX2L, HSP7C, TBB, and SLC31A2, SLC35B1, and SLC39A13. The last three are all members of the SLC superfamily and contribute to the transport of Zn^{2+} in the GA (SLC39A13), Cu^{2+} in the lysosome (SLC31A2), and sugars in the ER (SLC35B1). N-glycosylation (Fig. 11.1) has been proposed as a possible posttranslational modification of SLC41A1 (Goytain & Quamme, 2005a). The reported interaction between oligosaccharyl transferase (KCP2) and SLC41A1 (Nestler et al., 2013) further substantiates this assumption.

By amino acid sequence analysis (NetPhos 2.0) of mice SLC41A1, Goytain and Quamme (2005a) found four putative sites for cAMP-dependent protein kinase A (PKA) phosphorylation at residues

S-157, S-308, S-393, and T-508, and four putative sites for protein kinase C (PKC) at residues T-80, T-167, T-387, and S-407 (Fig. 11.1). These predictions are in agreement with several functional studies demonstrating that SLC41A1 phosphorylation mediated via cAMP-activated PKA (Cefaratti & Ruse, 2007; Schweigel et al., 2006) and PKC (Ferreira, Rivera, & Romero, 2004) plays a key role in the regulation of NME activity. Most importantly, our group (Kolisek et al., 2012) has shown that phosphorylation mediated by cAMP-activated PKA is a main determinant of SLC41A1-mediated NME activity. The N-terminal flanking sequence of SLC41A1 also possesses multiple putative phosphorylation hotspots for p38MAPK, cdc2, GSK3, cdk5, and DNAPK (NetPhos 1.0; Fleig, Schweigel-Röntgen, & Kolisek, 2013; Sponder et al., 2013) (Fig. 11.1). Involvement of these additional protein kinases in the regulation of SLC41A1 function remains to be explored.

Ubiquitination (Fig. 11.1) also plays a role in the regulation of SLC41A1 turnover. According to software prediction, ubiquitination hotspots are present at positions 4, 8, 58, 92, 146, and 339 (http://bdmpub.biocuckoo.org) (Fleig et al., 2013). The role of ubiquitination in the turnover/degradation of SLC41A1 has been experimentally substantiated by Nestler et al. (2013), who have demonstrated that a protein band with a molecular mass well exceeding 170 kDa comprises ubiquitinated SLC41A1, as well as ACCA1 and ubiquitin.

4. MODULATION OF SLC41A1-RELATED NME ACTIVITY

4.1. Inhibitors

No highly selective blockers for SLC41A1 are available. A reversible inhibition was observed after application of amiloride ($K_i = 0.6$ mM), quinidine ($K_i = 0.5$ mM), or imipramine ($K_i = 0.5$ mM) (Fagan & Romani, 2001; Günther, 2007; Günther & Vormann, 1995; Schweigel et al., 2006, 2000; Shaw et al., 1991; Tessman & Romani, 1998; Wolf, Di Francesco, Covacci, & Cittadini, 1997; Zhang & Melvin, 1995).

Imipramine acts on the extracellular Na^+-binding site of the NME and competition between these two compounds slows Mg^{2+} efflux. Using imipramine concentrations between 10 and 500 μM, most investigators achieved a 33–85% inhibition of Mg^{2+} efflux (Fagan & Romani, 2001; Günther & Vormann, 1995; Kubota et al., 2003; Standley & Standley, 2002; Tessman & Romani, 1998; Wolf et al., 1997, 2009).

In SLC41A1-overexpressing HEK293 cells, Mg^{2+} extrusion is blocked by >90% after imipramine (250 μM) application and is completely inhibited by 100 μM quinidine (Kolisek et al., 2012). An elevated imipramine effect has also been observed in cells either oversupplied with Mg^{2+} or long-term adapted to high extracellular [Mg^{2+}], and reflects increased NME expression (Schweigel et al., 2009; Watanabe, Konishi, Ohkido, & Matsufuji, 2005).

4.2. Regulation of SLC41A1 expression and activity

4.2.1 SLC41A1 activity depends on the free intracellular Mg^{2+} concentration

Basal NME activity is low in most cell types (Standley & Standley, 2002). Measurable Mg^{2+} efflux took place only when $[Mg^{2+}]_i$ was increased (K_m at half maximal rate: 1.0–4 mM) and it stopped when the normal Mg^{2+} content was achieved (Büttner et al., 1998; Frenkel et al., 1989; Günther et al., 1984; Handy et al., 1996; Kubota et al., 2003; Standley & Standley, 2002; Tashiro & Konishi, 1997; Tursun, Tashiro, & Konishi, 2005; Willis et al., 1992). Günther (1996) suggested that the transporter performs Mg^{2+}/Mg^{2+} exchange in cells with normal $[Mg^{2+}]_i$, and that binding of two Mg^{2+} to an intracellular modifier site is required for the allosteric transformation of the Mg^{2+}/Mg^{2+} to the Na^+/Mg^{2+} exchange mode. Hattori, Tanaka, Fukai, Ishitani & Nureki (2007) when investigating the crystal structure of MgtE found four putative Mg^{2+}-binding sites at the cytosolic domains of the transporter, and suggested that the sensing of $[Mg^{2+}]_i$ by these residues is involved in Mg^{2+}-dependent regulation of the transport activity. In agreement with this, Mandt et al. (2011) found that Mg^{2+}-dependent regulation of SLC41A1 total and surface expression is fully dependent on an intact cytoplasmic N-terminal domain.

4.2.2 Role of ATP

NME activity is obtained only with ATP or hydrolysable ATP analogs (Frenkel et al., 1989; Tashiro et al., 2009). In human erythrocytes, half saturation of Na^+-dependent Mg^{2+} efflux is reached at an intracellular [ATP] of 150 μM/l cells (Frenkel et al., 1989). Depletion of ATP by treatment with substances having an uncoupling effect on mitochondria (FCCP, nigericin) or inhibiting mitochondrial electron transport (KCN) results in a strong ~90% reduction of Na^+-dependent Mg^{2+} efflux (Tashiro et al., 2009). Nevertheless, no evidence for modulation by direct interaction between ATP and SLC41A1 has been found. However, the NME activity is directly dependent on extracellular Na^+ and indirectly dependent on the Na^+/K^+ pump (and thereby on ATP) (Günther & Vormann, 1995; Okorodudu et al., 2001; Yoshimura et al., 1995). Inhibition of the Na^+/K^+ ATPase is accompanied by elevated intracellular [Na^+] (Kolisek et al., 2012), resulting in a 50% inhibition of NME activity at concentrations of ~40 mM (Tashiro et al., 2005). Furthermore, metabolic inhibition is associated with intracellular acidification to pH_i 6.7–6.8 that also contributes to reduced NME activity (Tashiro et al., 2009). Another ATP-dependent mechanism involved in NME regulation is protein phosphorylation through the interplay of kinase–phosphatase systems.

4.2.3 The cAMP-PKA pathway as an important activator of SLC41A1

Functional data collected in a variety of tissues (e.g., heart, liver) and cell types (e.g., human erythrocytes, lymphocytes, thymocytes, cardiac myocytes, hepatocytes, ascites cells, ruminal epithelial cells) revealed that elevation of the intracellular cAMP concentration specifically up regulates Na^+-dependent Mg^{2+} efflux by 50–70% (Cefaratti & Ruse, 2007; Fatholati, LaNoue, Romani, & Scarpa, 2000; Günther & Vormann, 1992; Matsuura, Kanayama, Inoue, Takeda, & Morishima, 1993; Schweigel et al., 2006; Wolf, Di Francesco, Covacci, & Cittadini, 1994; Wolf et al., 1997). An increased affinity of the NME for intracellular Mg^{2+} induced by SLC41A1 phosphorylation via cAMP-dependent PKA has been postulated as the underlying mechanism (Günther & Vormann, 1992; Wolf et al., 1994). This is in agreement with PKA-phosphorylation sites within SLC41A1 (Goytain & Quamme, 2005a; Fig. 11.1). In our 2012 study (Kolisek et al., 2012) on HEK293 cells, we found a considerable amount of the hSLC41A1 protein to be present in the phosphorylated fraction, and application of Rp-cAMP—a substance that prevents PKA activation—reduced the SLC41A1-related Mg^{2+} extrusion significantly by 40%. Hormones that decrease cAMP production (e.g., insulin) inhibit Mg^{2+} efflux from cells (Romani, Matthews, & Scarpa, 2000). In hepatocytes and cardiomyocytes, insulin only abolishes the β-adrenergic Mg^{2+} extrusion (Keenan, Romani, & Scarpa, 1996; Romani et al., 2000). It is not clear if this inhibitory effect results from an insulin action on the β-adrenergic receptor, insulin stimulation of cytosolic phosphodiesterase, or direct modulation of the NME (Ferreira et al., 2004).

Besides acting on the antiport directly, cAMP and other agents could also induce the release of Mg^{2+} from intracellular stores that then activates the NME (Fagan & Romani, 2001; Günther & Vormann, 1992; Tessman & Romani, 1998; Wolf et al., 1994). In cardiomyocytes and hepatocytes, cAMP-induced intracellular Mg^{2+} mobilization (e.g., from mitochondria) has been shown to be responsible for the initiation of Mg^{2+} efflux

(Fatholati et al., 2000; Romani, Dowell, & Scarpa, 1991). We examined epithelial cells isolated from sheep rumen and in this cell type with general high-Mg^{2+} transport capacity (Schweigel et al., 1999), we found no indication for such a process to play a role in efflux initiation (Schweigel et al., 2006).

The degree of NME stimulation by cAMP depends on the cells' Mg^{2+} status and, for example, is more pronounced in nonloaded cells compared to artificially Mg^{2+}-loaded cells (e.g., by use of the ionophore A23187) (Günther, 2006; Günther & Vormann, 1992; Wolf et al., 1994). The stimulatory effect of cAMP amounts to 304% in rumen epithelial cells cultivated in medium with low Mg^{2+} (0.12 mM) compared to cells cultivated in media with normal (1.2 mM) or high (10 mM) Mg^{2+} (64% and 109%, respectively). It can be concluded that maximum or near maximum NME activity renders SLC41A1 insensitive to cAMP stimulation. This has also been shown in our recent work on characterization of a coding variant of SLC41A1 carrying substitution p.A350V. Mutation p.A350V led to increased NME activity of SLC41A1 accompanied by complete insensitivity to cAMP stimulation (Kolisek, Sponder, et al., 2013). Insensitivity of SLC41A1 to cAMP stimulation might be of general physiological importance because various hormones or mediators (e.g., catecholamines, prostaglandin E1 and E2, glucagon, and angiotensin II) use this pathway to induce a transient $[Mg^{2+}]_i$ decrease that directly or indirectly influences cellular transport mechanisms and physiological functions (Ferreira et al., 2004; Günther, 2006; Romani & Scarpa, 1990a,1990b; Schweigel et al., 2006; Vormann & Günther, 1987, Wolf et al., 1994, 1997).

Na^+-dependent Mg^{2+} extrusion from hepatocytes is alternatively activated by changes in intracellular Ca^{2+} content after α_1-adrenergic receptor stimulation and blocked by manipulations such as inhibition of IP_3-induced Ca^{2+} release from ER, chelation of intracellular Ca^{2+}, and inhibition of Ca^{2+} influx from the extracellular environment (Fagan & Romani, 2001). In rat sublingual acini, the Na^+-dependent Mg^{2+} efflux is inhibited by carbachol, a muscarinic agonist, but is not influenced by cAMP.

4.2.4 Regulation at the expression level

The expression of SLC41A1 has been shown to be quantitatively altered following changes in extracellular Mg^{2+} concentration and/or Mg^{2+} status of cells or tissues (Goytain & Quamme, 2005a; Mandt et al., 2011; Schweigel et al., 2009; Wolf et al., 2009). In rumen epithelial cells, NME protein expression and NME abundance per single cell were reduced (by 10%

and 25%) or increased (by 18% and 36%) after 24 h-exposure to Mg^{2+}-deficient medium (0.12 mM) or high-Mg^{2+} (5.0 mM) medium (Schweigel et al., 2009). In parallel, the Mg^{2+} efflux capacity was decreased by 32% and accelerated by 56% in response to Mg^{2+}-deficient and high-Mg^{2+} conditions, respectively. Very similar results were observed with the renal epithelial cell line NRK-52E (Ikari, Nakajima, Taki, & Suketa, 2002). Wolf et al. (2004) adapted mouse mammary epithelial cells (HC-11) to an environment containing very low (0.05 mM) or very high (45 mM) [Mg^{2+}] compared to controls grown in normal 0.5 mM Mg^{2+} medium. Total Mg^{2+} content (~20 nmol/mg protein) and growth rate (doubling time [DT]: ~26 h) were similar in control and high-Mg^{2+} adapted cells, but were reduced (total [Mg^{2+}]: 10 nmol/mg protein; DT: 43 h) in low-Mg^{2+} adapted HC11 cells. NME expression followed the order of low-Mg^{2+} cells < control cells < high-Mg^{2+} cells; the high-Mg^{2+} cells also showed the highest Mg^{2+} efflux activity. Thus, in the examined cell types, excess or deficient extracellular Mg^{2+} induced a homeostatic up- or downregulation of SLC41A1 accompanied by respective changes in the efflux capacity. This can be seen as the main mechanism to protect cells against Mg^{2+} overload or extensive Mg^{2+} loss. Indeed, in most cases, the functionally important $[Mg^{2+}]_i$ is unchanged with manipulations of the extracellular [Mg^{2+}] (Schweigel et al., 2009).

In agreement with the results described earlier, Feillet-Coudray et al. (2006) have observed a decreased Mg^{2+} efflux in erythrocytes obtained from mice fed an Mg^{2+}-deficient diet. However, Goytain and Quamme (2005a) showed that *SLC41A1* mRNA is upregulated 4.1-fold and 1.6-fold in the renal cortex of hypomagnesemic mice and MDCT cells, respectively. Bearing in mind the function of NME for Mg^{2+} transport across the basolateral membrane of renal distal tubule cells, *SLC41A1* transcript upregulation is not surprising and fits with its role in Mg^{2+} reabsorption (Quamme, 1989).

Another cell model of increased expression of SLC41A1 is chicken DT40 B-cells displaying disruption of *trpm7*—the most important Mg^{2+} influx pathway. These cells have a strongly reduced Mg^{2+} uptake capacity and a need for high extracellular Mg^{2+} concentrations of 10–15 mM for normal growth (Mandt et al., 2011). Interestingly, SLC41A1 expression was further upregulated in a TRPM7-deficient DT40 clone that has regained the ability to grow in media with normal [Mg^{2+}] of 0.5 mM. However, as no transport experiments were performed in this study, it is not possible to differentiate between a direct involvement of SLC41A1 in transmembrane Mg^{2+} uptake and other indirect effects of SLC41A1.

Mg^{2+} influx systems such as transient receptor potential 6 (TRPM6; and TRPM7) and MagT1 are rapidly inactivated at normal intracellular Mg^{2+} concentrations. Thus, an increased expression of SLC41A1 could be interpreted as a mechanism to promote Mg^{2+} uptake under Mg^{2+}-deficient conditions (He et al., 2005). Heterologous expression of SLC41A1 in TRPM7-deficient DT40 B-cells sufficiently restored their ability to grow at physiological extracellular [Mg^{2+}] of 0.5 mM (Mandt et al., 2011). Mutated variants of SLC41A1, p.D263A, and p.D487A lack the ability to complement growth of TRPM7-deficient cells. The SLC41A1 residues 263 and 487 correspond to highly conserved, and functionally important residues in the MgtE Mg^{2+} conducting pore (Hattori et al., 2009, 2007). Although the Mg^{2+} transport capacity is reduced in p.D263A and completely abolished in p.D487A mutants, they have no effect on protein folding, intracellular trafficking, or surface expression of SLC41A1 (Mandt et al., 2011).

Romanuik et al. (2009) have identified SLC41A1 expression as being stimulated by androgens.

Mandt et al. (2011) identified endosomal recycling as a main mechanism to regulate SLC41A1 function. The authors suggest a model in which SLC41A1 expression is posttranslationally regulated by Mg^{2+}-dependent translocation from or to the cell surface and by degradation in the endosomal/lysosomal pathway. Mg^{2+} dependency of these processes is lost after truncation of the entire N-terminal domain—a finding in agreement with the suggestion that this part of the protein has the capacity to sense the cellular Mg^{2+} status.

5. PHYSIOLOGICAL FUNCTIONS OF SLC41A1-RELATED NME ACTIVITY

The NME has been shown to be responsible for the maintenance of an optimal $[Mg^{2+}]_i$ (Schweigel et al., 2006) for a variety of processes critical for the cell such as bioenergetics, the regulation of cellular pH, volume, and total ion balance, and cell proliferation and differentiation (Jiang et al., 2007; Romani & Scarpa, 1992; Rubin, 2005; Wolf & Trapani, 2008; Yamaguchi & Ishikawa, 2008).

5.1. A role of NME in active transepithelial Mg^{2+} transport

As for other ions, transcellular Mg^{2+} transport is a three-step process consisting of (1) entry into the epithelial cell from the lumen, (2) transit through the cytosol, and (3) extrusion from the cell across the basolateral membrane. However, the mechanisms responsible for these three transport components are not well understood. The first step—entry into the cell across the apical membrane—occurs passively and is driven by a steep electrochemical gradient that results mainly from the inside-negative plasma membrane voltage. It has been postulated that the melastatin-related, TRPM6 ion channel, which is predominantly expressed in the apical membrane of renal and intestinal epithelia, represents the influx step in epithelial Mg^{2+} transport (Voets et al., 2004). In agreement with this idea, mutations in TRPM6 have been shown to cause a combined defect of renal magnesium conservation and intestinal magnesium absorption in patients suffering from primary hypomagnesaemia with secondary hypocalcemia (Paunier, Radde, Kooh, Conen, & Fraser, 1968; Schlingmann et al., 2002; Walder et al., 2002).

The NME has long been postulated to mediate Mg^{2+} transport across the basolateral membrane of intestinal, renal, and placental epithelia, although the available experimental data are not conclusive (Günther, Vormann, & Höllriegel, 1988; Hoenderup & Bindels, 2005; Quamme, 1989). There are very few studies showing a direct Na^{+}-dependent component of transepithelial Mg^{2+} transport. The forestomachs, in particular, the rumen, are the main and the most important sites of Mg^{2+} absorption in ruminants (Tomas & Potter, 1976), thereby making the rumen epithelium a good model to study the mechanisms of Mg^{2+} transport. By using of the Ussing-chamber technique, Leonhard-Marek et al. (2005) investigated the unidirectional $^{28}Mg^{2+}$ flux across the isolated ruminal epithelium and the effect of cAMP, imipramine, and amiloride on this transport. Decreasing the serosal $[Na^{+}]$ from 30 to 5 mM reduced net Mg^{2+} transport by 34%. The unidirectional mucosal to serosal $^{28}Mg^{2+}$ flux increased from 58 ± 10 nmol cm^{-2} h^{-1} in control tissues to 77 ± 17 nmol cm^{-2} h^{-1} in theophyllin (a phosphodiesterase inhibitor)-treated tissues—a finding consistent with the cAMP-dependent PKA stimulation of the NME seen in ruminal epithelial cells (Schweigel et al., 2006). The cAMP-mediated effect depended on the availability of mucosal Mg^{2+} and was reduced by low serosal $[Na^{+}]$ and serosal addition of imipramine or amiloride. Cyclic AMP regulation of transepithelial transport is of general importance, and the involvement of hormones and other signals needs to be investigated.

Shaw, Mughal, Maresh, and Sibley (1990, 1991) investigated the maternofetal Mg^{2+} transfer across the *in situ* perfused rat placenta. The authors found the transport to be energy dependent and temperature dependent. Perfusion of the fetal site of the placenta with Na^{+}-free solution or with solutions containing the Na^+/K^+-ATPase inhibitor ouabain or the NME blocker amiloride decreased the maternofetal Mg^{2+} transport markedly (Shaw et al., 1991). These experimental results provided reasonable evidence for NME-mediated Mg^{2+} efflux across the fetal-facing plasma membrane of the syncytiotrophoblast. In agreement with this finding, a functional NME was identified in human trophoblast cells (Standley & Standley, 2002). In addition, a recent study by Kolisek, Galaviz-Hernández, et al. (2013) demonstrated expression of SLC41A1 mRNA and protein in placenta tissue.

Most (about 80%) of the ultrafilterable Mg^{2+} is reabsorbed, and this process for the most part (60–70%) takes place in the cortical segment of the thick ascending loop of Henle (cTAL). Mg^{2+} reabsorption is predominantly mediated via passive, paracellular processes. The final regulation of renal Mg^{2+} excretion, however, occurs in the distal convoluted tubule by secondary active transport; the involvement of a basolateral NME has been proposed (De Rouffignac & Quamme, 1994). In agreement with this hypothesis, there is upregulation of SLC41A1 mRNA in renal cortex tissue from hypomagnesemic mice, and in MDCT cells exposed to low Mg^{2+} medium (Goytain & Quamme, 2005a).

5.2. A role of NME in cell signaling pathways

The possible existence of NME subtypes and of distinct modalities of NME activation/inhibition (e.g., by hormones) raises the question of the physiological significance of SLC41A1 for functions other than maintenance of Mg^{2+} homeostasis.

He et al. (2005) demonstrated that acute angiotensin II stimulation of smooth vascular muscle cells induced NME-mediated Mg^{2+} efflux followed by a reduction of $[Mg^{2+}]_i$ and vasoconstriction. TRPM7, which facilitates Mg^{2+} influx, may also be activated by decreased $[Mg^{2+}]_i$. In the long term, a TRPM7-mediated increase of intracellular Mg^{2+} is postulated to play a role in smooth vascular muscle cell growth (He et al., 2005).

In ruminal epithelial cells, a cAMP-induced reduction of $[Mg^{2+}]_i$ has been demonstrated to activate a nonselective cation conductance expressed in the apical membrane of these cells, and thereby to play an important role in the regulation of ruminal Na^+ transport/absorption (Leonard-Marek et al., 2005).

An important role of Na^+-dependent Mg^{2+} extrusion in the regulation of glucose transport and utilization by liver cells has been found. Hormones such as catecholamine, glucagon, or phenylephrine not only stimulate Mg^{2+}

efflux but also mobilize glucose, mainly via glycogenolysis (Fagan & Romani, 2000; Keenan et al., 1996). Overnight starvation reduces the total hepatic Mg^{2+} content and also depletes liver glycogen and glucose (Torres, Youngner, & Romani, 2005). This effect also occurs after stimulation of perfused livers with glucagon or catecholamine (Torres et al., 2005) and was described for livers obtained from type-I diabetic rats (Fagan, Cefaratti, & Romani, 2004). Conversely, NME inhibition by imipramine or amiloride inhibits glucose release from hepatocytes (Fagan & Romani, 2000). In addition, reduction of the glucose transport capacity by phloretin results in inhibition of Mg^{2+} release from liver cells (Fagan & Romani, 2000). Insulin-stimulated glucose accumulation in heart muscle cells and pancreatic β-cells is accompanied by a proportional increase of Mg^{2+} accumulation (Romani et al., 2000; Henquin, Tamagawa, Nenquin, & Cogneau, 1983), and extracellular Mg^{2+} deficiency leads to a reduction of glucose accumulation (Romani et al., 2000). These data clearly show a direct link between hormonal modulation of NME activity, glucose homeostasis/accumulation, and insulin sensitivity.

6. IMPLICATIONS FOR A ROLE OF DISTURBED SLC41A1-RELATED NME ACTIVITY IN DISEASE STATES

Evidence for an involvement of SLC41A1 in the pathophysiology of Parkinson's disease (PD), preeclampsia, and nephronophthisis (NPHP)-related ciliopathies has been reported recently (Hurd et al., 2013; Kolisek, Galaviz-Hernández, et al., 2013; Kolisek, Sponder, et al., 2013; Tucci et al., 2010; Yan et al., 2011).

SLC41A1 has been detected as being a part of a novel PD susceptibility locus named *PARK16* (Satake et al., 2009). Subsequently, rare PD-associated coding variants of SLC41A1, namely p.A350V (Tucci et al., 2010), p.L146G, p.P480P, and c.552+50G>A (Yan et al., 2011) have been identified. Our group (Kolisek, Sponder, et al., 2013) demonstrated that the conservative substitution p.A350V markedly enhanced (69%) NME activity and is therefore a gain-of-function mutation. Moreover, we found cells overexpressing the p.A350V variant of SLC41A1 to be insensitive to cAMP stimulation and to have a reduced proliferation rate. Therefore, we speculate that the p.A350V mutation may either augment the affinity of SLC41A1 for intracellular Mg^{2+}, change the Mg^{2+}-carrier-complex formation, or dysregulate transporter gating to facilitate the Mg^{2+} transport rate.

Another mutation in SLC41A1 (c.698G>T) has just been shown to cause the nephronophthisis-like phenotype (NPHLP) (Hurd et al., 2013). This mutation results in the skipping of exon 6 of *SLC41A1*, resulting in an in-frame deletion of one transmembrane helix responsible for the complete block of SLC41A1 function.

Kolisek, Galaviz-Hernández, et al. (2013) analyzed the transcription activities of various magnesium-responsive genes (*TRPM6, TRPM7, MagT1, TUSC3/N33, NIPA1, CNNM2, SLC41A1, SLC41A2, SLC41A3*) in human placenta samples from a Mexican Mestizo cohort of normoevolutive women and preeclamptic women. Among the tested genes, only *SLC41A1* had a deregulated transcription activity in placenta samples obtained from preeclamptic women, suggesting an important role of SLC41A1 in the pathophysiology of the disease. *SLC41A1* was significantly overexpressed in about 55% of tested samples from preeclamptic women in comparison to 9.5% in normoevolutive samples (Kolisek, Galaviz-Hernández, et al., 2013).

Taking into account earlier studies linking disturbed NME activity to various disease conditions such as hypertension (Kisters et al., 1999; Resnick, 1993), diabetes (Barbagallo & Dominguez, 2007; Reed, Cefaratti, Berti-Mattera, & Romani, 2008; Resnick et al., 1993), and cystic fibrosis (Vormann, 1992), it will not be surprising if the involvement of SLC41A1 in cardiovascular, metabolic, psychiatric, and neurological diseases other than PD will be demonstrated in the near future.

7. MOLECULAR AND FUNCTIONAL CHARACTERISTICS OF SLC41A2 AND SLC41A3

The SLC41 members A2 and A3 are products of the human 12q23.3 and 3q21.2 genes and their mRNA has been identified in various organs and tissues (reviewed in Sahni & Scharenberg, 2013; Schmitz, Deason, & Perraud, 2007). In a recent review, Sahni and Scharenberg (2013) summarized SLC41A2 expression data obtained from http://www.proteinatlas.org/ENSG00000136052. According to these data, *SLC41A2* shows the strongest expression in lymph nodes, stomach, lungs, testis, and skin, and is moderately expressed in spleen, intestine, heart, breast, and kidneys. In contrast to *SLC41A1, SLC41A2* mRNA expression was not upregulated in kidney cortical tissue from hypomagnesemic mice or in MDCT cells incubated in Mg^{2+}-free medium (Goytain & Quamme, 2005b). In addition, Sahni, Nelson, and Scharenberg (2007) found *SLC41A2* to be expressed in

Jurkat T-cells, as well as in wild-type and TRPM7-deficient DT40 B-cells. Bioinformatics data suggest different expression of human *SLC41A3* in organs such as cerebellum, testis, and prostate, as well as in some cell lines, for example, SW480, MOLT4, and MCF-7 (Sahni & Scharenberg, 2013).

SLC41A2 (573 amino acids) and SLC41A3 (500 amino acids) are 62.3- and 54.6-kDa proteins, respectively (Goytain & Quamme, 2005b; Sahni & Scharenberg, 2013). The human SLC41A2 is 70% similar to SLC41A1. Both SLC41A2 and SLC41A3 are plasma membrane proteins (Goytain & Quamme, 2005b; Quamme, 2010; Sahni et al., 2007). According to software prediction (TMPred at www.ch.embnet.org), the strongly preferred model of SLC41A2 possesses 10 TM domains with both termini oriented extracellularly (Fleig et al., 2013). However, Sahni et al. (2007) provided experimental data showing that SLC41A2 when overexpressed in DT40 chicken cells has 11 TM domains with an extracellular N terminus and an intracellular C terminus. In a recent review, Sahni and Scharenberg (2013) reinterpreted their results from SLC41A2 localization studies (Sahni et al., 2007) and hypothesized that SLC41A2 is normally targeted to membranes of intracellular compartments and involved in organellar Mg^{2+} transport (Sahni & Scharenberg, 2013). Sequence analysis gave no evidence for the existence of signal peptide-responsive elements and/or conserved domains in the long N terminus of SLC41A2 (Sahni & Scharenberg, 2013). An online-generated prediction of the SLC41A3 topology suggested an 11-TM-span "N-terminus-outside/C-terminus-inside" model (Fleig et al., 2013). So far, no information is available about the complex-forming abilities of SLC41A2 and SLC41A3.

Goytain and Quamme (2005b) performed the first functional characterization of SLC41A2 expressed in *X. laevis* oocytes using the two-microelectrode voltage-clamp technique. SLC41A2 carries Mg^{2+}, Ba^{2+}, Ni^{2+}, Co^{2+}, Fe^{2+}, and Mn^{2+}, but not Zn^{2+}, Cu^{2+}, or Ca^{2+}. Based on these data, SLC41A2 was suggested to represent a channel-like Mg^{2+} transporter (Goytain & Quamme, 2005b). However, overexpression of SLC41A2 in TRPM7-deficient B-cells induced no Mg^{2+} currents, but did support cell growth in normal, non-Mg^{2+}-supplemented media (Sahni et al., 2007). In another study on fibroblasts with TRPM7 knockout (Su et al., 2011), SLC41A2 expression restored normal cell morphology and motility. According to these results, it is unlikely that SLC41A2 functions as a Mg^{2+} channel, but rather, as a Mg^{2+} transporter.

No clear functional data are available for SLC41A3 (Sahni & Scharenberg, 2013). Distinct splice variants of SLC41A3 have to be investigated in their specific cellular context regarding intracellular binding partners or regulatory mechanisms. Interestingly, a conditional knockout mouse line, named *Slc41a3*$^{tm1a(KOMP)Wts}$ (http://www.knockoutmouse.org/martsearch/searchquery=Slc41a3), has been generated that shows a phenotype with abnormal locomotor coordination (http://en.wikipedia.org/wiki/SLC41A3).

8. CONCLUSION

Over the past decade, the field of Mg^{2+} homeostasis has undergone a dramatic expansion with the identification of several genes encoding putative Mg^{2+} transporters, among them the three members of the SLC41 family. At the molecular and physiological level, SLC41A1 is the best characterized protein, whereas not much is known about the exact function(s) of SLC41A2 and SLC41A3 (e.g., as channels or carriers). A possible role of SLC41A2 in organellar Mg^{2+} transport must be further validated. In contrast to the other family members, SLC41A3 may be more depended on cell type-specific factors, for example, partner proteins. This dependence could explain the existence of five SLC41A3 splice variants that should be characterized in their specific cell or tissue context. Cellular localization and membrane topology of SLC41A2 and SLC41A3 also await further examination. Nevertheless, based on their homology to SLC41A1 and initial functional studies, both appear to be Mg^{2+} transport proteins.

SLC41A1 has been shown to function as a NME—a mechanism that is long known to be the primary Mg^{2+} efflux system of mammalian and avian cells. Under specific conditions, the carrier also works in the reverse mode and mediates Mg^{2+} uptake. Mg^{2+}-dependent and hormonal regulation of SLC41A1 expression, localization, degradation, and/or activity has been described and is dependent on an intact intracellular N terminus. Sensing of intracellular Mg^{2+} and phosphorylation via cAMP-dependent PKA are important for the regulation of SLC41A1 NME activity. Multiple phosphorylation sites for other protein kinases (e.g., PKC, p38MAPK, cdc2, GSK3, cdk5, and DNAPK) are present at the N terminus, and their effects remain to be examined experimentally.

SLC41A1 has been demonstrated to form transient hetero-oligomeric complexes with proteins likely involved in its posttranslational maturation (modification), proper folding, anterograde transport, and vesicular sorting. The physiological relevance of interactions between SLC41A1 and other binding partners has to be clarified. Communication between various

Mg^{2+} transporters has been postulated to be essential for cellular Mg^{2+} homeostasis. Thus, the interplay between SLC41 family members and Mg^{2+} influx systems such as TRPM6/7 or MagT1 should be evaluated.

Mutations in SLC41A1 have been identified in patients with PD and others showing a NPHLP. Moreover, a link between preeclampsia and elevated placental expression of *SLC41A1* has been shown. Disturbed NME activity has also been associated with cardiovascular and metabolic diseases. Establishing the role of SLC41A1 in the pathophysiological processes underlying these and other diseases is an important direction for further research.

ACKNOWLEDGMENT

The work of the authors was supported by research grants from the German Research Foundation (DFG); Schw 642/1 to M. S. R. and KO-3586/3-2 to M. K. M. S. R. and M. K. are inventors on the pending patent EP2011/065979 "Na^+/Mg^{2+} exchanger."

REFERENCES

Altura, B. M., & Altura, B. A. (1995). Magnesium and cardiovascular biology: An important link between cardiovascular risk factors and atherogenesis. *Cellular & Molecular Biology Research*, *41*, 347–359.

Bara, M., & Guiet-Bara, A. (2001). Magnesium regulation of Ca^{2+} channels in smooth muscle and endothelial cells of human allantochorial placental vessels. *Magnesium Research*, *14*, 11–18.

Barbagallo, M., Belvedere, M., Di Bella, G., & Dominguez, L. J. (2011). Altered ionized magnesium levels in mild-to-moderate Alzheimer's disease. *Magnesium Research*, *24*, 115–121.

Barbagallo, M., & Dominguez, L. J. (2007). Magnesium metabolism in type 2 diabetes mellitus, metabolic syndrome and insulin resistance. *Archives of Biochemistry and Biophysics*, *458*, 40–47.

Büttner, S., Günther, T., Schäfer, A., & Vormann, J. (1998). Magnesium metabolism in erythrocytes of various species. *Magnesium-Bulletin*, *20*, 101–109.

Cefaratti, C., Romani, A., & Scarpa, A. (1998). Characterization of two Mg^{2+} transporters in sealed plasma membrane vesicles from rat liver. *American Journal of Physiology*, *275*, C2995–C.

Cefaratti, C., Romani, A., & Scarpa, A. (2000). Differential localization and operation of distinct Mg^{2+} transporters in apical and basolateral sides of rat liver plasma membrane. *Journal of Biological Chemistry*, *275*, 3772–3780.

Cefaratti, C., & Ruse, C. (2007). Protein kinase A dependent phosphorylation activates Mg^{2+} efflux in the basolateral region of the liver. *Molecular and Cellular Biochemistry*, *297*, 209–214.

De Rouffignac, Ch., & Quamme, G. (1994). Renal magnesium handling and its hormonal control. *Physiological Reviews*, *74*, 305–322.

Fagan, T. E., Cefaratti, C., & Romani, A. (2004). Streptozotocin-induced diabetes impairs Mg^{2+} homeostasis and uptake in rat liver cells. *American Journal of Physiology*, *286*, E184–E193.

Fagan, T. E., & Romani, A. (2000). Activation of Na^+- and Ca^{2+}-dependent Mg^{2+} extrusion by α_1- and β_1-adrenergic agonists in rat liver cells. *American Journal of Physiology*, *279*, G943–G950.

Fagan, T. E., & Romani, A. (2001). Alpha(1)-Adrenoceptor-induced Mg^{2+} extrusion from rat hepatocytes occurs via Na+-dependent transport mechanism. *American Journal of Physiology Gastrointestinal and Liver Physiology, 280*, G1145–G1156.

Fatholati, M., LaNoue, K., Romani, A., & Scarpa, A. (2000). Relationship between total and free cellular Mg^{2+} during metabolic stimulation of rat cardiac myocytes and perfused hearts. *Archives of Biochemistry and Biophysics, 374*, 395–401.

Feillet-Coudray, C., Trzeciakiewicz, A., Coudray, C., Rambeau, M., Chanson, A., Rayssiguier, Y., et al. (2006). Erythrocyte magnesium fluxes in mice with nutritionally and genetically low magnesium status. *European Journal of Nutrition, 45*, 171–177.

Feray, J. C., & Garay, R. (1986). An Na^{+}-stimulated Mg^{2+} transport system in human red blood cells. *Biochimica et Biophysica Acta, 856*, 76–84.

Féray, J. C., & Garay, R. (1988). Demonstration of Na^{+}: Mg^{2+} exchange in human red cells by its sensivity to tricyclic antidepressant drugs. *Naunyn-Schmiedeberg's Archives of Pharmacology, 338*, 332–337.

Ferreira, A., Rivera, A., & Romero, J. R. (2004). Na^{+}/Mg^{2+} exchange is functionally coupled to the insulin receptor. *Journal of Cellular Physiology, 199*, 434–440.

Flatman, P. W., & Smith, L. M. (1991). Sodium-dependent magnesium uptake by ferret red cells. *Journal of Physiology, 443*, 217–230.

Flatman, P. W., & Smith, L. M. (1996). Magnesium transport in magnesium-loaded ferret red blood cells. *Pflügers Archiv - European Journal of Physiology, 432*, 995–1002.

Fleig, A., Schweigel-Röntgen, M., & Kolisek, M. (2013). Solute carrier family SLC41, what do we really know about it? *WIREs Membrane Transport and Signaling, 2*, 227–239. http://dx.doi.org/10.1002/wmts.95.

Frenkel, E. J., Graziani, M., & Schatzmann, H. J. (1989). ATP requirement of the sodium-dependent magnesium extrusion from human red blood cells. *Journal of Physiology, 414*, 385–397.

Garban, Z. (1997). Implications of the chronobiochemistry-metabolism relationship in the induction of homeostasis changes. V. The action of Mg^{2+} and Zn^{2+} ions on DNA. In R. Smetana (Ed.), *Advances in magnesium research* (pp. 482–489). London: John Libbey & Company Ltd.

Goytain, A., & Quamme, G. A. (2005a). Functional characterization of human SLC41A1, a Mg^{2+} transporter with similarity to prokaryotic MgtE Mg^{2+} transporters. *Physiological Genomics, 21*, 337–342.

Goytain, A., & Quamme, G. A. (2005b). Functional characterization of the mouse solute carrier, SLC41A2. *Biochemical and Biophysical Research Communications, 330*, 701–705.

Günther, T. (1996). Putative mechanism of Mg^{2+}/Mg^{2+} exchange and Na^{+}/Mg^{2+} antiport. *Magnesium-Bulletin, 18*, 2–6.

Günther, T. (2006). Mechanisms, regulation and pathological significance of Mg^{2+} efflux from erythrocytes. *Magnesium Research, 19*, 190–198.

Günther, T. (2007). Na^{+}/Mg^{2+} antiport in non-erythrocyte vertebrate cells. *Magnesium Research, 20*, 89–99.

Günther, T., & Vormann, J. (1985). Mg^{2+} efflux is accomplished by an amiloride-sensitive Na^{+}/Mg^{2+} antiport. *Biochemical and Biophysical Research Communications, 130*, 540–545.

Günther, T., & Vormann, J. (1989). Characterization of Mg^{2+} efflux from human, rat and chicken erythrocytes. *FEBS Letters, 250*, 633–637.

Günther, T., & Vormann, J. (1990a). Characterization of Na^{+}-independent Mg^{2+} efflux from erythrocytes. *FEBS Letters, 271*, 149–151.

Günther, T., & Vormann, J. (1990b). Na^{+}-dependent Mg^{2+} efflux from Mg-loaded rat thymocytes and HL 60 cells. *Magnesium and Trace Elements, 9*, 279–282.

Günther, T., & Vormann, J. (1992). Activation of Na^{+}/Mg^{2+} antiport in thymocytes by cAMP. *FEBS Letters, 297*, 132–134.

Günther, T., & Vormann, J. (1995). Reversibility of Na^+/Mg^{2+} antiport in rat erythrocytes. *Biochimica et Biophysica Acta*, *1234*, 105–110.

Günther, T., Vormann, J., & Förster, R. (1984). Regulation of intracellular magnesium by Mg^{2+} efflux. *Biochemical and Biophysical Research Communications*, *119*, 124–131.

Günther, T., Vormann, J., & Höllriegel, V. (1988). Effects of amiloride and furosemide on ^{28}Mg transport into fetuses and maternal tissues of the rat. *Magnesium-Bulletin*, *10*, 34–37.

Handy, R. D., Gow, I. F., Ellis, D., & Flatman, P. W. (1996). Na-dependent regulation of intracellular free magnesium concentration in isolated rat ventricular myocytes. *Journal of Molecular and Cellular Cardiology*, *28*, 1641–1651.

Hartwig, A. (2001). Role of magnesium in genomic stability. *Mutation Research*, *475*, 113–121.

Hattori, M., Iwase, N., Furuya, N., Tanaka, Y., Tsukazaki, T., Ishitani, R., et al. (2009). Mg^{2+}-dependent gating of bacterial MgtE channel underlies Mg^{2+} homeostasis. *EMBO Journal*, *28*, 3602–3612.

Hattori, M., Tanaka, J., Fukai, S., Ishitani, R., & Nureki, O. (2007). Crystal structure of the MgtE Mg^{2+} transporter. *Nature*, *448*, 1072–1075.

He, Y., Yao, G., Savoia, C., & Touyz, R. M. (2005). Transient receptor potential melastatin 7 ion channels regulate magnesium homeostasis in vascular smooth muscle cells: Role of angiotensin II. *Circulation Research*, *96*, 207–215.

Henquin, J. C., Tamagawa, T., Nenquin, M., & Cogneau, M. (1983). Glucose modulates Mg^{2+} fluxes in pancreatic islets cells. *Nature*, *301*, 73–74.

Hintz, K., Günzel, D., & Schlue, W.-R. (1999). Na^+-dependent regulation of the free Mg^{2+} concentration in neuropile glial cells and P neurones of the leech *Hirudo medicinalis*. *Pflügers Archiv - European Journal of Physiology*, *437*, 354–362.

Hoenderup, G. J., & Bindels, R. J. M. (2005). Epithelial Ca^{2+} and Mg^{2+} channels in health and disease. *Journal of the American Society of Nephrology*, *16*, 15–26.

Hurd, T. W., Otto, E. A., Mishima, E., Gee, H. Y., Inoue, H., Inazu, M., et al. (2013). Mutation of the Mg^{2+} transporter SLC41A1 results in a Nephronophthisis-like phenotype. *Journal of the American Society of Nephrology*, *24*, 967–977.

Ikari, A., Nakajima, K., Taki, S., & Suketa, Y. (2002). Up-regulation of Na^+-dependent Mg^{2+} transport by nitric oxide and cyclic GMP pathway in renal epithelial cells. *European Journal of Pharmacology*, *451*, 133–139.

Jiang, J., Li, M., Inoue, K., Chu, X.-P., Seeds, J., & Xiong, Z.-G. (2007). Transient receptor potential melastatin 7-like current in human head and neck carcinoma cells: Role in cell proliferation. *Cancer Research*, *67*, 10929–10938.

Jóna, I., Szegedi, C., Sárközi, S., Szentisi, P., Csernoch, L., & Kovács, L. (2001). Altered inhibition of the rat skeletal ryanodine receptor/calcium release channel by magnesium in the presence of ATP. *Pflügers Archiv - European Journal of Physiology*, *441*, 729–738.

Keenan, D., Romani, A., & Scarpa, A. (1996). Regulation of Mg^{2+} homeostasis by insulin in perfused rat livers and isolated hepatocytes. *FEBS Letters*, *395*, 241–244.

Kisters, K., Krefting, E. R., Barenbrock, M., Spieker, C., & Rahn, K. H. (1999). Na^+ and Mg^{2+} contents in smooth muscle cells in spontaneously hypertensive rats. *American Journal of Hypertension*, *12*, 648–652.

Kolisek, M., Galaviz-Hernández, C., Vázquez-Alaniz, F., Sponder, G., Javaid, S., Kurth, K., et al. (2013). *SLC41A1* is the only magnesium responsive gene significantly overexpressed in placentas of preeclamptic women. *Hypertension in Pregnancy*, *32*, 378–389. http://dx.doi.org/10.3109/10641955.2013.810237.

Kolisek, M., Launay, P., Beck, A., Sponder, G., Serafini, N., Brenkus, M., et al. (2008). SLC41A1 is a novel mammalian Mg^{2+} carrier. *The Journal of Biological Chemistry*, *283*, 16235–16247.

Kolisek, M., Nestler, A., Vormann, J., & Schweigel-Röntgen, M. (2012). Human gene SLC41A1 encodes for the Na^+/Mg^{2+} exchanger. *American Journal of Physiology Cell Physiology*, *302*, C318–C326.

Kolisek, M., Sponder, G., Mastrototaro, L., Smorodchenko, A., Launay, P., Vormann, J., et al. (2013). Substitution p.A350V in Na^+/Mg^{2+} exchanger SLC41A1, putatively associated with Parkinson's disease, is a gain of function mutation. *PLoS ONE*, *8*(8), e71096. http://dx.doi.org/10.1371/journal.pone.0071096.

Kolisek, M., Zsurka, G., Samaj, J., Weghuber, J., Schweyen, R. J., & Schweigel, M. (2003). Mrs2p is an essential component of the major electrophoretic Mg^{2+} influx system in mitochondria. *EMBO Journal*, *22*, 1235–1244.

Kubota, T., Tokuno, K., Nakagawa, J., Kitamura, Y., Ogawa, H., Suzuki, Y., et al. (2003). Na^+/Mg^{2+} transporter acts as a Mg^{2+} buffering mechanism in PC12 cells. *Biochemical and Biophysical Research Communications*, *303*, 332–336.

Kucharski, L. M., Lubbe, W. J., & Maguire, M. E. (2000). Cation hexaammines are selective and potent inhibitors of the CorA magnesium transport system. *Journal of Biological Chemistry*, *275*, 16767–16773.

Leonhard-Marek, S., Stumpff, F., Brinkmann, I., Breves, G., & Martens, H. (2005). Basolateral Mg^{2+}/Na^+ exchange regulates apical nonselective cation channel in sheep rumen epithelium via cytosolic Mg^{2+}. *American Journal of Physiology. Gastrointestinal and Liver Physiology*, *288*, G630–G645.

Lovitt, B., Vanderporten, E. C., Sheng, Z., Zhu, H., Drummond, J., & Liu, Y. (2010). Differential effects of divalent manganese and magnesium on the kinase activity of leucine-rich repeat kinase 2 (LRRK2). *Biochemistry*, *49*, 3092–3100.

Mandt, T., Song, Y., Scharenberg, A. M., & Sahni, J. (2011). SLC41A1 Mg^{2+} transport is regulated via Mg^{2+}-dependent endosomal recycling through its N-terminal cytoplasmic domain. *Biochemical Journal*, *439*, 129–139.

Matsuura, T., Kanayama, Y., Inoue, T., Takeda, T., & Morishima, I. (1993). cAMP-induced changes of intracellular free Mg^{2+} levels in human erythrocytes. *Biochimica et Biophysica Acta*, *1220*, 31–36.

Mooren, F. C., Turi, S., Günzel, D., Schlue, W. R., Domschke, W., Singh, J., et al. (2001). Calcium–magnesium interactions in pancreatic acinar cells. *FASEB Journal*, *15*, 659–672.

Nestler, A., Sponder, S., Rutschmann, K., Mastrototaro, L., Weise, C., Vormann, J., et al. (2013). Nature of SLC41A1 complexes: Report on split-ubiquitin yeast two hybrid assay. *Magnesium Research*, *26*, 56–66.

Ödblom, M. P., & Handy, R. D. (1999). A novel DIDS-sensitive, anion-dependent Mg^{2+} efflux pathway in rat ventricular myocytes. *Biochemical and Biophysical Research Communications*, *264*, 334–337.

Okorodudu, A., Yang, H., & Elghetany, M. T. (2001). Ionized magnesium in the homeostasis of cells: Intracellular threshold for Mg^{2+} in human platelets. *Clinica Chimica Acta*, *303*, 147–154.

Paunier, L., Radde, I. C., Kooh, S. W., Conen, P. E., & Fraser, D. (1968). Primary hypomagnesemia with secondary hypocalcemia in an infant. *Pediatrics*, *41*, 385–402.

Quamme, G. A. (1989). Control of magnesium transport in the thick ascending limb. *American Journal of Physiology*, *256*, F197–F210.

Quamme, G. A. (2010). Molecular identification of ancient and modern mammalian magnesium transporters. *American Journal of Physiology Cell Physiology*, *298*, C407–C429.

Reed, G., Cefaratti, C., Berti-Mattera, L. N., & Romani, A. (2008). Lack of insulin impairs Mg^{2+} homeostasis and transport in cardiac cells of streptozotocin-injected diabetic rats. *Journal of Cellular Biochemistry*, *104*, 1034–1053.

Resnick, L. M. (1993). Ionic basis of hypertension, insulin resistance, vascular disease and related disorders. The mechanisms if "syndrome X" *American Journal of Hypertension*, *6*, 123S–134S.

Resnick, L. M., Altura, B. T., Gupta, R. K., Laragh, J. H., Alderman, M. H., & Altura, B. M. (1993). Intracellular and extracellular magnesium depletion in type 2 (non-insulin-dependent) diabetes mellitus. *Diabetologia*, *36*, 767–770.

Romani, A., Dowell, E., & Scarpa, A. (1991). Cyclic AMP induced Mg^{2+} release from rat liver hepatocytes, permeabilized hepatocytes and isolated mitochondria. *Journal of Biological Chemistry, 266*, 24376–24384.

Romani, A. M., Matthews, V. D., & Scarpa, A. (2000). Parallel stimulation of glucose and Mg^{2+} accumulation by insulin in rat hearts and cardiac ventricular myocytes. *Circulation Research, 86*, 326–333.

Romani, A., & Scarpa, A. (1990a). Hormonal control of Mg^{2+} in the heart. *Nature, 346*, 841–844.

Romani, A., & Scarpa, A. (1990b). Norepinephrine evokes a marked Mg^{2+} efflux from liver cells. *FEBS Letters, 269*, 37–40.

Romani, A., & Scarpa, A. (1992). Regulation of cell magnesium. *Archives of Biochemistry and Biophysics, 298*, 1–12.

Romanuik, T. L., Wang, G., Holt, R. A., Jones, S. J., Marra, M. A., & Sadar, M. D. (2009). Identification of novel androgen-responsive genes by sequencing of LongSAGE libraries. *BMC Genomics, 10*, 476. http://dx.doi.org/10.1186/1471-2164-10-476.

Rubin, H. (2005). The membrane, magnesium, mitosis (MMM) model of cell proliferation control. *Magnesium Research, 18*, 268–274.

Sahni, J., Nelson, B., & Scharenberg, A. M. (2007). SLC41A2 encodes a plasma-membrane Mg^{2+} transporter. *Biochemical Journal, 401*, 505–513.

Sahni, J., & Scharenberg, A. M. (2013). The SLC41 family of MgtE-like magnesium transporters. *Molecular Aspects of Medicine, 34*, 620–628.

Saris, N. E., Mervaala, E., Karppanen, H., Khawaja, J. A., & Lewenstam, A. (2000). Magnesium. An update on physiological, clinical and analytical aspects. *Clinica Chimica Acta, 294*, 1–26.

Satake, W., Nakabayashi, Y., Mizuta, I., Hirota, Y., Ito, C., Kubo, M., et al. (2009). Genome-wide association study identifies common variants at four loci as genetic risk factors for Parkinson's disease. *Nature Genetics, 41*, 1303–1307.

Schlingmann, K. P., Weber, S., Peters, M., Niemann Nejsum, L., Vitzthum, H., Klingel, K., et al. (2002). Hypomagnesemia with secondary hypocalcemia is caused by mutations in TRPM6, a new member of the TRPM gene family. *Nature Genetics, 31*, 166–170.

Schmitz, C., Deason, F., & Perraud, A.-L. (2007). Molecular components of vertebrate Mg^{2+}-homeostasis regulation. *Magnesium Research, 20*, 6–18.

Schweigel, M., Kuzinski, J., Deiner, C., & Kolisek, M. (2009). Rumen epithelial cells adapt magnesium transport to high and low extracellular magnesium conditions. *Magnesium Research, 22*, 133–150.

Schweigel, M., Lang, I., & Martens, H. (1999). Mg^{2+} transport in sheep rumen epithelium: Evidence for an electrodiffusive mechanism. *American Journal of Physiology, 277*, G976–G982.

Schweigel, M., & Martens, H. (2003). Anion-dependent Mg^{2+} influx and a role for a vacuolar H^{+}-ATPase in sheep ruminal epithelial cells. *American Journal of Physiology. Gastrointestinal and Liver Physiology, 285*, G45–G53.

Schweigel, M., Park, H.-S., Etschmann, B., & Martens, H. (2006). Characterization of the Na^{+}-dependent Mg^{2+} transport in sheep ruminal epithelial cells. *American Journal of Physiology. Gastrointestinal and Liver Physiology, 290*, G56–G65.

Schweigel, M., Vormann, J., & Martens, H. (2000). Mechanisms of Mg^{2+} transport in cultured ruminal epithelial cells. *American Journal of Physiology. Gastrointestinal and Liver Physiology, 278*, G400–G408.

Shaw, A. J., Mughal, M. Z., Maresh, M. J. A., & Sibley, C. P. (1990). Evidence for active transfer of magnesium across the in situ perfused rat placenta. *Pediatric Research, 27*, 622–625.

Shaw, A. J., Mughal, M. Z., Maresh, M. J. A., & Sibley, C. P. (1991). Sodium-dependent magnesium transport across in situ perfused rat placenta. *American Journal of Physiology. Regulatory, Integrative and Comparative Physiology, 30*, R369–R372.

Smith, R. L., Thompson, L. J., & Maguire, M. E. (1995). Cloning and characterization of MgtE, a putative new class of Mg^{2+} transporter from *Bacillus firmus* OF4. *Journal of Bacteriology*, *177*, 1233–1238.

Sponder, G., Rutschmann, K., & Kolisek, M. (2013). "Inside-in" or "inside-out"? The membrane topology of SLC41A1. *Magnesium Research*, *26*, 176–181.

Standley, P. R., & Standley, C. A. (2002). Identification of a functional Na^{+}/Mg^{2+} exchanger in human trophoblast cells. *American Journal of Hypertension*, *15*, 565–570.

Stout, A. K., Li-Smerin, Y., Johnson, J. W., & Reynolds, I. J. (1996). Mechanisms of glutamate-stimulated Mg^{2+} influx and subsequent Mg^{2+} efflux in rat forebrain neurones in culture. *Journal of Physiology*, *492*, 641–657.

Su, L. T., Liu, W., Chen, H. C., Gonzalez-Pagan, O., Habas, R., & Runnels, L. W. (2011). TRPM7 regulates polarized cell movements. *Biochemical Journal*, *434*, 513–521.

Tashiro, M., Inoue, H., & Konishi, M. (2009). Metabolic inhibition strongly inhibits Na^{+}-dependent Mg^{2+} efflux in rat ventricular myocytes. *Biophysical Journal*, *96*, 4941–4950.

Tashiro, M., & Konishi, M. (1997). Na^{+} gradient-dependent Mg^{2+} transport in smooth muscle cells of guinea pig tenia cecum. *Biophysical Journal*, *73*, 3371–3384.

Tashiro, M., Tursun, P., & Konishi, M. (2005). Intracellular and extracellular concentrations of Na^{+} modulate Mg^{2+} transport in rat ventricular myocytes. *Biophysical Journal*, *89*, 3235–3247.

Tessman, P. A., & Romani, A. (1998). Acute effect of EtOH on Mg^{2+} homeostasis in liver cells: Evidence for the activation of an Na^{+}/Mg^{2+} exchanger. *American Journal of Physiology. Gastrointestinal and Liver Physiology*, *275*, G1106–G1116.

Tomas, F. M., & Potter, B. J. (1976). The site of magnesium absorption from the ruminant stomach. *British Journal of Nutrition*, *36*, 37–45.

Torres, L. M., Youngner, J., & Romani, A. (2005). Role of glucose in modulating Mg^{2+} homeostasis in liver cells from starved rats. *American Journal of Physiology*, *288*, G195–G206.

Tucci, A., Nalls, M. A., Houlden, H., Revesz, T., Singleton, A. B., Wood, N. W., et al. (2010). Genetic variability at the PARK16 locus. *European Journal of Human Genetics*, *18*, 1356–1359.

Tursun, P., Tashiro, M., & Konishi, M. (2005). Modulation of Mg^{2+} efflux from rat ventricular myocytes studied with the fluorescent indicator furaptra. *Biophysical Journal*, *88*, 1911–1924.

Voets, T., Nilius, B., van der Kemp, A. W., Droogmans, G., Bindels, R. J., & Hoenderop, J. G. (2004). TRPM6 forms the Mg channel involved in intestinal and renal Mg^{2+} absorption. *Journal of Biological Chemistry*, *279*, 19–25.

Vormann, J. (1992). Mineral metabolism in erythrocytes from patients with cystic fibrosis. *European Journal of Clinical Chemistry and Clinical Biochemistry*, *30*, 193–196.

Vormann, J., & Günther, T. (1987). Amiloride sensitive net Mg^{2+} efflux from isolated perfused rat hearts. *Magnesium*, *6*, 220–224.

Wabakken, T., Rian, E., Kveine, M., & Aasheim, H. C. (2003). The human solute carrier SLC41A1 belongs to a novel eukaryotic subfamily with homology to prokaryotic MgtE Mg^{2+} transporters. *Biochemical and Biophysical Research Communications*, *306*, 718–724.

Walder, R. Y., Landau, D., Meyer, P., Shalev, H., Tsolia, M., Borochowitz, Z., et al. (2002). Mutation of TRPM6 causes familial hypomagnesemia with secondary hypocalcemia. *Nature Genetics*, *31*, 171–174.

Watanabe, Y., & Dreyfus, L. S. (1972). Electrophysiological effects of magnesium and its interactions with potassium. *Cardiovascular Research*, *6*, 79–88.

Watanabe, M., Konishi, M., Ohkido, I., & Matsufuji, S. (2005). Enhanced sodium-dependent extrusion of magnesium in mutant cells established from a mouse renal tubular cell line. *American Journal of Physiology Renal Physiology*, *289*, F742–F748.

Weaver, K. (1986). Pregnancy-induced hypertension and low birth weight in magnesium-deficient ewes. *Magnesium, 5*, 191–200.
Weglicki, W. B., & Phillips, T. M. (1992). Pathobiology of magnesium deficiency: A cytokine/neurogenic inflammation hypothesis. *American Journal of Physiology, 263*, R734–R737.
White, R. E., & Hartzell, H. C. (1989). Magnesium ions in cardiac function. Regulator of ion channels and second messengers. *Biochemical Pharmacology, 38*, 859–867.
Willis, J. S., Xu, W., & Zhao, Z. (1992). Diversities of transport of sodium in rodent red cells. *Comparative Biochemistry and Physiology, 102*, 609–614.
Wolf, F. I., Di Francesco, A., Covacci, V., & Cittadani, A. (1995). Regulation of Na-dependent magnesium efflux from intact tumor cells. In R. Smetana (Ed.), *Advances in magnesium research* (pp. 490–496). London: John Libbey & Company Ltd.
Wolf, F. I., Di Francesco, A., Covacci, V., & Cittadini, A. (1994). cAMP activates magnesium efflux via the Na/Mg antiporter in ascites cells. *Biochemical and Biophysical Research Communications, 202*, 1209–1214.
Wolf, F. I., Di Francesco, A., Covacci, V., & Cittadini, A. (1997). Regulation of magnesium efflux from rat spleen lymphocytes. *Archives of Biochemistry and Biophysics, 344*, 397–403.
Wolf, F. I., Fasanella, S., Tedesco, B., Torsello, A., Sgambato, A., Faraglia, B., et al. (2004). Regulation of magnesium content during proliferation of mammary epithelial cells (HC-11). *Frontiers in Bioscience, 9*, 2056–2062.
Wolf, F. I., & Trapani, V. (2008). Cell (patho)physiology of magnesium. *Clinical Science, 114*, 27–35.
Wolf, F. I., Trapani, V., Simonacci, M., Mastrototaro, L., Cittadini, A., & Schweigel, M. (2009). Modulation of TRPM6 and Na^{+}/Mg^{2+} exchange in mammary epithelial cells in response to variations of magnesium availability. *Journal of Cellular Physiology, 9999*, 1–8.
Xu, W., & Willis, J. S. (1994). Sodium transport through the amiloride-sensitive Na-Mg pathway of hamster red cells. *Journal of Membrane Biology, 141*, 277–287.
Yamaguchi, S., & Ishikawa, T. (2008). The electrogenic Na^{+}-HCO_3^{-} cotransporter NBCe1-B is regulated by cellular Mg^{2+}. *Biochemical and Biophysical Research Communications, 376*, 100–104.
Yan, Y., Tian, J., Mo, X., Zhao, G., Yin, X., Pu, J., et al. (2011). Genetic variants in the RAB7L1 and SLC41A1 genes of the PARK16 locus in Chinese Parkinson's disease patients. *International Journal of Neuroscience, 121*(11), 632–636.
Yoshimura, M., Oshima, T., Matsuura, H., Watanabe, M., Higashi, Y., Ono, N., et al. (1995). Effect of the transmembrane gradient of magnesium and sodium on the regulation of cytosolic free magnesium concentration in human platelets. *Clinical Science (London), 89*, 293–298.
Zhang, G. H., & Melvin, J. E. (1995). Regulation by extracellular Na^{+} of cytosolic Mg^{2+} concentration in Mg^{2+}-loaded rat sublingual acini. *FEBS Letters, 371*, 52–56.

INDEX

Note: Page numbers followed by "*f*" indicate figures and "*t*" indicate tables.

A

B

C

O

P

R

Z

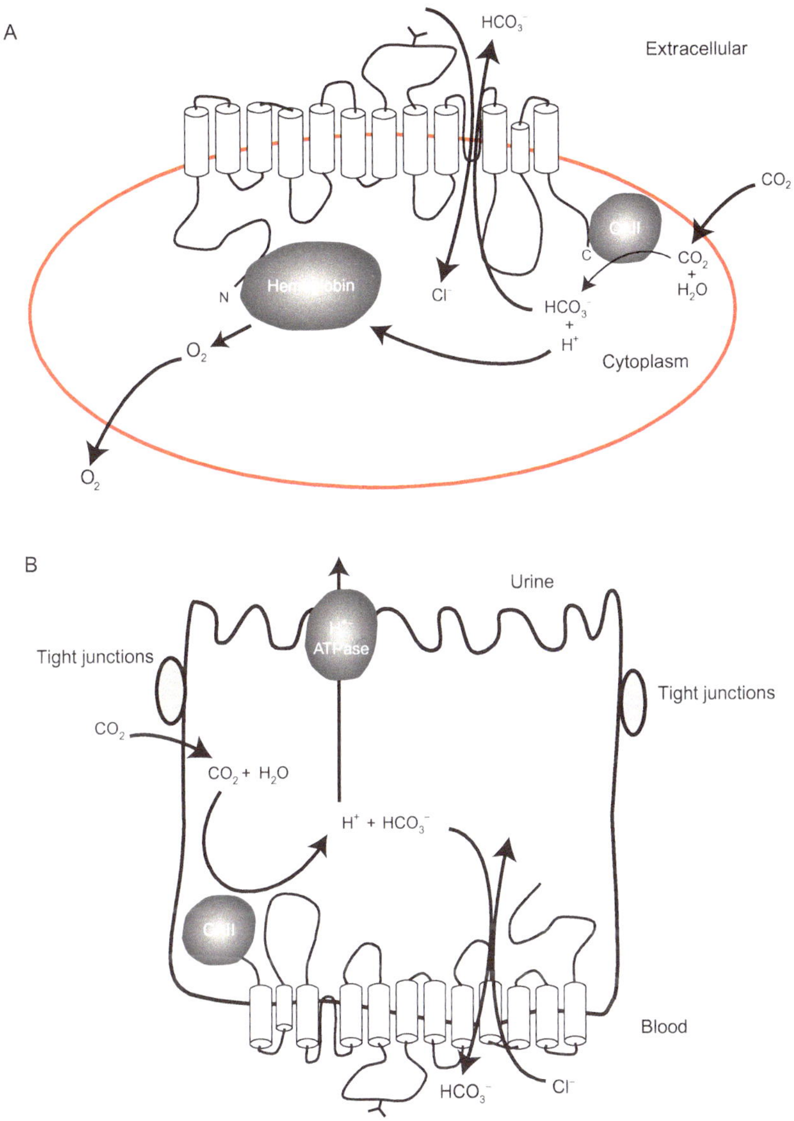

Please refer to Fig. 1.1 in text for figure legend.

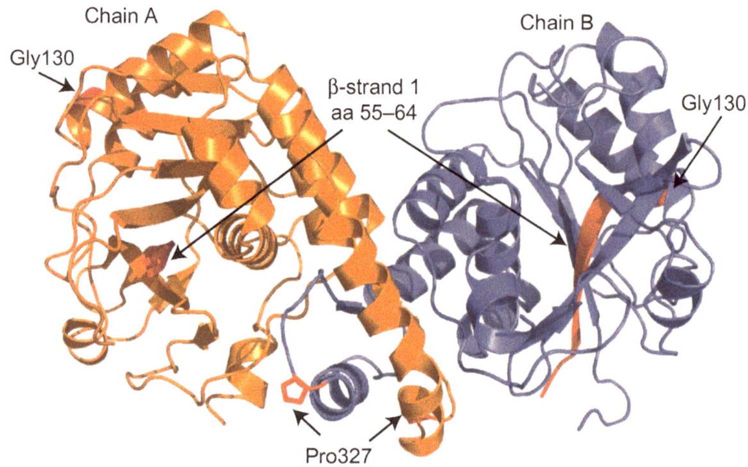

Please refer to Fig. 1.2 in text for figure legend.

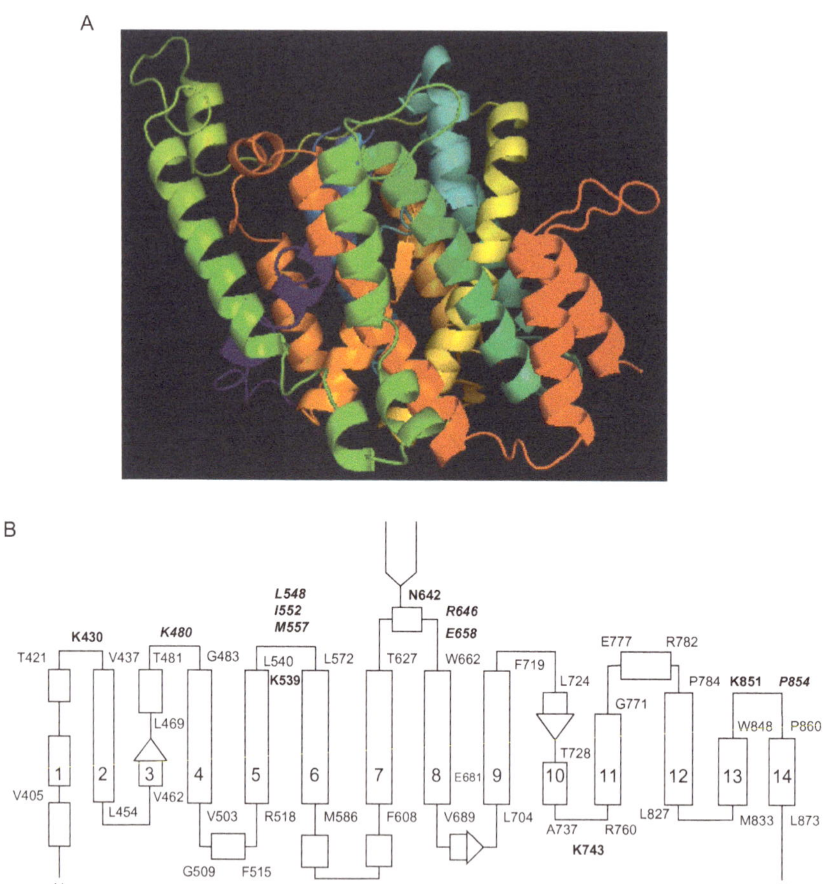

Please refer to Fig. 1.3 in text for figure legend.

Overall structure and topology of the SLC9A family

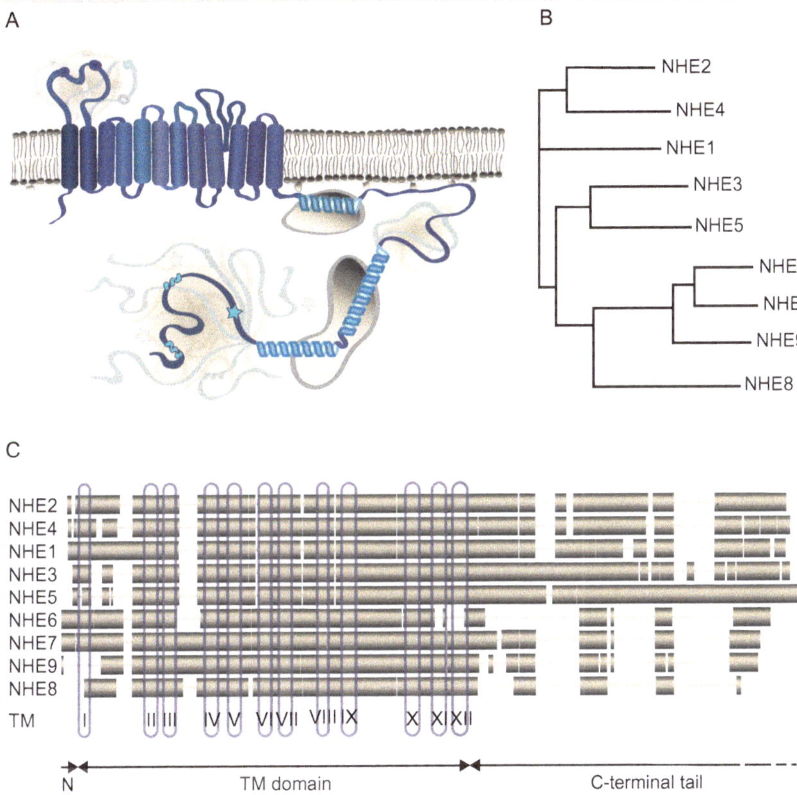

Please refer to Fig. 2.1 in text for figure legend.

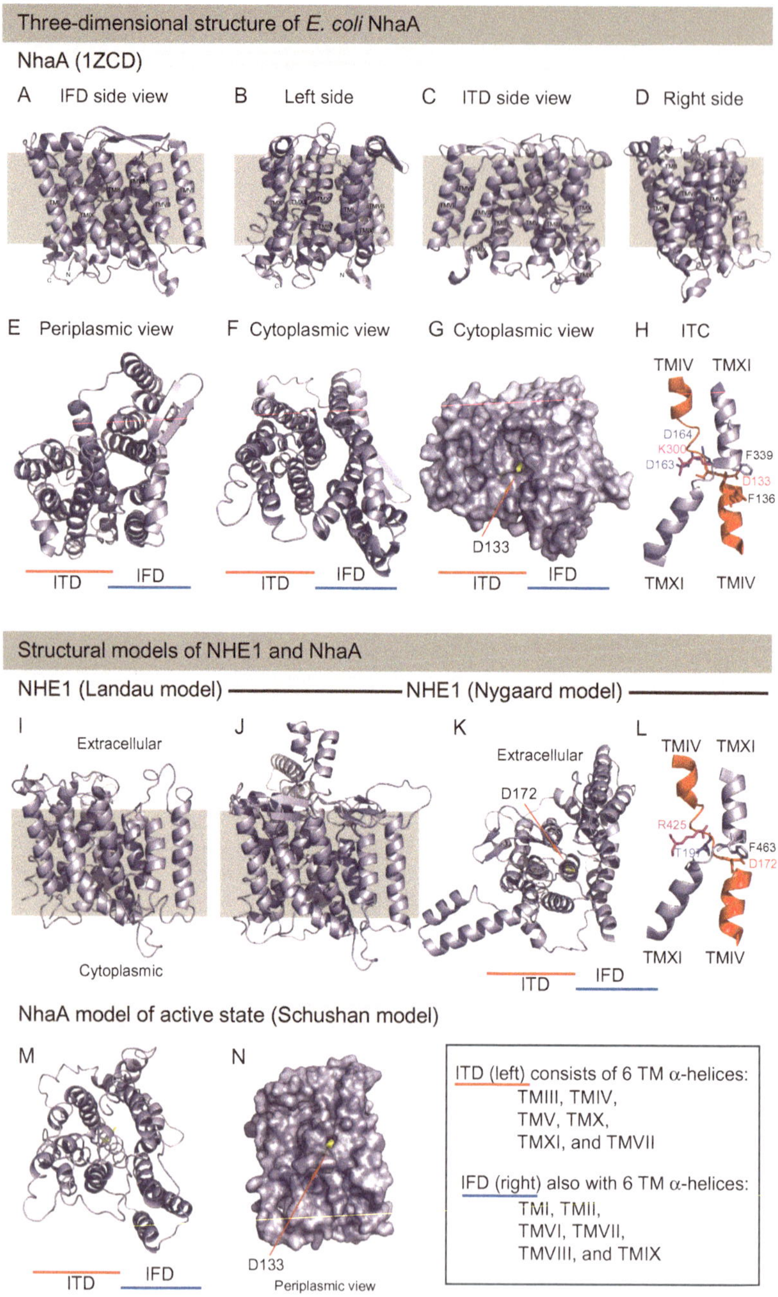

Please refer to Fig. 2.2 in text for figure legend.

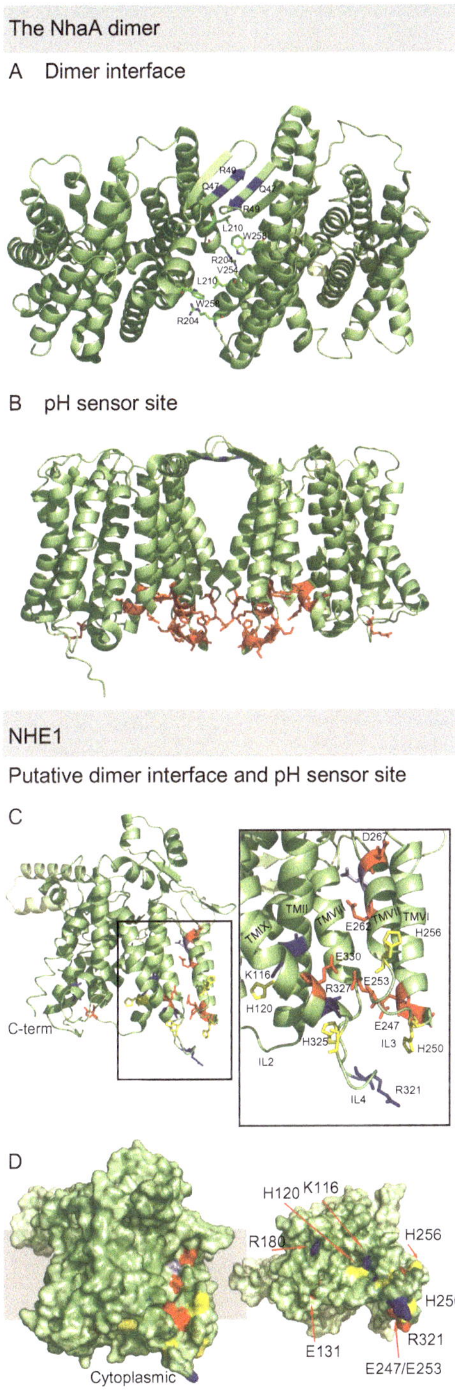

Please refer to Fig. 2.3 in text for figure legend.

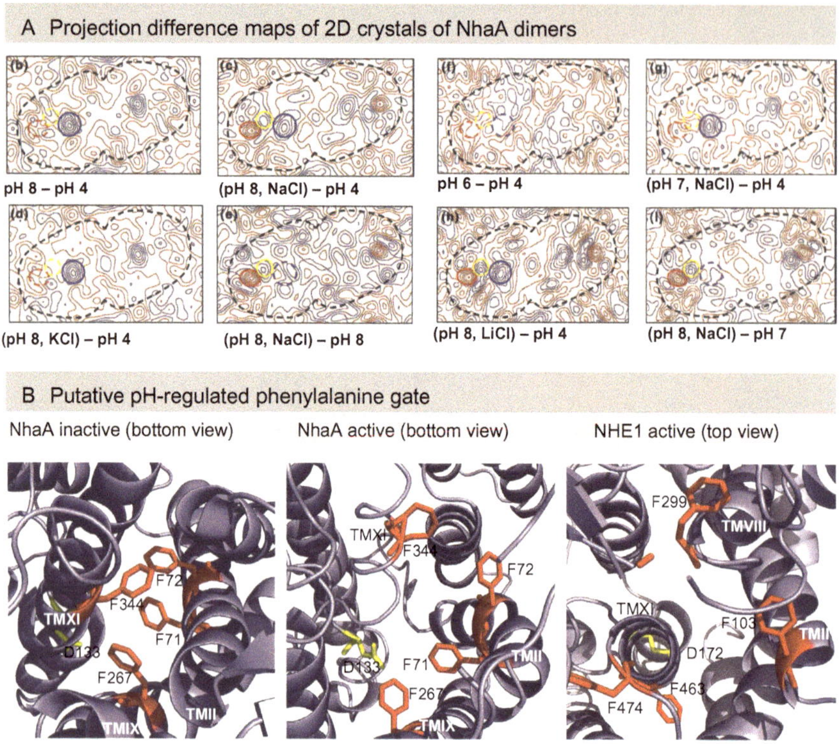

Please refer to Fig. 2.4 in text for figure legend.

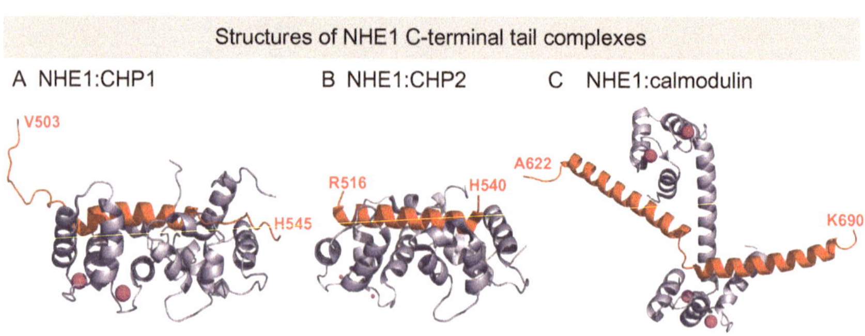

Please refer to Fig. 2.5 in text for figure legend.

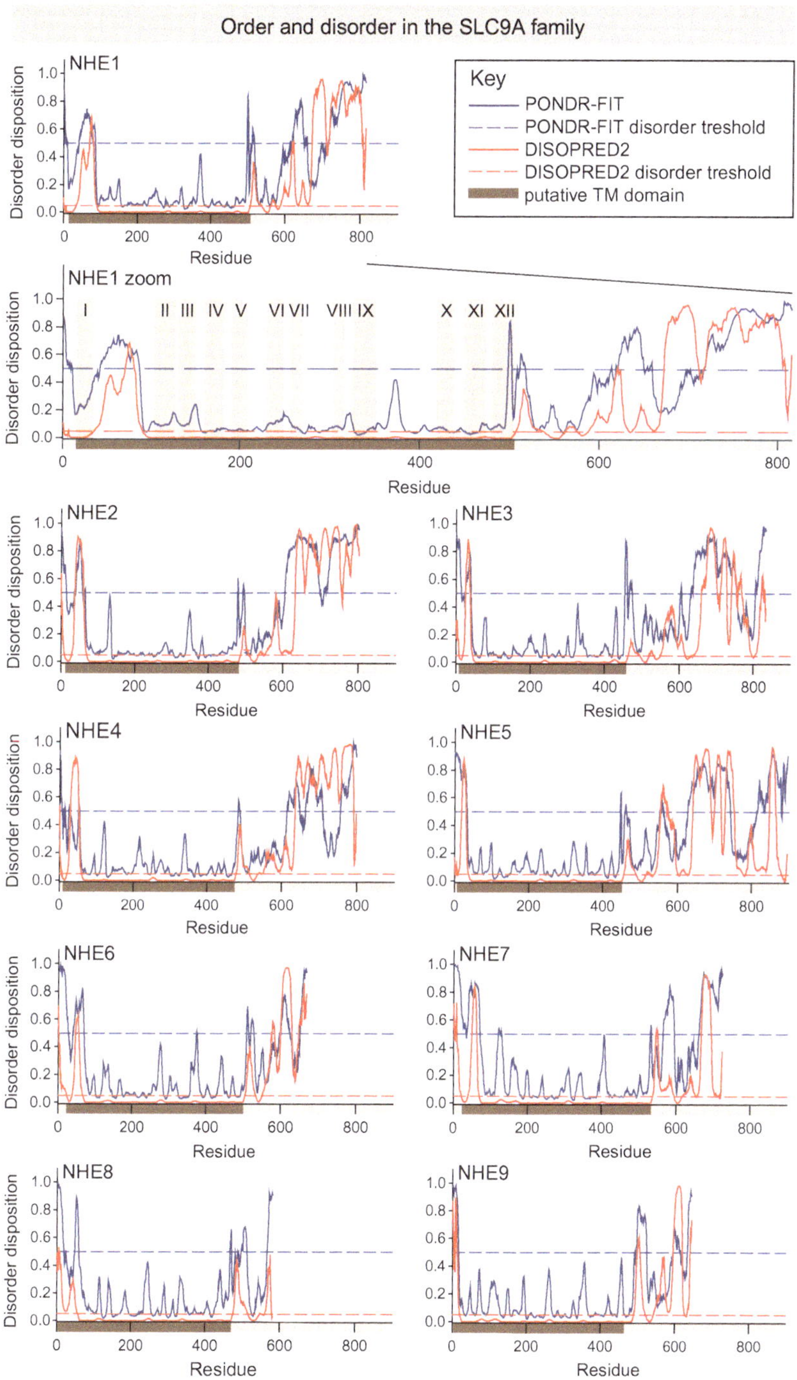

Please refer to Fig. 2.6 in text for figure legend.

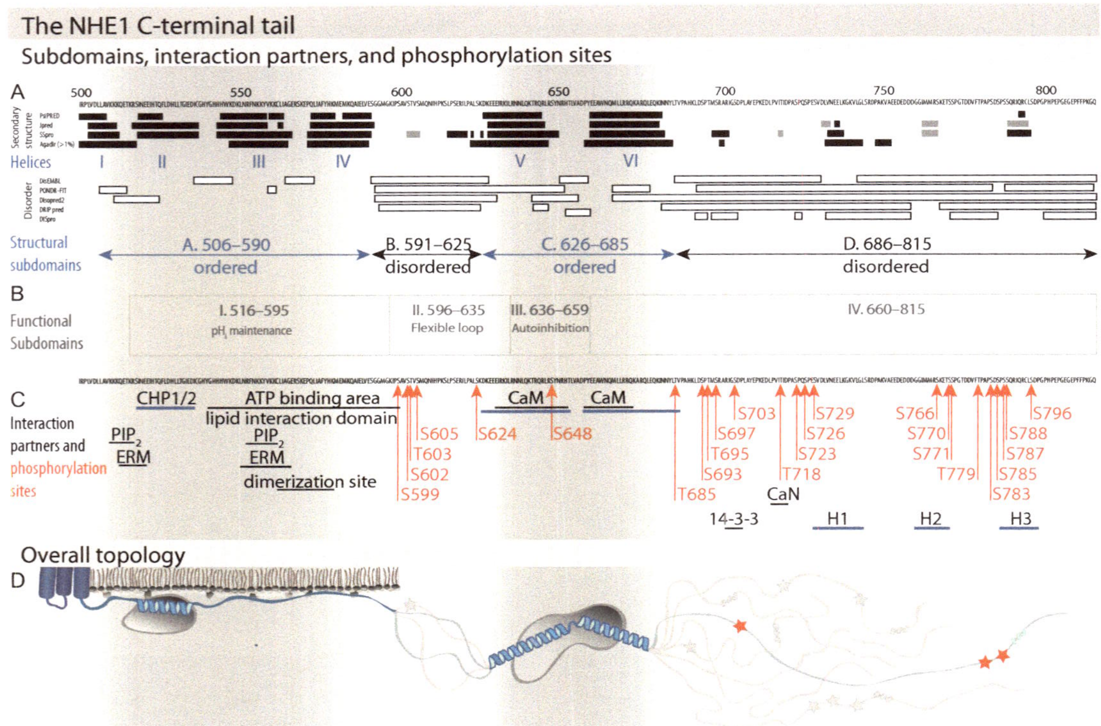

Please refer to Fig. 2.7 in text for figure legend.

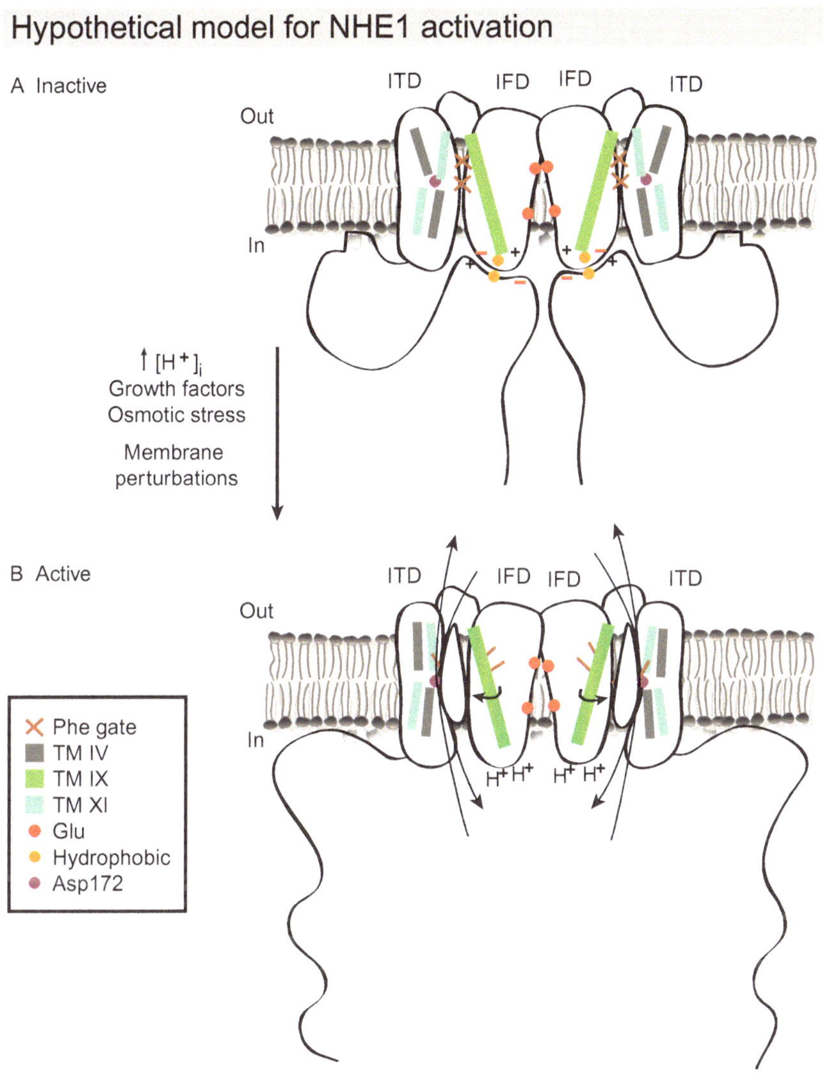

Please refer to Fig. 2.8 in text for figure legend.

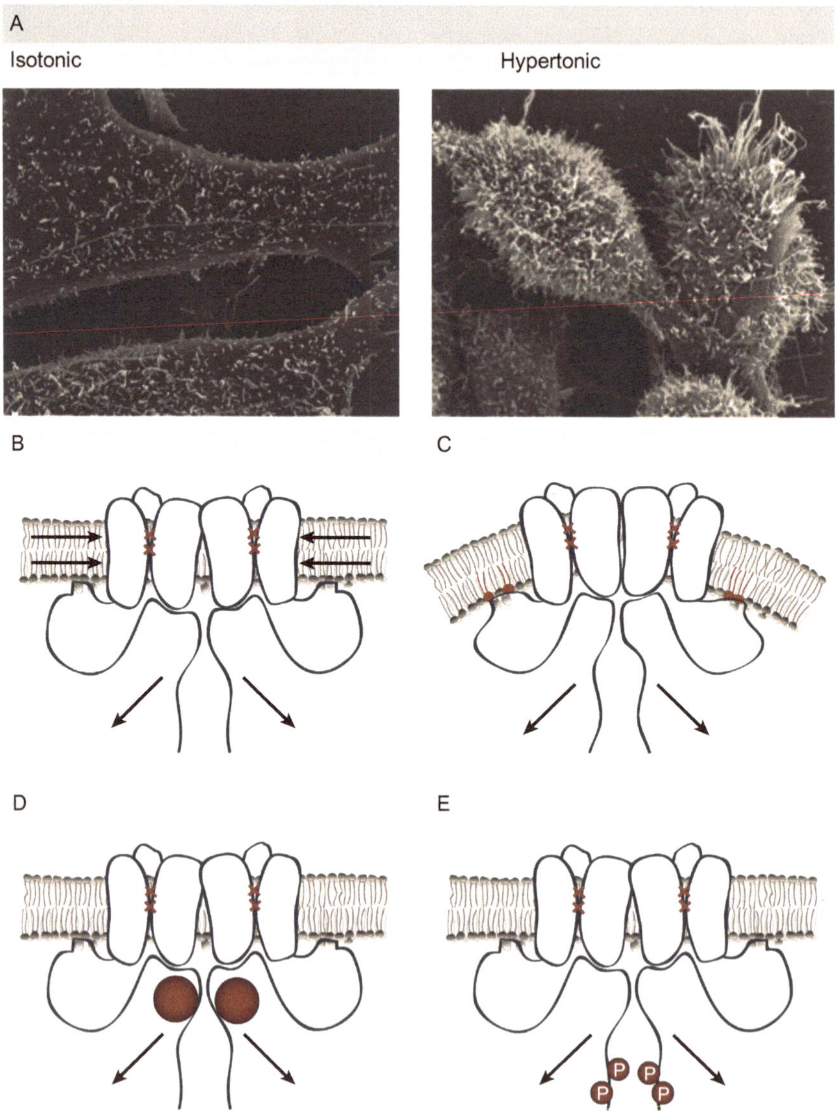

Please refer to Fig. 2.9 in text for figure legend.

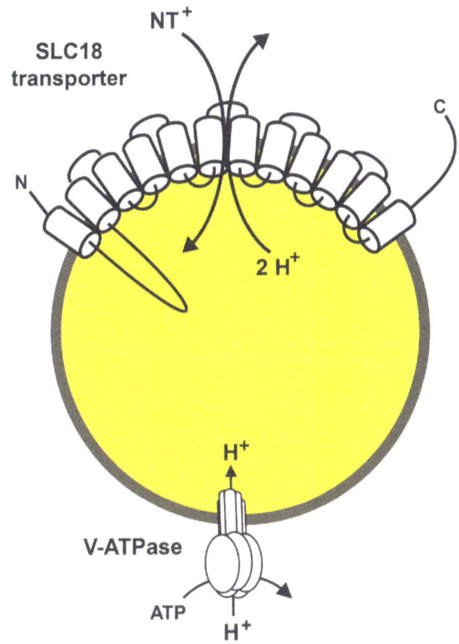

Please refer to Fig. 3.1 in text for figure legend.

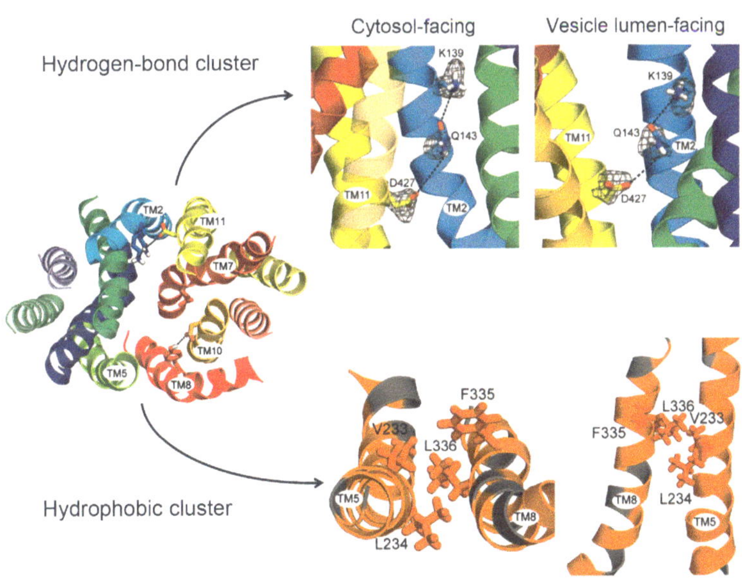

Please refer to Fig. 3.2 in text for figure legend.

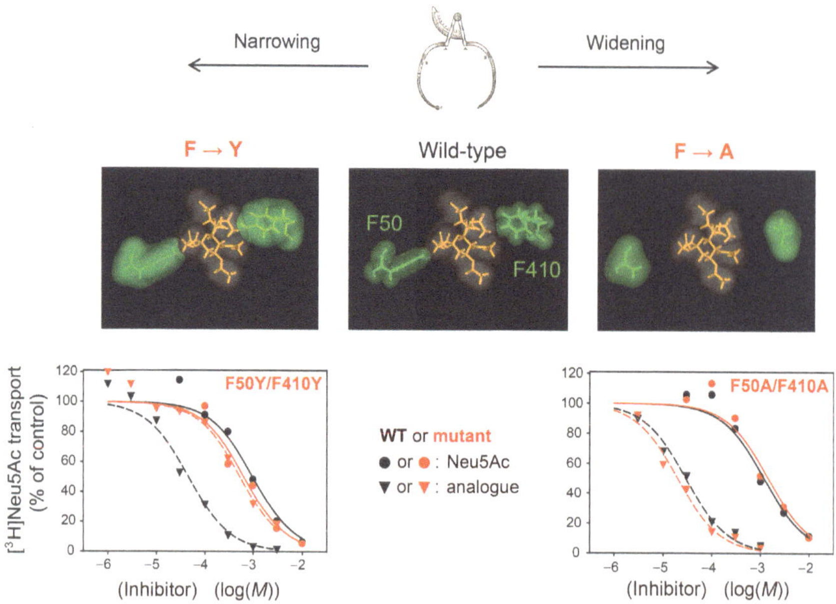

Please refer to Fig. 3.3 in text for figure legend.

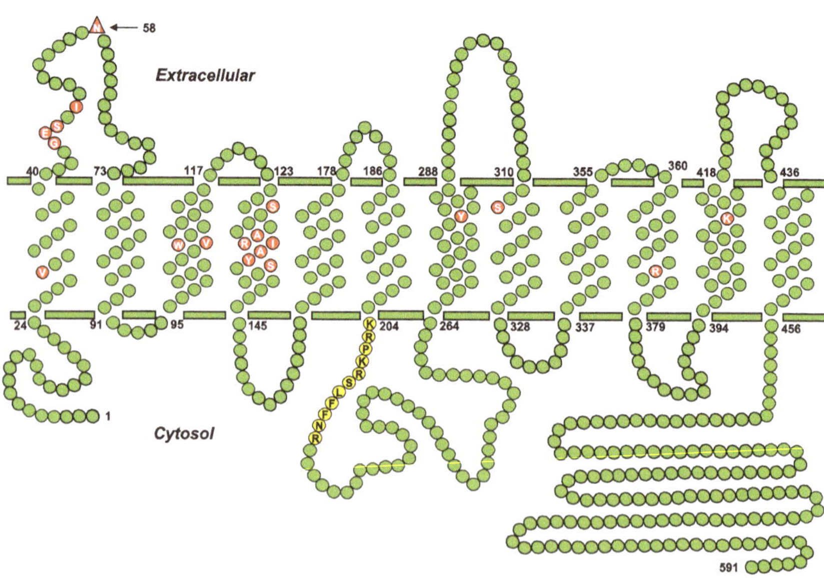

Please refer to Fig. 4.2 in text for figure legend.

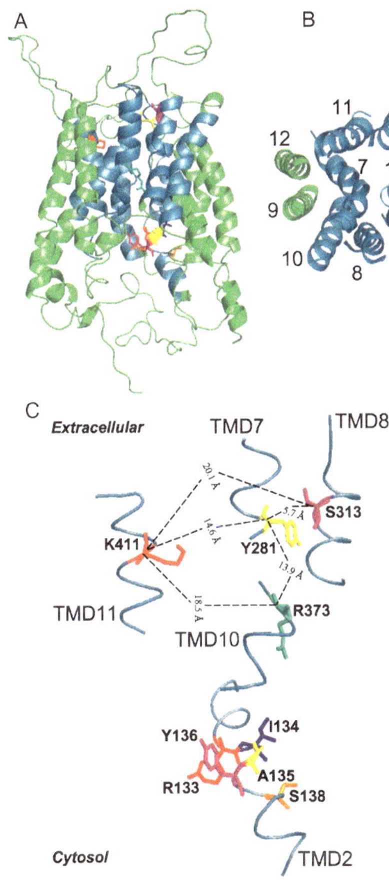

Please refer to Fig. 4.3 in text for figure legend.

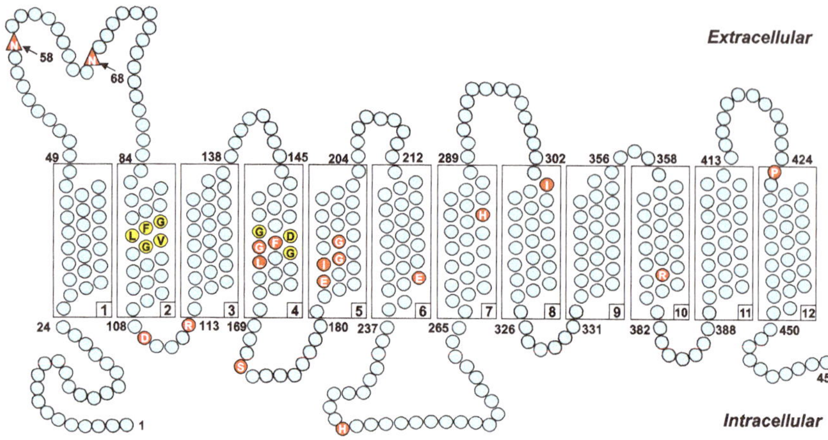

Please refer to Fig. 4.4 in text for figure legend.

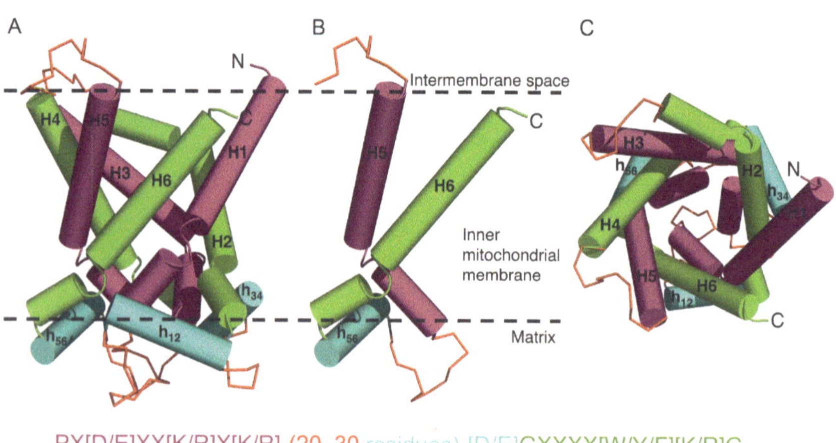

Please refer to Fig. 8.1 in text for figure legend.

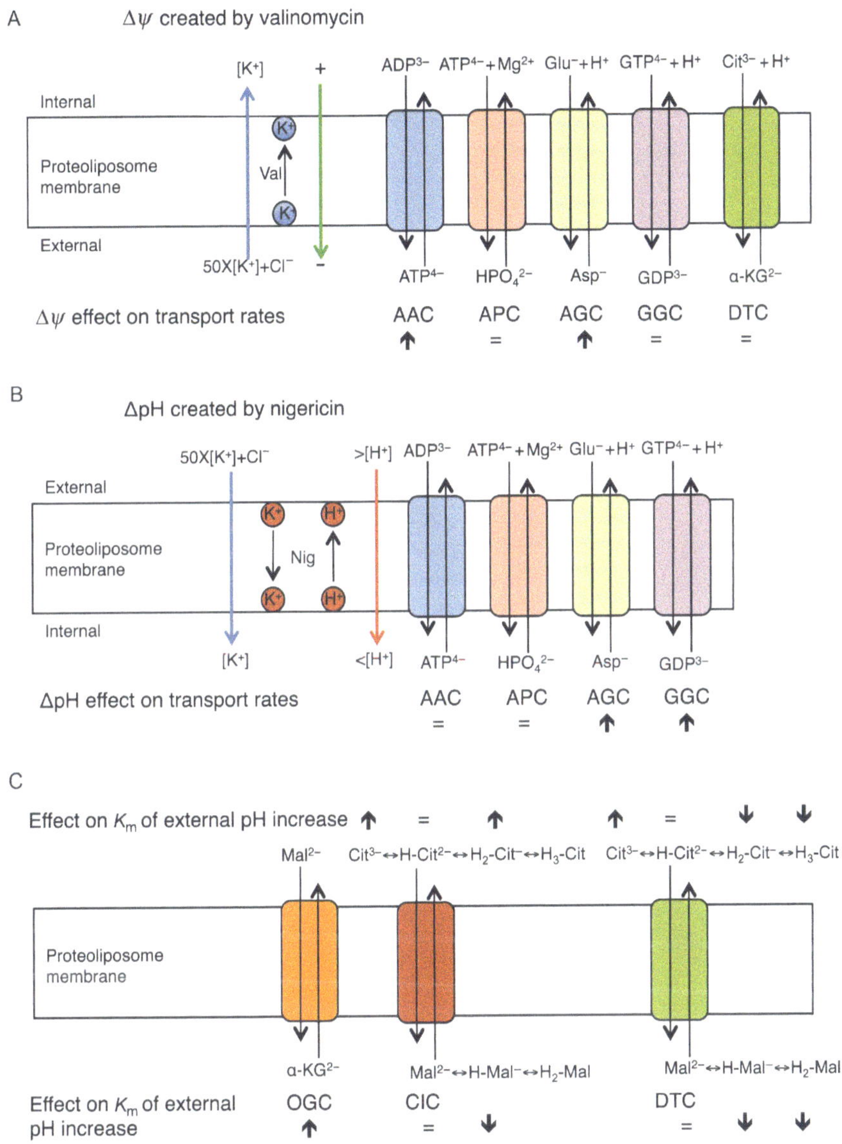

Please refer to Fig. 8.2 in text for figure legend.

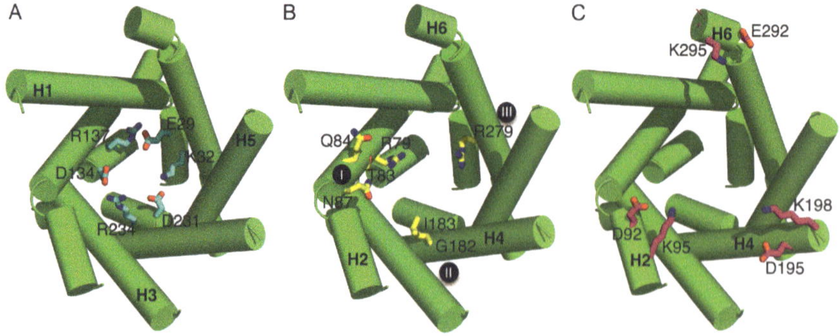

Please refer to Fig. 8.3 in text for figure legend.

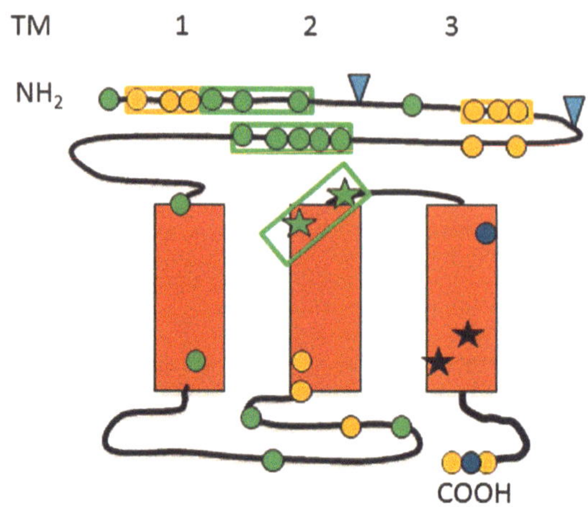

Please refer to Fig. 9.1 in text for figure legend.

Gene	Organ/tissue distribution		Subcellular localization							Link to disease
	Widespread	Predominant expression	PM	SG	Vesicles	GA	Mito	C	N	
SLC30										
A1	× (ubiquitous)	Tissues involved in Zn^{2+} acquisition such as small intestine, renal tubule, placenta	×							Embryonic death, Alzheimer's disease (AD)
A2		Mammary gland, prostate, retina, placenta, testis, seminal vesicles, pancreas, small intestine, kidney	×	×	Endosomal, lysosomal		×			Breast cancer
A3		Brain, testis, pancreas			Synaptic					AD
A4	×	Mammary gland, placenta, prostate, brain, kidney	×	×	Endosomal					Lethal milk syndrome, prostate, cancer, AD
A5	×	Heart, placenta, prostate, ovary, testis, small intestine, thymus, bone	×	×	×	×				
A6	×	Brain, lung, small intestine, liver			×	×				AD
A7	×	Intestine, stomach, pancreas, prostate, testis, retina, muscle			×	×				
A8		Pancreas, thyroid, testis, adrenal gland		×						Diabetes (type 1 and 2)
A9	×							×	×	
A10		Brain, retina, liver	Unknown							

Abbreviations: PM = plasma membrane, SG = secretory granules, GA = Golgi apparatus, Mito = mitochondria, C = cytosol, N = nucleus

Please refer to Table 9.1 in text for Table legend.

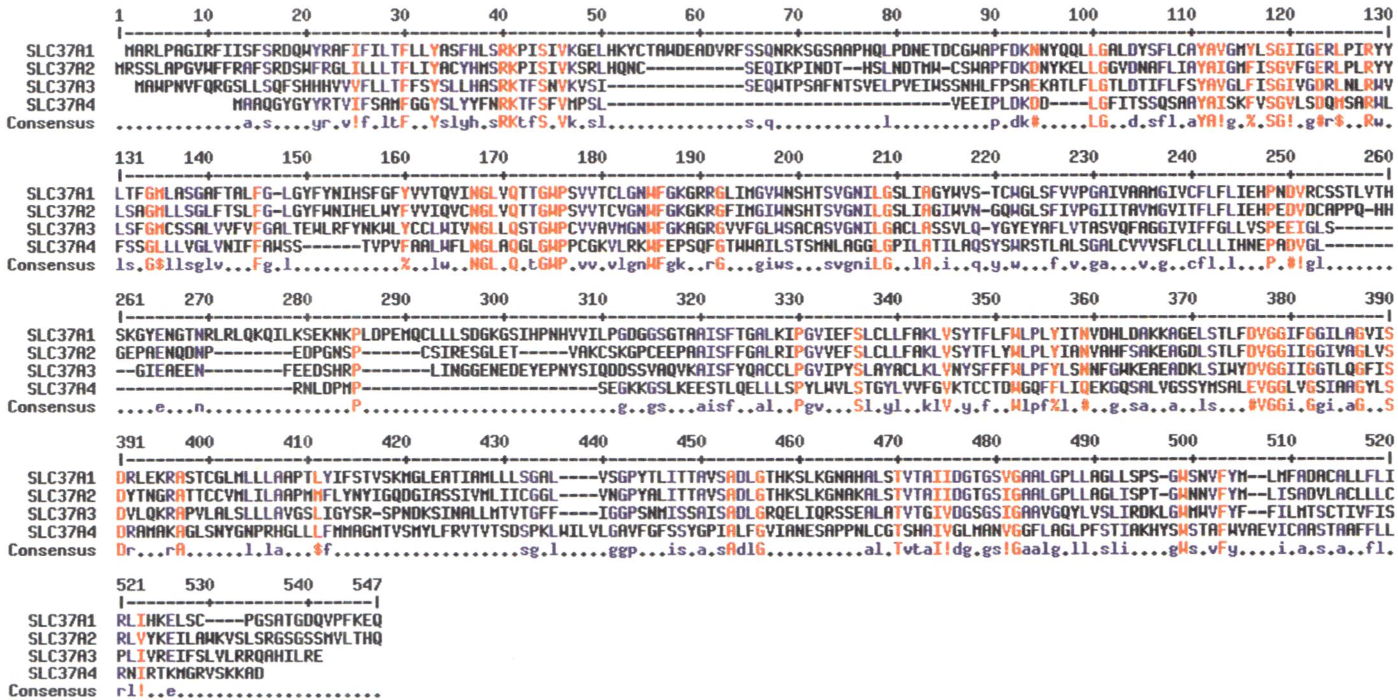

Please refer to Fig. 10.1 in text for figure legend.

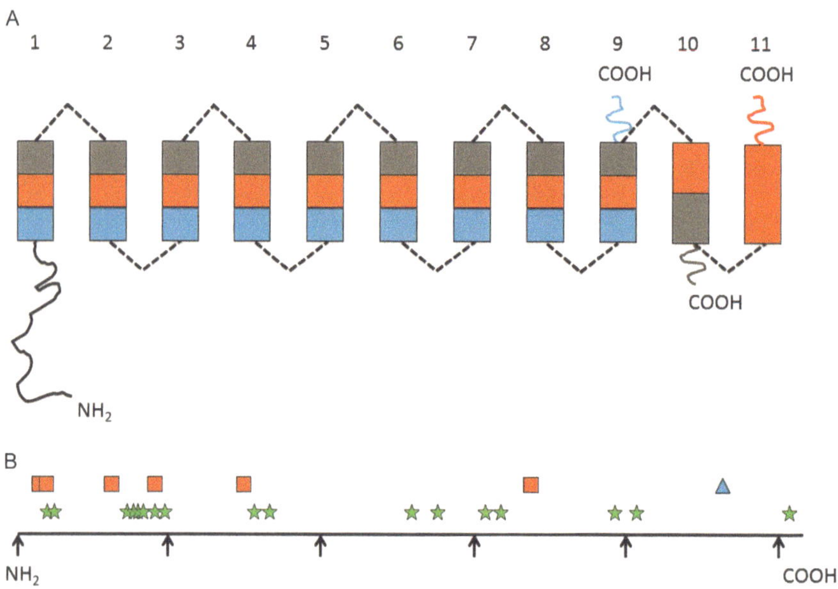

Please refer to Fig. 11.1 in text for figure legend.

MIX
Papier | Fördert gute Waldnutzung
FSC® C083411

Kolibri 360 GmbH
Ferdinand-Jühlke-Straße 7
99095 Erfurt, Deutschland
produktsicherheit@kolibri360.de